JOHN H. NELSON

UNIVERSITY OF NEVADA, RENO

Nuclear Magnetic Resonance Spectroscopy

Prentice Hall

Pearson Education, Inc.
Upper Saddle River, NJ 07458

Library of Congress Cataloging-in-Publication Data

Nelson, John H. (John Henry)
  Nuclear magnetic resonance spectroscopy / by John H. Nelson.
    p. cm.
  Includes bibliographical references and index.
  ISBN 0-13-033451-0
  1. Nuclear magnetic resonance spectroscopy. I. Title.

QD96.N8 N45 2003
543′.0877--dc21   00-069869

Senior Editor: Nicole Folchetti
Assistant Managing Editor, Science: Beth Sweeten
Editorial Assistants: Nancy Bauer/Eliana Ortiz
Manufacturing Manager: Trudy Pisciotti
Assistant Manufacturing Manager: Michael Bell
Editor in Chief, Physical Science: John Challice
Art Studio: Preparé, Inc.
Art Director: Jayne Conte
Cover Designer: Bruce Kenselaar
Production Services/Composition: Preparé, Inc.

© 2003 by Pearson Education, Inc.
Pearson Education, Inc.
Upper Saddle River, NJ 07458

Printed in the United States of America
10   9   8   7   6   5   4   3   2   1

ISBN 0-13-033451-0

Pearson Education LTD., *London*
Pearson Education Australia PTY, Limited, *Sydney*
Pearson Education Singapore, Pte. Ltd.
Pearson Education North Asia Ltd., *Hong Kong*
Pearson Education Canada, Ltd., *Toronto*
Pearson Educación de Mexico, S.A. de C.V.
Pearson Education—Japan, *Tokyo*
Pearson Education Malaysia, Pte. Ltd.

CONTENTS

CHAPTER 1

Basic Theory of NMR Spectroscopy

CHAPTER 2

Measurement of the Spectrum: Instrumentation

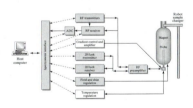

C H A P T E R 3

The Chemical Shift and Examples for Selected Nuclei

C H A P T E R 4

Symmetry and NMR Spectroscopy

C H A P T E R 5

Spin–Spin Coupling and NMR Spin Systems

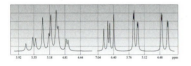

CHAPTER 6

Typical Magnitudes of Selected Coupling Constants

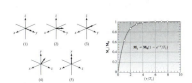

CHAPTER 7

Nuclear Spin Relaxation

CHAPTER 8

The Nuclear Overhauser Effect

$$\frac{M_C(^1H_{irrad})}{M_C^0} = 1 + \left[\frac{6J_2(\omega_H + \omega_C) - J_0(\omega_H - \omega_C)}{J_0(\omega_H - \omega_C) + 3J_1\omega_C + 6J_2(\omega_C + \omega_H)} \right] \cdot \frac{\gamma_H}{\gamma_C} = 1 + \eta_{CH}$$

$$\frac{M_C(^1H_{irrad})}{M_C^0} = 1 + 0.5\frac{\gamma_H}{\gamma_C} = 2.988 \quad \text{or} \quad \eta_{CH} = 1.988$$

$$\frac{1}{T_1(\text{other})} = \frac{1}{T_1(\text{obs})}\left[1 - \frac{\eta_{\text{obs}}}{\eta_0} \right]$$

CHAPTER 9

Editing ^{13}C NMR Spectra

CHAPTER 10

Two-Dimensional NMR Spectroscopy

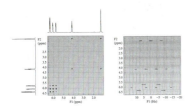

CHAPTER 11

Dynamic NMR Spectroscopy

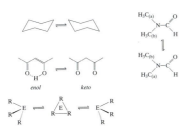

CHAPTER 12

Lanthanide-Shift Reagents

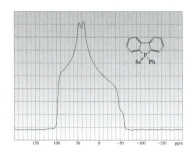

CHAPTER 13

NMR of Solids

CHAPTER 14

Problems

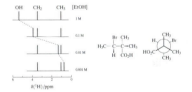

APPENDIX

Selected Values of Longitudinal Relaxation Times for Some Nuclei

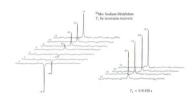

PREFACE

It is probably safe to say that NMR spectroscopy is the most broadly utilized spectroscopic technique presently available for obtaining detailed information on chemical systems at the molecular level. Although there are a number of excellent books on NMR spectroscopy, most of the textbooks focus on ^{1}H and ^{13}C NMR spectroscopy and discuss other nuclei only briefly or not at all. Because Fourier Transform (FT) spectrometers are now generally available in both academic and industrial settings, the synthetic chemist can, and often does, routinely investigate the spectroscopy of a variety of nuclides. Moreover, some of the fundamental principles are best illustrated by considering the properties and spectra of nuclei other than ^{1}H and ^{13}C.

This book was written as a textbook for senior undergraduate and graduate students who need a reasonably thorough discussion of the subject at an introductory level. It contains a fair amount of chemical-shift and coupling-constant data for seven of the more widely studied nuclides (^{1}H, ^{11}B, ^{13}C, ^{15}N, ^{19}F, ^{31}P, and ^{195}Pt), but is not a compendium of such data. References to compendia are given in the appropriate places in the text. The book is largely empirical, but the theoretical basis for most of the more important aspects is presented. I have attempted to explain each of the various experiments at a level such that the reader can ascertain which among several alternatives would be the best experiment to use. The practitioner should always ask, "What information do I want to obtain?" Then the best, and hopefully easiest, experiments can be chosen to provide that information most expeditiously.

Since students often learn this type of material best by solving problems, a large number of practice problems have been included. The majority of them involve actual spectra. The problems have been chosen to represent the concepts discussed in the book, but, of course, are biased by my own interests. An answer book containing detailed solutions is available.

Some figures have been taken from the literature, and acknowledgments are given in the appropriate captions. I wish to thank Dr. George A. Gray of Varian Associates, Mr. Lewis W. Cary and Dr. Stephan E. Boiadjiev of the University of Nevada, and Professor Roderick E. Wasylishen of Dalhousie University for several of the spectra reproduced herein. I am also grateful to the many students at UNR who have endured the various forms of this book in my classes. I appreciate their comments, their insightful and probing questions, and their encouragement to complete the book.

The following individuals reviewed portions of this textbook at various stages during its development:

Samuel O. Grim	*University of Maryland*
Joseph M. O'Connor	*University of California, La Jolla*
Richard Keiter	*Eastern Illinois University*
Robert T. Paine, Jr.	*University of New Mexico*
Thomas B. Rauchfuss	*University of Illinois*
Lawrence R. Sita	*University of Chicago*
Nancy S. True	*University of California, Davis*

I am particularly grateful for their detailed written comments and suggestions for improvement. I also wish to acknowledge the help of the Prentice Hall editorial staff, especially John Challice. I am grateful as well to Mrs. Jenny Johnson and Mrs. Nancy Ball, who expertly typed the manuscript. I welcome your comments, criticisms, suggestions, and, most importantly, your corrections.

John H. Nelson

PHYSICAL AND CHEMICAL CONSTANTS

Avogadro's number $\quad N = 6.02214199 \times 10^{23} \text{ mol}^{-1}$

Planck's constant $\quad h = 6.62606876 \times 10^{-34} \text{ Js}$
$$= 6.62606876 \times 10^{-27} \text{ erg}$$
$$= 6.62606876 \times 10^{-34} \text{ Nms}$$

reduced Planck's constant $\quad \hbar = \dfrac{h}{2\pi} = 1.054589 \times 10^{-34} \text{ Js}$

Boltzman's constant $\quad k = 1.3806503 \times 10^{-23} \text{ JK}^{-1}$

magnetic permeability of vacuum $\dfrac{\mu_0}{4\pi} = 10^{-7} \text{ mkg C}^{-2}$

electron charge $\quad e = 4.8030 \times 10^{-10} \text{ abs esu}$
$$= 1.602176462 \times 10^{-19} \text{ C}$$

electron mass $\quad m_e = 9.10938188 \times 10^{-31} \text{ kg}$
$$= 5.485799 \times 10^{-4} \text{ amu}$$
$$= 0.5110 \text{ MeV}$$

proton mass $\quad m_H = 1.67262158 \times 10^{-27} \text{ kg}$
$$= 1.0072765 \text{ amu}$$

neutron mass $\quad m_n = 1.0086649 \text{ amu}$
$$= 1.67492716 \times 10^{-24} \text{ g}$$

gas constant $\quad R = 8.31441 \text{ J mol}^{-1} \text{ K}^{-1}$
$$= 1.9872 \text{ cal mol}^{-1} \text{ K}^{-1}$$
$$= 0.08206 \text{ Latm mol}^{-1} \text{ K}^{-1}$$

Bohr magneton $\quad \mu_B = 9.274078 \times 10^{-24} \text{ JT}^{-1}$

nuclear magneton $\quad \mu_N = 5.050824 \times 10^{-27} \text{ JT}^{-1}$

proton moment $\quad \mu_H = 1.4106171 \times 10^{-26} \text{ JT}^{-1}$

gyromagnetic ratio of protons $\quad \gamma_H = 2.651987 \times 10^{-8} \text{ rads T}^{-1} \text{ s}^{-1}$

reduced gyromagnetic ration of proton $\quad \dfrac{\gamma_H}{2\pi} = 4.257711 \times 10^7 \text{ T}^{-1} \text{ s}^{-1}$

electric permittivity of vacuum $\quad \varepsilon_0 = \left(\dfrac{1}{36\pi}\right) 10^{-9} \text{ m}^{-3} \text{ kg}^{-1} \text{ s}^2 \text{ C}^2$

Other constants $\quad \Pi = 3.14159\ldots$
$$e = 2.7183\ldots$$
$$\ln 10 = 2.3026\ldots$$

SYMBOLS AND ABBREVIATIONS

a %	natural abundance
ADC	analog-to-digital converter
APT	attached–proton test
AQT	acquisition time
$\mathbf{B}_0$	static magnetic field (flux density)
$\mathbf{B}_1, \mathbf{B}_2$	radio frequency field with frequencies ν_1 and ν_2
$\mathbf{B}_{\text{eff}}, \mathbf{B}_{\text{loc}}$, or $\mathbf{B}_N$	effective field at position of nucleus
C	Coulomb
COSY	correlated spectroscopy
CSA	chemical-shift anisotropy
CW	continuous wave
χ	magnetic susceptibility
$^{13}\text{C}\{^1\text{H}\}$	observation of ^{13}C resonance while ^1H is decoupled
δ	chemical shift relative to a standard
2D	two-dimensional
DD	dipole–dipole
ΔE	energy difference between two states
D_{ij}	direct coupling constant
DEPT	distortionless enhancement by polarization transfer
DMSO	dimethylsulfoxide
DNMR	dynamic NMR
DPM	dipivalomethane (2,2,6,6-tetramethylheptanedione)
$\Delta\delta$	chemical-shift change
e	electron charge
E_A	Arrhenius activation energy
E_X	electronegativity of substituent
η	fractional enhancement in NOE
FID	free-induction decay
FOD	heptafluoro-7,7-dimethyl-4,6-octanedione
FT	Fourier transform
ϕ	phase difference between two vectors
F_1, F_2	frequency axes in a 2D NMR spectrum
$\Delta G^{\ddagger}$	Gibbs free energy of activation
ΔG°	standard-state Gibbs free energy
γ	gyromagnetic ratio
$\gamma\!\!\!/ = \dfrac{\gamma}{2\pi}$	reduced gyromagnetic ratio
h	Planck's constant
$\hbar = \dfrac{h}{2\pi}$	reduced Planck's constant
$\Delta H^{\ddagger}$	enthalpy of activation
HETCOR	heteronuclear chemical-shift correlation
ΔH°	standard-state enthalpy

HMBC	heteronuclear multiple-bond correlation
HMQC	heteronuclear multiple quantum coherence
I	nuclear angular-momentum quantum number (spin)
I	nuclear-spin operator
Ix, Iy, Iz	eigenvalue of the component of **I**
INADEQUATE	incredible natural-abundance double quantum transfer
INEPT	insensitive nuclei enhanced by polarization transfer
$^nJ_{AX}$	indirect spin–spin coupling constant
$^nK_{AX}$	reduced indirect coupling constant
K_{eq}	equilibrium constant
k	rate constant
k_c	rate constant at the coalescence temperature T_c
k_B	Boltzman constant
k_0	frequency factor
LSR	lanthanide-shift reagent
μ	magnetic moment of nucleus
$\mu_0/4\pi$	magnetic permeability of vacuum
μ_z	component of μ along the static-field direction (z-axis)
m	magnetic quantum number
m	meter
$\mathbf{M}_0$	macroscopic magnetization of the sample
$\mathbf{M}_{eff}$	total magnetization along $\mathbf{B}_0 + \mathbf{B}_1$
MO	molecular orbital
$\mathbf{M}_{X'} \mathbf{M}_{Y'}$	transverse magnetization components in the x' and y' directions
$\mathbf{M}_z$	longitudinal magnetization in the z direction (static-field direction)
$\mathbf{M}_x$	magnetization for the x nuclei
$\mathbf{M}_H^{C\alpha}, \mathbf{M}_H^{C\beta}$	^{1}H magnetization in a two-spin spin system with the ^{13}C nuclei in the α and β states
$\mathbf{M}_C^{H\alpha}, \mathbf{M}_C^{H\beta}$	^{13}C magnetization vectors in a two-spin spin system with the protons in the α and β states
N, n	number of scans
N	total number of nuclei in a sample
N	newton (unit of force)
N_α, N_β	numbers of nuclei in the α and β states
N_i	number of nuclei in level i
ν	frequency
ν_L	Larmor frequency
ν_i	resonance frequency of nucleus i
ν_1	frequency of RF generator (observing frequency)
ν_2	decoupling frequency
$\Delta\nu_{1/2}$	line width at half-height
NOE	nuclear Overhauser effect/enhancement
NOESY	nuclear Overhauser effect spectroscopy
ω	angular velocity
ppm	parts per million
P	angular momentum of nucleus
P_Z	component of P in the Z direction
eQ	electric quadrupole moment
R	universal gas constant

R_A	receptivity of nucleus A
RF	radio frequency field
r_{ij}	interatomic (or internuclear) distance
σ	shielding (screening) constant
$S(t), S(f)$	signal as a function of time or of frequency
$\Delta S^{\ddagger}$	entropy of activation
ΔS°	standard-state entropy
S/N	signal-to-noise ratio
SW	Sweep width also called spectral width
τ	time interval between pulses
τ_c	correlation time
τ_l	lifetime of a nucleus in a particular spin state or magnetic environment
τ_p	pulse duration
T_1	variable time in a 2D experiment; usually incremented in regular steps
Δ	fixed time interval in a 2D pulse sequence
T	tesla (unit of magnetic flux density)
T	absolute temperature (in K)
THF	tetrahydrofuran
T_1	spin–lattice, or longitudinal, relaxation time
$T_{1\rho}$	relaxation time in the rotating frame
T_2	spin–spin, or transverse, relaxation time
Θ	Pulse angle
TMS	Tetramethylsilane
W_0, W_1, W_2	Transition probabilities for zero-quantum, single-quantum, and double-quantum relaxation processes

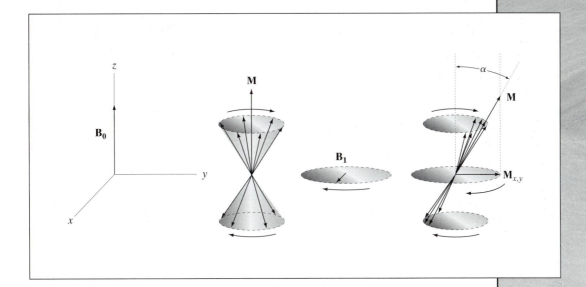

CHAPTER 1

Basic Theory of
NMR Spectroscopy

Nuclear magnetic resonance (NMR) spectroscopy involves the magnetic energy of nuclei when they are placed in a magnetic field. For fields of appropriate strength, we shall learn that transitions between nuclear energy levels occur in the radio-frequency (RF) region of the electromagnetic spectrum. The first NMR experiments were conducted in 1945, but the technique did not become useful to chemists until 1949, when it was discovered that transitions between the energy levels were environment dependent. This was the birth of the chemical shift. NMR spectroscopy has developed very rapidly since the introduction of the first commercial spectrometer in 1953. Despite the relatively high costs of modern spectrometers, it is safe to say that NMR spectroscopy has become the spectroscopic technique most widely used by chemists.

A ll forms of spectroscopy give spectra that may be described in terms of the frequency, intensity, and shape of spectral lines or bands. These observable properties, which, for NMR spectroscopy, are found to be the shielding constants and coupling constants of the nuclei, and the lifetimes of the energy levels are all a function of molecular structure. The NMR spectroscopist is concerned with (a) obtaining the optimum spectrum and learning how to do so, (b) extracting the nuclear magnetic resonance parameters from the spectrum, (c) interpreting or predicting these parameters in terms of appropriate theoretical models, and, (d) for chemists, using the information gained from the spectra to ascertain molecular structure or obtain kinetic and thermodynamic data. These topics will be discussed in succeeding chapters.

1.1 Nuclear Spin

The basic theory of NMR spectroscopy is common to all experiments and all NMR-active nuclei independent of the nature of the spectrometer. The nuclear spin (I), which may have values of $0, \frac{1}{2}, 1, \frac{3}{2}$, etc., in units of $h/2\pi = \hbar$, is the nuclear property responsible for the NMR phenomenon. The value of the nuclear spin for a given nucleus depends upon its mass (number of protons plus neutrons) and atomic number (number of protons). I is obtained by the vector addition of individual proton and neutron spin quantum numbers of $\frac{1}{2}$ each, with the restriction that neutron spins can cancel only neutron spins and proton spins only proton spins. Three cases result:

1. Zero spin for both neutrons and protons (e.g., ^{12}C, ^{16}O, ^{32}S). Nuclei in this category neither give rise to NMR spectra nor couple to other nuclei.

2. Those with half-integral spin in which either the number of protons or the number of neutrons is odd (e.g., ^{1}H, ^{11}B, ^{19}F, ^{31}P, ^{35}Cl, ^{79}Br).

3. Those with integral spin in which both the number of neutrons and the number of protons is odd (e.g., ^{2}H, ^{14}N).

Nuclei with nuclear spins of $I = \frac{1}{2}$, called **dipolar nuclei**, act as though they were spherical bodies possessing a uniform charge distribution, which circulates over their surfaces. When a probing electrical charge approaches such a nucleus, it experiences an electrostatic field whose magnitude is independent of the direction of approach. Nuclei of this sort are ^{1}H, ^{13}C, ^{15}N, ^{19}F, ^{29}Si, ^{31}P, ^{57}Fe, ^{77}Se, ^{89}Y, ^{103}Rh, ^{107}Ag, ^{109}Ag, ^{111}Cd, ^{113}Cd, ^{115}Sn, ^{117}Sn, ^{119}Sn, ^{123}Te, ^{125}Te, ^{129}Xe, ^{169}Tm, ^{183}W, ^{187}Os, ^{195}Pt, ^{203}Tl, ^{205}Tl, and ^{207}Pb. (See Table 1.1.)

Nuclei with $I > \frac{1}{2}$, called **quadrupolar nuclei** (see Table 1.2) have a nonspherical distribution of spinning charge, resulting in nonsymmetrical electric and magnetic fields. This imparts an electric quadrupole (Q) to the nucleus, a property that can complicate its NMR behavior. The most frequently studied nuclei are those with a nuclear spin of $\frac{1}{2}$.

Table 1.1 NMR Properties of the Spin-1/2 Nuclei[a]

Isotope[b]	Natural abundance[c] C/%	Magnetic moment[d] μ/μ_N	Magnetogyric ratio[d] $\gamma/10^7$ rad $T^{-1}s^{-1}$	NMR frequency[e] Ξ/MHz	Standard	Relative receptivity[f] D^P	D^C
^{1}H	99.985	4.83724	26.7519	100.000 000	Me$_4$Si	1.000	5.67×10^3
^{3}H[g]	–	5.1596	28.535	(106.664)	Me$_4$Si-t	–	–
^{3}He	1.3×10^{-4}	−3.6851	−20.380	(76.182)	–	5.75×10^{-7}	3.26×10^{-3}
^{13}C	1.108	1.2166	6.7283	25.145 004	Me$_4$Si	1.76×10^{-4}	1.00
^{15}N	0.37	−0.4903	−2.712	10.136 783	MeNO$_2$ or [NO$_3$]$^-$	3.85×10^{-6}	2.19×10^{-2}
^{19}F	100	4.5532	25.181	94.094 003[i]	CCl$_3$F	0.834	4.73×10^3
^{29}Si	4.70	−0.96174	−5.3188	19.867 184	Me$_4$Si	3.69×10^{-4}	2.10
^{31}P	100	1.9602	10.841	40.480 737	85%H$_3$PO$_4$	0.0665	3.77×10^2
^{57}Fe	2.19	0.1566	0.8661	(3.238)	Fe(CO)$_5$	7.43×10^{-7}	4.22×10^{-3}
^{77}Se	7.58	0.925	5.12	19.071 523	Me$_2$Se	5.30×10^{-4}	3.01
^{89}Y	100	−0.23786	−1.3155	(4.917)	Y(NO$_3$)$_3$aq.	1.19×10^{-4}	0.675
^{103}Rh	100	−0.153	−0.846	3.172 310	mer-[RhCl$_3$(SMe$_2$)$_3$]	3.16×10^{-5}	0.179
(^{107}Ag)	51.82	−0.1966	−1.087	4.047 897	Ag$^+$aq.	3.48×10^{-5}	0.197
^{109}Ag	48.18	−0.2260	−1.250	4.653 623	Ag$^+$aq.	4.92×10^{-5}	0.279
(^{111}Cd)	12.75	−1.0293	−5.6926	21.215 478	CdMe$_2$	1.23×10^{-3}	6.97
^{113}Cd[h]	12.26	−1.0768	−5.9550	22.193 173	CdMe$_2$	1.35×10^{-3}	7.67
(^{115}Sn)	0.35	−1.590	−8.792	(32.86)	Me$_4$Sn	1.24×10^{-4}	0.705
(^{117}Sn)	7.61	−1.732	−9.578	35.632 295	Me$_4$Sn	3.49×10^{-3}	19.8
^{119}Sn	8.58	−1.8119	−10.021	37.290 662	Me$_4$Sn	4.51×10^{-3}	25.6
(^{123}Te[h])	0.87	−1.275	−7.049	(26.35)	Me$_2$Te	1.59×10^{-4}	0.903
^{125}Te	6.99	−1.537	−8.498	31.549 802	Me$_2$Te	2.24×10^{-3}	12.7
^{129}Xe	26.44	−1.345	−7.441	(27.81)	XeOF$_4$	5.69×10^{-3}	32.3
^{169}Tm	100	−0.400	−2.21	(8.27)	–	5.66×10^{-4}	3.21
^{171}Yb	14.31	0.8520	4.712	(17.61)	–	7.82×10^{-4}	4.44
^{183}W	14.40	0.2025	1.120	4.161 733	WF$_6$	1.06×10^{-5}	5.99×10^{-2}
^{187}Os	1.64	0.111	0.616	2.282 343	OsO$_4$	2.00×10^{-7}	1.14×10^{-3}
^{195}Pt	33.8	1.043	5.768	21.414 376	[Pt(CN)$_6$]$^{2-}$	3.39×10^{-3}	19.2
^{199}Hg	16.84	0.87072	4.8154	17.910 841	Me$_2$Hg	9.82×10^{-4}	5.57
(^{203}Tl)	29.50	2.7912	15.436	(57.70)	TlNO$_3$aq.	0.0567	3.22×10^2
^{205}Tl	70.50	2.8187	15.589	57.633 833	TlNO$_3$aq.	0.140	7.91×10^2
^{207}Pb	22.6	1.002	5.540	20.920 597	Me$_4$Pb	2.01×10^{-3}	11.4

[a] A complete list, excluding most radioactive nuclei.

[b] Nuclei in parentheses are considered to be not the most favorable for the element concerned.

[c] Data from *Handbook of Chemistry and Physics*, 55th edition, CRC Press (1974–5), pages B248–332, except for the value of ^{13}C (which is taken from page E69).

[d] Data derived from the compilation of G. H. Fuller, *J. Phys. Chem. Ref. Data* **5**, 835 (1976), which lists values of $\mu_{max} = \gamma\hbar I$, corrected for diamagnetic shielding.

[e] Resonance frequency in a magnetic field such that the protons of TMS resonate at exactly 100 MHz. The values quoted are for the resonances of the standards listed in the next column and are taken (except where otherwise stated) from *NMR and the Periodic Table*, ed. R. K. Harris & B. E. Mann, Academic Press, New York (1978). Values in brackets are calculated from the magnetogyric ratios given in the preceding column, and are therefore relative to the resonant frequency of bare protons.

[f] D^P is the receptivity relative to that of ^{1}H, whereas D^C is relative to ^{13}C.

[g] Radioactive (half-life 12y).

[h] Long-lived radioactive isotope.

[i] S. Brownstein & J. Bornais, *J. Magn. Reson.* **38**, 131 (1980).

Table 1.2   The Spin Properties of Quadrupolar Nuclei[a]

Isotope[b]	Spin[c]	Natural abundance[c] C/%	Magnetic moment[d] μ/μ_N	Magnetogyric ratio[c] $\gamma/10^7$ rad T^{-1}s^{-1}	Quadrupole moment $10^{28}Q$/m^2	NMR frequency[f] Ξ/MHz	Linewidth factor[g] $10^{56}\ell$/m^4	Relative receptivity[f] D^P	D^C
^{2}H	1	0.015	1.2126	4.1066	2.8×10^{-3}	15.351	3.9×10^{-5}	1.45×10^{-6}	8.21×10^{-3}
^{6}Li	1	7.42	1.1625	3.9371	-8×10^{-4}	14.717	3.2×10^{-6}	6.31×10^{-4}	3.58
^{7}Li	3/2	92.58	4.20394	10.3975	-4×10^{-2}	38.866	2.1×10^{-3}	0.272	1.54×10^{3}
^{9}Be	3/2	100	-1.52008	-3.7598	5×10^{-2}	14.054	3.3×10^{-3}	1.39×10^{-2}	78.7
^{10}B	3	19.58	2.0792	2.8746	8.5×10^{-2}	10.746	1.4×10^{-3}	3.93×10^{-3}	22.3
^{11}B	3/2	80.42	3.4708	8.5843	4.1×10^{-2}	32.089	2.2×10^{-3}	0.133	7.52×10^{2}
^{14}N[i]	1	99.63	0.57099	1.9338	1×10^{-2}	7.228	5.0×10^{-4}	1.00×10^{-3}	5.69
^{17}O	5/2	0.037	-2.2407	-3.6279	-2.6×10^{-2}	13.561	2.2×10^{-4}	1.08×10^{-5}	6.11×10^{-2}
^{21}Ne	3/2	0.257	-0.85433	-2.1130	9×10^{-2}	7.899	1.1×10^{-2}	6.33×10^{-6}	3.59×10^{-2}
^{23}Na	3/2	100	2.86265	7.08013	0.10	26.466	1.3×10^{-2}	9.27×10^{-2}	5.26×10^{2}
^{25}Mg	5/2	10.13	-1.012	-1.639	0.22	6.126	1.5×10^{-2}	2.72×10^{-4}	1.54
^{27}Al	5/2	100	4.3084	6.9760	0.15	26.077	7.2×10^{-3}	0.207	1.17×10^{3}
^{33}S	3/2	0.76	0.8308	2.055	-5.5×10^{-2}	7.681	4.0×10^{-3}	1.72×10^{-5}	9.77×10^{-2}
^{35}Cl	3/2	75.53	1.0610	2.6240	-0.10	9.809	1.3×10^{-2}	3.56×10^{-3}	20.2
^{37}Cl	3/2	24.47	0.88313	2.1842	-7.9×10^{-2}	8.165	8.3×10^{-3}	6.66×10^{-4}	3.78
^{39}K	3/2	93.1	0.50533	1.2498	4.9×10^{-2}	4.672	3.2×10^{-3}	4.75×10^{-4}	2.69
(^{41}K)	3/2	6.88	0.27740	0.68608	6.0×10^{-2}	2.565	4.8×10^{-3}	5.80×10^{-6}	3.29×10^{-2}
^{43}Ca	7/2	0.145	-1.4936	-1.8025	0.2[j]	6.738	5.4×10^{-3}	8.67×10^{-6}	4.92×10^{-2}
^{45}Sc	7/2	100	5.3927	6.5081	-0.22	24.328	6.6×10^{-3}	0.302	1.72×10^{3}
^{47}Ti	5/2	7.28	-0.93292	-1.5105	0.29	5.646	2.7×10^{-2}	1.53×10^{-4}	0.867
^{49}Ti	7/2	5.51	-1.25198	-1.51093	0.24	5.648	7.8×10^{-3}	2.08×10^{-4}	1.18
(^{50}V)[k]	6	0.24	3.6152	2.6717	$\pm6\times10^{-2}$	9.987	1.4×10^{-4}	1.34×10^{-4}	0.759
^{51}V	7/2	99.76	5.8379	7.0453	-5×10^{-2}	26.336	3.4×10^{-4}	0.383	2.17×10^{3}
^{53}Cr	3/2	9.55	-0.6113	-1.512	-0.15	5.651	1.2×10^{-3}	8.62×10^{-5}	0.489
^{55}Mn	5/2	100	4.081	6.608	0.4	24.70	5.1×10^{-2}	0.176	9.97×10^{2}
^{59}Co	7/2	100	5.234	6.317	0.38	23.61	2.0×10^{-2}	0.277	1.57×10^{3}
^{61}Ni	3/2	1.19	-0.9680	-2.394	0.16	8.949	3.4×10^{-2}	4.06×10^{-5}	0.231
^{63}Cu	3/2	69.09	2.8696	7.0974	-0.211	26.530	5.94×10^{-2}	6.45×10^{-2}	3.66×10^{2}
^{65}Cu	3/2	30.91	3.0741	7.6031	-0.195	28.421	5.07×10^{-2}	3.55×10^{-2}	2.01×10^{2}
^{67}Zn	5/2	4.11	1.0356	1.6768	0.16	6.2679	8.2×10^{-3}	1.18×10^{-4}	0.670
(^{69}Ga)	3/2	60.4	2.6007	6.4323	0.19	24.044	4.8×10^{-2}	4.19×10^{-2}	2.38×10^{2}
^{71}Ga	3/2	39.6	3.3046	8.1731	0.12	30.551	1.9×10^{-2}	5.65×10^{-2}	3.20×10^{2}
^{73}Ge	9/2	7.76	-0.97197	-0.93574	-0.18	3.498	2.4×10^{-3}	1.10×10^{-4}	0.622
^{75}As	3/2	100	1.858	4.595	0.29	17.18	0.11	2.53×10^{-2}	1.44×10^{2}
(^{79}Br)	3/2	50.54	2.7182	6.7228	0.37	25.130	0.18	4.01×10^{-2}	2.28×10^{2}
^{81}Br	3/2	49.46	2.9300	7.2468	0.31	27.089	0.13	4.92×10^{-2}	2.79×10^{2}
^{83}Kr	9/2	11.55	-1.073	-1.033	0.26	3.860	5.0×10^{-3}	2.19×10^{-4}	1.24
(^{85}Rb)	5/2	72.15	1.6002	2.5909	0.26	9.685	2.2×10^{-2}	7.65×10^{-3}	43.4

	I								
^{87}Rb[k]	3/2	27.85	3.5502	8.7807	0.13	32.823	2.3×10^{-2}	4.92×10^{-2}	2.79×10^{2}
^{87}Sr	9/2	7.02	−1.208	−1.163	0.3	4.349	6.7×10^{-3}	1.91×10^{-4}	1.08
^{91}Zr	5/2	11.23	−1.5415	−2.4959	−0.21[l]	9.3298	1.4×10^{-2}	1.06×10^{-3}	6.04
^{93}Nb	9/2	100	6.818	6.564	−0.22	24.54	3.6×10^{-3}	0.487	2.77×10^{3}
^{95}Mo	5/2	15.72	1.081	1.750	−0.022	6.542	4.6×10^{-3}	5.14×10^{-4}	2.92
(^{97}Mo)	5/2	9.46	−1.104	−1.787	0.255	6.679	0.39	3.29×10^{-4}	1.87
^{99}Tc[k]	9/2	—	6.281	6.046	0.3	22.60	6.7×10^{-3}	—	–
^{99}Ru	5/2	12.72	−0.7623[m]	−1.234[m]	7.6×10^{-2}	4.614[m]	1.8×10^{-3}	1.46×10^{-4}	0.827
^{101}Ru	5/2	17.07	−0.8544[m]	−1.383[m]	0.44	5.171[m]	6.2×10^{-2}	2.75×10^{-4}	1.56
^{105}Pd	5/2	22.23	−0.760	−1.23	0.8	4.60	0.20	2.52×10^{-4}	1.43
(^{113}In)	9/2	4.28	6.1058	5.8782	0.82	21.973	5.0×10^{-2}	1.50×10^{-2}	85.0
^{115}In[k]	9/2	95.72	6.1190	5.8908	0.83	22.020	5.1×10^{-2}	0.337	1.91×10^{3}
^{121}Sb	5/2	57.25	3.9747	6.4355	−0.28	24.056	2.5×10^{-2}	9.30×10^{-2}	5.27×10^{2}
(^{123}Sb)	7/2	42.75	2.8876	3.4848	−0.36	13.026	1.8×10^{-2}	1.98×10^{-2}	1.13×10^{2}
^{127}I	5/2	100	3.3238	5.3817	−0.79	20.117	0.20	9.20×10^{-2}	5.39×10^{2}
^{131}Xe[i]	3/2	21.18	0.8918	2.206	−0.12	8.245	1.9×10^{-2}	5.94×10^{-4}	3.37
^{133}Cs	7/2	100	2.9231	3.5277	-3×10^{-3}	13.187	1.2×10^{-6}	4.82×10^{-2}	2.73×10^{2}
(^{135}Ba)	3/2	6.59	1.080	2.671	0.18	9.984	4.3×10^{-2}	3.28×10^{-4}	1.86
^{137}Ba	3/2	11.32	1.208	2.988	0.28	11.17	0.10	7.89×10^{-4}	4.47
^{139}La	7/2	99.911	3.150	3.801	0.22	14.210	6.6×10^{-3}	6.02×10^{-2}	3.42×10^{2}
^{177}Hf	7/2	18.50	0.8960	1.081	4.5	4.042	2.8	2.57×10^{-4}	1.46
^{179}Hf	9/2	13.75	−0.705	−0.679	5.1	2.54	1.9	7.42×10^{-4}	0.421
^{181}Ta	7/2	99.988	2.66	3.22	3	12.0	1.2	3.65×10^{-2}	2.07×10^{2}
(^{185}Re)	5/2	37.07	3.753	6.077	2.3	22.72	1.7	5.13×10^{-2}	2.91×10^{2}
^{187}Re[k]	5/2	62.93	3.791	6.138	2.2	22.94	1.5	8.81×10^{-2}	5.00×10^{2}
^{189}Os[l]	3/2	16.1	2.096	0.8475	0.8	7.836	0.85	3.87×10^{-4}	2.20
(^{191}Ir)	3/2	37.3	0.1877	0.4643	1.1	1.735	1.6	9.77×10^{-6}	5.54×10^{-2}
^{193}Ir	3/2	62.7	0.2044	0.5054	1.0	1.889	1.3	2.11×10^{-5}	0.120
^{197}Au	3/2	100	0.18701	0.46254	0.59	1.729	0.46	2.58×10^{-5}	0.147
^{201}Hg[i]	3/2	13.22	−0.71871	−1.7776	0.44	6.645	0.26	1.94×10^{-4}	1.10
^{209}Bi[k]	9/2	100	4.511	4.342	−0.38	16.23	1.1×10^{-2}	0.141	8.01×10^{2}

[a] Excluding the lanthanides.

[b] See footnote b to Table 1.1.

[c] Loc. cit. in footnote d to Table 1.1, except where otherwise stated.

[d] Loc. cit. in footnote c to Table 1.1.

[e] It should be noted that reported values of Q may be in error by as much as 20–30%.

[f] Calculated from the quoted value of γ. (Therefore, diamagnetic corrections are included and the frequency quoted is not with respect to TMS.)

[g] $\ell = (2I + 3)Q^2/[I^2(2I - 1)]$.

[h] See footnote (f) to Table 2.1.

[i] A useful isotope of $I = \frac{1}{2}$ exists.

[j] R. Neumann, F. Träger, J. Kowalski, & G. zu Putlitz, Z. Physik **A279**, 249 (1976).

[k] Radioactive, with a long half-life. Other radioactive nuclei (e.g., ^{40}K, ^{41}Ca) have been examined by NMR, but are unimportant for chemical studies.

[l] S. Büttgenbach, R. Dicke, H. Gebauer, R. Kuhnen, & F. Träber, Z. Physik **A286**, 125 (1978).

[m] Derived from data in C. Brevard & P. Granger, J. Chem. Phys. **75**, 4175 (1981).

[n] S. Büttgenbach, R. Dicke, H. Gebauer & M. Herschel, Z. Physik **A280**, 217 (1977).

1.2 Interaction of Nuclear Spins with Magnetic Fields

The nuclear magnetic moment (μ) is directly proportional to the spin. That is,

$$\mu = \gamma I \hbar, \tag{1.1}$$

where the proportionality constant is called the **magnetogyric** or **gyromagnetic ratio** and is a constant for each particular nucleus. Table 1.1 lists the spin properties for the dipolar $\left(I = \frac{1}{2}\right)$ nuclei, and Table 1.2 lists the spin properties of the quadrupolar $\left(I > \frac{1}{2}\right)$ nuclei.

When a nucleus is placed in a magnetic field, the nuclear moments orient themselves with only certain allowed orientations, as we are considering a quantum mechanical system. A nucleus with spin I has $2I + 1$ possible orientations, given by the values of the magnetic quantum number M_I, which may take values of $-I, -I + 1, -I + 2, \ldots, I$. Thus, for a nucleus with $I = \frac{5}{2}$, M_I has the values $-\frac{5}{2}, -\frac{3}{2}, -\frac{1}{2}, \frac{1}{2}, \frac{3}{2}$, and $\frac{5}{2}$.

The energy of interaction is proportional to the nuclear magnetic moment and the applied field strength, and from Equation 1.1, we may write the equation

$$E = -\gamma \hbar M_I \mathbf{B_0}, \tag{1.2}$$

where $\mathbf{B_0}$ is the applied field strength. In the CGS system of units, the magnetic field $\mathbf{B_0}$ is measured in gauss, the nuclear moment (μ) has units of erg/gauss and is often given in nuclear magnetons (1 nuclear magneton equals 5.05×10^{-24} erg/gauss), and the magnetogyric ratio (γ) has the units of radians/gauss-sec. In SI units, $\mathbf{B_0}$ is in tesla (1 tesla $= 10^4$ gauss), μ is in ampere-meter^2, and γ is in radians/tesla-sec.

The energy levels for a nucleus with $I = \frac{5}{2}$ are shown schematically in the figure on the left, which is obtained directly from Equation 1.2. Because the sign of μ for a neutron is negative, the lower energy orientation corresponds to $M_I = +\frac{5}{2}$.

The selection rule for NMR transitions requires that M_I can change only by one unit (only single quantum transitions are allowed); that is, $\Delta M_I = \pm 1$. Thus, the transition energy is given by

$$\Delta E = \gamma \hbar \mathbf{B_0}. \tag{1.3}$$

In order to detect a transition with this energy, radiation given by $\Delta E = h\upsilon$ must be applied. Combining this with Equation 1.3, we have the fundamental resonance condition for all NMR experiments:

$$h\upsilon = \gamma \hbar \mathbf{B_0} \rightarrow \upsilon = \frac{\gamma \mathbf{B_0}}{2\pi}. \tag{1.4}$$

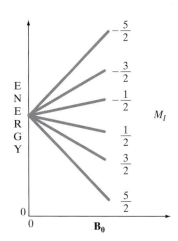

SAMPLE PROBLEM 1.1

Calculate the resonance frequency for a proton in a field of 14,092 gauss $=$ 1.4092 tesla.

Solution

From Equation 1.4 and Table 1.1, we have

$$\upsilon = \frac{\gamma \mathbf{B_0}}{2\pi} = \left(\frac{26.7519 \times 10^7 \text{ rad T}^{-1}\text{sec}^{-1}}{2\pi} \right) \left(\frac{10^{-4}\text{T}}{\text{G}} \right) (14,092\,\text{G})$$

$$= 59999467.84\,\text{s}^{-1} \cong 60\,\text{MHz}.$$

When a nucleus with magnetogyric ratio γ is placed in a magnetic field with field strength $\mathbf{B_0}$, the resonance condition is satisfied when the frequency of the applied radiation, υ, is given by Equation 1.4. Note, in particular, the relationship between field strength and frequency in this equation. As the field strength increases, the resonance frequency increases. At the magnetic fields currently obtainable in the laboratory (10–176 kgauss, or 1–17.6 tesla), the resonance frequencies υ of most nuclei are in the radio-frequency region (10–750 MHz).

1.3   Precession and the Larmor Frequency

Nuclei with $I \neq 0$, when placed in a magnetic field, adopt $2I + 1$ spin orientations with respect to the z-axis of the field. Each orientation has a different energy. Before these nuclei can absorb photons from an applied radio-frequency field, they must be oscillating with a uniform periodic motion with a frequency that corresponds to that of the radio-frequency field. The magnetic moments are not statically aligned parallel or antiparallel to $\mathbf{B_0}$. Rather, they remain at a certain angle with respect to $\mathbf{B_0}$ and **precess** like a top around the z-axis of $\mathbf{B_0}$ at a fixed frequency called the **Larmor frequency**, given by

$$\omega = \gamma \mathbf{B_0}. \qquad (1.5)$$

The angular Lamor frequency, in units of radians/sec, is related to the linear frequency υ, in reciprocal seconds or Hz, by the equation

$$\upsilon_{\text{precession}} = \frac{\omega}{2\pi} = \frac{\gamma \mathbf{B_0}}{2\pi}. \qquad (1.6)$$

Compare Equations 1.4 and 1.6.

The precessional motion causes the tip of the magnetic-moment vectors to trace out circular paths as shown in Figure 1.1. The precession frequency is independent of the value of M_I so that all spin orientations of a given nucleus precess at the same frequency in a fixed magnetic field.

1.4   Nuclear Energy Levels and Relaxation

There are two possible orientations of the nuclear spin for a dipolar nucleus $\left(I = \frac{1}{2}\right)$. In the absence of a magnetic field, these have the same energy, but in a magnetic field their degeneracy is lifted, due to the Zeeman effect, and the $M_I = \frac{1}{2}$ becomes the lower energy state (often symbolized by α). The higher energy state, $M_I = -\frac{1}{2}$, is frequently symbolized by β. The energy separation between these two states increases as the magnetic field strength $\mathbf{B_0}$ increases (see Equation 1.3), as illustrated in Figure 1.2.

There are two allowed transitions ($\Delta I = \pm 1$): (a) $\alpha \rightarrow \beta$, which corresponds to an absorption of energy, and (b) $\beta \rightarrow \alpha$, which corresponds to induced emission. The probabilities of absorption and induced emission are equal for NMR spectroscopy, and, therefore, there would be no net transfer of energy from the applied radiation to the sample if the populations of the two states were equal.

However, because the sample is in thermal equilibrium, the population of the two states is governed by the Boltzman distribution (Equation 1.7). If N_α and N_β are the number of spins in the α and β states, then

$$\frac{N_\beta}{N_\alpha} = e^{-(h\upsilon/kT)} = e^{-(\gamma\hbar\mathbf{B_0}/kT)} \qquad (1.7)$$

$$= 1 - \frac{h\upsilon}{kT} \qquad \text{when } h\upsilon < kT.$$

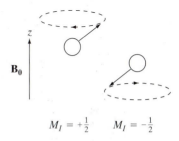

$M_I = +\frac{1}{2}$ $M_I = -\frac{1}{2}$

▲ FIGURE 1.1
Precession of the magnetic moments for the $M_I = \frac{1}{2}$ and $M_I = -\frac{1}{2}$ spin states in an external magnetic field $\mathbf{B_0}$.

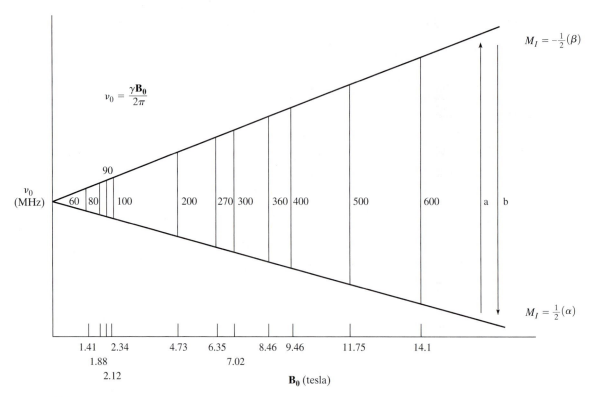

▲ FIGURE 1.2
The splitting of magnetic energy levels of protons, expressed as resonance frequency, as a function of magnetic field strength, expressed in tesla (T).

Thus, $N_\alpha > N_\beta$, and a net absorption of energy will occur, giving rise to an NMR signal. The amplitude of the signal is roughly proportional to the field strength $\mathbf{B}_0^{3/2}$, so strong magnetic fields are desirable.

With frequencies ν in the radio-frequency region, $h\nu \cong 10^{-2}$ cal (4.2×10^{-2} Joules); thus, at ambient temperature, the excess population of spins in the lower energy state is only approximately 1×10^{-5}. This very small population difference is the basic reason for the low sensitivity of NMR compared with infrared (IR) and especially electronic absorption spectroscopy. We do, however, get some compensation for this, as the coefficient of absorption is a constant for any nucleus, and thus, **the NMR signal obtained is directly proportional to the number of nuclei producing it**.

Obviously, if a constant RF field were applied with sufficient power (actually, very little power, in the milligauss range), the populations in the two energy states would equalize, and no more absorption of energy would occur, thus rendering the experiment useless. This phenomenon occurs frequently and is termed **saturation**. It is desirable then to use the minimum-power RF field necessary for observation of resonance consistent with maximum signal-to-noise ratio.

In order for the experiment to be useful, there must be a mechanism for return of the system from the higher energy state to the lower energy state. This is accomplished by a phenomenon known as **relaxation**, and the time required is called the **relaxation time**. (See also Chapter 7.) This may be a long time—several seconds or sometimes minutes for some nuclei. A simple

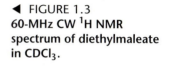

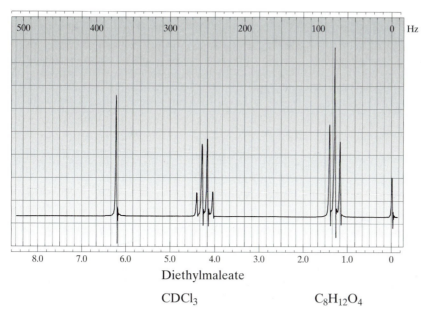

Diethylmaleate

CDCl$_3$ C$_8$H$_{12}$O$_4$

manifestation of the relaxation time is the ringing or wiggles observed immediately after a sharp signal has been detected in continuous wave operation. (See Chapter 2.) A typical example is illustrated in Figure 1.3. When the spectrometer is swept sufficiently slowly (slow passage conditions) so that the nuclei are in thermal equilibrium at all times, these wiggles are not observed. There is a more important consequence of the long nuclear relaxation times in pulsed spectrometers (in which those nuclei with the longest relaxation times produce the smallest signals if the pulsing frequency is too rapid to allow relaxation back to thermal equilibrium between each pulse). In such cases, the signal intensity is related to the relaxation times rather than to the number of nuclei.

There are two fundamental mechanisms of relaxation:

a. **Spin–spin relaxation (T$_2$).** Here, a nucleus of one atom imparts its energy to another atom in a low-energy state, with the result that no overall net change in populations occurs.
b. **Spin–lattice relaxation (T$_1$).** Here, energy is transferred to the lattice (whatever that may be), solvent, electrons in the system, or other kinds of atoms or ions. This energy is converted into translational or rotational energy, and the nucleus returns to the lower energy state. Energy transfer is essential for the observation of an NMR spectrum.

The spin–lattice (or **longitudinal**) relaxation time T_1 involves an exponential decay process given by

$$\frac{P_{eq} - P_t}{P_{eq} - P_o} = \exp\left(-\frac{t}{T_1}\right). \qquad \textbf{(1.8)}$$

In Equation 1.8, $P_{eq} - P_t$ is the difference between the equilibrium population (P_{eq}) of a given state, such as the α state of a proton, and the population after time $t(P_t)$, and $P_{eq} - P_o$ is the difference between the equilibrium population and that at time zero (P_o).

SAMPLE PROBLEM 1.2

If $T_1 = 1.0$ sec for a given set of ^{1}H nuclei at 25°C in a 7.0-tesla magnetic field, how long will it take for an initially equal distribution of ^{1}H spin states to progress 95% of the way toward equilibrium?

Solution

From Equation 1.3, we have

$$\Delta E = \gamma \hbar B_0 = \frac{(26.7519 \times 10^7 \, radT^{-1}s^{-1})(6.63 \times 10^{-34} \, Js)(7.02 \, T)}{2(3.14 \, rad)}$$

$$= 1.98 \times 10^{-25} \, J.$$

From Equation 1.7, we have

$$\frac{P_\beta}{P_\alpha} = \exp\left(\frac{-\Delta E}{kT}\right) = \exp\left(\frac{-1.98 \times 10^{-25} \, J}{1.381 \times 10^{-23} \, JK^{-1} \times 298 \, K}\right) = 0.9995.$$

Or since $P_\alpha + P_\beta = 1$, $P_\alpha = 0.5001$ at equilibrium. At 95% of equilibrium, $P_{eq} - P_t = 0.05(P_{eq} - P_o)$, or

$$\frac{0.05(P_{eq} - P_o)}{(P_{eq} - P_o)} = \exp\left(-\frac{t}{T_1}\right),$$

whence ($0.05 = \exp(-t/T_1)$). Taking the natural logarithm of both sides of this equation gives $-3.00 = -t/T_1$, and since $T_1 = 1 \, s$, $t = 3(1 \, s) = 3.0 \, s$.

In Chapter 2, we will discuss how an NMR signal is actually generated. Figure 1.4 illustrates a collection of $I = \frac{1}{2}$ nuclei at equilibrium in a magnetic field $\mathbf{B_0}$. Before RF irradiation begins (Figure 1.4a), nuclei in both the α and β

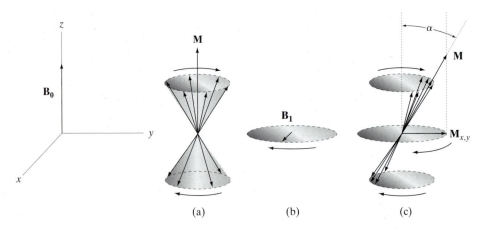

▲ FIGURE 1.4
Precession of a collection of $I = \frac{1}{2}$ nuclei around an external magnetic field $\mathbf{B_0}$. M represents the vector sum of the individual nuclear magnetic moments.

(a) Before irradiation by $\mathbf{B_1}$. (b) The orientation of the rotating magnetic component of the RF field $\mathbf{B_1}$. (c) During irradiation by $\mathbf{B_1}$.

spin states precess with their Larmor frequency, but they are completely **out of phase**, or randomly oriented, around the z-axis. The *net* nuclear magnetization **M** is the vector sum of the individual nuclear magnetic moments, with its magnitude determined by the excess population of α over β spins. **M** is aligned parallel to $\mathbf{B_0}$ and has no precessional motion and no component in the xy-plane.

When a weak magnetic field $\mathbf{B_1}$, *perpendicular* to $\mathbf{B_0}$, is applied that precesses in the xy-plane at the Larmor frequency, then all the individual magnetic moments *focus*, or become *phase coherent*. They track the oscillating magnetic field $\mathbf{B_1}$ and form a precessing bundle (Figure 1.4c). If the $\mathbf{B_1}$ field is sufficiently weak such that saturation does not occur, this phase coherence also requires that **M** tip away from the z-axis and begin to precess around that axis with the characteristic Larmor frequency. But now, since **M** is no longer aligned along the z-axis, it has a component in the xy-plane ($\mathbf{M_{x,y}}$) that oscillates with the same frequency. The flip angle α that **M** makes with the z-axis gives the magnitude of $\mathbf{M_{x,y}}$ according to the equation

$$\mathbf{M_{x,y}} = \mathbf{M} \sin \alpha. \qquad \textbf{(1.9)}$$

The angle α is determined by the power and duration of $\mathbf{B_1}$ with a maximum at $\alpha = 90°$ because $\sin 90° = 1$.

After irradiation by the $\mathbf{B_1}$ field stops, not only do the α and β states revert to their Boltzman distribution, but also, the individual nuclear magnetic moments begin to lose their phase coherence and return to a random orientation around the z-axis (Figure 1.4a). The latter process, called *spin–spin* (or *transverse*) *relaxation*, causes a decay of $\mathbf{M_{x,y}}$ at a rate that is controlled by T_2, the *spin–spin relaxation time*. Normally, T_2 is much shorter than T_1.

1.5   The Rotating Frame of Reference

We often want to focus our attention on the net nuclear magnetic moment **M**, rather than on the individual nuclear spins. Because **M** usually precesses around the z-axis of $\mathbf{B_0}$ when the weak RF field $\mathbf{B_1}$ is applied, we seek a more convenient way than the dotted ellipses of Figure 1.4 to depict this motion. We will adopt the **rotating frame of reference** as a convention to more easily visualize the effects of $\mathbf{B_1}$ on **M**.

Figure 1.5 illustrates four representations of **M** in what is termed the **laboratory frame of reference**, an xyz coordinate system as viewed by a stationary observer in the laboratory.

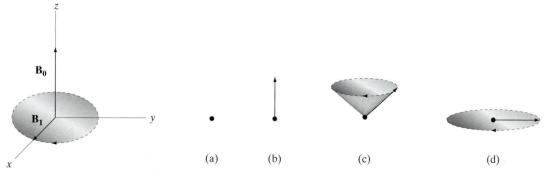

▲ FIGURE 1.5
Depiction of M in the laboratory frame.

▶ FIGURE 1.6
Axes of the rotating frame of reference, as viewed (a) by a stationary observer in the laboratory frame and (b) by an observer precessing in the rotating frame at the Larmor frequency.

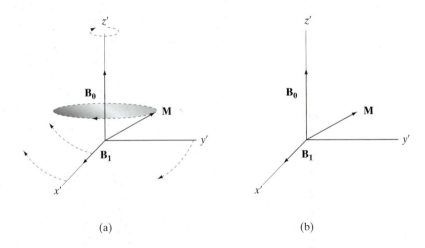

(a)   (b)

In part (a) of the figure, $\mathbf{B_0}$ and $\mathbf{B_1}$ are turned off so that the populations of the α and β spin states are equal and the magnitude of $\mathbf{M}$ is zero. In (b), the Boltzman equilibrium distribution of $\mathbf{M}$ has been achieved by turning on the $\mathbf{B_0}$ field. But remember that the individual nuclear magnetic moments are now precessing around the z-axis of $\mathbf{B_0}$ with their Larmor frequency. Part (c) shows $\mathbf{M}$ tipped to an angle of 45° through interaction with the $\mathbf{B_1}$ field with the appropriate strength and duration, and the resulting precession of $\mathbf{M}$ about the z-axis of $\mathbf{B_0}$ describes a cone. Finally, in (d), the flip angle is 90° such that precession of $\mathbf{M}$ now describes a disk in the xy-plane.

If we now let the x- and y-axes move at the Larmor frequency and we, as observers, also move at this same frequency, then $\mathbf{M}$ will appear to us to be stationary. We will designate this *rotating frame of reference* by the axes labels x', y', and z' to distinguish them from the stationary laboratory axes, x, y, and z. The essence of the rotating frame is illustrated in Figure 1.6.

Viewing the motion of $\mathbf{M}$ in the rotating frame of reference simplifies our drawings. But it is important to remember that whenever $\mathbf{M}$ is anywhere except directly along the z'-axis, it has a component that oscillates in the $x'y'$-plane, and it is this component that gives rise to an NMR signal.

1.6 The Bloch Equations

Felix Bloch* treated the bulk magnetization $\mathbf{M}$ of a collection of nuclei by what has become to be called the *Bloch equations*. When $\mathbf{M}$ is not collinear with any of the x-, y-, or z-axes, as in Figure 1.6, and $\mathbf{B_0}$ is along the z-axis, then $\mathbf{M}$ precesses about z with an angular frequency of ω_0. If no relaxation were to occur and the $\mathbf{B_1}$ field were not turned on, then the projection of $\mathbf{M}$ on the z-axis, $\mathbf{M_z}$, would remain constant, and

$$\frac{d\mathbf{M}_z}{dt} = 0. \tag{1.10}$$

The magnitudes of the x and y projections, $\mathbf{M_x}$ and $\mathbf{M_y}$, vary with time as $\mathbf{M}$ precesses about the z-axis. $\mathbf{M_x}$ and $\mathbf{M_y}$ are 180° out of phase with each other, since $\mathbf{M_x}$ is at a maximum when $\mathbf{M_y}$ is zero and vice versa. The time dependencies of $\mathbf{M_x}$ and $\mathbf{M_y}$ are given by

---

*F. Bloch, *Phys. Rev.* **1946**, *70*, 460.

$$\frac{d\mathbf{M}_x}{dt} = \gamma\mathbf{M}_y \cdot \mathbf{B}_0 = \omega_o \mathbf{M}_y \tag{1.11}$$

and

$$\frac{d\mathbf{M}_y}{dt} = \gamma\mathbf{M}_x \cdot \mathbf{B}_0 = -\omega_o \mathbf{M}_x. \tag{1.12}$$

When the $\mathbf{B}_1$ field is turned on and it rotates in the xy-plane with frequency ω (ω_o only exactly on resonance), Equations 1.10 to 1.12 become

$$\frac{d\mathbf{M}_z}{dt} = \gamma[\mathbf{M}_x(\mathbf{B}_1)_y - \mathbf{M}_y(\mathbf{B}_1)_x], \tag{1.13}$$

$$\frac{d\mathbf{M}_x}{dt} = \gamma[\mathbf{M}_y \mathbf{B}_0 - \mathbf{M}_z(\mathbf{B}_1)_y], \tag{1.14}$$

and

$$\frac{d\mathbf{M}_y}{dt} = -\gamma[\mathbf{M}_x \mathbf{B}_0 + \mathbf{M}_z(\mathbf{B}_1)_x]. \tag{1.15}$$

In these equations, $(\mathbf{B}_1)_x$ and $(\mathbf{B}_1)_y$ are the components of $\mathbf{B}_1$ along the x- and y-axes and are given by

$$(\mathbf{B}_1)_x = \mathbf{B}_1 \cos \omega t \tag{1.16}$$

and

$$(\mathbf{B}_1)_y = \mathbf{B}_1 \sin \omega t. \tag{1.17}$$

The relaxation of $\mathbf{M}_z$ to its equilibrium value, $\mathbf{M}_0$, is given by

$$\frac{d\mathbf{M}_z}{dt} = \frac{-(\mathbf{M}_z - \mathbf{M}_0)}{\mathbf{T}_1}. \tag{1.18}$$

The transverse relaxation similarly involves $\mathbf{T}_2$ as the time constant:

$$\frac{d\mathbf{M}_x}{dt} = -\frac{\mathbf{M}_x}{\mathbf{T}_2}; \tag{1.19}$$

$$\frac{d\mathbf{M}_y}{dt} = -\frac{\mathbf{M}_y}{\mathbf{T}_2}. \tag{1.20}$$

The transverse magnetization relaxes to zero rather than to equilibrium values. Adding the relaxation terms to equations 1.13 through 1.17 gives the complete Bloch equations:

$$\frac{d\mathbf{M}_z}{dt} = -\gamma(\mathbf{M}_x \mathbf{M}_1 \sin \omega t + \mathbf{M}_y \mathbf{B}_1 \cos \omega t) - \frac{\mathbf{M}_z - \mathbf{M}_0}{\mathbf{T}_1}; \tag{1.21}$$

$$\frac{d\mathbf{M}_y}{dt} = -\gamma(\mathbf{M}_x \mathbf{B}_0 - \mathbf{M}_z \mathbf{B}_1 \cos \omega t) - \frac{\mathbf{M}_y}{\mathbf{T}_2}; \tag{1.22}$$

$$\frac{d\mathbf{M}_x}{dt} = -\gamma(\mathbf{M}_y \mathbf{B}_0 - \mathbf{M}_z \mathbf{B}_1 \sin \omega t) - \frac{\mathbf{M}_z}{\mathbf{T}_2}. \tag{1.23}$$

It is easier to visualize the experimental effects of these equations in the rotating frame, in which both $\mathbf{B}_0$ and $\mathbf{B}_1$ are stationary. We call the projections of $\mathbf{M}$ on the xy-plane u and v, which are along and perpendicular to $\mathbf{B}_1$, respectively. Now, u and v will be in phase and out of phase, respectively, with $\mathbf{B}_1$ as shown in Figure 1.7.

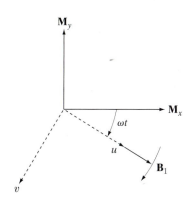

▲ FIGURE 1.7
Orientation of u and v in the xy-plane of the laboratory frame.

To transform to the rotating frame, we note that

$$\mathbf{M}_x = u \cos \omega t - v \sin \omega t \tag{1.24}$$

and

$$\mathbf{M}_y = -u \sin \omega t - v \cos \omega t \tag{1.25}$$

and recall that $\gamma \mathbf{B}_0 = \omega_0$. Then, equations 1.21 through 1.23 are replaced, respectively, by

$$\frac{du}{dt} + u/\mathbf{T}_2 + (\omega_0 - \omega)v = 0, \tag{1.26}$$

$$\frac{dv}{dt} + v/\mathbf{T}_2 - (\omega_0 - \omega)u + \gamma \mathbf{B}_1 \mathbf{M}_z = 0, \tag{1.27}$$

and

$$\frac{d\mathbf{M}_z}{dt} \frac{(\mathbf{M}_z - \mathbf{M}_0)}{\mathbf{T}_1} - \gamma \mathbf{B}_1 v = 0. \tag{1.28}$$

The term $\omega_0 - \omega$ is a measure of how far we are from being on resonance. From Equation 1.28, we see that changes in $\mathbf{M}_z$, or changes in the energy of the spin system, are associated only with v, the out-of-phase component of the macroscopic magnetization, and not with u. A result of this is that absorption signals will be associated with the measurement of v. The component u will be associated with *dispersion-mode* signals. These methods of observation will be considered further in Chapter 2.

Under specific conditions of experimental observation, we obtain a steady state in which u, v, and $\mathbf{M}_z$ are constant in the rotating frame. We pass through resonance by varying ω (or $\mathbf{B}_0$) at such a slow rate that u, v, and $\mathbf{M}_z$ always remain at these steady-state values. Under these so-called *slow-passage* conditions, one obtains the following equations from Equations 1.26 through 1.28:

$$u = \mathbf{M}_0 \left(\frac{\gamma \mathbf{B}_1 \mathbf{T}_2^2 (\omega_0 - \omega)}{1 + \mathbf{T}_2^2 (\omega_0 - \omega)^2 + \gamma^2 \mathbf{B}_1^2 \mathbf{T}_1 \mathbf{T}_2} \right); \tag{1.29}$$

$$v = -\mathbf{M}_0 \left(\frac{\gamma \mathbf{B}_1 \mathbf{T}_2}{1 + \mathbf{T}_2^2 (\omega_0 - \omega)^2 + \gamma^2 \mathbf{B}_1^2 \mathbf{T}_1 \mathbf{T}_2} \right); \tag{1.30}$$

$$\mathbf{M}_z = \mathbf{M}_0 \left(\frac{1 + \mathbf{T}_2^2 (\omega_0 - \omega)^2}{1 + \mathbf{T}_2^2 (\omega_0 - \omega)^2 + \gamma^2 \mathbf{B}_1^2 \mathbf{T}_1 \mathbf{T}_2} \right). \tag{1.31}$$

Now, when $\gamma^2 B_1^2 T_1 T_2 \ll 1$, which occurs when $\mathbf{B}_1$ is a few milligauss and $\mathbf{T}_1$ and $\mathbf{T}_2$ are no greater than a few seconds, the absorption, or *v-mode*, signal should be proportional to $\gamma \mathbf{B}_1 \mathbf{T}_2 / [1 + \mathbf{T}_2^2 (\omega_0 - \omega)^2]$. The latter function is known as a *Lorentzian line shape*, as shown in Figure 1.8(a), and is the type of signal normally observed in high-resolution spectra. At the center, when we are exactly on resonance, $\omega_0 = \omega$ and the signal height is proportional to $\gamma \mathbf{B}_1 \mathbf{T}_2$. The signal width is inversely proportional to $\mathbf{T}_2$. The peak width at half the maximum height, $\delta\omega$ or δv, is then given by

$$\delta\omega = \frac{2}{\mathbf{T}_2} \tag{1.32}$$

and

$$\delta v = \frac{1}{\pi \mathbf{T}_2}. \tag{1.33}$$

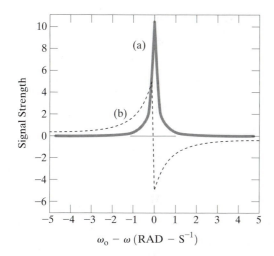

◀ FIGURE 1.8
v-mode [absorption (a)] and
u-mode [dispersion (b)]
Lorentzian line shapes.

For some purposes, it is advantageous to observe the *u*-mode, or *dispersion*, signal [Figure 1.8 (b)], which, under the same conditions, will be proportional to $\gamma B_1 T_2 (\omega_o - \omega)/[1 + T_2^2(\omega_o - \omega)^2]$. The maximum and minimum of the dispersion signal are separated by $1/\pi T_2$ for Lorentzian signals.

Additional Reading

1. An excellent computer tutorial entitled *The Basics of NMR Spectroscopy*, written by Joseph P. Hornak, is available through him at the Department of Chemistry, Rochester Institute of Technology, Rochester, NY 14623. This software employs realistic graphic animations that illustrate such processes as spin equilibrium, absorption, and relaxation.

2. ABRAHAM, R. J., FISHER, J., and LOFTUS, P., *Introduction to NMR Spectroscopy*, Wiley, New York, 1988.

3. ATTA-UR-RAHMAN, *One- and Two-Dimensional NMR Spectroscopy*, Springer-Verlag, New York, 1989.

4. ATTA-UR-RAHMAN, and CHOUDHARY, M. I., *Solving Problems with NMR Spectroscopy*, Academic, San Diego, 1996.

5. AKITT, J. W. and MANN, B. E. "NMR and Chemistry," Stanley Thornes, Cheltenham, U.K., 2000.

6. BECKER, E. O., *High Resolution NMR*, 2d ed., Academic, New York, 1980.

7. BOVEY, F. A., *Nuclear Magnetic Resonance Spectroscopy*, 2d ed., Academic, New York, 1988.

8. BREITMAIER, E. and VOELTER, W., *Carbon-13 NMR Spectroscopy*, 3d ed., VCH publishers, New York, 1990.

9. CLARIDGE, T. D. W., "High-Resolution Techniques in Organic Chemistry," Elsevier, London, 1999.

10. DEROME, E., *Modern NMR Techniques for Chemistry*, Pergamon, Oxford, 1987.

11. FRIEBOLIN, H., *Basic One- and Two-Dimensional NMR Spectroscopy*, 2d ed., VCH Publishers, New York, 1993.

12. GÜNTHER, H., *NMR Spectroscopy*, 2d ed., Wiley, New York, 1995.

13. HARRIS, R. K., *Nuclear Magnetic Resonance Spectroscopy*, Pitman, London, 1983.

14. JACKMAN, L. M. and STERNHELL, S., *Applications of NMR Spectroscopy in Organic Chemistry*, Pergamon, Oxford, 1969.

15. KEMP, W., *NMR in Chemistry: A Multinuclear Introduction*, Macmillan, London, 1986.

16. MARTIN, M. L., DELPUECH, J.-J., and MARTIN, G. J., *Practical NMR Spectroscopy*, Heyden, London, 1980.

17. SANDERS, J. K. M. and HUNTER, B. K., *Modern NMR Spectroscopy*, Oxford University Press, Oxford, 1987.

18. WERHLI, F. W. and WIRTHLIN, T., *Interpretation of Carbon-13 NMR Spectra*, Heyden, London, 1976.

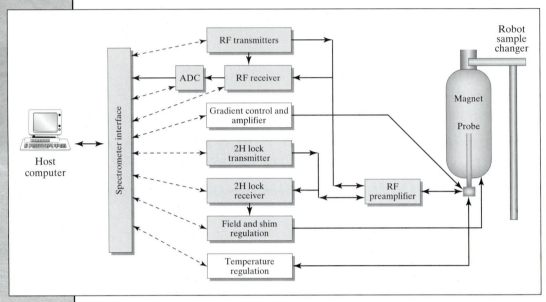

Components of a modern FT NMR Spectrometer.

Measurement of the Spectrum: Instrumentation

2.1   Fundamental Components
of an NMR Spectrometer

The essential components of all NMR spectrometers are a magnet to generate B_0; a radio frequency oscillator to generate B_1 in the transmitter coil; a receiver coil to pick up the signal; the electronics, usually including a computer and a plotter, to convert the signal into a spectrum; and a sample to be analyzed. Each of these components is discussed in some detail in this chapter.

2.2   The Magnet

It is imperative in NMR spectroscopy that the magnetic field have sufficient strength to bring the separation of nuclear energy states into the radio-frequency region. It is also critically important that both the stability and homogeneity of the B_0 field be very high. Because the precessional or resonance frequency of a nucleus is directly proportional to the strength of B_0 (Equation 1.4), the difference in resonance frequencies of nonidentical nuclei is also directly proportional to B_0. Therefore, it is often a considerable advantage to use the highest field-strength magnet available in order to obtain the greatest separation (dispersion) between NMR signals. Figure 2.1 illustrates this point. The higher field strength also results in greater energy gaps between spin states and, consequently, a greater population of the lower energy spin state (Equation 1.7). Thus, the intensity of the signal increases as the field strength increases.

There are three general types of magnets, and each type has advantages and disadvantages. **Permanent magnets** are the least expensive; they have relatively stable fixed magnetic fields, which generally do not exceed 1.4 tesla, and they require no electric current to generate the field. Their field strength is quite temperature dependent, however, such that the temperature in the environment in which they are used needs to be regulated. In contrast, **electromagnets** are huge, heavy, and more costly to build and operate, but their field strength may range up to 34 tesla. In practice, however, because electrical resistance to the high current necessary to generate strong magnetic fields generates considerable heat, they are usually limited to about 2.34 tesla. The excess heat generated by this resistance must be dissipated to assure a stable field. This is accomplished by circulating cooling water from a very large constant-temperature bath, through the pole pieces of the magnet.

The cost of electricity (up to 10 kw/h) and chilled water (about 10 liters/h) can be considerable, making electromagnets expensive to operate and maintain.

Superconducting magnets are a special subcategory of electromagnets. The cylindrical solenoid of a superconducting magnet is made of a special niobium–tin or niobium–tin–tantalum alloy that becomes superconducting at very low temperatures. When immersed in liquid helium (4.2 K), solenoids wound from these alloys are able to maintain sufficiently high current densities to provide *persistent* magnetic fields of 12 tesla or greater. A cross section of a superconducting magnet is shown in Figure 2.2.

The solenoid is bathed in liquid helium. The helium dewar is thermally isolated by means of a vacuum chamber and thermal baffles (not illustrated) from an outer dewar that contains liquid nitrogen (77 K). Once a superconducting magnet is energized, no additional electrical current is required, as

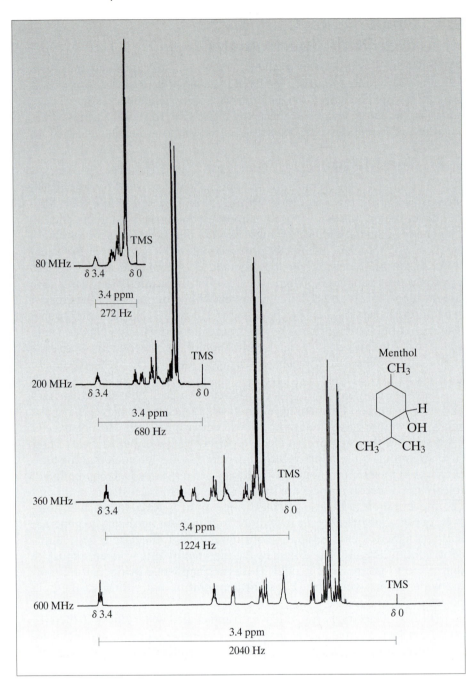

▲ FIGURE 2.1
^{1}H NMR spectra of a CDCl$_3$ solution of menthol containing a trace of TMS as a function of field strength.

See Figure 1.2 for the relation between ^{1}H observation frequency and **B$_0$** (tesla).

long as the magnet is kept at liquid helium temperature. If not, the field **quenches**, and it might not be possible to restore it. Most superconducting magnets must be filled with liquid nitrogen every 7 to 10 days and with liquid helium every 8 to 12 weeks. Still, the cost of these cryogens—liquid helium and

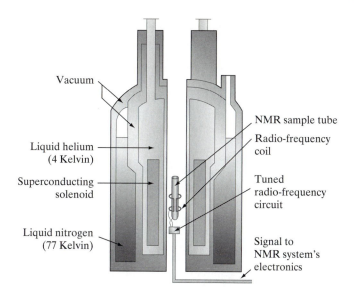

Vacuum

NMR sample tube

Liquid helium
(4 Kelvin)

Radio-frequency
coil

Superconducting
solenoid

Tuned
radio-frequency
circuit

Liquid nitrogen
(77 Kelvin)

Signal to
NMR system's
electronics

◀ FIGURE 2.2
**Cross section of a
superconducting magnet for
NMR spectroscopy.**

The magnet and sample tube are
not drawn to scale. The magnet
diameter is approximately 70 cm
and that of the sample tube is 5 to
20 mm.

nitrogen—is considerably less than the cost of electricity and coolant required for an electromagnet, such that over time superconducting magnets are the most desirable of the three types of magnets.

Superconducting magnets have fringe fields that extend from the magnet by more than a meter. The intensity of the fringe fields depends on the magnetic field strength and bore size. Magnetic materials, such as common tools and paper clips, are strongly attracted to these magnets and should be kept away from them. Also, the magnetically encoded information on credit cards and floppy disks is erased by these fringe fields and pacemakers may stop functioning by coming too close to them.

It is necessary, regardless of the type of magnet, to provide a means by which the stability of the field is monitored and controlled. This is accomplished by a process called **locking the magnetic field**. In order to lock the field and keep it from drifting, a substance with nuclei that give rise to a strong NMR signal (the **lock signal**), which occurs at a different frequency from those of the nuclei of interest, is chosen. If this substance is kept physically apart from, but close to, the sample, it is termed an **external lock**. More commonly, the lock substance is used as the solvent for the sample and is then termed an **internal lock**. In either case, the resonance frequency of the lock signal is continuously monitored and electronically compared with a fixed **reference frequency** oscillator. A difference between the lock and reference frequencies causes a direct microcurrent to pass through a secondary coil (known as the $Z°$ gradient coil, aligned with and inside the solenoid). This results in the generation of a small secondary magnetic field that is either aligned with or opposed to $\mathbf{B_0}$ and thus changes $\mathbf{B_0}$ until the lock and reference frequencies match. When this occurs, the magnet or spectrometer is said to be *locked*. Deuterium is the nucleus that is most commonly used for the lock signal, and this is one reason to use deuterated solvents.

Two techniques are used to improve the magnetic field homogeneity. First, the shape of the field is corrected by using auxiliary coils with highly specific geometries, called **shim coils**, and the process is called *shimming the magnet*. The shim coils are wound to approximate the spherical harmonic functions

Table 2.1 NMR Shims

Shim name	Function
Z^0	$1 - $ main field
Z^1	Z
Z^2	$2Z^2 - (X^2 + Y^2)$
Z^3	$Z[2Z^2 - 3(X^2 + Y^2)]$
Z^4	$8Z^2[Z^2 - (X^2 + Y^2)] + 3(X^2 + Y^2)^2$
Z^5	$48Z^3[Z^2 - 5(X^2 + Y^2)] + 90Z(X^2 + Y^2)^2$
X	X
Y	Y
ZX	ZX
ZY	ZY
XY	XY
$X^2 - Y^2$	$X^2 - Y^2$
Z^2X	$X[4Z^2 - (X^2 + Y^2)]$
Z^2Y	$Y[4Z^2 - (X^2 + Y^2)]$
ZXY	ZXY
$Z(X^2 - Y^2)$	$Z(X^2 - Y^2)$
X^3	$X(X^2 - 3Y^2)$
Y^3	$Y(3X^2 - Y^2)$

listed in Table 2.1. Many of these shims interact with each other, and multiple iterations through the different shims are usually necessary to arrive at a highly homogeneous field.*

In Table 2.1, the letters x, y, and z refer to the three axes of the laboratory frame of reference with z being along the magnetic field direction. For permanent magnets or electromagnets, z is between the two pole pieces, horizontal or parallel to the floor. For superconducting magnets, z is vertical or perpendicular to the floor.

The differences in these directions are important in understanding the effects of **sample spinning**, the second technique used to improve the field homogeneity. The sample is placed in a high-precision glass tube of uniform diameter, which in turn is placed in the center of the field in a region called the **probe**. The sample is then spun about its long axis at about 10 to 30 Hz. This spinning helps to average out slight inhomogeneities of the field in the sample region, as long as the spinning rate is fast compared with the field gradient across the sample. Spinning does not average out inhomogeneities along the axis of spin, which is the x-axis for permanent magnets or electromagnets and the z-axis for superconducting magnets. Sample spinning modulates the field and produces **sidebands** called spinning side bands (SSBs), which can usually be minimized by careful shimming.

The higher order shims are usually optimized for each probe during the field-mapping process that is done on installation. These shim files are stored in the computer and recalled when probe changes occur. Normally only low-order shims are optimized upon changing samples. A properly

---

*See W. W. Conover in *Topics in Carbon-13 NMR Spectroscopy*, G. C. Levy, Ed. Vol. **4**, 1984, pp. 38ff, for excellent step-by-step instructions on how to shim a magnet.

shimmed sample affects both sensitivity (usually measured by peak height) and resolution (usually measured by line width at half height).

Modern spectrometers have computer-controlled shims. The computer monitors the lock level to iterate to optimum values for the shim settings. An **auto shim** feature tracks the lock level and makes small corrections to Z^1, keeping the homogeneity at its optimum level. This is especially useful for long-term spectral accumulations where small temperature changes cause changes in the magnetic field homogeneity. Controlling the temperature of the probe and maintaining it slightly above ambient temperature is useful for the same purpose.

2.3   The Transmitter Coil

In order to tip **M** off the z-axis so that it has a component in the xy-plane, an RF field, $\mathbf{B_1}$, is applied through a transmitter coil (Figure 2.3). This field is generated by passing an alternating current through the transmitter coil. The magnetic field, $\mathbf{B_1}$, generated in this way is *linearly polarized* and oscillates back and forth along the x-axis. It can be viewed as if it were the vector sum of two oppositely phased *circularly polarized* rotating fields, $\mathbf{B_1}$ and $\mathbf{B_1'}$ (Figure 2.4). Resonance occurs when the frequency of $\mathbf{B_1}$ (the clockwise rotating field) matches the Larmor frequency. Hence, in the rotating frame, $\mathbf{B_1}$ remains stationary along the x'-axis. The transmitter coil needs to be **tuned** to the Larmor frequency of the nucleus of interest. This is usually done by varying the capacitance of the circuit, as the resonance frequency ω is related to the impedance of the inductor L and capacitor C by the following equation:

$$\omega = \left[\frac{1}{LC}\right]^{1/2}. \tag{2.1}$$

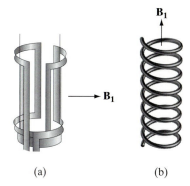

▲ FIGURE 2.3
Typical NMR coils.

(a) A Helmholtz coil used in superconducting magnets. (b) A solenoid coil used in permanent magnets and electromagnets and for solid-state NMR in superconducting magnets.

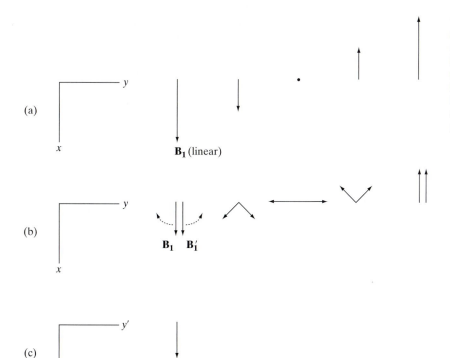

◄ FIGURE 2.4
(a) Linearly polarized field $\mathbf{B_1}$ oscillating along the x-axis of the laboratory frame. (b) Oppositely rotating circularly polarized fields $\mathbf{B_1}$ and $\mathbf{B_1'}$, both rotating at the Larmor frequency. (c) Orientation of $\mathbf{B_1}$ along the x'-axis of the rotating frame.

2.4   The Receiver Coil

A current is induced in a coil of wire by the movement of a magnetic field through the center of the coil. The receiver coil design is similar to that of the transmitter coil. In some spectrometers, a single coil serves both purposes. The receiver coil is normally placed along the y-axis, orthogonal to the transmitter coil. Both coils are housed in the *probe* (Figure 2.5). These coils act as inductors in circuits that are *tuned* for maximum absorption of the desired frequency. Tuning is accomplished by adjusting capacitors, which usually have extension rods that can be adjusted from outside the magnet. A probe is tuned when there is an exact balance between the inductance and capacitance of the circuit. (See Equation 2.1.) The exact tuning of a probe will be dependent on the sample, which is part of the circuit. There are many kinds of probes. The most common ones are called **single-frequency** or **dedicated probes** and **broadband probes**. A single-frequency probe is actually double tuned to the observation frequency and the lock frequency. A broadband probe is triple tuned to the observation frequency, the decoupling frequency, and the lock frequency. There are two designs for broadband probes: the *normal* and *inverse* designs. In the normal design, the broadband receiver coil is on the inside, closer to the sample, and the ^{1}H receiver coil is on the outside, further

▶ FIGURE 2.5
A typical probe design for a superconducting magnet and a cutaway view of the magnet showing the locations of the shim coils and solenoid.

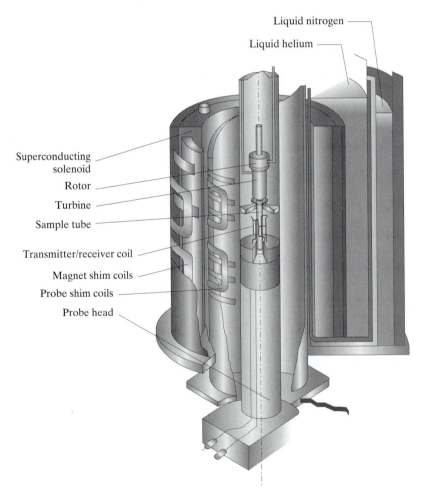

Liquid nitrogen

Liquid helium

Superconducting solenoid

Rotor

Turbine

Sample tube

Transmitter/receiver coil

Magnet shim coils

Probe shim coils

Probe head

from the sample. The inverse probe has inverse locations for these two receiver coils. The strength of the detected signal is a function of the distance of the receiver coil from the sample, and it decreases with increasing distance. In broadband experiments in which a nucleus other than ^{1}H is detected, the normal probe design is employed. In experiments in which ^{1}H is directly detected, the inverse probe design is generally used. These considerations are most important in two-dimensional NMR spectroscopy. (See Chapter 11.)

2.5   The Sample

Since solids generally give broad resonance signals, unless CP/MAS techniques (Chapter 14) are used, the sample is generally a liquid. The compound under investigation is therefore dissolved in a suitable solvent, and a small amount (0.1% to 1%) of a reference compound is added (or sometimes the residual solvent peak may be used as a reference) to obtain the spectrum.

The standard size of the sample tube for most ^{1}H spectrometers is 5 mm OD (outside diameter), which requires about 0.4 mL of solution for normal operation. The depth of the solution is determined by two factors. Above a certain height, the top part of the sample is not acted upon by an RF field (it is outside the region of the receiver and transmitter coils); therefore, it does not contribute to the signal. However, if the solution is too shallow, the vortex produced by spinning extends into the volume "sensed" by the coils, which results in poor resolution and spurious peaks. For the same reason, the sample should not be spun too rapidly. Spinning the sample always produces some spurious signals, due to modulation of the signal by the spinning tube, which never spins perfectly. As a consequence, small spinning sidebands (SSBs) are produced on either side of any large resonance at a separation from the central resonance (in Hz) equal to the spinning rate. This is of importance, as these spinning sidebands may be differentiated from satellites (due to coupling of the observed nucleus to another nucleus with less than 100% natural abundance) by varying the spinning rate of the sample, thus altering the position and intensity of the sidebands (Figure 2.6). The faster the spinning rate, the further apart are the spinning sidebands and the lower is their intensity.

For insensitive nuclei, more of the sample is required and a larger sample tube is used. (Up to 20-mm OD sample tubes are commercially available.) It is, however, pertinent to note that larger sample tubes are of no value if the quantity of the sample is limited. For example, if one has 100 mg of a compound that will dissolve easily in 0.4 mL of solvent, then there is no advantage to be gained in dissolving the compound in the 2 to 4 mL of solvent required for a 10-mm OD sample tube. On the contrary, if the 100-mg sample will not dissolve in 0.4 mL, but will dissolve in 2 to 4 mL, then the larger diameter tube does have an advantage. For highly soluble samples, the minimum-size sample tube is best, and for poorly soluble samples, the maximum-size sample tube is best. Thus, one wants to maximize what is called the **volume-filling factor** by placing the maximum amount of sample in the region "sensed" by the coils, without producing a highly viscous solution.

It is also important to obtain a clear mobile solution, as any solid particles or a viscous solution will seriously impair the resolution of the spectrum. A few minutes spent in finding the best solvent and in filtering the solution will save considerable spectrometer time. Since we generally wish to observe the

▶ FIGURE 2.6
(a) ^{1}H NMR spectrum of CHCl$_3$ showing ^{13}C satellites (arrows), and spinning sidebands (SSBs). (b) ^{1}H NMR spectrum of trans-1, 2-dichloroethylene showing ^{13}C satellites (arrows) and spinning sidebands (SSBs).

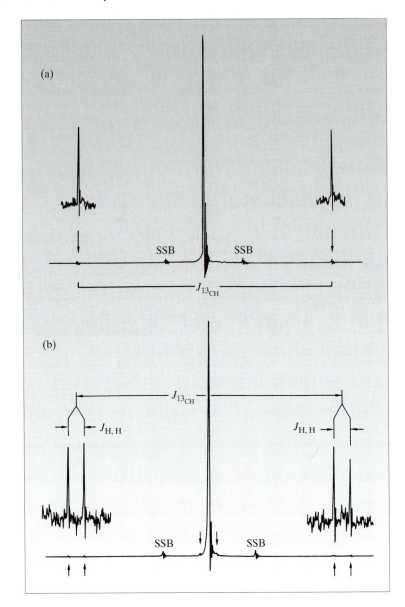

spectrum of the solute and not the solvent, we choose a solvent that does not contain nuclei that resonate in the same spectral region as those of the solute. Consequently, deuterated solvents are typically used when one observes a ^{1}H NMR spectrum. Some common solvents with their chemical shifts and liquid ranges are given in Table 2.2.

If a protonated solvent were used for ^{1}H observation, then approximately a 0.5- to 1.0-ppm region would be obscured by each solvent resonance. Using deuterated solvents will alleviate this problem, although a small residual proton resonance, depending upon the percentage deuteration of the solvent, will remain. The most commonly used solvent is deuteriochloroform (CDCl$_3$). This is a much better solvent than CCl$_4$ or CS$_2$, both of which are also suitable for ^{1}H or ^{13}C studies. They are all "transparent" in the proton spectrum; however, commercial CDCl$_3$ always contains some CHCl$_3$ (1% or less), which does not normally cause a problem. Since the chemical shift dis-

Table 2.2 Some Useful Solvents for NMR Spectroscopy[a],*

Solvent	Dielectric constant	Liquid temperature range (°C)	Chemical shifts		
			δ_H	δ_{HOD}[b]	δ_C
Acetone-d_6	20.7	−95–56	2.05	2.8	29.92, 206.68
Acetonitrile-d_3	37.5	−44–82	2.00	2.1	1.3, 117.7
Benzene-d_6	2.28	6–80	7.16	0.4	128.0
Cyclohexane-d_{12}	2.01	6–81	1.43		26.4
CCl_4	2.24	−23–77	–		96.7
CS_2	2.64	−112–46	–		192.8
$CDCl_3$	4.8	−64–61	7.26	1.5	77.0
CD_2Cl_2	8.9	−95–40	5.33	1.5	53.1
$CDCl_2CDCl_2$	8.2	−44–146	5.94	1.5	75.5
CF_2BrCl	–	−140–−25	–		109.2[c]
$CFCl_3$	2.3	−111–24	–		117.6[d]
Dioxane-d_8	2.2	12–101	3.53	2.4	66.5
D_2O	78.5	0–100	4.82		–
DMSO-d_6	46.7	19–189	2.50	3.3	39.4
DMF	36.7	−60–153	2.74, 2.91, 8.02	3.5	30.1, 35.2, 162.7
Hexamethyl phosphoramide	30.0	7–233	2.53		35.8
Ethanol-d_6		−130–79	1.1, 3.55, 5.26[e]	5.3	17.2, 56.8
Methanol-d_4	32.7	−98–65	3.30, 4.84[e]	4.9	59.05
Nitrobenzene-d_5	34.8	6–211	8.2, 7.6, 7.5		149, 134, 129, 124
Nitromethane-d_3	35.87	−29–101	4.33		61.4
Pyridine-d_5	12.40	−42–115	7.20, 7.57, 8.73	5.0	123.4, 135.4, 149.8
Tetrahydrofuran-d_8	7.6	−108–66	1.72, 3.57	2.5	25.3, 67.4
Trifluoroacetic acid-d	8.6	−15–72	11.50[e]		116.6, 164.2[f]
1,2,4-Trichlorobenzene	3.9	17–214	7.1, 7.3, 7.4		133.3, 132.8, 130.7, 130.0, 127.6

[a] The physical constants are given for the protonated solvents.
[b] [1]H chemical shift of HOD in the solvent.
[c] $J(CF) = 342$ Hz.
[d] $J(CF) = 337$ Hz.
[e] The OH protons' chemical shift may vary, depending on the solute concentration, etc.
[f] $J(CF) = 283$ Hz; $J(CCF) = 43$ Hz
* A very useful tabulation of NMR chemical shifts of common laboratory solvents as trace impurities is given in Gottleib, H.E.; Kotlyar, V., Nudleman, A., *J. Org. Chem*, **1997**, *62*, 7512.

persion range is much greater than for protons (about 300 ppm versus 25 ppm) in ^{13}C NMR spectra, the possibility of any resonance being obscured by the solvent is correspondingly much less. However, the deuterated solvent is usually vital, as the majority of spectrometers use the solvents' deuterium resonance to stabilize and lock the spectrometer. Also, the large signal of the protonated solvent in both ^{1}H and ^{13}C NMR spectra causes digitization problems in FT spectrometers. An interesting recent development is the commercial availability of ^{12}C enriched solvents, which give very weak residual ^{13}C resonances, and this can be of great use in some cases.

A common problem in NMR is to obtain suitable solvents for both high-temperature and low-temperature studies. For temperatures up to 140°C, 1,1,2,2-tetrachloroethane is an excellent solvent, having solvent properties very similar to those of $CHCl_3$. For higher temperatures, dimethylsulfoxide is often used, but it has disadvantages, as it is a very reactive compound, and the sample often cannot be recovered from it after the experiment.

Other useful solvents for high-temperature studies are nitrobenzene, o-dichlorobenzene, and 1,2,4-trichlorobenzene.

For low-temperature work, dichloromethane, acetone, methanol, and dichloromethane–toluene mixtures may be used (down to −140°C for the last). Tetrahydrofuran–carbon-disulfide mixtures may be used down to about −150°C. Obviously, some thought should be given to the choice of solvent.

2.6   Obtaining a Spectrum

There are two main methods used to achieve a resonant condition in NMR spectroscopy: CW (continuous wave) and FT (Fourier transform). The actual method for achieving resonance is more complicated than it appears. Because most samples generally have more than one type of magnetically equivalent sets of nuclei, they have more than one distinct chemical shift. Such a sample will have multiple Larmor frequencies, each corresponding to a different chemical shift. Any excitation method must cover all of the Larmor frequencies in the sample. In the CW method, the resonance condition is met for each magnetically equivalent group by sweeping either the $\mathbf{B_1}$ frequency (frequency sweep) or the $\mathbf{B_0}$ field (field sweep). Rather than exciting one group at a time as in the CW method, the FT method provides a short, but powerful, pulse so that all of the groups of the same nuclei in the sample are simultaneously excited. In either case, the nuclei are upset from their original Boltzman distribution.

2.7   Continuous-Wave Experiments

Each group in a sample can successively be brought into resonance by sweeping either the frequency or the field. The applied radio-frequency field can be decomposed into two counterrotating vectors in the $x'y'$-plane (Figure 2.4b). One vector rotates in the same direction as the rotating frame, and if its frequency is ω_o, it will be stationary in that frame. This component is the $\mathbf{B_1}$ field. The other vector has a negligible effect, because it is rotating in the opposite direction, and in the rotating frame it is at twice the Larmor frequency.

Since the radio frequency coil in a cross-coil probe is wound so that it is perpendicular to the main field, $\mathbf{B_0}$, we must consider the effect that a slow sweep of either $\mathbf{B_0}$ or $\mathbf{B_1}$ has on the x and y components of the magnetization. The magnetization vector, $\mathbf{M}$, tries to follow the effective field, but its motion is balanced out by spin–spin and spin–lattice relaxation processes.

In a **cross-coiled** system, the exciting and monitoring coils are perpendicular to each other. When a particular nucleus comes into resonance as a consequence of the exciting $\mathbf{B_1}$ field, a voltage is induced in the perpendicular coil. This signal, microvolts or less in magnitude, is amplified and detected directly in the **frequency domain**. The shape of this signal depends upon a number of factors, including the scan rate and T_2, the former of which is illustrated in Figure 2.7.

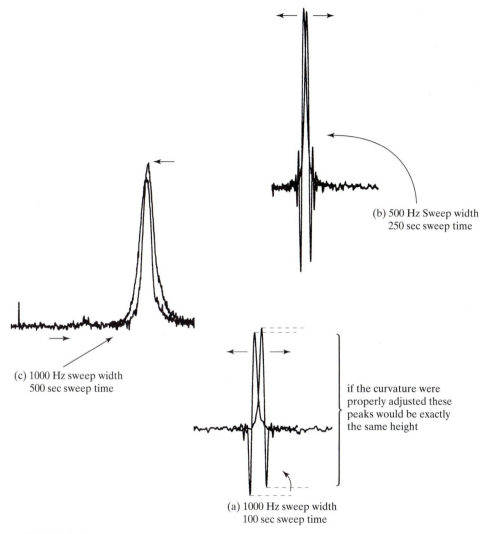

(b) 500 Hz Sweep width
250 sec sweep time

(c) 1000 Hz sweep width
500 sec sweep time

if the curvature were
properly adjusted these
peaks would be exactly
the same height

(a) 1000 Hz sweep width
100 sec sweep time

▲ FIGURE 2.7
Sweep rate dependence of the 60-MHz ^{1}H NMR resonance of $(CH_3)_4$ Si*.

*The resonances were swept in both directions, as indicated by the horizontal arrows.

The correct choice of the scan rate is important and complicated because the following requirements contradict one another:

a reasonably short recording time (several minutes),
a maximum signal height H (and signal-to-noise ratio), and
a minimum signal distortion.

The first requirement necessitates fast sweep rates, which, in turn, cause the following:

1. *a displacement of the resonance frequency.* Note in Figure 2.7 that the faster the sweep rate, the greater is the separation of the two resonances obtained by sweeping in opposite directions.

▶ FIGURE 2.8

Resolution test. CW ^{1}H NMR spectrum of a degassed 10% (v/v) *ortho*-dichlorobenzene solution in carbon tetrachloride at 25°C (left half of the spectrum only). Standard conditions: frequency sweep, 60 Hz; sweep time, 2000 s (sweep rate, 0.03 Hz s^{-1}; filter bandwidth, 1 Hz; 5-mm OD sample tube. Line width of the third transition (from the left-hand side), 0.15 Hz. The 0.15-Hz resolution thus obtained is also evidenced by the splitting of *ca.* 0.15 Hz, which is observed between transitions 5 and 6.

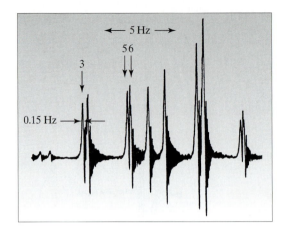

2. *broadening of the signal and therefore, a loss of resolution.* The line width, $\Delta\nu_{1/2}$, of the resonance in Figure 2.7c is actually one-fifth that in Figure 2.7a.

3. *the appearance of "wiggles" or "ringing."* These oscillations, which accompany the signal, may cause difficulties when there are two resonances that are close together, as in Figure 2.8.

4. *a decrease in the signal-to-noise (S/N) ratio due in part to the necessity of reducing the filter time constant.*

It appears that the best procedure for recording single-scan spectra consists of using the lowest possible sweep rate, determined mainly by the maximum allowed recording time, and the corresponding optimum saturation parameter. Although the CW method was the primary means for signal detection in the early days of NMR spectroscopy, it has largely been supplanted by FT methods in all but the most simple spectrometers.

Conceptually, FT NMR is rather different from CW spectroscopy. A major difference is that detection is performed in the **time domain** and the signal is Fourier transformed into the **frequency domain**.

2.8 Fourier Transform Experiments

In a Fourier transform experiment, a short burst or pulse of a radio-frequency signal, nominally monochromatic at ν_0, is used to simultaneously excite all the Larmor frequencies in a sample. This short pulse is ideally a square wave, which can be approximated by orthogonal functions of higher and higher frequencies. An illustration is given in Figure 2.9, where, on the left, a positive and a negative square wave are approximated by 1,3,5, and 7 orthogonal functions. The difference between these approximations and a square wave (the error) is plotted at the right of the figure. Because the square wave is like a superposition of many frequencies, it can excite all of the frequencies in the spectrum if the pulse is sufficiently powerful.

The length of the pulse (t_p) determines the range of frequencies that are excited; that is,

$$\Delta\nu = \nu_0 \pm \frac{1}{t_p},$$

where ν_0 is the center frequency.

The duration of the pulse necessary to tip the magnetization vector from the z'-axis onto the $x'y'$-plane, termed a **90° pulse**, is inversely related to the

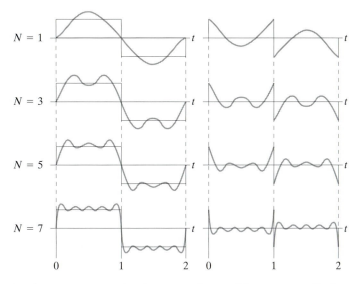

◀ FIGURE 2.9
Approximation of two square waves by orthogonal functions (left) and the instantaneous error remaining after each approximation (right).

power of the radio-frequency signal. The 90° pulse width, as it is called, is generally between 5 and 30 μs in duration. Its length depends upon the RF power level, the frequency of the measurement, and the physical characteristics of the probe.

The flip angle θ is related to the strength of the exciting field $\mathbf{B_1}$ and to t_p by the equation

$$\theta = \frac{\gamma}{2\pi}\mathbf{B_1}t_p. \tag{2.2}$$

In terms of spin populations, the effect of a 90° pulse can be visualized as equalizing the populations of the nuclear spin energy levels. Following the pulse, the spins undergo a free induction decay (FID) to reestablish their equilibrium populations. The FID is collected as a function of time and represents an interferogram in the time domain. The frequency domain information, or normal spectrum, is obtained by Fourier transformation of the FID. The detector is placed in the $x'y'$-plane, and hence, if only a single pulse is applied, then it is desirable to perform a 90° pulse so that the maximum magnetization is detected. The effects of 90°, 45°, 180°, and 270° pulses on signal amplitudes are compared in Figure 2.10. The detector samples only the projection of the

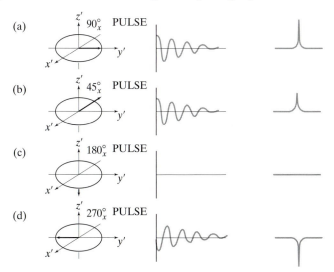

◀ FIGURE 2.10
The magnetization vector (left column), the resultant FID (center column), and the transformed spectrum (right column) after a (a) 90° pulse, (b) 45° pulse, (c) 180° pulse, and (d) 270° pulse. In all cases, the $\mathbf{B_1}$ field is applied along the x direction.

▶ FIGURE 2.11
Dependence of the signal amplitude on the pulse angle θ.

Note that the first null point occurs at $\theta = 180°$ and the second occurs at $\theta = 360°$.

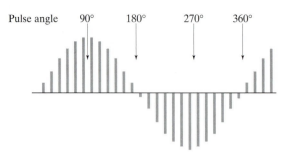

Pulse angle   90°   180°   270°   360°

magnetization that is in the $x'y'$-plane, so that a 45° pulse gives only 0.707 the intensity of a 90° pulse, a 180° pulse gives no signal, and a 270° pulse gives a signal −1.0 times the intensity of a 90° pulse. The 90° pulse width must be determined experimentally. It is best determined from a null spectrum or one produced by a 180° pulse. The length of the 180° pulse is then halved to obtain the proper length of the 90° pulse (Figure 2.11).

The optimum pulse width is not the 90° pulse when multiple FIDs are acquired, as then the relaxation time is of importance. The optimum pulse for multiple acquisitions is given by

$$\cos \theta_{optimum} = e^{-\tau/T_1}, \tag{2.3}$$

where τ is the acquisition time plus the delay time (**duty cycle**) and T_1 is the spin–lattice relaxation time.

SAMPLE PROBLEM 2.1

What is the optimum pulse width for ^{13}C observation of a typical medium-sized organic molecule (formula weight 300 or above) for which the average T_1 is about 4 s if the acquisition time is 2 s and the delay time zero?

Solution

$\theta = \text{arc cos } e^{-2s/4s} = 52.7°$.

In addition to having a magnitude or strength, the pulse in an FT experiment also has phase coherence. In all of the examples in Figure 2.10, the RF pulse was *along* the x direction, which rotated the magnetization from z' to y'. A 90° pulse, phase shifted by 90°, would be along the $-y$ direction. Figure 2.12

▶ FIGURE 2.12
The magnetization vector (left column), the resultant FID (center column), and the transformed spectrum (right column) after a (a) $90°_x$ pulse, (b) $90°_{-y}$ pulse, (c) $90°_{-x}$ pulse, and (d) $90°_y$ pulse.

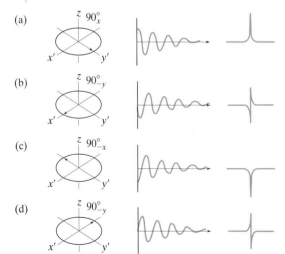

shows the phases of the spectra resulting from 90° pulses along the x, $-x$, y, and $-y$ directions. Some forms of two-dimensional NMR spectroscopy (Chapter 11) require application of RF pulses along directions other than x. Experimentally, a 90°-phase-shifted pulse is produced by introducing a delay equal to 0.25 of the radio-frequency cycle.

2.9   Detecting the Signal and Digitizing the Data

The NMR signal that is sensed by the receiver coil is extremely weak and must be amplified prior to detection. This function is performed by the **preamplifier**, which is normally placed very close to the probe in order to minimize signal loss. The preamplifier can be either **broadbanded** (to amplify signals over a very large frequency range) or **tuned** (to amplify signals over a narrow frequency range and reject all others). A tuned preamplifier or an associated **band-pass** (or **high-pass**) filter are used to eliminate unwanted noise from an NMR spectrum.

The amplified signal is in the megahertz, or radio-frequency, region. Audio signals occur over a fairly narrow bandwidth and are easier to work with than RF signals at different frequencies. Accordingly, the high-frequency signal is converted to a common audio signal before detection. In this process, the radio-frequency carrier, ν_{o}, is stripped off, leaving only the frequency-modulated signal, $\Delta \nu$. In this process, the phase of the radio-frequency carrier is used for *phase sensitive detection*. The audio signal FID then carries information about its offset ($\Delta \nu$) from the carrier or transmitter frequency (ν_{o}).

The audio signal is then passed through audio band-pass filters to remove noise and spurious signals outside the spectral bandwidth. These audio filters attenuate the noise or unwanted signals that occur outside the region of interest. It is the audio signal that is filtered, but it is easier to visualize the effects of filtering by looking at a spectrum in the frequency domain (Figure 2.13).

The filter widths are always set to be slightly wider than the spectral widths so that roll-off and end effects do not affect the appearance of the spectrum. In most spectrometers with quadrature detection, the audio filter bandwidth is automatically set by the software to be slightly greater than twice the sweep width.

The audio signal, which is in continuously varying analog format, is converted to a suitable digital representation by an A-to-D converter, which stores the incoming signal in discrete time bins. There are three important factors to be considered in the digitization process: the **digitization rate**, **digital resolution**, and **dynamic range**.

The rate that the data should be sampled or converted into a digital representation is governed by the sampling theorem. According to this theorem, a sine wave must be described by at least two points per period in order to be faithfully represented. A sampling rate of two points per cycle (Figure 2.14a)

◀ FIGURE 2.13
Superposition of a filter cutoff function on a transformed spectrum.

The filter prevents signals outside the cutoff region from being folded into the spectrum.

▶ FIGURE 2.14
Illustration of the Nyquist frequency.

(a) To be faithfully represented, a signal of frequency ν must be sampled twice per cycle. (b) When sampled at this rate, a signal of frequency $\nu + \Delta\nu$ cannot be distinguished from (c) a signal of frequency $\nu - \Delta\nu$.

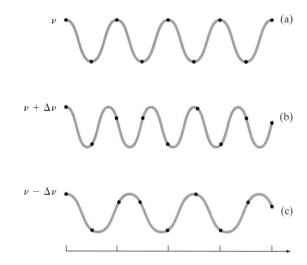

for a sine wave of frequency ν is known as the **Nyquist frequency**. If, instead of ν, the frequency of the sine wave were somewhat greater (say, $\nu + \Delta\nu$; Figure 2.14b), then that frequency could not be accurately represented by sampling the data twice per period ν, nor could that signal be distinguished from one whose frequency was $\nu - \Delta\nu$ (Figure 2.14c). Consequently, a signal with a frequency greater than the Nyquist frequency will appear in the spectrum at $\nu + \Delta\nu$. This signal is said to be **folded in**, also called **folded over**, or **aliased**. Figure 2.15 illustrates the influence of sweep width and transmitter frequency on foldover. The Nyquist frequency dictates the rate at which the FID must be sampled. For example, if the sweep width is 1 kHz, the signal must be sampled at a 2-kHz rate, or the data must be sampled every 500 μs. This sampling time is called the **dwell time** and is automatically set by the spectrometer when the sweep width is entered.

The product of the dwell time and the total number of data points determines the **digital resolution**. For example, if 8,192 data points were used to acquire a spectrum with a 1-KHz sweep width (500 μs dwell time), the **acquisition time** would be 8,192 points $\times$ 500 μs/point = 4×10^6 μs = 4 s. The digital resolution would then be 1 KHz/8,192 points = 0.12 Hz/point. This calculation applies to a single channel of data. Most current spectrometers employ digital quadrature detection, or two data channels. Then,

$$\text{acquisition time} = \frac{\text{number of data points}}{2 \times \text{sweep width (Hz)}} \qquad \textbf{(2.4)}$$

and

$$\text{digital resolution} = \frac{2 \times \text{sweep width (Hz)}}{\text{number of data points}}. \qquad \textbf{(2.5)}$$

The number of data points used to acquire the spectrum affects the acquisition time (more data points yield a longer acquisition time), the digital resolution (more data points yield a greater digital resolution), and the computing time (more data points yield a longer computing time). There is, therefore, a compromise that needs to be made, and spectra are not normally obtained with more than 32K data points for ^{1}H NMR observation and 64K for the observation of other nuclei. It should also be clear from equations 1.4,

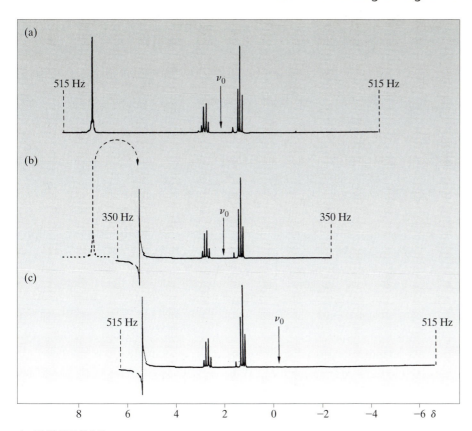

▲ FIGURE 2.15
Frequency folding in FT NMR spectroscopy.

(a) An 80-MHz ^{1}H NMR spectrum of ethylbenzene observed with quadrature detection with Nyquist frequency +515 Hz. (b) Nyquist frequency +350 Hz, but sweep width too small. (c) Nyquist frequency +515 Hz, but transmitter offset incorrectly set. The folded signals can usually be recognized by their incorrect phase and a relative chemical shift that does not track linearly with ν_0. Foldover may occur to either side of the spectrum, depending upon the particular spectrometer, but the distance from the edge of the spectrum is the same on either side as in (b).

2.4, and 2.5 that $\mathbf{B_0}$, or the field strength, influences the sweep width (Hz) and the number of data points to be used. As $\mathbf{B_0}$ increases, more data points should be used.

The **dynamic range** is a function of the word size of the computer used to process the data. When the FID is digitized, the voltage gain must be adjusted so that the signal is properly represented by the number of available bits, or the **word size**. The strongest part of the signal must not overflow the computer word size, and the weakest part must be represented by at least 1 bit. Thus, a large word size (20, 24, or 32 bits) or the ability to perform double-precision acquisition (linking together two words to represent each data point) is desirable.

In FT NMR spectroscopy, the FID is usually acquired in **quadrature**. This technique uses two phase-sensitive detectors that are out of phase with each other by 90°. The two FIDs that are also 90° out of phase with each other are digitized by the A-to-D converter in even (0, 2, 4, etc.) and odd (1, 3, 5, etc.) time bins. These two FIDs are often called the "real" (cosine function) and "imaginary"

▶ FIGURE 2.16
(a) Vector diagram in the rotating frame of two signals, $v + \Delta v$ and $v - \Delta v$, that have been tipped onto the $x'y'$-plane. The two quadrature detectors (A and B) lie along the y'- and x'-axes. Detection by A alone yields identical signals, but when detected by B, the signals are 180° out of phase.

(a)

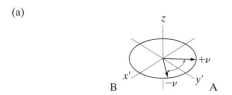

(b)

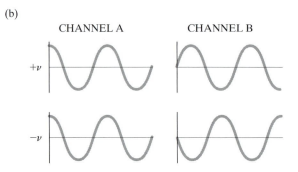

(sine function) signals. Complex Fourier transformation of these signals produces the **absorption** and **dispersion** spectra, respectively (Section 2.10).

There are several advantages of quadrature detection. With it, signals above and below the carrier frequency may be distinguished. Without quadrature detection, if the transmitter frequency (carrier frequency) were in the center of the spectrum, it would be impossible to distinguish $v + \Delta v$ from $v - \Delta v$ signals (Figure 2.16).

Alternatively, the carrier could be placed at one edge of the spectrum and the sweepwidth could be doubled. In this case, noise from the unwanted part of the spectrum would be folded in, thereby decreasing the S/N ratio. Also, the power required to cover twice the sweep width would be much greater. Finally, with the carrier in the center of the spectrum, the acquisition time can be twice as long, as it is inversely proportional to the sweep width.

The signal-to-noise ratio can also be improved by phase cycling the transmitter and receiver by some permutation of 0°, 90°, 180°, and 270°, while co-adding or co-subtracting the FIDs. The phase of the signal from the sample varies as the transmitter phase is changed, but the phase of a signal from a coherent noise source does not. Thus, sample signals co-add in a coherent fashion, but noise signals do not, when phase cycling is employed. For this reason, spectra should be acquired in multiples of 4, 8, or 16 acquisitions, depending upon the particular experiment.

2.10   Processing the Signal

The digitized signal is typically mathematically manipulated to increase the *apparent* resolution, improve the *apparent* signal-to-noise ratio, or emphasize some aspect of the spectrum. One cannot increase the *actual* information content beyond what is already present, but rather, some of the information can be enhanced at the expense of some other information. For example, S/N can be enhanced at the expense of resolution and vice versa. The early parts of the FID contain information related to S/N, while later parts contain information related to resolution.

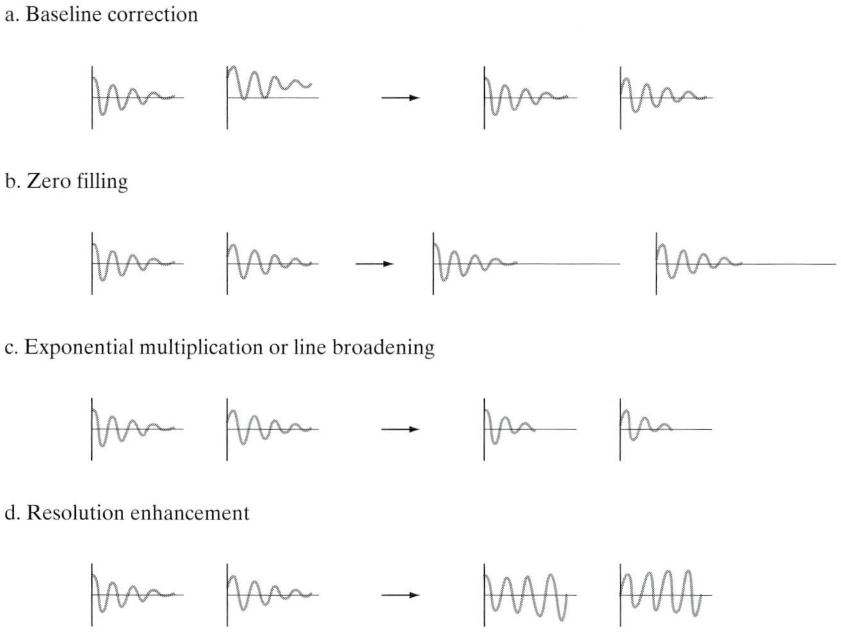

a. Baseline correction

b. Zero filling

c. Exponential multiplication or line broadening

d. Resolution enhancement

◀ FIGURE 2.17
The effects of various weighting functions on the FID.

When the signal is acquired in quadrature, it is common practice to **baseline correct** the data. This process removes any residual effect from the DC offset in the two quadrature channels (Figure 2.17a). If a baseline correction were not applied, a spike (or "glitch") would normally be observed in the spectrum at the carrier frequency, the height of which decreases with increasing S/N.

Following baseline correction, it is often desirable to **zero fill** the data, which entails adding zeros to the end of an FID. Normally an FID of N data points is zero filled to $2N$ data points, or an FID that has been acquired with a number of data points that is not an even power of two is zero filled up to the next power of two. Zero filling results in an interpolation between more data points in a spectrum and provides a better representation of line shapes and a better measurement of line widths. Figure 2.17b shows the effect on the FID of zero filling.

One of the more common weighting functions is **exponential multiplication** or **line broadening**. This is accomplished by multiplying the FID by a decaying exponential function, $\exp(-t/k)$, where k is a time constant that optimally matches the natural T_2 decay of the FID (Figure 2.18). The resultant

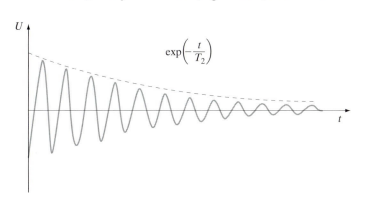

$$\exp\left(-\frac{t}{T_2}\right)$$

◀ FIGURE 2.18
Time dependence of the voltage μ induced in the receiver coil (FID) and a matched exponential weighting function.

▶ FIGURE 2.19
The effects of various weighting functions on the FID of an AMX spin system and the resulting transformed spectra.

(a) FID and spectrum with 8K data points. (b) Exponential multiplication of (a) with 1-Hz line broadening. Note that the S/N is improved over (a), but small splittings are lost. (c) Resolution enhancement. (d) and (e) 2 K data points result in a truncated FID, oscillations in the baseline, and "clipped" lines in (e) the high-field expansion. (f) Zero filling (d) and (e) to 10 K data points removes the clipping and reveals the small coupling effects.

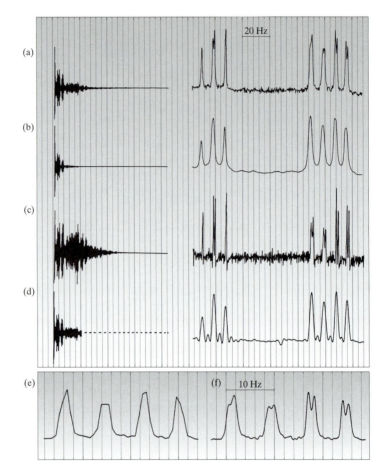

spectrum has an increased S/N ratio at the expense of resolution, where the amount of line broadening is given by $\pm 1/\pi k$. When k is negative, the process is called **resolution enhancement**, and it occurs at the expense of S/N. The most effective function for resolution enhancement is the product function, often called a double exponential, viz.,

$$\exp\left(\frac{t}{k}\right)\exp\left(-\frac{t^2}{l}\right),$$

for which the parameters k and l are determined empirically for each spectrum. Figure 2.19 illustrates the effects of various weighting functions on the FID (left) and the resulting transformed spectra (right).

2.11 The Fourier Transformation Process

The FID, which is a signal in the time domain, $f(t)$, is converted into a signal in the frequency domain, $F(\omega)$, by Fourier transformation. Figures 2.10, 2.12, and 2.19 show several examples of FIDs (the time-domain spectra) and their Fourier transformations (the frequency-domain spectra). The time $f(t)$ and frequency $F(\omega)$ functions are Fourier inverses that are related to one another by

$$f(t) = \frac{1}{2\pi}\int_{-\infty}^{+\infty} F(\omega)e^{i\omega t}\,d\omega \tag{2.6}$$

and

$$F(\omega) = \int_{-\infty}^{+\infty} f(t)e^{-i\omega t}\, d\omega, \tag{2.7}$$

and since

$$e^{ix} = \cos(x) + i\sin(x),$$

Eq. 2.7 becomes

$$F(\omega) = \int_{-\infty}^{+\infty} f(t)\cos \omega_0 t - \int_{-\infty}^{+\infty} f(t)i\sin \omega_0 t\, dt. \tag{2.8}$$

Equation 2.6 can be expanded as a series of sines and cosines because $f(t)$ can be expressed as a series of sines and cosines. This results in the following equation:

$$F(\omega) = \int \big[A_1 \cos(\omega_1 t) + A_2 \cos(2\omega_2 t) + \cdots \tag{2.9}$$

$$+ A_n \cos(n\omega_n t)\big] \cos(n\omega_0 t)\, dt$$

$$- i \int \big[A_1 \cos(\omega_1 t) + A_2 \cos(2\omega_2 t) + \cdots$$

$$+ A_n \cos(n\omega_n t)\big] \sin(n\omega_0 t)\, dt$$

$$- i \int \big[B_1 \sin(\omega_1 t) + B_2 \sin(2\omega_2 t) + \cdots$$

$$+ B_n \sin(n\omega_n t)\big] \cos(n\omega_0 t)\, dt$$

$$- i^2 \int \big[B_1 \sin(\omega_1 t) + B_2 \sin(2\omega_2 t) + \cdots$$

$$+ B_n \cos(n\omega_n t)\big] \sin(n\omega_0 t)\, dt.$$

Equation 2.9 applies to a continuous signal, but because the data are digitized, discrete signals occurring at discrete data points must be considered. The integrals in that equation must be replaced by sums, and the Fourier coefficients A_k must be found so that they can be summed to give $F(\omega)$. This is done in the computer with the fast Fourier transform (FFT) algorithm developed some time ago by Cooley and Tukey:

$$F(\omega) = \sum_{-t}^{+t} f(t)\exp\{i\omega t\}\, dt. \tag{2.10}$$

In today's modern microcomputers, the transformation takes only a few seconds and the operation is transparent to the user. Thus, the user need not worry about the foreboding appearance of equations 2.6 through 2.10.

2.12   Double and Triple Resonance

Double resonance involves the simultaneous or sequential application of two different RF frequencies to a sample in a magnetic field. The decoupling field, often called $\mathbf{B_2}$, must be larger in frequency units than the interaction to be decoupled. The $\mathbf{B_2}$ field "drives" the spins, causing their z components to be flipped rapidly, effectively removing the interaction from the spectrum. Many different types of double-resonance experiments are possible. Here we introduce the terminology and the basic concepts. Whenever decoupling experiments are performed, more RF power is transferred to the sample, which is thereby heated; the chemical shifts, which are temperature dependent, may change significantly. The effect of large decoupling powers on chemical shifts is called a **Bloch–Siegert shift**. It is thus important to decouple with the minimum possible RF power. Recently devised schemes called GARP, WALTZ-16, and MLEV-64, among others, allow the decoupling frequency to be distributed over a broad band (up to 10,000 Hz) with minimum power, thereby minimizing heating problems in the sample.

2.13 Homonuclear Decoupling

Double-resonance experiments for solution NMR spectroscopy are usually performed by irradiating the sample at a single frequency while observing the entire spectrum by pulse methods. This is accomplished by time-sharing the transmitter coil for the two purposes. Homonuclear double resonance is most commonly employed for proton spectra, as illustrated in Figure 2.20. One can vary the power of the irradiating RF fields so that the spins are merely "tickled" or so that they are completely saturated. When the spins are saturated, they are effectively **decoupled**, and the effects of spin–spin coupling to adjacent nuclei are removed, as in Figure 2.20b. This usually produces sub-

▶ FIGURE 2.20
500-MHz ^{1}H NMR spectra of $C_6H_5CH_2CH_3$ in $CDCl_3$.

(a) FID of ^{1}H{^{1}H, CH_3} and (b) Fourier transform of (a). (c) FID of normal spectrum and (d) Fourier transform of (c). Note that the "beat patterns" in (c) are separated by $1/^3J(HH)$.

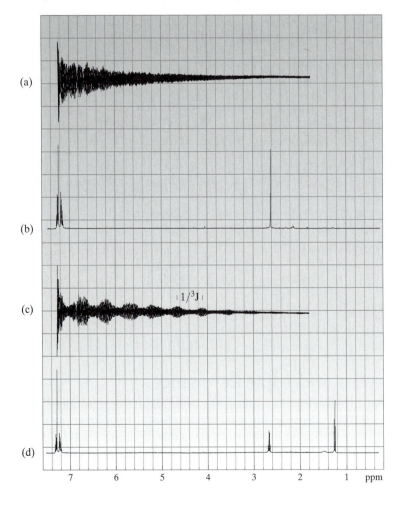

stantial spectral simplification and can be used as an aid in assigning resonances or determining the values of coupling constants in proton NMR spectra. In Chapter 11 we shall see **two-dimensional** NMR methods that can be
used in place of single-frequency homonuclear decoupling to determine coupling pathways among all coupled nuclei, but generally not the values of
coupling constants.

2.14 Heteronuclear Decoupling

Double-resonance experiments are also used in solution NMR spectroscopy
to remove the multiplicities caused by spin–spin couplings between protons
and the heteronucleus. In a heteronuclear double-resonance experiment, the
protons are saturated during the acquisition of the free-induction decay signal. This causes the heteronuclear spin multiplets to collapse into a single
peak. Continuous saturation of the proton spin system (as opposed to saturation only during acquisition) produces a signal enhancement caused by the
nuclear Overhauser effect (NOE), to be further discussed in Chapter 9. A proton-decoupled spectrum is therefore simple and of enhanced signal-to-noise
ratio. Figures 2.21 and 2.22 illustrate two examples. In Figure 2.21, both

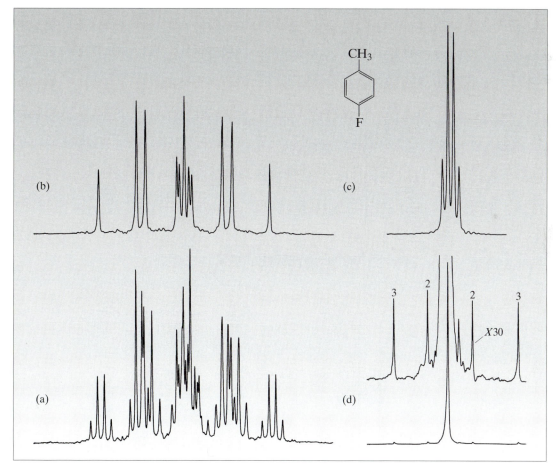

▲ FIGURE 2.21
56.4 MHz ^{19}F and ^{19}F{^{1}H} NMR spectra of 4-F-C$_6$H$_4$CH$_3$ in CDCl$_3$.

(a) Normal ^{19}F spectrum, (b) ^{19}F{^{1}H} spectrum with CH$_3$ protons decoupled (X resonance of $AA'BB'X$ spin system), (c) ^{19}F{^{1}H} spectrum
with C$_6$H$_4$ protons decoupled (X resonance of A_3X spin system), and (d) ^{19}F{^{1}HBB} completely decoupled ^{19}F spectrum; the
superposition X30 shows the ^{13}C satellites due to ^2J(CF) and ^3J(CF), which are marked with 2 and 3, respectively.

selective and broadband $^{19}F\{^1H\}$ experiments are illustrated, whereas Figure 2.22 illustrates a $^{13}C\{^1H\}$ broadband decoupling experiment. In both figures, the spectral simplification and the increased sensitivity are readily apparent. In some cases—particularly when one observes carbon NMR spectra of molecules containing both protons and fluorine or protons and phosphorus—it is sometimes desirable to perform triple-resonance experiments, as illustrated in Figure 2.23.

▶ FIGURE 2.22
22.5 MHz ^{13}C NMR spectra of (CH$_3$O)$_2$CH$_3$P=O in CDCl$_3$.
(a) $^{13}C\{^1H\}$ spectrum, (b) ^{13}C spectrum with proton coupling. Note that for ^{31}P, $1 = \frac{1}{2}$, 100% natural abundance, and $^{13}C—^{31}P$ coupling is present in both spectra.

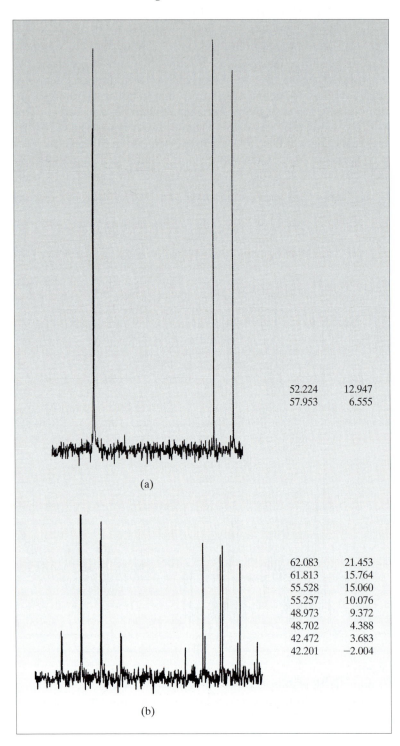

52.224	12.947
57.953	6.555

(a)

62.083	21.453
61.813	15.764
55.528	15.060
55.257	10.076
48.973	9.372
48.702	4.388
42.472	3.683
42.201	−2.004

(b)

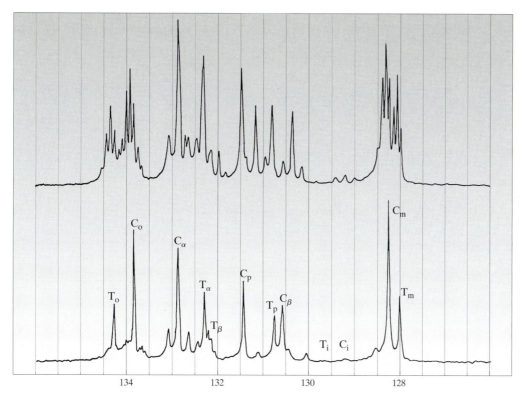

▲ FIGURE 2.23
75-MHz ^{13}C{^{1}H} (top) and ^{13}C{^{1}H, ^{31}P} (bottom) NMR spectra of (C) *cis*- and
(T) *trans*-diiodobis (divinylphenylphosphine)platinum (II) in CDCl$_3$ at 30°C
(o = ortho, p = para, m = meta, i = ipso phenyl carbons; α, β = vinyl
carbons). Holt, M. S.; Nelson, J. H.; Alcock, N. W. *Inorg. Chem.* **1986**, *25*, 2288.

2.15   Pulse Sequences

A pulse sequence consists of a precisely timed series of radio-frequency pulses, whose frequencies, phases, and duration are used to produce the desired
information. Figure 2.24 shows some examples, illustrating several variations
on heteronuclear decoupling and NOE experiments. For example, the pulse
diagram in Figure 2.24a is read as follows: A 90° pulse (of unspecified phase)
is applied at the ^{13}C frequency, the carbon signal is detected during the acquisition time, and there is an acquisition delay before the sequence is repeated. Simultaneously, the proton-irradiating frequency is applied
continuously during all three periods. This produces a heteronuclear decoupled spectrum with NOE. (See Chapter 9.)

   The pulse diagram of Figure 2.24b shows what is known as a "gated decoupling experiment." The resultant spectrum will be decoupled, but the signal intensities will not be perturbed by differential NOEs. This type of
experiment is useful for obtaining quantitative carbon NMR spectra. In a gated
decoupling experiment, the proton decoupling frequency is turned on only
during the time that the free-induction decay signal is detected. This pulse
sequence works because the spin–spin coupling is destroyed essentially instantaneously when the proton $\mathbf{B_2}$ field is applied, but the NOE takes some
time to build up to its maximum value. Furthermore, only the magnetization
in the $x'y'$-plane is detected, and hence, any Overhauser enhancement that

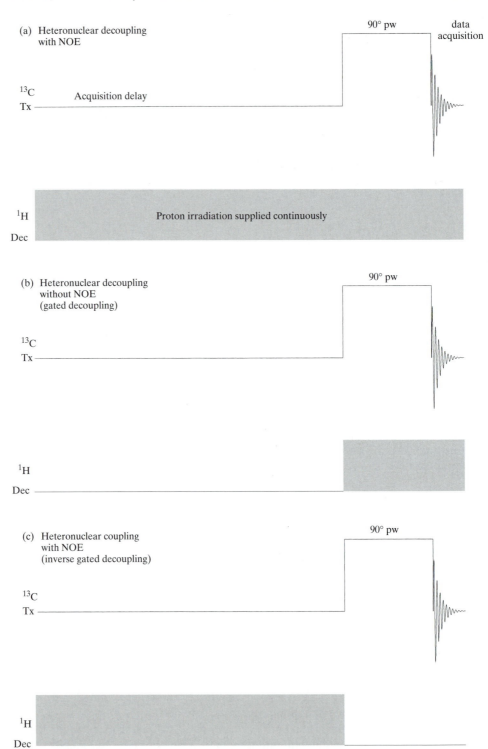

▲ FIGURE 2.24
Heteronuclear pulse sequences.

(a) ${}^1H$ decoupling with NOE, (b) gated decoupling, and (c) heteronuclear, coupled, with NOE (inverse gated decoupling).

builds up will be for the magnetization that has already relaxed back into the z' direction. This built-up Overhauser enhancement will be lost during the acquisition delay.

An additional variation of this straightforward double-resonance experiment is to obtain a heteronuclear NMR spectrum that is fully coupled, yet has the enhanced signal intensity caused by the NOE. The pulse sequence for such an experiment is shown in Figure 2.20c. It is called an inverse gated decoupling experiment.

2.16   Signal-to-Noise Ratio

NMR spectroscopy is inherently an insensitive spectroscopic technique. (See Section 1.4.) The signal-to-noise ratio can be improved by acquiring many spectra and adding them together. With a CW instrument, this can be accomplished by using a small computer called a multichannel analyzer to digitize and store each frequency-domain spectrum and add them together. This is a time-consuming and difficult process, for several reasons. Each spectrum consists of less signal than baseline, but both must be digitized, and the acquisition time is several minutes for each. The typical multichannel analyzer does not have a large number of channels, and as a result, the data-point resolution is not good. Also, in order to add properly, each spectrum must start at exactly the same point, a difficult task to accomplish. With an FT spectrometer, each spectrum is digitized in a few seconds in the time domain, and the FIDs can be added together with much higher data-point resolution. In either case, the signal-to-noise ratio increases as the square root of the number of spectra that are added together. Thus, a doubling of S/N requires a quadrupling of the number of spectra. A common misconception about this technique is that an infinite number of spectra will always give rise to a high S/N ratio. In practice, if no signals are observed after some reasonably large number of spectra (say, x) are added together, then it is usually found that no signals or exceedingly weak signals are observed even after 10x or 100x spectra are added together. If the sample is too dilute, a spectrum will not be observed even with the most sensitive high-field spectrometer.

Additional Reading

1. Consult the references given at the end of Chapter 1.

2. KING, R. W. and WILLIAMS, K. R., have published a four-part series entitled, "The Fourier Transform in Chemisty" in the *Journal of Chemical Education*: Part 1, *66*, A213 (1989); Part 2, *66*, A243 (1989); Part 3 (which also has a useful glossary of terms), *67*, A125 (1990), Part 4, *67*, A125, (1990).

3. There is an excellent computer tutorial entitled, *13th BCCE Workshop FT NMR Simulations*, written by Harold Bell of the Virginia Polytechnic Institute and State University, Blacksburg, VA 24061, which is available from him. This software produces FIDs from input spectral parameters and graphically illustrates data-processing procedures. It also simulates various one- and two-dimensional spectra.

4. FARRAR, T. C. and BECKER, E. D., *Pulse and Fourier Transform NMR*, Academic, New York, 1971.

5. MÜLLEN, K. and PREGOSIN, P. S., *Fourier Transform NMR Techniques: A Practical Approach*, Academic, London, 1976.

6. SHAW, D., *Fourier Transform NMR Spectroscopy*, 2d ed., Elsevier, Amsterdam, 1984.

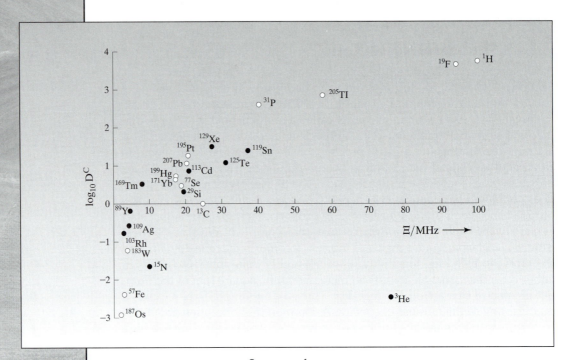

CHAPTER 3

Relative NMR receptivities, D^C, for spin-$\frac{1}{2}$ nuclei as a function of the standardised NMR frequencies Ξ. For each element only the most favourable nuclide has been included. The open circles represent nuclei with positive magnetogyric ratios; filled circles indicate nuclei with negative γ. See Table 1.1.

The Chemical Shift and Examples for Selected Nuclei

3.1   General Considerations

$\mathbf{T}$he NMR phenomenon is useful to chemists because the energy of the res-onance (i.e., the field strength required to bring about resonance at a con-stant frequency) is dependent upon the **electronic environment** about the nucleus. The electrons around the nucleus shield the nucleus so that the mag-nitude of the field acting on the nucleus, $\mathbf{B}_N$, is reduced from the applied field $\mathbf{B}_0$ by a screening or shielding constant according to the equation

$$\mathbf{B}_N = \mathbf{B}_0(1 - \sigma).\qquad(3.1)$$

The screening constant σ depends upon several things, among which are hy-bridization and electronegativity of groups attached to the nucleus. For chem-ically different nuclei of the types in a molecule, the resonance frequencies are

$$\nu_i = \frac{\beta_p \mathbf{B}_0 g_i(1 - \sigma_i)}{h},\qquad(3.2)$$

for $i = 1, 2, \ldots, r$.

Identical nuclei in chemically equivalent environments (to be defined shortly) in a molecule have the same values of g_i and σ_i and thus have the same NMR transition frequency (chemical shift). Identical nuclei in chemi-cally different environments have the same values of g_i, but different values of σ_i and, since σ_i is generally small compared to unity, have NMR transitions that lie close together. Nonidentical nuclei have different values of g_i, and their transitions are well separated.

If we consider a molecule such as ethanol (CH_3CH_2OH), the ^{16}O and ^{12}C nuclei therein have $I = 0$ and may be ignored. There are three kinds of chem-ically different protons in this molecule (we shall show later that we would predict, from symmetry arguments, three sets of equivalent protons), and each kind has its own value of σ. It might be argued that the three CH_3 protons are not quite in chemically equivalent environments, and this is true in a frozen conformation. However, the barrier to rotation about the C—C bonds is so low that, at ambient temperature, the CH_3 and CH_2 protons each become av-eraged into equivalent environments by this rotation. (See Chapter 11.) The OH proton is bonded to a highly electronegative atom, and we might expect the electron density to be lower about this proton than about the other protons in the molecule. This means that the OH proton has a lower diamagnetic shield-ing and a lower value of σ than do the CH_3 and CH_2 protons. Experimental-ly, it is found that, for pure liquid CH_3CH_2OH, $\sigma_{OH} < \sigma_{CH_2} < \sigma_{CH_3}$, in apparent accord with the electronegativity argument. However, in the gas phase and in dilute CCl_4 solutions, σ of the OH proton is greater than σ for the CH_2 and CH_3 protons. Thus, there must be other factors besides electronega-tivity that determine the chemical shifts in the isolated ethanol molecule. The variation in position of the OH resonance of ethanol is due to intermolecular hydrogen bonding.

From Equation 3.2 we see that at a fixed value of $\mathbf{B}_0$ the resonance fre-quency is *highest* for the OH proton (which has the lowest σ) in pure ethanol, and we would expect the ^1H NMR spectrum of pure ethanol to appear as in Figure 3.1. Most commonly, NMR spectra are observed with fixed RF frequency ($\nu_{spectrometer} = 40, 60, 80, 100, 200, \ldots$ MHz), and $\mathbf{B}_0$ is varied.

▶ FIGURE 3.1
**Low resolution ^{1}H NMR
spectrum of pure CH$_3$CH$_2$OH.**

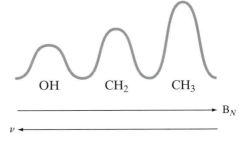

Under these conditions, the value of the field strength required to bring about resonance of the i nucleus is

$$\mathbf{B}_{0,i} = \frac{h\nu_{\text{spectrometer}}}{\beta_p g_i (1 - \sigma_i)}. \tag{3.3}$$

Hence, $\mathbf{B}_{0,i}$ is lowest for the OH proton in pure ethanol. A sweep from low to high $\mathbf{B}_0$ at fixed RF frequency corresponds to a sweep from high to low RF frequency at fixed $\mathbf{B}_0$. The change in NMR transition frequency at fixed $\mathbf{B}_0$ (or equivalently, the change in $\mathbf{B}_0$ at fixed $\nu_{\text{spectrometer}}$) due to the shielding of a nucleus by the electrons of the molecule is called the **chemical shift**. An isotropic chemical shift is observed in liquid and gaseous samples due to rapid movement of the molecules, but in solid samples an anisotropic chemical shift is observed, because the chemical shift is really a tensor quantity. (A tensor is a three-dimensional vector; see Chapter 14.)

The shielding tensor for a linear molecule has only three nonzero components, two of which are independent. The electronic circulations induced in such a molecule will be different, depending on whether the molecule is oriented parallel or perpendicular to the magnetic field. In this case, the two independent $\sigma_\parallel$ and $\sigma_\perp$ characterize the nuclear shielding. In a nuclear site with no symmetry at all, there are nine components of σ, but the off-diagonal components have negligible effect, and only the diagonal components, σ_{11}, σ_{22}, and σ_{33}, have great importance. In fluid phases, only the isotropic average $(\sigma_{\text{avg}} = \frac{1}{3}[\sigma_{11} + \sigma_{22} + \sigma_{33}])$ is observed.

The shielding constant σ is numerically small compared to unity. For protons in diamagnetic molecules, σ_i ranges from 1×10^{-5} to 4×10^{-5}, while for heavier nuclei, it may be as large as 10^{-4} to 10^{-3}. Actually, σ_i itself cannot be determined with high accuracy, because it requires accurate measurements of both $\nu_{\text{spectrometer}}$ and $\mathbf{B}_0$, the latter of which is difficult to measure accurately. Because of this difficulty, chemical shifts are measured *relative* to the shift of some standard compound.

It follows from Equation 3.3 that the separation δ_i between the resonances of two nuclei with different shielding constants σ_i is proportional to ν. (In an experiment at constant $\mathbf{B}_0$ and variable RF frequency, the separation is proportional to $\mathbf{B}_0$.) In order to express chemical shifts in a form independent of ν or $\mathbf{B}_0$, we define

$$\delta_i \equiv (\sigma_{\text{ref}} - \sigma_i) \times 10^6 \text{ ppm}, \tag{3.4}$$

where σ_{ref} and σ_i are the shielding constants for the standard reference nucleus and for nucleus i, respectively. This is the origin of the chemical shift unit in ppm. The definition makes δ_i positive for nuclei less shielded than the

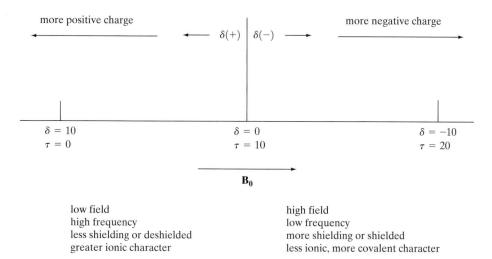

▲ FIGURE 3.2
Chemical shift terminology.

reference. δ ^{1}H ranges from 0 to 16 ppm in most organic compounds and to −30 ppm for transition metal hydrides. $\delta = 31$ for the bare proton. Another chemical shift unit, τ (most frequently encountered in the older literature) is defined as

$$\tau \equiv 10.000 - \delta. \tag{3.5}$$

τ ^{1}H ranges from 10 to −10 for most organic compounds and up to 40 for transition metal hydrides. These terms are more clearly defined in Figure 3.2.

Common reference standards are given in Table 1.1. For ^{1}H NMR spectroscopy, the recommended reference is tetramethylsilane [$(CH_3)_4Si$], usually termed TMS, for which $\delta = 0$ and $\tau = 10$. Because TMS is not soluble in water or dimethylsulfoxide, it is not recommended for these solvents. An alternative for these solvents is DSS, sodium 2,2-dimethyl-2-silapentane-5-sulfonate [$(CH_3)_3SiCH_2CH_2CH_2SO_3Na$], for which the CH_3 protons resonate at $\delta = 0.0$. A disadvantage of this reference compound is that it contains a number of other protons, which give signals in the region of interest. TMS is also the recommended reference for ^{13}C NMR spectroscopy, giving rise to only one resonance under normal operation. For TMS, $\delta = 0.0$, and downfield (less shielded) resonances are again positive. In the earlier literature, some ^{13}C chemical shifts were measured relative to CS_2 ($\delta_c = 192.3$ ppm). To convert, we have $\delta_c = 192.3 - \delta(CS_2)$. The usual ^{13}C reference for aqueous solutions is dioxane ($\delta_c = 66.5$); external TMS has also been used. External references are not recommended for ^{1}H measurements, but for other nuclei, the errors involved are small.*

We may understand the shielding in a molecule in the following way: Consider the s-electrons in a molecule. (Other electrons have zero probability of residing at the nucleus, although they do have an effect on chemical

---

* See Akitt, J. W. and Mann, B. E. "NMR and Chemistry," Stanley Thornes, Chettenham, U.K., 2000 for a discussion of the relative merits of various referencing techniques.

shifts.) The *s* electrons are spherically symmetric and circulate or precess in the applied magnetic field as illustrated in the following diagram:

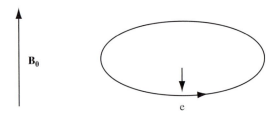

A circulating electron is an electron current, which produces a magnetic field at the nucleus opposing the applied field B_0. Thus, in order to obtain the resonance condition (Equation 1.4), it is necessary to increase the applied field over that for the bare nucleus. If B_0 is the applied field and B_i is the field at nucleus *i*, then the shielding (ΔB) is given by $\Delta B = B_0 - B_i$. This upfield shift (or shielding) of the nucleus is called a **diamagnetic shift**, and the phenomenon, called **diamagnetism**, is a universal contribution, as every molecule has *s* electrons. For electrons in *p* orbitals and *d* orbitals, and for all nonspherical molecules, there is no spherical symmetry. These electrons produce large magnetic fields at the nucleus, which, when averaged over the molecular motions, give a **low-field shift**. This deshielding is called the **paramagnetic shift**.

3.2   Proton Chemical Shifts

The proton (^{1}H) is a special case, as it is the only nucleus with only *s* electrons. This is the fundamental reason for its small chemical-shift dispersion range. The direct influence of the diamagnetic term may be seen in ^{1}H NMR spectroscopy. For example, in substituted methanes, CH_3X, as *X* becomes more electronegative, the electron density around the protons decreases, and they resonate at lower fields ie., increasing δ ^{1}H values. Indeed (Table 3.1), there is a reasonable correlation between δ ^{1}H and the electronegativity of *X* (E_X). We also find that acidic protons resonate at low field, (e.g., CO_2H or C(O)H groups have δ ≈ 11–12). Figure 3.3 shows the δ ^{1}H ranges for some commonly occuring functional groups. Again, the general pattern of

CH₃X	δ_H	δ_c	E_X
Table 3.1 Chemical Shifts (δ) of CH₃X Compounds and the Electronegativity of X, Denoted E_X			
X = SiMe₃	0.0	0.0	1.90
H	0.23	−2.6	2.20
Me	0.88	5.7 ⎫	
CN	1.97	1.3 ⎬	2.60
COCH₃	2.08	29.2 ⎭	
NH₂	2.36	28.3	3.05
I	2.16	−20.7	2.65
Br	2.68	10.0	2.95
Cl	3.05	25.1	3.15
OH	3.38	49.3	3.50
F	4.26	75.4	3.90

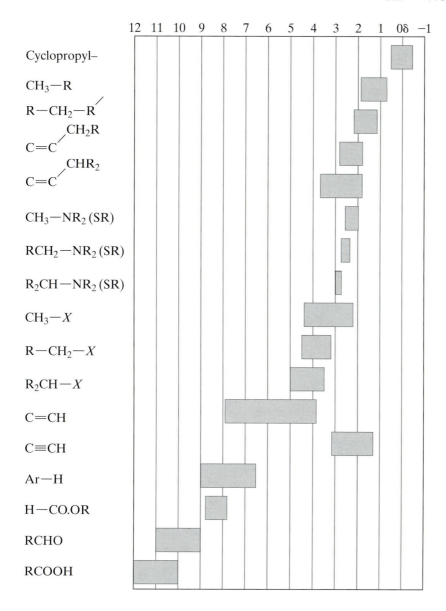

^{1}H chemical shifts of some common functional groups.

X = halogen, —OR, —NHCOR, —OCOR (R = alkyl).

increasing δ value for protons with diminished electron density is observed. There are many exceptions, however, as ^{1}H chemical shifts are affected by the diamagnetic circulations of valence electrons as well as by the neighboring p electrons and other anisotropic effects. (See later.)

More extensive tables of chemical shifts are available. One very simple and useful set of rules, known as Shoolery's rules,* enables the prediction of the chemical shifts of any CH_2XY or $CHXYZ$ proton. The chemical shift is given by the chemical shift of methane ($\delta = 0.23$), plus the sum of the substituent contributions in Table 3.2. The predictions are generally good to within 0.3 ppm for CH_2 groups, but are less accurate for CH groups. With the inclusion of the value for X=H, predictions can be extended to CH_3X groups by regarding them as H—CH_2—X groups.

---

* Shoolery, J. N., *Varian Associates Technical Information Bulletin*, Vol. 2, No. 3, Varian Associates, Palo Alto, CA, 1959.

Table 3.2 Additive Contributions to the Chemical Shifts of CH$_2$ and CH Groups: δH = 0.23 + Σ Contributions

Group	Contribution	Group	Contribution	Group	Contribution
H	0.17	NR$_2$	1.57	Br	2.33
CH$_3$	0.47	CONR$_2$	1.59	OR	2.36
CH$_2$R	0.67	SR	1.64	Cl	2.53
CF$_3$	1.14	CN	1.70	OH	2.56
C=C	1.32	COR	1.70	O.CO.R	3.13
C≡CR	1.44	I	1.82	O.Ph	3.33
				F	3.60
CO$_2$R	1.55	Ph	1.85	RCONH	2.36

SAMPLE PROBLEM 3.1

Predict the chemical shifts for the indicated protons in CH$_3$C$\underline{H}_2$Cl, CH$_3$C$\underline{H}_2$NR$_2$, and C$\underline{H}_2$Cl$_2$, using Shoolery's rules.

Solution

				Experimental
For	CH$_3$C$\underline{H}_2$Cl,	δ = 0.23 + 0.47 + 2.53 = 3.23 ppm		3.57
	CH$_3$C$\underline{H}_2$NR$_2$,	δ = 0.23 + 0.47 + 1.57 = 2.27 ppm		2.45
	C$\underline{H}_2$Cl$_2$,	δ = 0.23 + (2)(2.53) = 5.29 ppm		5.32

Shoolery's rules have recently been modified* to include contributions of not only directly attached substituents (α parameters), but also substituents one carbon removed (β parameters) and two carbons removed (γ parameters). These substituent parameters are listed in Table 3.3.

Table 3.3 Substituent Parameters for the Calculation of ^{1}H Chemical Shifts: δ CH$_3$ = 0.9 + α; δ CH$_3$ = 0.9 + Σ (β + γ); δ CH$_2$ = 1.2 + Σ (α + β + γ); δ CH = 1.5 + Σ (α + β + γ)

X	α	β	γ	X	α	β	γ
R—	0.0	0.0	0.0	H$_2$N—	1.5	0.2	0.1
R$_2$C=CR—	0.8	0.2	0.1	RCONH—	2.1	0.3	0.1
RC≡C—	0.9	0.3	0.1	O$_2$N—	3.2	0.8	0.1
Ar—	1.4	0.4	0.1	HS—	1.3	0.4	0.1
F—	3.2	0.5	0.2	RS—	1.3	0.4	0.1
Cl—	2.2	0.5	0.2	OHC—	1.1	0.4	0.1
Br—	2.1	0.7	0.2	RCO—	1.2	0.3	0.0
I—	2.0	0.9	0.1	ArCO—	1.7	0.3	0.1
HO—	2.3	0.3	0.1	HO$_2$C—	1.1	0.3	0.1
RO—	2.1	0.3	0.1	RO$_2$C—	1.1	0.3	0.1
R$_2$C=CRO—	2.5	0.4	0.2	H$_2$NOC—	1.0	0.3	0.1
ArO—	2.8	0.5	0.3	ClOC—	1.8	0.4	0.1
RCO$_2$—	2.8	0.5	0.1	N≡C—	1.1	0.4	0.2
ArCO$_2$—	3.1	0.5	0.2	RSO—	1.6	0.5	0.3
ArSO$_3$—	2.8	0.4	0.0	RSO$_2$—	1.8	0.5	0.3

*Beauchamp, P. S., and Marquez, R., *J. Chem. Educ.* **1997**, *74*, 1483.

SAMPLE PROBLEM 3.2

Predict the ^1H chemical shifts of the CH_3 and CH_2 protons for $CH_3C(O)CH_2CH_2CH_2CO_2H$, $ClCH_2CH_2CH_2C{\equiv}CH$, and $Br\ CH_2CH_2CH_2C(O)Cl$.

Solution

For $CH_3C(O)CH_2CH_2CH_3CO_2H$,   **Experimental**
          $a\qquad b\quad c\quad d$

$\delta\ CH_3 = 0.9 + 1.2 = 2.1$ 2.1
$\delta\ CH_2(b)\ = 1.2 + 1.2 + 0.1 = 2.5$ 2.6
$\delta\ CH_2(c)\ = 1.2 + 0.3 + 0.3 = 1.8$ 1.9
$\delta\ CH_2(d)\ = 1.2 + 0.0 + 1.1 = 2.3$ 2.4

For $ClCH_2CH_2CH_2C{\equiv}CH$,
          $a\quad b\quad c$

$\delta\ CH_2(a)\ = 1.2 + 2.2 + 0.1 = 3.5$ 3.7
$\delta\ CH_2(b)\ = 1.2 + 0.5 + 0.3 = 2.0$ 2.0
$\delta\ CH_2(c)\ = 1.2 + 0.2 + 0.9 = 2.3$ 2.4

For $Br\ CH_2CH_2CH_2C(O)Cl$,
          $a\quad b\quad c$

$\delta\ CH_2(a)\ = 1.2 + 2.1 + 0.1 = 3.4$ 3.4
$\delta\ CH_2(b)\ = 1.2 + 0.7 + 0.4 = 2.3$ 2.2
$\delta\ CH_2(c)\ = 1.2 + 0.2 + 1.8 = 3.2$ 3.1

The effect of substituents on the ^1H chemical shifts of olefinic and aromatic protons has also been investigated in detail. Tables 3.4 and 3.5 give the appropriate substituent effects. The additivity relations for olefins predict experimental chemical shifts to within 0.3 ppm over a range of about 5 ppm. Note the alternation in the substituent effects of some groups (OR, NR, F) between the geminal $C{=}C\overset{\displaystyle H}{\underset{\displaystyle X}{\diagdown}}$ and vicinal ($H{-}C{=}C{-}X$) effects. This is a measure of the difference in the inductive electron-withdrawing effect, which largely affects the nearest proton, and the resonance electron-withdrawing effect, which affects mainly the vicinal protons.

There is an analogy between Z_{cis} in olefins and the ortho-substituent effect in benzenes. These effects generally have the same sense, but Z_{cis} is almost twice as large, which would be expected on the basis of simple valence-bond principles.

A significant contribution to proton chemical shifts in aromatic compounds is due to the aromatic ring current. Let a molecule of benzene be oriented with its plane perpendicular to the applied magnetic field B_0, as shown in the following diagram:

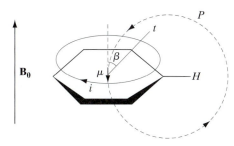

Table 3.4 Additive Shielding Increments for Olefins

$$\delta C{=}C{\diagdown}_{H} = 5.25 + Z_{gem} + Z_{cis} + Z_{trans}$$

$$R_{cis}{\diagdown}_{R_{trans}}C{=}C{\diagup}^{H}_{R_{gem}}$$

Substituent R	Z_i for R (ppm)		
	Z_{gem}	Z_{cis}	Z_{trans}
H	0.0	0.0	0.0
alkyl	0.45	−0.22	−0.28
Alkyl (cyclic)	0.69	−0.25	−0.28
CH₂OH	0.64	−0.01	−0.02
CH₂SH	0.71	−0.13	−0.22
CH₂X(X=F, CL, Br)	0.70	0.11	−0.04
CH₂N	0.58	−0.10	−0.08
C=C (isolated)	1.00	−0.09	−0.23
C=C (conjugated)	1.24	0.02	−0.05
C≡N	0.27	0.75	0.55
C≡C	0.47	0.38	0.12
C=O (isolated)	1.10	1.12	0.87
C=O (conjugated)	1.06	0.91	0.74
COOH (isolated)	0.97	1.41	0.71
COOH (conjugated)	0.80	0.98	0.32
COOR (isolated)	0.80	1.18	0.55
COOR (conjugated)	0.78	1.01	0.46
CF₃	0.66	0.61	0.32
CO.H	1.02	0.95	1.17
CO.N	1.37	0.98	0.46
CO.Cl	1.11	1.46	1.01
OR (R, aliphatic)	1.22	−1.07	−1.21
OR (R, conjugated)	1.21	−0.60	−1.00
O.CO.R	2.11	−0.35	−0.64
CH₂.CO; CH₂.CN	0.69	−0.08	−0.06
CH₂AR	1.05	−0.29	−0.32
Cl	1.08	0.18	0.13
Br	1.07	0.45	0.55
I	1.14	0.81	0.88
NR (R, aliphatic)	0.80	−1.26	−1.21
NR (R, conjugated)	1.17	−0.53	−0.99
N.CO	2.08	−0.57	−0.72
Ar	1.38	0.36	−0.07
Ar(o-subs)	1.65	0.19	0.09
SR	1.11	−0.29	−0.13
SO₂	1.55	1.16	0.93
F	1.54	−0.40	−1.02

The increments "R conjugated" are to be used instead of "R isolated" when either the substituent or the double bond is conjugated with further substituents. The increment alkyl (cyclic) is to be used when both the substituent and the double bond form part of a ring. (Data for compounds containing 3- and 4-membered rings have not been considered.)

Table 3.5   Substituent Chemical Shifts ($\Delta\delta$) in Benzenes[a]

| Substituent | $\Delta\delta_H$ | | | $\Delta\delta_C$ | | | |
	ortho	meta	para	C_1	ortho	meta	para
NO_2	0.95	0.26	0.38	20.0	−4.8	0.9	5.8
$CO.OCH_2$	0.71	0.11	0.21	1.8	1.0	−0.2	5.3
$COCH_3$	0.62	0.14	0.21	9.1	0.1	0.0	4.2
CHO	0.56	0.22	0.29	8.6	1.3	0.6	5.5
CN	0.36	0.18	0.28	−15.4	3.6	0.6	3.9
F	−0.29	−0.02	−0.23	34.8	−12.9	1.4	−4.5
Cl	0.03	−0.02	−0.09	6.2	0.4	1.3	−1.9
Br	0.18	−0.08	−0.04	−5.5	3.4	1.7	−1.6
I	0.39	−0.21	0.00	−34.1	8.7	1.4	−1.4
OH	−0.56	−0.12	−0.45	26.9	−12.7	1.4	−7.3
OCH_3	−0.48	−0.09	−0.44	31.4	−14.4	1.0	−7.7
$O.CO.CH_3$	−0.25	0.03	−0.13	19.3	−9.8	−2.2	−6.9
CH_3	−0.20	−0.12	−0.22	8.9	0.7	−0.1	−2.9
NH_2	−0.75	−0.25	−0.65	18.0	−13.3	0.9	−9.8
NMe_2	−0.66	−0.18	−0.67	22.6	−15.6	1.0	−11.5
$C\equiv CH$	–	–	–	−5.8	+6.9	0.1	0.4

[a] In ppm from benzene (δ_H 7.16; δ_c 128.0).

Then the π electrons precess in exactly the same way, and for the same reason, as the s electrons mentioned earlier. Now, however, there is a molecular circulation of electrons, rather than the atomic circulation of the s electrons, and the resulting ring current is symbolized by i. (Remember that the current flows in the direction opposite that of the electrons). The induced current gives rise to a magnetic moment (μ) along the sixfold proper rotation axis (C_6) of the benzene ring. The extra magnetic field produced by the ring current *opposes* the applied field, giving a *high-field shift*. Conversely, at the proton on the benzene ring, the ring current field *adds* to B_0, giving a *low-field* shift. The ring current is induced only when the applied magnetic field is perpendicular to the benzene ring. In a fluid solution, the benzene molecules rotate rapidly, and the chemical shift is the average over all orientations.

Many calculations of this ring current have been attempted. The simplest involves a calculation of the magnetic field due to the equivalent dipole (μ), and for any point P, this is given by

$$\Delta\delta(\text{ppm}) = \frac{\mu(1 - 3\cos^2\theta)}{r^3}, \tag{3.6}$$

where r and θ are as shown above. Thus, when $\theta = 0°$, (that is, above the benzene plane), $\Delta\delta$ is negative and there is a high-field shift. Conversely, when $\theta = 90°$, $\Delta\delta$ is positive and there is a low-field shift. At $\theta = 54.7°$, $1 - 3\cos^2\theta$ goes to zero, and there is no ring-current effect on chemical shifts.

Schematic representation of the 220-MHz ^{1}H NMR spectrum of coproporphyrin 1 methyl ester in CDCl$_3$ solution (M=Me, P=CH$_2$CH$_2$CO$_2$Me).

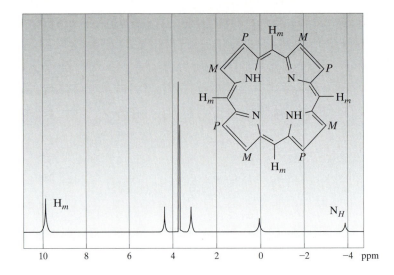

With more refined calculations that use the values $\mu = 27.0$, $r = 2.5$ Å, and $\theta = 90°$, $\Delta\delta = 27.0(2.5)^3 = 1.7$ ppm. For C$_6$H$_6$, δ ^{1}H is 7.16 ppm, which, when compared with $\delta = 5.86$ ppm for the olefinic protons of cyclohexa-1,3-diene, gives a ring-current shift of $\Delta\delta = 7.16 - 5.86 = 1.30$ ppm for the ring protons. Further from the benzene ring, the equivalent dipole model is more accurate.

An interesting manifestation of the ring current occurs for [10]-paracyclophane, for which the δ ^{1}H values of the various CH$_2$ groups reflect their positions relative to the benzene ring:

[10]-paracyclophane δ_H values

Those directly over the ring resonate at the highest field.

Another example is provided by the ^{1}H NMR spectrum of coproporphyrin 1 methyl ester (Figure 3.4). The large macrocycle of the porphyrin ring is aromatic (18 π electrons; $4n + 2$, $n = 4$) and gives rise to a large ring current. As a consequence, the mesoprotons (H$_m$) on the periphery of the porphyrin ring are shifted to low field ($\sim$10 δ), and the NH protons in the middle of the ring are shifted to high field ($\sim$−4 δ).

Because of its large effect on chemical shifts, the presence of a ring current is sometimes used as a test for aromaticity. For example, in their low-temperature ^{1}H NMR spectra, [16]-annulene has proton chemical shifts of 10.3 δ (inner protons) and 5.28 δ (outer protons), whereas [18]-annulene has shifts of −4.22 δ (inner) and 10.75 δ (outer):

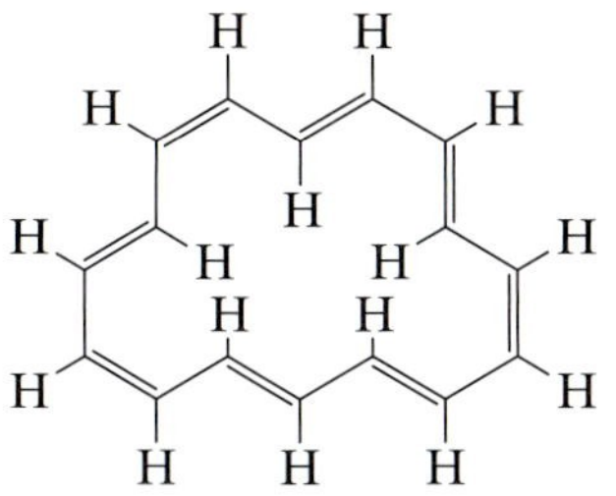

[16]-annulene 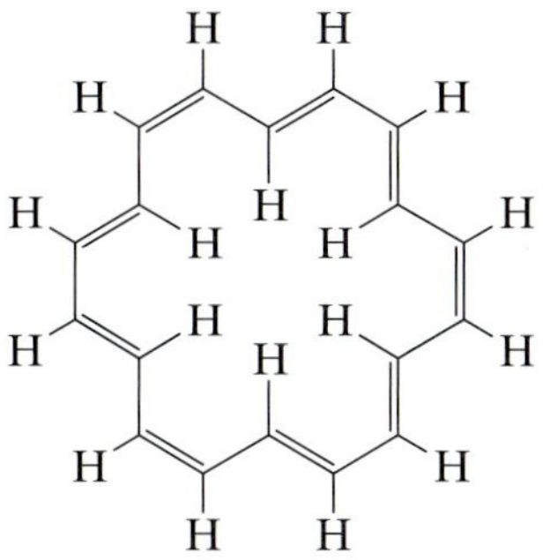   [18]-annulene

This suggests that the $4n + 2$ annulene is aromatic, whereas the $4n$ annulene is not.

The ring current is a magnetic property and its effect on chemical shifts should be exactly the same (in ppm) for any nucleus. But since all nuclei other than protons have much larger chemical-shift dispersion ranges, the ring-current effects are less noticeable for other nuclei.

The circulation of electrons giving rise to the diamagnetic effects in spherically symmetric atoms and in the benzene ring can also occur around the axis of any linear molecule when the axis is parallel to the applied field. This will produce an induced magnetic moment and magnetic effects on neighboring nuclei in exactly the same manner as for s electrons and the π electrons of C_6H_6. Two such groups are found in acetylene and nitriles. In these two cases, the diamagnetic circulation around the linear axis will produce high-field shifts along the molecular axis (e.g., for the $C\equiv CH$—proton) and low-field shifts perpendicular to the axis. This explains why acetylenic protons resonate at such high fields ($H-C\equiv C-H$, 1.48 δ, compared with C_2H_6, 0.88 δ and C_2H_4, 5.31 δ) when, on the basis of carbon hybridization, we would expect the order of chemical shifts to be δ acetylene $> \delta$ ethylene $> \delta$ ethane.

There are several other functional groups in which the circulation of electrons is less restricted about one molecular axis than the others. This produces a **magnetic anisotropy**. Protons near a magnetically anisotropic group will experience both high-field and low-field shifts, depending upon their position relative to the anisotropic group.

One such group is the carbonyl group. For this group, the anisotropy causes deshielding of protons, which lie in a cone whose axis is along the $C=O$ bond. Shielding occurs outside this cone, as illustrated in the following figure:

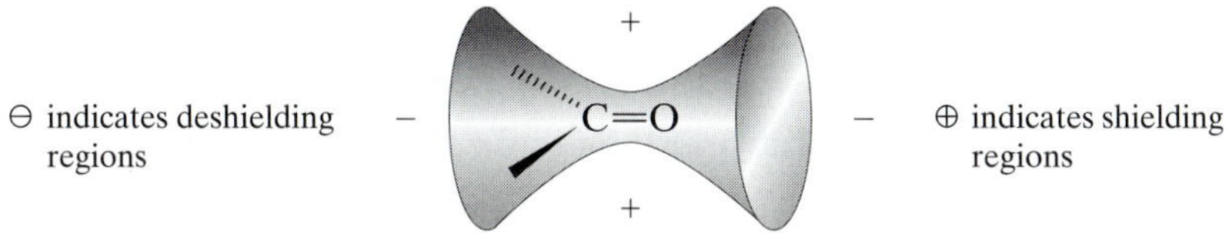

⊖ indicates deshielding
regions

⊕ indicates shielding
regions

Thus, an aldehyde proton within the cone experiences a low-field shift due to the anisotropy and resonates at low fields (δ 9.5–10.0). The alkene and nitro groups are also anisotropic, with deshielding regions in the planes of the double bonds and shielding regions above and below these planes,* as illustrated in the following diagram:

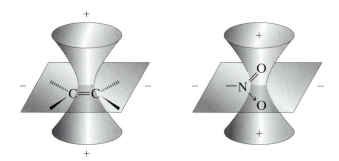

3.3   Carbon Chemical Shifts

The normal chemical shift dispersion range for ^{13}C (200 ppm) is about 10 times that for protons. Figure 3.5 gives the ^{13}C chemical shift ranges for a number of common functional groups. Tables 3.6 and 3.7 give both ^{1}H and ^{13}C chemical shifts for some typical ring systems. Comparing Figures 3.3 and 3.5, we see that there is a parallel between ^{1}H and ^{13}C chemical shifts. Downfield from the reference (TMS in both cases), the order alkanes, substituted alkanes, alkenes and aromatics, and ketones and aldehydes are the same for both nuclei. The same parallel is evident in Table 2.2, which gives the proton and carbon chemical shifts for some common solvents. However, the analogy between ^{1}H and ^{13}C chemical shifts should not be carried too far. Comparing the ^{1}H and ^{13}C chemical shifts of substituted methanes (Table 3.1) shows that, although for the first-row atoms deshieldings follow the order of the electronegativities of the substituent, this is not the case for the Cl, Br, and I substituents, in which large upfield shifts of the methyl carbon are observed. This "heavy atom" effect has no counterpart in ^{1}H chemical shifts.

Furthermore, the effect of substituents on δ ^{13}C is not confined to the nearest atom, as in Shoolery's rules for ^{1}H chemical shifts: Substituents two, three, or even four bonds away may have an appreciable effect, as is noted in Beauchamp and Marquez's modification of Shoolery's rules. A set of simple additivity rules has been proposed by Grant and Paul[†] for the ^{13}C chemical shifts of alkanes, in which these effects are considered explicitly. The chemical shift (δ) of the ith carbon atom in a hydrocarbon is given by

$$\delta_i = -2.6 + 9.1n_\alpha + 9.4n_\beta - 2.5n_\gamma + 0.3n_\delta, \tag{3.7}$$

where n_α is the number of carbons bonded directly to the ith carbon atom and n_β, n_γ, and n_δ are, respectively, the number of carbon atoms two, three,

---

*But see Martin, N. H., Allen, N. W., Minga, E. K., Ingrassia, S. T., Brown, J. D., *J. Am. Chem. Soc.*, **1998**, *120*, 11510.

[†] Grant, D. M.; Paul, E. G., *J. Am. Chem. Soc.*, **1964**, *86*, 2984.

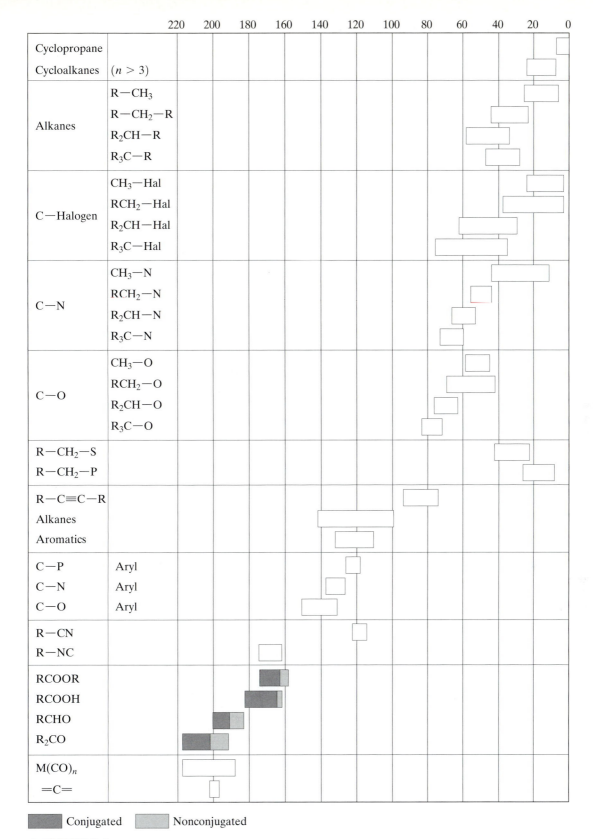

		220	200	180	160	140	120	100	80	60	40	20	0

Cyclopropane
Cycloalkanes $(n > 3)$

Alkanes
R—CH₃
R—CH₂—R
R₂CH—R
R₃C—R

C—Halogen
CH₃—Hal
RCH₂—Hal
R₂CH—Hal
R₃C—Hal

C—N
CH₃—N
RCH₂—N
R₂CH—N
R₃C—N

C—O
CH₃—O
RCH₂—O
R₂CH—O
R₃C—O

R—CH₂—S
R—CH₂—P

R—C≡C—R
Alkanes
Aromatics

C—P   Aryl
C—N   Aryl
C—O   Aryl

R—CN
R—NC

RCOOR
RCOOH
RCHO
R₂CO

M(CO)ₙ
=C=

■ Conjugated   ■ Nonconjugated

▲ FIGURE 3.5
¹³C chemical shifts of some common functional groups (R = alkyl).

Table 3.6 Proton and ^{13}C Chemical Shifts (δ) of Some Unsaturated Cyclic Systems

Molecule	1	2	3	4	5	6	7	8	9
Furan (positions 2,1; O)	7.42 (142.6)	6.37 (109.6)							
Pyridine (positions 3,2,1; N)	8.5 (150.6)	7.1 (124.5)	7.5 (136.4)						
Naphthalene (positions 5,4,3,2,1)	7.8 (128.1)	7.5 (125.9)			(133.7)				
Quinoline (positions 6,5,4,3,2,1,7,8,9,N)	8.8 (150.9)	7.3 (121.5)	8.0 (136.0)	(128.7)	7.7 (128.3)	7.4 (126.8)	7.6 (129.7)	8.0 (130.1)	(140.0)
Pyrrole (positions 2,1; N–H)	6.68 (118.5)	6.22 (108.2)							
Thiophene (positions 2,1; S)	7.30 (125.4)	7.10 (127.2)							
Pyridazine (positions 2,1; N,N)	9.21 (152.8)	7.50 (127.6)							
Pyrimidine (positions 2,3,1; N)	9.2 (159.5)	8.6 (157.5)	7.1 (122.1)						
Pyrazine (positions 2,1; N,N)	8.5 (145.6)								
Isoquinoline (positions 4,5,3,2,6,1,7,8,9; N)	8.5 (143.8)	7.5 (120.8)	(136.0)	7.7 (126.8)	7.6 (130.5)	7.5 (127.5)	7.9 (127.9)	(129.0)	9.1 (153.1)
Indole (positions 4,5,3,2,6,1,7,8; N–H)	6.5 (125.2)	6.3 (102.6)	(128.8)	(121.3)	(122.3)	(120.3)	(111.8)	(136.1)	
Imidazole (positions 2,3,1; N,N–H)	7.7 (136.2)	7.2 (122.3)	7.2 (122.3)						
Thiazole (positions 2,3,1; N,S)	8.84 (152.7)	7.97 (143.2)	7.41 (118.6)						
Anthracene (positions 6,5,4,3,2,1)	7.9 (130.1)	7.4 (125.5)			(132.2)	8.3 (132.6)			
Coumarin (positions 5,4,3,2,6,7,8,9; O,=O)	(160.4)	6.42 (116.4)	7.72 (143.4)	(118.7)	(127.9)	(124.3)	(131.7)	(116.5)	(153.8)
2-Pyridone (positions 4,3,2,5,1; N–H,=O)	(165.3)	6.6 (120.1)	7.3 (134.8)	6.1 (106.7)	7.3 (120.1)				

Figures in parentheses are ^{13}C shifts.

Table 3.7 Proton (δ_H) and ^{13}C (δ_C) Chemical Shifts of Some Saturated Heterocyclic Systems

Molecule	1	2	3	4
(tetrahydrofuran, O, positions 1,2)	3.7 (67.9)	1.8 (25.8)		
(pyrrolidine, NH, positions 1,2)	2.7 (47.1)	1.6 (25.7)		
(tetrahydrothiophene, S, positions 1,2)	2.8 (31.2)	1.9 (31.4)		
(sulfolane, SO₂, positions 1,2)	3.00 (51.5)	2.23 (22.8)		
(tetrahydropyran, O, positions 1,2,3)	3.6 (68.6)	1.6 (27.2)	1.6 (24.2)	
(piperidine, NH, positions 1,2,3)	2.7 (47.9)	1.5 (27.8)	1.5 (25.9)	
(thiane, S, positions 1,2,3)	2.6	1.8	1.8	
(ethylene oxide, O, position 1)	2.6 (39.7)			
(cyclopropane, position 1)	0.3 (−2.6)			
(aziridine, NH, position 1)	1.6 (N—H at 0.0) (28.7 for N—Me)			
(thiirane, S, position 1)	2.3 (18.9)			
(2-pyrrolidinone, NH, positions 1,2,3,4)	(179.4)	2.2 (30.3)	2.2 (20.8)	3.4 (N—H at 7.7) (42.4)
(γ-butyrolactone, O, positions 1,2,3,4)	(177.9)	2.5 (27.7)	2.2 (22.2)	4.3 (68.6)

Figures in parentheses are ^{13}C shifts.

Table 3.8 Substitutent Effects on ^{13}C Chemical Shifts ($\Delta\delta_c$) in 1-Substituted Pentanes[a]

Substituent	C_1	C_2	C_3	C_4	C_5
F	70.1	8.0	−6.7	0.1	0.0
Cl	30.6	10.0	−5.3	−0.5	−0.1
Br	19.3	10.1	−4.1	−0.7	0.0
I	−7.4	10.5	−2.1	−1.1	−0.1
CH_3	9.3	9.4	−2.5	0.4	0.2
NH_2	29.7	11.2	−5.0	0.1	0.0
OH	48.3	10.1	−6.0	0.3	0.2
CHO	31.4	0.7	−1.9	0.8	0.5
$CO.CH_3$	30.7	2.1	−1.2	1.4	1.2
COOH	20.5	2.3	−2.7	0.2	0.3
$C{\equiv}N$	3.7	3.2	−2.9	−0.4	−0.8
$C{\equiv}CH$	5.0	5.8	−3.0	0.4	–
$CH{=}CH_2$	20.3	6.2	−2.8	0.0	−0.1

[a] In ppm from pentane (C_1, 13.7; C_2, 22.6; C_3, 34.5).

and four bonds removed. The constant ($-2.6\ \delta$) is the ^{13}C chemical shift for methane. For *n*-octane, these rules would predict

$$C_1-C_2-C_3-C_4-C_5-C_6-C_7-C_8;$$

for $C_{1,8}$, $\quad \delta = -2.6 + 9.1 + 9.4 - 2.5 + 0.3 = 13.7;$

for $C_{2,7}$, $\quad \delta = -2.6 + 2(9.1) + 9.4 - 2.5 + 0.3 = 22.8;$

for $C_{3,6}$, $\quad \delta = -2.6 + 2(9.1) + 2(9.4) - 2.5 + 0.3 = 32.2;$

for $C_{4,5}$, $\quad \delta = -2.6 + 2(9.1) + 2(9.4) + 2(-2.5) + 0.3 = 29.7.$

These values compare favorably with the experimental values (14.0, 23.0, 32.4, and 29.7 δ). Table 3.8 gives substituent effects for a variety of substituents in 1-substituted pentanes. Note that a substituent in the α or β position generally deshields the carbon nucleus (the heavy atom *I* in the α position is an exception), while one in the γ position is shielding. This γ effect is of importance in conformational-analysis studies and will be considered in more detail later. In acyclic alkanes the effects of substituents in the δ or more remote positions are very small, but in cyclic compounds these long-range effects (γ, δ, etc.) may be large, depending upon the distance of the substituent from the carbon under consideration. For this reason, Equation 3.7 and Table 3.8 are not so accurate for cyclic compounds. However, the substituent parameters may be used in acyclic compounds to predict chemical shifts for multiply substituted compounds in the same way that Shoolery's rules for ^1H chemical shifts were used, and where the substituents do not interact directly, a reasonable additivity relationship holds.

SAMPLE PROBLEM 3.3

Predict the ^{13}C chemical shifts of 4-chlorobutan-1-ol.

Solution

Using Equation 3.7 we predict the following chemical shifts of butane $C_1-C_2-C_3-C_4$:

$$\delta_{1,4} = -2.6 + 9.1 + 9.4 - 2.5 = 13.4$$

$$\delta_{2,3} = -2.6 + 2(9.1) + 9.4 = 25.0$$

Observed

13.0

24.8

Then we use Table 3.8 to predict values for the chemical shifts of HO C_1—C_2—C_3—C_4 Cl:

Observed

$$\delta_{C1} = 13.4 + 48.3 - 0.5 = 61.2 \qquad 62.0$$

$$\delta_{C2} = 25.0 + 10.1 - 5.3 = 29.8 \qquad 29.7$$

$$\delta_{C3} = 25.0 - 6.0 + 10.0 = 29.0 \qquad 30.0$$

$$\delta_{C4} = 13.4 + 0.3 + 30.6 = 44.3 \qquad 45.4$$

The agreement between the predicted and observed values is often sufficiently good to immediately assign the spectrum, which is an important consideration in such spectra. However, we should note that these additivity relationships may break down for strongly interacting subsitutents. This will be particularly true for geminal (CR_1R_2) and, to a lesser extent, vicinal (CR_1CR_2) substituents.

SAMPLE PROBLEM 3.4

Use the substituent effects of Table 3.5 to assign the $^{13}C\{^1H\}$ NMR spectrum of

for which the following chemical shifts were observed: 117.1, 117.8, 118.2, 128.5, 132.4, 159.0, and 171.8 δ.

Solution

171.8 δ is the carbonyl carbon. By using the CO_2—CH_3 group to approximate the

$(CH_3)_2N\overset{\overset{\textstyle O}{\|}}{C}$ group, the other six carbon chemical shifts are predicted to be as follows:

	Observed	$\Delta\delta = \delta_{obs} - \delta_{cal}$	Percent difference
$\delta_{C1} = 128.0 + 26.9 + 1.0 = 155.9$	159.0	+3.1	1.9
$\delta_{C2} = 128.0 - 12.7 - 0.2 = 115.1$	117.1	+2.0	1.7
$\delta_{C3} = 128.0 + 1.4 + 5.3 = 134.7$	132.4	−2.3	−1.7
$\delta_{C4} = 128.0 - 7.3 - 0.2 = 120.5$	118.2	−2.3	−1.9
$\delta_{C5} = 128.0 + 1.4 + 1.0 = 130.4$	128.5	−1.9	−1.5
$\delta_{C6} = 128.0 - 12.7 + 1.8 = 117.2$	117.8	+0.6	0.5

Considering that we have approximated the $(CH_3)_2NC(O)$ functionality by the CO_2CH_3 functionality and that hydrogen bonding (strong interaction) occurs between the amide and OH groups, the agreement is not too bad.

An analysis analogous to that of Equation 3.7 has been developed for olefinic carbons. In this case, we need to consider the location of the substituent relative to the olefinic carbon under consideration, as indicated in the following chemical formula:

$$\overset{\gamma'}{C}-\overset{\beta'}{C}-\overset{\alpha'}{C}-\overset{\kappa'}{C}=\overset{\kappa}{C}-\overset{\alpha}{C}-\overset{\beta}{C}-\overset{\gamma}{C}$$

Then, the olefinic carbon chemical shift may be calculated from the following equation, with the correction terms being 0 (α, α' *trans*), -1.1 (α, α' *cis*), -4.8 (α, α), $+2.5$ (α', α'), and $+2.3$ (β, β):

$$\delta_c(k) = 123.3 + 10.6n\alpha + 7.2n\beta - 1.5n\gamma - 7.9n\alpha' - 1.8n\beta' + 1.5n\gamma'$$
$$+ \text{ correction term.} \tag{3.8}$$

For example, consider the following C_1 and C_2 chemical shifts of 1-butene, 2-methyl-1-butene, and 3,3-dimethyl-1-butene:

1-butene: $\underset{1}{C}=\underset{2}{C}-\underset{3}{C}-\underset{4}{C}$

$\delta_{C1} = 123.3 - 7.9 - 1.8 = 113.6$ vs. 112.8 observed

$\delta_{C2} = 123.3 + 10.6 + 7.2 = 141.1$ vs. 140.2 observed

2-methyl-1-butene: $\underset{1}{C}=\overset{\displaystyle C \atop |}{\underset{2}{C}}-C-C$

$\delta_{C1} = 123.3 + 2(-7.9) - 1.8 + 2.5 = 108.2$ vs. 107.7 observed
$\delta_{C2} = 123.3 + 2(10.6) + 7.2 - 4.8 = 146.9$ vs. 146.4 observed

3,3-dimethyl-1-butene: $\underset{1}{C}=\underset{2}{C}-\overset{\displaystyle C \atop |}{\underset{\displaystyle | \atop C}{C}}-C$

$\delta_{C1} = 123.3 - 7.9 + 3(-1.8) = 110.0$ vs. 108.4 observed
$\delta_{C2} = 123.3 + 10.6 + 3(7.2) + 2.3 = 157.8$ vs. 148.2 observed

Table 3.9 lists typical substituent effects on the ^{13}C chemical shifts of the olefinic carbons of ethylene.

Comparing these data with those given in Table 3.8 and the corresponding data for olefinic proton chemical shifts (Table 3.4) is illuminating. Whereas, for alkanes, the α- and β-substituent effects usually have the same sign, both giving rise to a downfield shift, for many substituents on olefins, a strong alternating effect causes a downfield shift for C_1 and an upfield shift for C_2. This is the case for F, CH_3, and CH_3O, paralleling the substituent effects of these groups on proton chemical shifts. Note, however, the upfield shifts produced by the heavy atom (I), which is similar to, but larger than, the corresponding upfield shifts produced by this substituent in alkanes.

As expected, the substituent chemical shifts (SCSs) in arenes (Table 3.5) are intermediate between those of alkanes and alkenes. The SCS of C_1, the ipso carbon, are very similar to those of the corresponding carbon in ethylene

Table 3.9   Substituent Effects on the ^{13}C Chemical Shifts ($\Delta\delta_c$) in 1-Substituted Ethylenes[a]

Substituent	C_1	C_2
F	24.9	−34.3
Cl	3.3	−5.4
Br	−7.2	−0.7
I	−37.4	7.7
CH_3	10.3	−7.8
OCH_3	30.3	−37.3
CHO	13.6	13.2
COOH	5.2	9.1
C(O)R	13.8	4.7
CN	−15.1	15.0

[a] In ppm from ethylene (δ_c = 123.3).

and generally much less than the SCS produced at C_1 in pentane. However, the SCSs of the *ortho* carbon atom are roughly what one would expect on the basis of an aromatic C—C bond having a bond order intermediate between a double and single bond. These SCSs may be used with caution to predict chemical-shift values and provide probable assignments for multiply substituted aromatics, as was illustrated in Sample Problem 3.4. As another example, consider the ^{13}C{^{1}H} NMR spectrum of methyl salicylate, shown below, for which

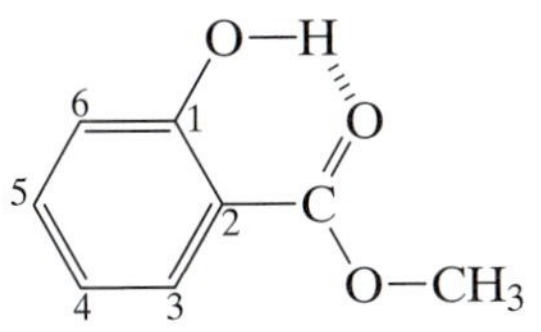

the following chemical shifts were observed in CDCl$_3$ at 22.5 MHz: δ = 170.6, 161.9, 135.7, 130.0, 119.1, 117.6, 112.5 and 52.1. The resonances at δ 170.6 (C=O) and 52.1 (OCH$_3$) are readily assigned. By using the SCS (Table 3.5), the following are predicted: δ C_1 (156.4), C_2 (117.6), C_3 (130.9), C_4 (121.0), C_5 (135.2), C_6 (115.6), providing a reasonable assignment, except for C_2 and C_6. These latter two carbons are readily distinguished on the basis of the intensities of their resonances. Since C_2 does not have an attached hydrogen, its resonance will be less intense than that of C_6. (We will discuss why this is so later.)

The shielding effect of substituents on the γ-carbon chemical shifts in alkanes has been noted previously. In acyclic compounds, rapid rotation about C—C bonds averages all substituent effects over many possible rotamers. Thus, it is necessary to obtain substituent chemical shifts for rigid molecules in order to ascertain their geometric or spatial dependence.

Table 3.10   Substituent Chemical Shifts ($\Delta\delta$) for Cyclohexanes[a]

Substituent		C_α	C_β	C_γ	C_δ
CH₃	eq	5.0	9.0	0.0	−0.2
	ax	1.4	5.4	−6.4	0.0
CN	eq	1.4	2.8	−1.9	−1.9
	ax	0.1	−0.9	−4.5	−1.4
OH	eq	42.6	8.0	−2.7	−2.1
	ax	37.8	5.5	−6.8	−0.7
NH₂	eq	23.8	9.9	−1.6	−1.0
F	eq	64.5	5.6	−3.4	−2.5
	ax	61.1	3.1	−7.2	−2.0
Cl	eq	32.3	10.5	−0.6	−2.2
	ax	32.3	6.7	−7.1	−1.4
Br	eq	24.6	11.2	0.3	−2.5
	ax	27.5	7.2	−6.5	−1.5
I	eq	2.0	13.7	2.0	−2.1
	ax	9.1	9.1	−4.6	−1.3

[a] In ppm from cyclohexane ($\delta_c = 27.6$).

An extensive compilation of ^{13}C chemical shifts for transition-metal organometallic compounds may be found in Mann B. E., Taylor, B. E., *^{13}C NMR Data for Organometallic Compounds*, Academic, New York, 1981. An excellent treatise on ^{13}C NMR is Breitmaier, E., Volter, W., *Carbon-13 NMR Spectroscopy*, VCH, New York, 1990. This book contains a wealth of data.

One such series of molecules—the cyclohexanes—has been studied extensively, and Table 3.10 gives substituent chemical shifts for the cyclohexane ring. The shifts may be used, with caution, to predict the carbon chemical shifts of variously substituted cyclohexanes. Again, where geminal substituents or 1,3-diaxial interactions are present, these additivity rules break down. It is interesting to note the very different effects of any given substituent in the axial and equatorial positions. In general, an equatorial substituent gives shifts for all the cyclohexane carbons, which are to low field of the corresponding axial substituent. In the particular case of the γ carbons, axial substituents produce a sizable upfield shift (about 4 to 7 ppm), whereas the corresponding equatorial substituents produce a much smaller upfield shift (about 0 to 3 ppm). As would be expected, the γ effect in the open chain compounds (Table 3.8) falls between these values.

3.4   Boron Chemical Shifts

Two boron nuclei are available for investigation: ^{10}B ($I = 3$) and ^{11}B $\left(I = \frac{3}{2}\right)$. There is, however, very little interest in ^{10}B NMR spectroscopy, because of the lower natural abundance of ^{10}B (19.5%, compared with 80.42% for ^{11}B) and its inherently low sensitivity. ^{10}B has a receptivity of 22.1 relative to ^{13}C, while the relative receptivity of ^{11}B is 754. Although, in principle, ^{10}B and ^{11}B nuclei in the same compound have different chemical shifts, this is not likely to cause much practical concern. Thus, for the pair of ions $[BH_4^-]$ and $[BF_4^-]$, there is a primary isotope effect of 0.11 ± 0.03 ppm. For the vast majority of boron compounds, 0.11 ppm is negligible compared to the line width of the resonance. The generally accepted reference standard for boron NMR is $F_3B \cdot OEt_2$ as an external reference. There is a great deal of chemical-shift data for boron compounds.[*] For simple compounds, there is a close relationship between $\delta\,^{11}$B and $\delta\,^{13}$C for isoelectronic species. Boron-11 chemical shifts are proportional to the electronegativity and π-bonding character of the ligand attached to the boron. Thus, for a series of R_nBF_{3-n} compounds, there are good relationships between $\delta\,^{11}$B and the boron $2p$ orbital population and between $\delta\,^{11}$B and boron p charge densities. Some representative ^{11}B chemical shifts are given in Table 3.11.

The ^{11}B chemical-shift range covers some 200 ppm (similar in dispersion to ^{13}C), with four coordinate sp^3 species ($\delta = +20$ to -128 ppm) generally occurring to high field of three coordinate sp^2 species ($\delta + 92$ to -8 ppm), as shown schematically in Figure 3.6.

The chemical shifts of representative binary boron hydrides are given in Figures 3.7 to 3.9.

Table 3.11   Some Representative ^{11}B Chemical Shifts Referenced to $F_3B \cdot OEt_2$ (R, R^1, R^2 = alkyl)

	δ/ppm		δ/ppm
RBF_2	28 to 30	R_2BF	59 to 60
$RBCl_2$	62 to 64	R_2BCl	76 to 78
$RBBr_2$	62 to 66	R_2BBr	79 to 82
$PhBF_2$	24.8	Ph_2BF	47.4
$PhBCl_2$	54.8	Ph_2BCl	61.0
$RB(NMe_2)_2$	33.5 to 34.2	R_2BNMe_2	44 to 46
$RB(OMe)_2$	29 to 32	$R_2B(OMe)$	53 to 54
$[B(OMe)_4]^-$	3	$[B(OH)_4]^-$	1.1
$[BR_4]^-$	-16 to -20	$[BPh_4]^-$	-6.3
$[BF_4]^-$	-2.2	$[BCl_4]^-$	6.6
$[BBr_4]^-$	-24.1	$[BI_4]^-$	-1.28
$[B(NO_3)_4]^-$	-86.6	$[B(OPh)_4]^-$	2
$(R^1BNR^2)_3$	23 to 37	$B(NR_2)_2X$	20 to 30

[*] Nöth, H. and Wrackmeyer, B., "NMR Spectroscopy of Boron Compounds," in Diehl, P., Fluck, F., Kosfeld, R. Eds., *NMR—Basic Principles and Progress*, Springer–Verlag, Berlin 1978; Wrackmeyer, B., *Annu. Rep. NMR Spectrosc.*, **1988**, *20*, 61; Siedle, A. R., *Annu. Rep. NMR Spectrosc.*, **1988**, *20*, 205.

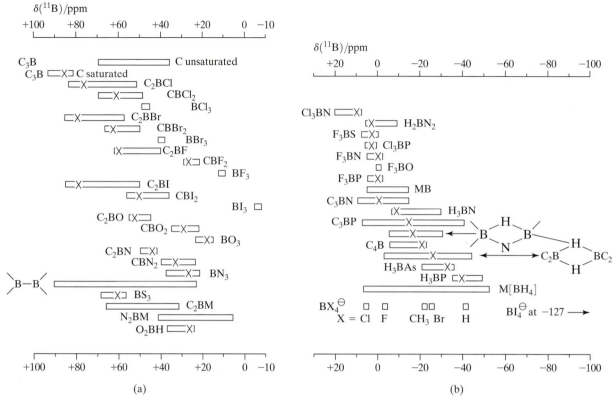

▲ FIGURE 3.6
Chemical-shift ranges for three-coordinate, BXYZ (a) and four-coordinate, B$WXYZ$ (b) compounds. X indicates the shift of the simplest CH$_3$ derivative. Note that in donor solvents most BXYZ compounds will be four coordinate.

▶ FIGURE 3.7
Chemical shifts for selected arachno boranes.

B$_3$H$_8$$^-$ and B$_9$H$_{14}$$^-$ are fluxional in solution. See Muettertics, E. L., *The Chemistry of Boron and Its Compounds*, Wiley, New York, 1967, for a discussion of these structures.

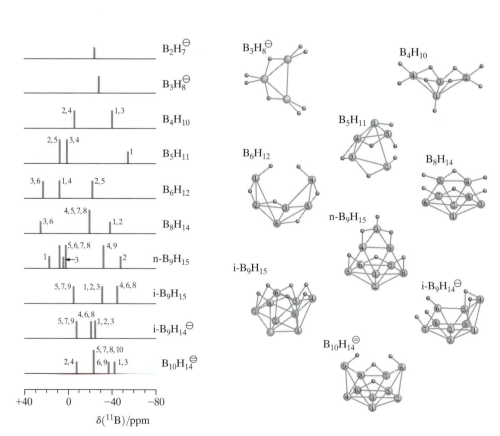

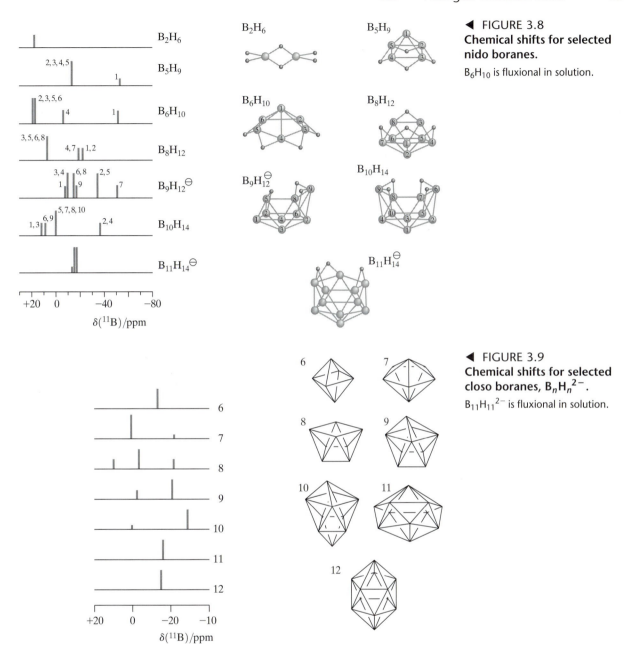

◀ FIGURE 3.8
Chemical shifts for selected nido boranes.

B_6H_{10} is fluxional in solution.

◀ FIGURE 3.9
Chemical shifts for selected closo boranes, $B_nH_n^{2-}$.

$B_{11}H_{11}^{2-}$ is fluxional in solution.

3.5   Nitrogen Chemical Shifts

Nitrogen possesses two isotopes of interest to NMR spectroscopists. Both present considerable problems. Nitrogen-14 is 99.63% naturally abundant, with a receptivity of 5.69 relative to that of ^{13}C, but because $I = 1$, this nucleus possesses a quadrupole moment. Thus, ^{14}N signals are easily detected, but the lines are generally broad, with line widths often between 100 and 1000 Hz (14 to 140 ppm at 2.35 tesla). Consequently, resolution is poor. Although the chemical-shift dispersion range is 1000 ppm, the range for similar types of nitrogen is only about 50 ppm. Nitrogen-15 NMR spectroscopy would appear to be better for structural investigations, as $I = \frac{1}{2}$, but unfortunately, it is only 0.37% abundant, with a consequent low receptivity, and γ is negative, leading to a negative NOE. (See Chapter 14.) The presence of two isotopes again raises the possibility of a primary isotope effect giving rise to fundamentally different chemical shifts, but usually the difference between the ^{14}N and ^{15}N chemical shifts of a nucleus in the same sample is no greater than the experimental error (0.2 ppm). Reference compounds are not fully agreed upon, but most commonly, NO_3^- is used for aqueous solutions and CH_3NO_2 for organic solvents. The relative shift between these two references is very small. Some typical chemical shifts are given in Figures 3.10 and 3.11. Amines and ammonium ions resonate at highest field, and double bonding and electronegative substituents cause a shift to a low field, with nitroso compounds occurring at lowest field. It should be noted that most chemical shifts are upfield of the reference and are therefore negative.

Nitrogen-15 and carbon-13 are similar in electronic structure and bonding, and their chemical shifts generally show a parallel dependence on chemical structure. It is found, in general, that shielding decreases in the order $sp^3 > sp > sp^2$ for varying nitrogen hybridization. Protonation of alkylamines causes strong deshielding, whereas protonation of pyridines causes strong shielding. For example, for CH_3NH_2, $\delta = -377.3$, and for $CH_3NH_3^+Cl^-$, $\delta = -361.4$, but for pyridine, $\delta = -63.2$, and for pyridine hydrochloride, $\delta = -161.6$.

Substituents on alkyl amines give rise to chemical-shift changes of the same type previously discussed for the substituent effects on δ ^{13}C for alkanes. Substituents on pyridine rings have an additive effect on δ ^{15}N, and some substituent constants are given in Table 3.12. The SCSs in Table 3.12 may be used with caution to predict the chemical shifts of multiply substituted pyridines.

Table 3.12 Substituent Effects (Δ_i) on Nitrogen-15 Chemical Shifts in Monosubstituted Pyridines; δ $^{15}N = -69.2 + \Delta_i$

Substituent	Δ_{C-2}	Δ_{C-3}	Δ_{C-4}
—CH$_3$	−0.4	0.3	−8.0
—CH$_2$CH$_3$	−1.8		−6.6
—CH(CH$_3$)$_2$	−5.1		5.9
—C(CH$_3$)$_3$	−2.5		−5.8
—CN	−0.9	−0.8	10.6
—CHO	10	11	29
—CO—CH$_3$	−9	15	11
—CO—OCH$_2$CH$_3$	11.8		−5
—OCH$_3$	−49	0	−23
—OH	−126	−2	−118
—NO$_2$	−23	1	22
—NH$_2$	−45	10	−46
—F	−42	−18	
—Cl	−4	4	−6
—Br	2	8	7

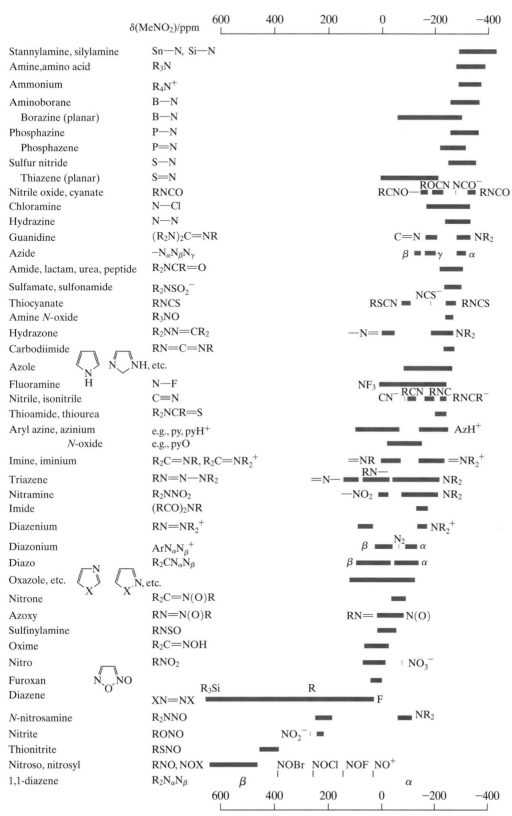

▲ FIGURE 3.10
Nitrogen chemical-shift ranges for main group compounds.

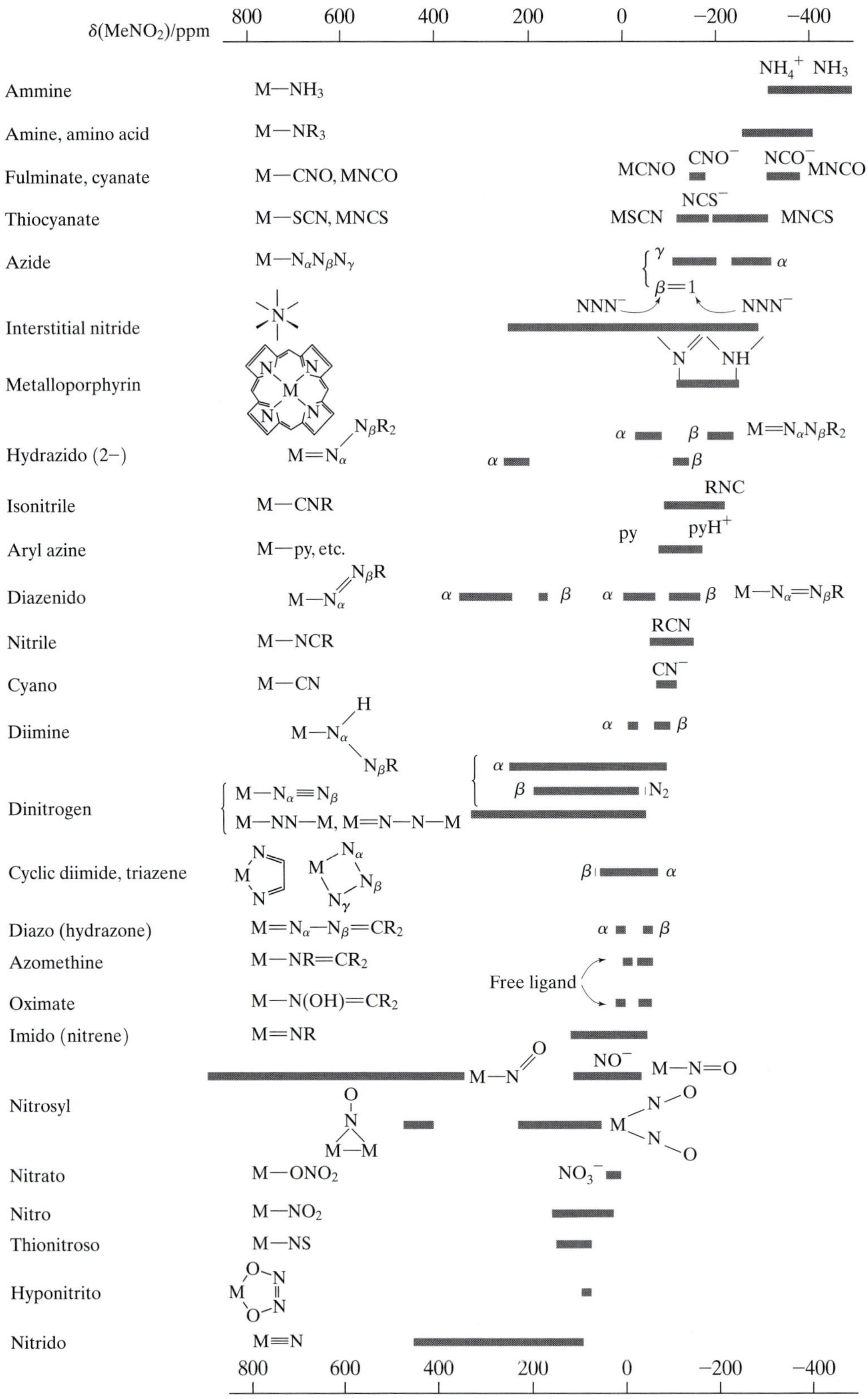

▲ FIGURE 3.11
Nitrogen chemical-shift ranges for metal complexes.

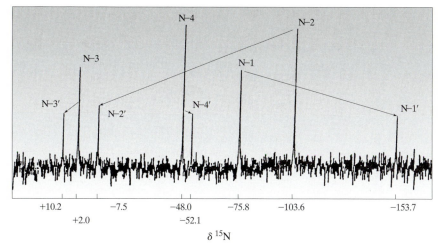

◀ FIGURE 3.12
**10.13 MHz ^{15}N NMR
spectrum of a mixture
of 1,5-dimethyl (1)
and 2,5-dimethyltetrazole (2)
in CDCl$_3$.^{a}**

^{a} Sveshnikov, N. N.; Nelson, J. H.
Magn. Reson. Chem., **1997**,
35, 209.

In heterocyclic rings, δ ^{15}N is strongly dependent on the ring position. This is strikingly illustrated by the spectra of the isomeric 1,5-dimethyltetrazole and 2,5-dimethyltetrazole shown in Figure 3.12.

Predict the relative ^{15}N chemical shifts of 4-hydroxy-, 2-methyl, and 4-nitro-pyridine.

Solution

By using the SCSs in Table 3.12, we make the following predictions:

Experimental

$$\delta\ ^{15}N = -69.2 - 118 = -187.2 \qquad -201$$

$$\delta\ ^{15}N = -69.2 - 0.4 = -69.6 \qquad -72$$

$$\delta\ ^{15}N = -69.2 + 22 = -47.2 \qquad -35$$

Although the actual values compare rather poorly, the relative values are clearly established.

Coordinating a nitrogen donor ligand to a transition metal can cause a large change in the chemical shift, and the change is often to high field. This is well illustrated by the ^{15}N NMR spectrum of the cobalt complex shown in the following diagram:

^{15}N chemical shifts of $Co(DMGH)_2(N\text{---}CH_3IM)(N_3)$

The chemical shifts of the free azide ion move from -128 (β) and -277 (α) up-field by 68.7 ppm for N_α and downfield by 68 ppm for N_γ upon coordination. The change for N_β (-1.95 ppm) is small. Coordinating N-methylimidazole causes an upfield shift of 112.8 ppm for the donor nitrogen and an upfield shift of 4.21 ppm for the N-methylnitrogen.*

3.6   Fluorine Chemical Shifts

Fluorine-19 ($I = \frac{1}{2}$, 100% abundant) has a chemical-shift dispersion range of over 1200 ppm (δ from -419.4 to $+825$), which causes considerable difficulty in accurately measuring chemical shifts, particularly at high fields. For this reason, in addition to the primary reference, $CFCl_3$, several secondary standards (Table 3.13) are commonly used. Further, problems are encountered in reproducing literature values, since, as with early data on ^{15}N and ^{31}P, both sign conventions (upfield either positive or negative) have been used, often without designation. Typical ranges for ^{19}F chemical shifts are illustrated in Figures 3.13 and 3.14. There is a large solvent effect on ^{19}F chemical shifts, as illustrated in Table 3.14. The resonances are generally shifted to low fields by solvents of high polarizability, suggesting that the solvent effect is due to intermolecular van der Waals interactions. Fluorine-19 chemical shifts are sensitive not only

---

*A useful compilation of nitrogen NMR data is Witanowski, M., Webb, G. A., *Nitrogen NMR*, Plenum, New York, 1973. The old sign convention (with downfield shifts negative) is used in that book. Chemical-shift and coupling constant data for azoles are reviewed in Claramunt, R. M., Sanz, D., Lopez, C., Jimenez, J. A., Jimeno, M. L., Elguero, J., Fruchier, A. *Magn. Reson. Chem.* **1997**, *35*, 35. Chemical shifts and coupling constants for a wide variety of compounds are given in Berger, S., Braun, S., Kalinowski, H.-O., *NMR Spectroscopy of the Non-Metallic Elements*, Wiley, New York, 1997.

Table 3.13   δ ^{19}F of Various Secondary Standards Relative to CFCl$_3$

Compound	δ ^{19}F		δ ^{19}F
F$_2$	+422.9		
C$_6$H$_5$—SO$_2$F	+65.5	(fluorobenzene structure)	−113.1
CCl$_2$F$_2$	−6.9		
CF$_4$	−63.3		
C$_6$H$_5$—CF$_3$	−63.9		
CCl$_2$F—CCl$_2$F	−67.3	(hexafluorobenzene structure)	−162.9
CF$_3$—CO$_2$CH$_3$	−74.2		
CBrF$_2$CBrF$_2$	−63.4		
CF$_3$—COOH	−78.5	(tetrafluorodichlorocyclobutane structure)	−114.1
CF$_3$—CCl$_3$	−82.2		
CF$_3$—CO—CF$_3$	−84.6	(octafluorocyclobutane structure)	−136.0
CF$_3$—C(OH)$_2$—CF$_3$	−92.7		

Table 3.14   Illustrative Solvent Effects on ^{19}F Chemical Shifts Relative to CFCl$_3$

Solvent	C$_6$H$_5$CF$_3$	Cl$_2$CCCl$_2$F	C$_5$H$_{11}$F	(CH$_3$)$_3$CF
CH$_2$I$_2$	69.48	67.22	71.03	72.20
CHBr$_3$	67.73	66.41	68.23	69.65
CS$_2$	65.72	65.17	69.05	67.58
CCl$_4$	65.40	64.75	65.58	66.25
CH$_2$Cl$_2$	65.12	64.13		
C$_6$H$_6$	65.01	63.73	64.94	
C$_{16}$H$_{34}$	64.47			
C$_2$H$_5$OH	63.87			
C$_7$H$_{16}$	63.70	63.70	63.70	63.70
(C$_2$H$_5$)$_2$O	63.62	62.92		
C$_7$F$_{16}$	61.25	61.94	60.70	

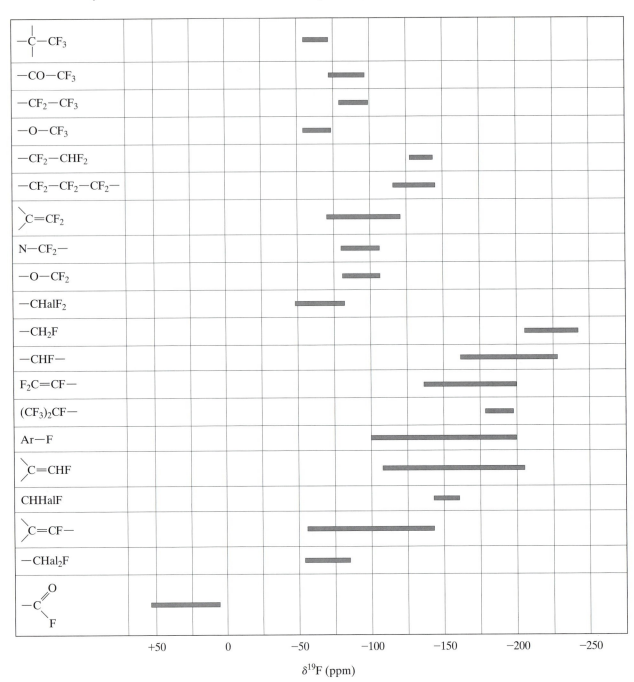

▲ FIGURE 3.13
Typical ^{19}F Chemical-shift ranges for compounds with CF bonds.

to directly attached substituents, but also to long-range effects. Table 3.15 gives some representative chemical-shift values relative to the common reference, $CFCl_3$. Electronegative substituents cause a low-field shift, as does increasing the oxidation state of the element to which fluorine is bound.

There is also a marked stereochemical effect on ^{19}F chemical shifts, such as that for the square-based pyramidal (C_{4V}) symmetry BrF_5. The axial fluorine

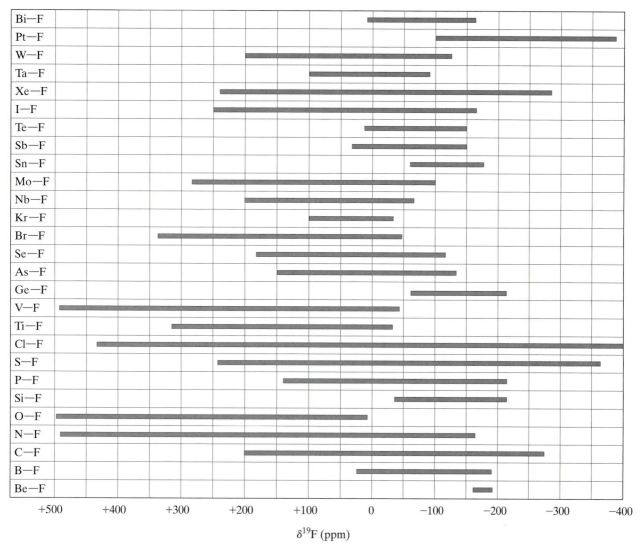

▲ FIGURE 3.14
Typical ^{19}F chemical-shift ranges for compounds with metal and nonmetal fluorine bonds.

($\delta = 132$) has a considerable chemical-shift difference from the equatorial fluorines ($\delta = 270$) ppm. Charge has a much smaller effect—for example, SiF_4 at $\delta - 164$, SiF_5^- at $\delta - 136$, and $[SiF_6]^{2-}$ at $\delta - 127$ ppm; and PF_3 at $\delta - 34$, PF_5 at $\delta - 71.5$, and PF_6^- at $\delta - 71$. Even within the substantially more restricted range of organic fluorine compounds, ^{19}F chemical shifts cover a very wide range.* The great sensitivity of ^{19}F chemical shifts has made them very useful tools to probe changes in molecular structure. For example, if WCl_6 and WF_6 are mixed in solution, the ^{19}F NMR spectrum shows well-resolved resonances for the 12 different fluorine environments in the 9 fluorine-containing species $WF_{6-n}Cl_n$,

---

* See Emsley, J. W., Phillips, L., *Prog. in Nucl. Magn. Reson. Spectrosc.*, **1971**, 7, 1, for an extensive compilation of ^{19}F chemical shifts. See also Berger, Braun, and Kalinowski referenced in Section 3.5.

Table 3.15 Some Representative ^{19}F Chemical Shifts Referenced to $CFCl_3$

NF_3	146.9	$[SiF_6]^{2-}$	−127	$FCH{=}CH_2$	−114
BrF_3	−16.3	XeF_2	−199.6	$F_2C{=}CH_2$	−81.3
PF_3	−34	BF_4^-	−151.3	$F_2C{=}CF_2$	−135.2
AsF_3	−41.5	BrF_5	270, 132	C_6F_6	−162.9
SbF_3	−52.6	$CFBr_3$	7.4	$[BeF_4]^{2-}$	−163.5
SF_4	88.4, 34.1	CF_2Br_2	7	ClF_5	259.8, 428.8
SeF_4	37.7, 12.1	CFH_2Ph	−207	ReF_7	347
TeF_4	24.7	CF_2Cl_2	−6.9	SbF_5	−114.0
SiF_4	−164	$[AsF_6]^-$	−64.3	WF_6	+165
GeF_4	−173.5	ClF_3	118.1, 7.7	XeF_6	118.3
MeF	−267.9	MoF_6	+278	PF_5	−71.5
EtF	−211.5	$[SbF_6]^-$	−121.7	SF_6	56.5
CF_2H_2	−143.6	TeF_6	−57.7	$XeOF_4$	101.6
CF_3R	−60 to −70	CF_4	−63.3	$CClF_3$	−33
AsF_5	−65.2	XeF_4	−15.7	$SiClF_3$	−134.6
BF_3	−126.8	$[PF_6]^-$	71.0	$GeClF_3$	−144.1
IF_7	168	IF_5	50.1, 1.4	BeF_3^-	−166.6
SeF_6	49.6	SiF_5^-	−136.0	BeF_2	−168.8
				SiF_4	−164.0

$n = 0$ to 5, present in solution. Fluorine-19 chemical shifts are also sensitive to isotope effects: 0.086 ppm for ^{12}C versus ^{13}C in flourobenzene, 0.05 ppm for $[^{10}BF_4^-]$ compared with $[^{11}BF_4^-]$ (Figure 3.15) and 0.009 ppm for $C^{35}Cl_3F$ compared with $C^{35}Cl^{37}Cl_2F$.

Recent ^{19}F NMR spectroscopy studies of the coordinating ability of the weakly coordinating anions BF_4^-, SiF_6^{2-}, PF_6^-, AsF_6^-, and SbF_6^- have established that each of these anions can coordinate to transition metals.[*] Coordinating to a transition metal causes an upfield shift, sometimes of considerable magnitude, as illustrated in Figure 3.16.

The additivity relationships found for proton and carbon chemical shifts are not as general for the fluorine chemical shifts of fluorocarbons, but substituent chemical shifts have been determined for fluorobenzenes and fluoropyridines.[**] For a saturated fluorocarbon, the terminal CF_3 group resonance occurs substantially downfield of those of the internal CF_2 groups. Proton substitution on a CF_3 group to give a CF_2H group usually causes an upfield shift, while halogen substitution, such as CF_3 to CF_2X, usually causes a downfield shift.

The chemical shifts of meta- and para-substituted fluorobenzenes correlate roughly with Hammet σ constants of the substituent. This fact has been used to assess the trans[†] influence of various ligands.[‡]

---

[*] Strauss, S. H., *Chem. Rev.* **1993**, *93*, 927.

[**] See Emsley and Phillips.

[†] Parshall, G. W., *J. Am. Chem. Soc.*, **1964**, *86*, 5367; **1966**, *88*, 704.

[‡] Appleton, T. G., Clark, H. C., Manzer, L. E., *Coord. Chem. Rev.* **1973**, *10*, 335.

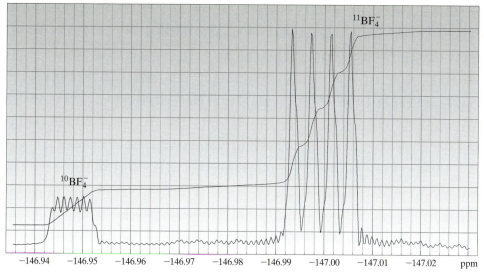

▲ FIGURE 3.15
470-MHz fluorine-19 NMR spectrum of aqueous NaBF₄.

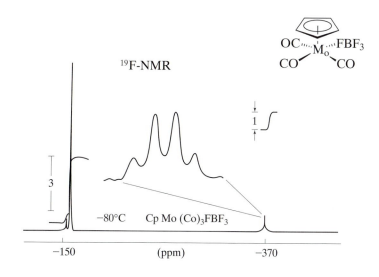

◀ FIGURE 3.16
Fluorine-19 NMR spectrum of (η^5—C₅H₅)Mo(CO)₃FBF₃ in CD₂Cl₂ at −80C.*

*Sunkel, K.; Urban, G.; Beck, W. *J. Organometallic Chem.*, **1983**, *252*, 187.

3.7 Phosphorus Chemical Shifts

The ^{31}P ($I = \frac{1}{2}$, 100% abundant) nucleus is very sensitive and has a very wide chemical-shift dispersion range of ~4000 ppm (δ = 3471 to −461 ppm, relative to the common reference 85% H₃PO₄). Typical ^{31}P chemical-shift ranges are illustrated in Figure 3.17. The chemical-shift range depends upon the phosphorus oxidation state. Thus, for λ^3 compounds, the range is δ = −488 for P₄ to δ = 245 for CH₃PF₂, showing that the chemical shifts are highly dependent on the substituents. Electronegative substituents, such as (CH₃)₂N, CH₃O, and halogens, generally cause a downfield shift. For P(IV) compounds, the range is smaller (δ = −50 to 100 ppm). In going from (CH₃)₃P to (Buᵗ)₃P, the chemical shift changes from δ = −62 to δ = 63 (i.e., by 125 ppm), but on going

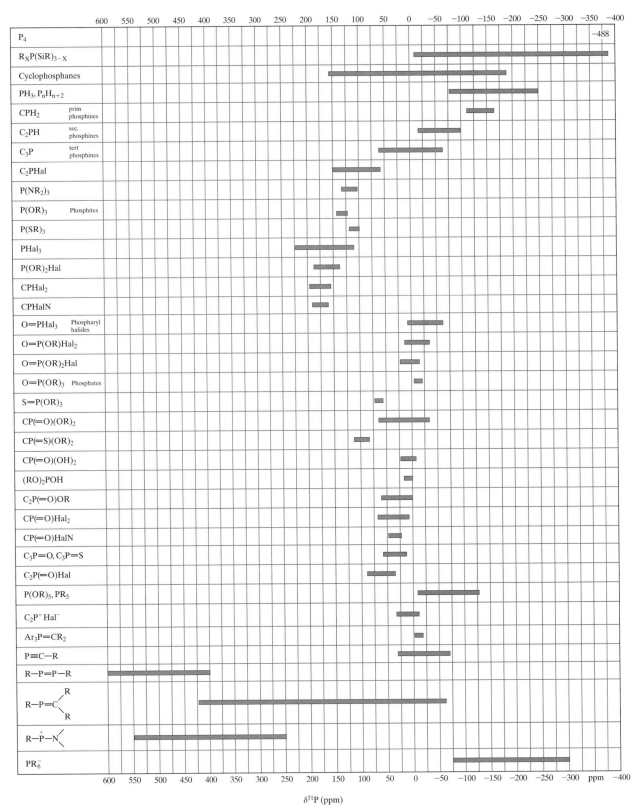

▲ FIGURE 3.17
Typical ^{31}P chemical-shift ranges.

from $(CH_3)_4P^+$ to $[Bu_3{}^t PH^+]$, the chemical shift only changes from $\delta = 25$ to $\delta = 58.3$.

Van Wazer and coworkers* related the chemical shifts to a combination of three parameters: bond angles, the electronegativity of the substituents, and the π-bonding character of the substituents. When only one parameter is changed at a time, useful correlations exist. Thus, for alkyl substituents, the order $CH_3 < C_2H_5 < i\text{-Pr} < Bu^t$ is found with chemical shifts moving downfield in the same order. Additivity relationships, which are found for alkyl subsituents, fail for π-bonding and for electronegative substituents, such as PCl_3 ($\delta = 220$), PCl_2F ($\delta = 224$), $PClF_2$ ($\delta = 176$), and PF_3 ($\delta = 97$). Table 3.16 gives some representative ^{31}P chemical shifts relative to 85% H_3PO_4 (external), the most commonly used reference. Older literature for ^{31}P NMR gives low-field shifts as negative, whereas the opposite is the present convention. This poses some difficulties, as the

Table 3.16 Representative ^{31}P Chemical Shifts Relative to External 85% H_3PO_4, Downfield Positive

Structure	CH_3	C_2H_5	$n\text{-}C_4H_9$	$t\text{-}C_4H_9$	$C_6H_5CH_2$	C_6H_5
			R			
PH_2R	−163.5	−127	−135	−82	−121.8	−122
PHR_2	−99	−55.5	−69.5	20.1	−47.2	−41
PR_3	−62.2	−20	−32.5	63	−12.9	−6
PCl_2R	191.2	196.3	190	198.6		165
$PClR_2$	92	119	118.8	145		81.5
PBr_2R	184	194	194.5		173.3	152
$PBrR_2$	91	116.2		151.4		70.8
$P(OR)Cl_2$	180	177	178.8		174.8	173
$P(OR)_2Cl$	169	165	165	170		157
$P(OR)_3$	141	137.6	139	138.3	124.8	128.2
$P(SR)Cl_2$	206	210.7	209.7		205.5	204.2
$P(SR)_3$	125.6	115.6	116.1		112.1	130.5
$O{=}PR_3$	36.2	48.3	43.2	66.5	40.7	28
$S{=}PR_3$	−28.9	54.5	48	90.6	45.0	42.6
$O{=}P(OR)_3$	+2.1	−1.0	14.0	−13.3	0.9	−17.3
$RP(C_6H_5)_3{}^+$	22.6	26.2	24.0	34.7	21.2	22.1

	H	F	Cl	Br	I	CN	NCO	NCS
				X				
PX_3	−239	97	218	227	178	−136	97	86
$O{=}PX_3$		−35.5	2	−102.9				

*Crutchfield, M. M., Dungan, C. H., Letcher, J. H., Mark, V., Van Wazer, J. R., in *Topics in Phosphorus Chemistry*, Vol. 6; Grayson, M., and Griffiths, E. J., Eds., Wiley-Interscience, New York, 1967, Chap 4.

convention often is not stated. The ^{31}P chemical shifts may be predicted to within 1 to 2 ppm by additivity relationships given in the following equations for primary, secondary, and tertiary phosphines and quarternary phosphonium salts, in that order:*

Primary phosphines RPH_2: $\delta = -163.5 + 2.5\,\sigma^P$;   **(3.9)**

Secondary phosphines R_2PH: $\delta = -99 + 1.5 \displaystyle\sum_{n=1}^{2} \sigma_n^P$;   **(3.10)**

Tertiary phosphines R_3P: $\delta = -62 + \displaystyle\sum_{n=1}^{3} \sigma_n^P$;   **(3.11)**

Quarternary
phosphonium salts R_4P^+: $\delta = +21.5 + 0.26 \displaystyle\sum_{n=1}^{2} \sigma_n^P - 3.22\,m - 5.5\,l.$

                                                            (3.12)

In these equations, σ^P are constants characteristic of the individual groups (Table 3.17). In the equation for the quarternary phosphonium salts, m is the number of allyl, benzyl, or cyclohexyl groups, and l is the number of phenyl groups.

Table 3.17   Empirical σ^P Constants for Equations 3.9 to 3.12[†]

Group	σ^P	Group	σ^P	Group	σ^P
NC	−24.5	n-C_nH_{2n+1} (N > 3)	+10	c-C_5H_9	+21
		$CH_2C{\equiv}C{-}H$	+13		
Et_2N	−1.0	$C{\equiv}C{-}H$	−10	c-C_6H_{11}	+23
		$C{\equiv}CCH_3$	−6		
CH_3	0	$NCCH_2CH_2$	+13	sec-Bu	+24
Me_3CCH_2	+3	C_2H_5	+14	i-pr	+27
i-Bu	+6	$PhCH_2$	+17	t-bu	+42
allyl	+9	Ph	+18	⎯〈 CH$_3$ / CH$_3$	−6
					+9
n-pr	+10	vinyl	+15	t-amyl	+42
				dibenzophosphole	+37

[†] See footnote below.

---

*Grim, S. O., McFarlane, W., *Nature (London)*, **1965**, *208*, 995. Maier, L., *Helv. Chim. Acta*, **1966**, *49*, 1718. Fluck, E., Lorenz, J., *Z. Naturforsch.* **1967**, *22B*, 1095.

SAMPLE PROBLEM 3.6

Predict the ^{31}P chemical shifts for the following series of phosphines: Me$_3$P, Me$_2$PPh, MePPh$_2$, and Ph$_3$P.

Solution

		Experimental
Me$_3$P: $\delta = -62 + 3(0) = -62$ ppm		-62 ppm
Me$_2$PPh: $\delta = -62 + 2(0) + 18 = -44$ ppm		-48 ppm
MePPh$_2$: $\delta = -62 + 0 + 2(18) = -26$ ppm		-28 ppm
Ph$_3$P: $\delta = -62 + 0 + 3(18) = -8$ ppm		-6 ppm

Phosphorus-31 chemical shifts are highly solvent and temperature dependent, and a mixture of Ph$_3$P and Ph$_3$P=O in toluene can be used as an NMR thermometer.*

Coordinating a phosphine to a transition metal brings about a low-field shift in the phosphorus resonance called a *coordination chemical shift* and defined as $\Delta\delta\,^{31}$P $= \delta\,^{31}$P(complex) $- \delta\,^{31}$P(ligand). This coordination chemical shift is often linearly related to the free-ligand chemical shift by relations of the form $\Delta\delta\,^{31}$P $= A\delta\,^{31}$P(ligand) $+ B$. These equations have found utility in determining the geometry of complexes and in the evaluation of comparative donor abilities of ligands. Table 3.18 gives some typical coordination chemical shifts for Cr, Mo, W, and Pt complexes. The coordination chemical shifts are a function of the transition metal, its coordination geometry, and the other ancillary ligands within the coordination sphere. This relationship is illustrated by the data for Pd(II) complexes given in Table 3.19.

One should note from the data in Table 3.18 that the coordination chemical shifts for phosphorus donors, which bear electronegative substituents (P(OR)$_3$, PX_3), are often negative. For ligands of these types, coordination to a transition metal often causes an upfield shift in the ^{31}P resonance. Coordination chemical-shift equations have been determined for several complex types (Table 3.20).

Chelation of a diphosphine brings about a large chemical-shift difference compared with the coordination of two similar model monodentate phosphines to the same metal center.† This chemical-shift difference, called a ring contribution (Δ_R), is a function of the chelate-ring size and the metal. For four- and five-membered chelate rings, Δ_R ranges from -12 to -50 ppm and from $+20$ to $+40$ ppm, respectively. For six-membered and larger chelate rings, Δ_R is small and positive or negligible.

---

*Dickert, F. L., Hellman, S. W., *Anal. Chem.*, **1980**, *52*, 996.

†Garrou, P. E., *Chem. Rev.*, **1981**, *81*, 229.

Extensive compilations of ^{31}P chemical-shift data may be found in the following references: Crutchfield, M. M., Dungan, C. H., Letcher, J. H., Mark, V., Van Wazer, J. R., in Grayson, M. and Griffiths, E. J. Eds., *Topics in Phosphorus in Chemistry*, Vol. 5, Wiley-Interscience, New York, **1967**. Gorenstein, D. G., *Prog. NMR Spectrosc*, **1983**, *16*, 1. Gorenstein, D. G., Ed., *Phosphorus-31 NMR Principles and Applications*, Academic, New York, **1984**. Pregosin, P. S., Kuntz, R. W., *^{31}P and ^{13}C NMR of Transition Metal Phosphine Complexes*, Springer–Verlag, Berlin, **1979**. Verkade, J. G., Quin, L. D., Eds., *Phosphorus-31 NMR Spectroscopy in Stereochemical Analysis, Organic Compounds and Metal Complexes*, VCH Publishers, Deerfield Beach, Florida, **1987**. Verkade, J. G., Quin, L., Eds. *Phosphorus-31 NMR Spectral Properties in Compound Characterization and Structural Analysis*, VCH Publishers, Deerfield Beach, Florida, **1994**. Berger, S., Braun, S., Kalinowski, H.-O., *NMR Spectroscopy of the Non-Metallic Elements*, Wiley, New York, **1997**.

Table 3.18 Ligand Chemical Shifts (ppm) and Coordination Chemical Shifts (ppm)[a] for Cr, Mo, W, and Pt Complexes

Ligand	Ligand shift	$Cr(CO)_5L$ $\Delta\delta$	$Mo(CO)_5L$ $\Delta\delta$	$W(CO)_5L$ $\Delta\delta$	$cis\text{-}PtCl_2L_2$ $\Delta\delta$
PH_3	−239	109	74	51	
$P(CH_2O)_3CBu$	−81		40		
PMe_3	−62	71	45	24	37
PMe_2Ph	−48		41		32
PBu_3^n	−32.5	63	45	26	34
PPr_3^n	−33				33
$PMePh_2$	−28	63	43	24	27
PPr_2^nPh	−27				25
PBu_2^nPh	−26	61	44	25	24
PEt_3	−20		40		30
PPr^nPh_2	−18		44		25
PEt_2Ph	−17				21
PBu^nPh_2	−17	62	44	25	24
$PEtPh_2$	−12	60	42	24	22
PPh_3	−6	61	44	27	14
PPr^iPh_2	0	59	43	26	
PBu^iPh_2	17	55	40	24	
$PClPh_2$	81.5				−8
$P[N(Me)CH_2\}_3Cpen$	87			25	
$P(OCH_2)_3Cpen$	93			22	
$P(OCH_2)_3CEt$	93		44		
PF_3	97		50		−27
$PF_2(OPh)$	110				−41
$Me_2Si\{(CH_2)_3OPPh_2\}_2$	111				−28
$Ph_2Si\{(CH_2)_3OPPh_2\}_2$	112				−30
$PPh_2(OMe)$	116		28		
$P(NC_5H_{10})_3$	117				−61
$P(NMe_2)_3$	123		23		
$PF(OPh)_2$	123				−57
$P(SMe)_3$	125		5		
$P(OPh)_3$	128.2		17		−61
$P(O\text{-}o\text{-}tol)_3$	130				−71
$P(OEt)_3$	137.6		22		
$P(OMe)_3$	141	39	26	−4	−68
$PF_2(NMe_2)$	144				−56
$PF(NMe_2)_2$	151				−65
$P(OEt)_2Ph$	159				−66

(continues on next page)

Table 3.18 *(cont.)*

Ligand	Ligand shift	Cr(CO)$_5$L $\Delta\delta$	Mo(CO)$_5$L $\Delta\delta$	W(CO)$_5$L $\Delta\delta$	*cis*-PtCl$_2$L$_2$ $\Delta\delta$
P(OMe)$_2$Ph	159		23		
[P(F)N(Bu^t)]$_2$	164				−79
PCl(OCH$_2$)$_2$	167				−51
PCl(OC$_3$H$_2$)$_2$	168				−57
TPPB^b	178	20	2	−22	
PI$_3$	179	−291			
PCl$_2$Me	191				−82
[P(Cl)N(Bu^t)]$_2$	211				−119
PCl$_3$	218	−31	−66	−120	−102
PBr$_3$	227	−132	−163	−231	
(Mes) P=CPh$_2$	233				−105
DMPP^b	−3	51	30	10	11
PDBP^c	−11	59	37	20	16

^{a} 2,4,6-triphenylphosphabenzene.
^{b} 3,4-dimethyl-1-phenylphosphole.
^{c} phenyldibenzophosphole.

Table 3.19 ^{31}P Coordination Chemical Shifts for *cis*- and *trans*-L$_2$PdX$_2$

L	X	$\delta\ ^{31}P_{ligand}$	$\Delta\delta\ ^{31}P$ *cis*	$\Delta\delta\ ^{31}P$ *trans*
PMe$_2$Ph	Cl	−48	54.4	42.8
PMePh$_2$	Cl	−28	47.1	35.8
PMe$_2$(p-MeOPh)	Cl	−39.1	44.7	33.0
PMe$_2$(p-MePh)	Cl	−38.6	44.2	32.9
PMe$_2$(p-ClPh)	Cl	−37.9	43.6	33.0
PMe(p-MePh)$_2$	Cl	−21.0	38.9	27.3
PMe(p-ClPh)$_2$	Cl	−20.2	38.0	27.6
PMe$_2$Ph	N$_3$	−48	53.2	46.0
PMePh$_2$	N$_3$	−28	45.1	37.1
PMe$_2$(p-MeOPh)	N$_3$	−39.1	43.2	34.6
PMe$_2$(p-MePh)	N$_3$	−38.6	43.0	34.6
PMe$_2$(p-ClPh)	N$_3$	−37.9	42.5	35.1
PMe(p-MePh)$_2$	N$_3$	−21.0	37.2	27.8
PMe(p-ClPh)$_2$	N$_3$	−20.2	36.6	27.7

Table 3.20   Coordination Chemical-Shift Equations
$\Delta\delta\ ^{31}P = A\delta\ ^{31}P(ligand) + B$

Compound	A	B
$LCr(CO)_5$	−0.153	60.58
$LMo(CO)_5$	−0.142	40.19
$LW(CO)_5$	−0.174	19.29
$cis\text{-}L_2Cr(CO)_4$	−0.196	57.92
$cis\text{-}L_2Mo(CO)_4$	−0.136	43.90
$cis\text{-}L_2W(CO)_4$	−0.138	19.35
$fac\text{-}L_3Cr(CO)_3$	−0.216	64.60
$fac\text{-}L_3Mo(CO)_3$	−0.105	36.49
$fac\text{-}L_3W(CO)_3$	−0.094	17.69
$mer\text{-}L_3Cr(CO)_3$[a]	−0.111	61.66
	−0.142	72.71
$mer\text{-}L_3Mo(CO)_3$[a]	−0.093	35.94
	−0.091	48.92
$mer\text{-}L_3W(CO)_3$[a]	−0.109	17.33
	−0.102	22.33
$cis\text{-}L_2PdCl_2$	−0.202	38.63
$trans\text{-}L_2PdCl_2$	−0.304	26.79
$cis\text{-}L_2Pd(N_3)_2$	−0.440	28.25
$trans\text{-}L_2Pd(N_3)_2$	−0.353	23.47
$cis\text{-}L_2Pd(NCO)_2$	−0.899	10.04
$cis\text{-}L_2PtCl_2$	−0.326	18.83
$trans\text{-}L_2PtCl_2$	−0.481	21.41
$trans\text{-}L_4RuCl_2$	−0.348	37.65
all $cis\text{-}L_2Ru(CO)_2Cl_2$	−0.416	26.84
$trans\text{-}L_2Rh(CO)Cl$	−0.335	35.89
$mer\text{-}L_3RhCl_3$	−0.654	10.36
	−0.741	9.95
$trans\text{-}L_2Ir(CO)Cl$	−0.357	29.11
$mer\text{-}L_2Ir(CO)Cl_3$	−0.377	0.70
$mer\text{-}L_3IrCl_3$	−0.799	−30.80
	−0.722	−32.47

[a] The first entry corresponds to the pair of mutually *trans* phosphines. See Verstuyft, A. W., Nelson, J. H., Cary, L. W., *Inorg. Nucl. Chem. Lett.*, **1976**, *12*, 53. Mann B. E., Masters, C., Shaw, B. L., Slade, R. M., Stainbank, R. E., *Inorg. Nucl. Chem. Lett.*, **1971**, 7, 881. Maun, B. E., Shaw, B. L., Slade, R. M., *J. Chem Soc., A*, **1971**, 2976.

3.8   Platinum Chemical Shifts

The only naturally occurring isotope of platinum with nonzero spin is ^{195}Pt ($I = \frac{1}{2}$, 33.8% abundant). This isotope was not widely studied prior to 1968. Because coupling constants to platinum are large and ligand exchange is generally slow, ^{195}Pt *satellites* are often visible for a wide range of compounds, so that double-resonance techniques such as INDOR are still widely used, although direct observation of a ^{195}Pt resonance is no longer difficult. Solvent effects on $\delta\,^{195}$Pt are quite large so that a coordinatively saturated reference such as $[PtCl_6]^{2-}$ is desirable. The chemical shift of this substance has a considerable temperature dependence (1.1 ppm K^{-1}). This fact, coupled with the very large chemical-shift dispersion range ($\sim$13,000 ppm) has led to the use of an absolute frequency of 21.4 MHz (at 2.35 tesla) as the reference. On this scale, $\delta\,[PtCl_6]^{2-}$ is 4533 ppm. Some representative data are given in Tables 3.21 to 3.23.

A striking feature of the data in Table 3.22 is the spread of the values for $[PtX_6]^{2-}$, X = F, Cl, Br, and I, which encompasses nearly the complete ^{195}Pt chemical-shift dispersion range. There is no clear division of the ^{195}Pt chemical shifts on the basis of oxidation state; resonances for platinum (0) complexes are found in the same region as those for platinum (II) complexes.

Table 3.21   Variation of ^{195}Pt Chemical Shift^a with the Neutral Ligand in $[NR_4][PtX_3L]$

L	X=Cl	Br	I
$C{\equiv}O^a$	1264	563	−953
$C{\equiv}NMe$	1487	806	−730
$H_2C{=}CH_2$	1748	1060	
$NH_2(p\text{-}C_6H_4CH_3)$	2717	2099	
$NHMe_2$	2670	2114	529
NMe_3	2818	2282	
$MeN{=}CMe_2$			526
4-Etpy	2772	2181	522
$N{\equiv}CMe$	2511	1827	
PMe_3	1033	414	−973
PPh_3	1020		
$P(OMe)_3$	1037	371	−1060
PF_3	907	183	−1370
$AsMe_3$	1360	664	−913
$SbMe_3$	1390	605	−1109
OH_2	3330		
SMe_2	1776	1118	−440
$SeMe_2$	1764	1057	−596
$TeMe_2$	1474	1057	−995
$SOMe_2$	1535	892	
X	3055	2066	

^{a} In ppm to high frequency of $\Xi\,(^{195}$Pt$) = 21.4$ MHz at 2.35 tesla.

Table 3.22   The ^{195}Pt Chemical Shifts[a] of Some Complex Ions of Platinum

Compound	Medium	δ (^{195}Pt)
[PPh$_3$CH$_2$Ph]$_2$[PtF$_6$]	CH$_2$Cl$_2$(0.1 M)	11,847
fac-[PtCl$_3$F$_3$]$^{2-}$	H$_2$O	7605
[PtCl$_6$]$^{2-}$		4533
[PtBrCl$_5$]$^{2-}$		4236
trans-[PtBr$_2$Cl$_4$]$^{2-}$		3942
cis-[PtBr$_2$Cl$_4$]$^{2-}$		3941
mer-[PtBr$_3$Cl$_3$]$^{2-}$	aq. HCl/HBr	3636
fac-[PtBr$_3$Cl$_3$]$^{2-}$	(total Pt 2 M)	3635
trans-[PtBr$_4$Cl$_2$]$^{2-}$		3320
cis-[PtBr$_4$Cl$_2$]$^{2-}$		3319
[PtBr$_5$Cl]$^{2-}$		2990
[PtBr$_6$]$^{2-}$		2651
[NBu$_4$]$_2$[PtCl$_6$]	CH$_2$Cl$_2$	4783
H$_2$[PtI$_6$]	aq. HI	−1528
Na$_2$[Pt(CN)$_6$]	H$_2$O	672
[Pt(en)$_3$][ClO$_4$]$_4$	H$_2$O (sat'd)	3579
[PtCl$_4$]$^{2-}$		2887
[PtBrCl$_3$]$^{2-}$		2661
trans-[PtBr$_2$Cl$_2$]$^{2-}$	H$_2$O	2418
cis-[PtBr$_2$Cl$_2$]$^{2-}$	(total Pt 0.5 M)	2408
[PtBr$_3$Cl]$^{2-}$		2138
[PtBr$_4$]$^{2-}$		1843
[NBu$_4$]$_2$[PtCl$_4$]	CH$_2$Cl$_2$(0.3 M)	3055
[NBu$_4$]$_2$[PtCl$_6$]	CH$_2$Cl$_2$	3325
[NBu$_4$]$_2$[PtBr$_4$]	CH$_2$Cl$_2$(0.3 M)	2066
[NBu$_4$]$_2$[Pt$_2$Br$_6$]	CH$_2$Cl$_2$	2233
H$_2$[PtI$_4$]	aq. HI	−993
[NBu$_4$]$_2$[Pt$_2$I$_6$]	CH$_2$Cl$_2$	−600
Na$_2$[Pt(CN)$_4$]	H$_2$O	−204
K$_2$[Pt(SCN)$_4$]	H$_2$O(2 M)	604
Na$_2$[Pt(NO$_2$)$_4$]	H$_2$O(sat'd)	2353
[Pt(NH$_3$)$_4$][ClO$_4$]$_2$	H$_2$O(ca. 310K)	1900
[Pt(en)$_2$]Cl$_2$	H$_2$O(sat'd)	1493

[a] See footnote to Table 3.21.

However, the resonances for platinum (II) complexes often appear at higher field than those for analogous platinum (IV) complexes, as illustrated by the following data:

Pt(II)	δ ^{195}Pt	Pt(IV)	δ ^{195}Pt
PtCl$_4$$^{2-}$	2887	PtCl$_6$$^{2-}$	4533
PtBr$_4$$^{2-}$	1843	PtBr$_6$$^{2-}$	2651
Pt(CN)$_4$$^{2-}$	−204	Pt(CN)$_6$$^{2-}$	672
[Pt(en)$_2$]$^{2+}$	1493	[Pt(en)$_3$]$^{4+}$	3579
PtI$_4$$^{2-}$	−993	PtI$_6$$^{2-}$	−1528

Table 3.23   Variation of ^{195}Pt Chemical Shift^a with Complex Structure

L	cis-PtCl$_2$L$_2$	trans-PtCl$_2$L$_2$	[PtClL$_3$]$^+$	[PtL$_4$]$^{2+}$
CNMe	665		203	−318
NMe$_3$		2647		
PMe$_3$	125	583	−146	−358
AsMe$_3$	242	753	−211	
SbMe$_3$	−79		−646	
SMe$_2$	982	1109	523	
SeMe$_2$	798	1029	241	

	[PtCl$_5$L]$^-$	cis-PtCl$_4$L$_2$	trans-PtCl$_4$L$_2$	mer-[PtCl$_3$L$_3$]$^-$
CNMe	3323	2447		
PMe$_3$	3060	1921	2367	1348
AsMe$_3$	1359	2126	2502	1449
SMe$_2$	3656	2788	2850	

^{a} See footnote to Table 3.21.

Within a family of compounds [e.g., PtX$_n$Y$_{6-n}$ (X = F, Cl, Br, and I combinations), there are additivity relationships for δ ^{195}Pt.

In a series of closely related chloride complexes, δ moves downfield in the order O > N > S > As > P donor. Thus, δ [PtCl$_3$(H$_2$O)]$^-$ (3330) > δ [PtCl$_3$(NMe$_2$)]$^-$ (2818) > δ [PtCl$_3$(SMe$_2$)]$^-$ (1776) > δ [PtCl$_3$(AsMe$_3$)]$^-$ (1360) > δ [PtCl$_3$(PMe$_3$)]$^-$ (1033). As one descends a family in the periodic table, the chemical shift usually (but not always) moves upfield with increasing mass, as illustrated by the following data:

trans-PtCl$_2$(SMe$_2$)$_2$	trans-PtBr$_2$(SMe$_2$)$_2$	trans-PtI$_2$(SMe$_2$)$_2$
1109	634	−598
trans-[PtI(CH$_3$)(PEt$_3$)$_2$]	trans-[PtI(SiH$_3$)(PEt$_3$)$_2$]	trans-[PtI(GeH$_3$)(PEt$_3$)$_2$]
−292	−737	−784
[PtCl$_3$SMe$_2$]$^-$	[PtCl$_3$SeMe$_2$]$^-$	[PtCl$_3$TeMe$_2$]$^-$
1776	1764	1474

but

[PtCl$_3$NMe$_3$]$^-$	[PtCl$_3$PMe$_3$]$^-$	[PtCl$_3$AsMe$_3$]$^-$	[PtCl$_3$SbMe$_3$]$^-$
2818	1033	1360	1390

Complex geometry also affects δ ^{195}Pt. For L_2PtX$_2$ complexes, the resonance for the *cis* isomer usually occurs 200 to 500 ppm upfield of that for the *trans* isomer (Table 3.22). The difference in the chemical shift decreases as the donor abilities of L and X become more similar. (Cf cis-[(SMe$_2$)$_2$PtBr$_2$] (654) vs. trans-[(SMe$_2$)$_2$PtBr] (634).

Platinum compounds participate in redistribution equilibria of both anionic and neutral ligands, *viz.*,

$$(R_3P)_2PtX_2 + (R_3P)_2PtY_2 \rightleftharpoons 2(R_3P)_2PtXY$$

and

$$(R_3P)_2PtX_2 + (R_3'P)_2PtX_2 \rightleftharpoons 2(R_3P)(R_3'P)PtX_2,$$

with the former being much more general than the latter. ^{195}Pt NMR spectroscopy is a very useful way of monitoring these equilibria, as shown by the spectra in Figures 3.18 and 3.19.

▶ FIGURE 3.18
21.28 MHz ^{195}Pt{^{1}H} NMR
spectrum of an equilibrium mixture
of (A) *trans*-(Bzl$_3$P)$_2$PtCl$_2$, (B) *trans*-
(Bzl$_3$P)$_2$PtBr$_2$, and (C) *trans*-
(Bzl$_3$P)$_2$PtBrCl in CDCl$_3$ at 300 K.
Reference is PtCl$_6^{2-}$.* ^1J(PtP) is
indicated in each triplet.

*Rahn, J. A., Holt, M. S., Nelson,
J. H., *Polyledron*, **1989**, *8*, 897.

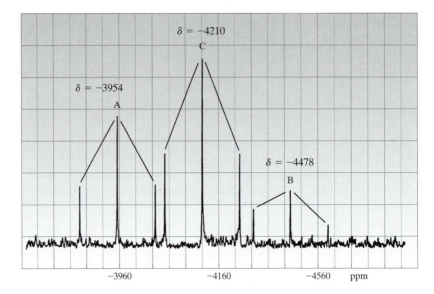

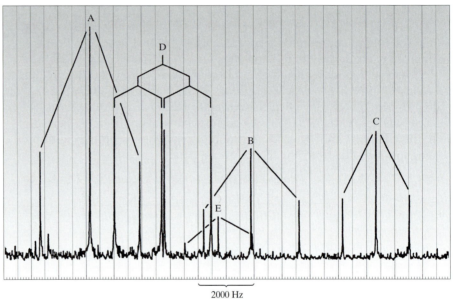

▲ FIGURE 3.19
21.28 MHz ^{195}Pt{^{1}H} NMR spectrum of an equilibrium mixture of (A) *cis*-(MePh$_2$P)$_2$PtBr$_2$
(δ = −4685 ppm), (B) *cis*-(MePh$_2$P)$_2$PtI$_2$ (δ = −5230 ppm), (C) *trans*-(MePh$_2$P)$_2$PtI$_2$
(δ = −5648 ppm) (D) *cis*-(MePh$_2$P)$_2$PtBrI (δ = −4931 ppm), and (E) *trans*-(MePh$_2$P)$_2$PtBrI
(δ = −5120 ppm) in CDCl$_3$ at 300 K. Reference is PtCl$_6^{2-}$.* ^1J(PtP) is indicated in each triplet.
* See reference to Figure 3.19.

The data in these figures illustrate the halide additivity relationships as follows: for *trans*-(Bzl$_3$P)$_2$PtBrCl, one expects a δ ^{195}Pt of $(-3954 - 4478)/2 =$ −4216 versus −4210 observed, and for *cis*-(MePh$_2$P)$_2$PtBrCl, one expects a δ ^{195}Pt of $(-5230 - 4685)/2 = -4958$ versus −4931 observed. The chemical shift of a given nucleus is given by Equation 3.2 in general form, *viz.*,

$$\nu_i = \frac{\gamma \mathbf{B}_0 (1 - \sigma_i)}{2\pi}$$

and

$$\sigma_i = \sigma_p + \sigma_d + \sigma_x,$$

where σ_i represents the total screening constant and σ_p, σ_d, and σ_x are the paramagnetic, diamagnetic, and extraneous (e.g., ring current and other anisotropic contributions) expressions, respectively. The diamagnetic term is produced by electrons in the ground state of a molecule and the paramagnetic term by electronic transition into excited singlet electronic configurations of a molecule. For most heavy nuclei, and especially for metals, the paramagnetic term dominates. It has recently been shown[*] for d^6 transition metals such as Pt(IV) that

$$\sigma_p = \frac{-8\mu_0 \mu_B^2}{\pi} \langle r_d^{-3} \rangle \frac{\eta_\Sigma \eta_{\sigma\pi}}{\Delta E} \qquad (3.13)$$

where μ_B is the Bohr magneton, $\langle r_d^{-3} \rangle$ is the d-orbital radial expectation value, η_Σ is a configuration interaction term, $\eta_{\sigma\pi}$ is a covalency ratio, and ΔE is a transition energy from the ground state to an excited state. For most d^6 complexes, η_Σ differs only slightly from unity, and typical values of $\eta_{\sigma\pi}$ depend upon the nature of the donor atom, the ligand, and the metal. Equation 3.13 predicts that σ_p should be inversely proportional to electronic transition energies and that ν and δ should increase with increasing electronic transition energy.

For a series of Pt(IV) complexes, one finds the following data:

Compound	$\delta\,^{195}$Pt(ppm)	ΔE(nm)
PtF_6^{2-}	11,847	317
$PtCl_6^{2-}$	4533	380
$PtBr_6^{2-}$	2651	435
$Pt(SCN)_6^{2-}$	1073	370

These data demonstrate the effects of the $\eta_{\sigma\pi}$ covalency ratio (compare $\delta\,^{195}$Pt for $PtCl_6^{2-}$ and $Pt(SCN)_6^{2-}$ and ΔE^{-1}). Ligands have been ordered in a **magnetochemical series** by Juranic according to increasing magnetic shielding of the metal nucleus. The magnetochemical series is comparable to the spectrochemical series determined from electronic spectroscopy. The spectrochemical series is

$$CO > P(OR)_3 > CN^- > en > NH_3 > F^- > N_3^- > Cl^- > Br^- > I^-.$$

The magnetochemical series is

$$\text{Pt(IV)} \quad I^- > CH_3 > CN^- > Br^- > en > NH_3 > Cl^- > NO_2^- > OH_2 > F^-.$$

---

[*] Juranic, N., *Coord. Chem. Rev.*, **1989**, *96*, 253, discusses a modification of the original Ramsey equation.

An extensive compilation of ^{195}Pt spectral data may be found in Pregosin, P. S. *Coord. Chem. Rev.*, **1982**, *44*, 247.

To date, NMR spectra have been obtained on nearly every possible nuclide. Additional data may be found in Harris, R. K., and Mann, B. E., *NMR and the Periodic Table*, Academic Press, New York, **1978**; Mason, J., *Multinuclear NMR*, Plenum, New York, **1987**, and Brevard, C. and Granger P., *Handbook of High Resolution Multinuclear NMR*, Wiley-Interscience, New York, **1981**. See also Grant, D. M.; Harris, R. K.; Eds., *Encyclopedia of Nuclear Magnetic Resonance*, Wiley, New York, **1996**.

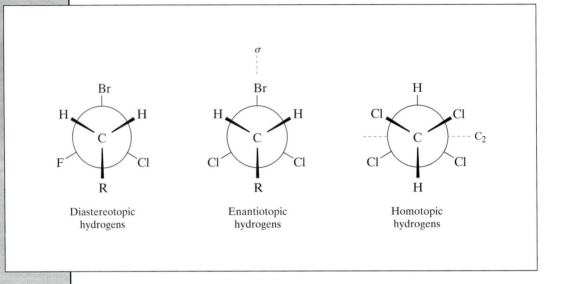

CHAPTER 4

Diasteretopic
hydrogens

Enantiotopic
hydrogens

Homotopic
hydrogens

Symmetry and NMR Spectroscopy

4.1   Introduction

In this chapter, chemical shifts are considered from a symmetry point of view. Two atoms are **chemical shift equivalent** or **isochronous** if there is a symmetry element in the molecule, C_n, S_n, or a σ relating them. Two nuclei are chemically equivalent if there is a C_n axis relating them. Chemically equivalent nuclei are isochronous, but the converse is not necessarily true. Nuclei related by a C_n axis are termed **homotopic**, and they are chemical-shift equivalent (isochronous) in both chiral and achiral environments. Two nuclei related *only* by a mirror plane are termed **enantiotopic**. They are isochronous in an achiral environment and anisochronous in a chiral environment. Thus, in asymmetric molecules, which do not possess any symmetry operations, all nuclei must be anisochronous. That is, chiral molecules may contain **diastereotopic**, but not homotopic or enantiotopic, nuclei. Diastereotopic nuclei are not related by any symmetry operation.

Two atoms are **magnetically equivalent** if they are isochronous and if all the coupling constants (J) for couplings to any other nucleus are equal for each nucleus (**isogamous coupling**). Examples of these terms and some representative molecules illustrating them are given in Table 4.1. In the tetrahedral CH_2F_2 molecule, the two hydrogen and two fluorine atoms are each related by a C_2 and a σ symmetry operation. They are thus homotopic pairs

Table 4.1   Table of Equivalencies

Molecule							
Symmetry element for the nuclei considered	C_3	C_2	C_2	C_2	σ	σ	none
Protons	homotopic	homotopic	homotopic	homotopic	enantiotopic	enantiotopic	diastereotopic
Chemical shifts	isochronous	isochronous	isochronous	isochronous	isochronous	isochronous	anisochronous
Coupling constants	isogamous	isogamous	anisogamous	anisogamous	isogamous	anisogamous	anisogamous
Spin System	A_4	A_2X_2	$AA'XX'$	$AA'XX'$	A_2X	$AA'XX'$	ABX

of nuclei, and there will be only one chemical shift for each pair of (isochronous) nuclei. If we consider the geometric relationships among the hydrogens and the fluorines in the figure

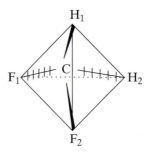

we note that the H_1CF_1 angle is equal to each of the H_2CF_1, H_2CF_2, and H_1CF_2 angles. Thus, each hydrogen nucleus sees a geometrically equivalent pair of fluorines, and the magnetic interactions between each hydrogen and each fluorine are equal. The hydrogens and fluorines are said to be magnetically equivalent (isogamous). For 1,1-difluoroethylene, that is,

$$F_1 \diagdown \quad \diagup H_1$$
$$\quad C = C$$
$$F_2 \diagup \quad \diagdown H_2$$

although H_1 and H_2 are related by C_2 and σ symmetry operations and are therefore chemical shift equivalent, they are not magnetically equivalent, because the geometric relationship of H_1 to F_1 (*cis*) is not the same as that of H_1 to F_2 (*trans*). As a consequence, the ^1H and ^{19}F NMR spectra of these two molecules, CH_2F_2 and $F_2C=CH_2$, will be very different, as shown in Figure 4.1.

Nuclei that are not related by a symmetry operation can be chemical shift equivalent only by chance. All four protons of propyne, $CH_3C\equiv CH$, in dilute chloroform solution are an example. This is called **accidental equivalence**. They can also become chemical shift equivalent by virtue of a time-averaging process, which places the individual nuclei in equivalent environments. Typical processes capable of doing this are conformational interconversion, such as for cyclohexane, or rapid rotation about single bonds, such as in methyldiphenylphosphine. (See Chapter 11.) Rapid rotation about the P—C bonds in $(PhCH_2)_2PPh$, shown in the following figure, cannot bring the CH_2 protons into equivalent environments, as it can for CH_3PPh_2, yet the ^1H NMR spectrum of this compound shows that the CH_2 protons are accidentally chemical shift equivalent, even though they are diastereotopic:

$$CH_3PPh_2 \qquad (PhCH_2)_2PPh$$

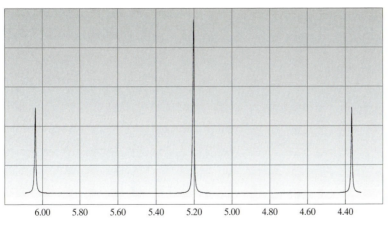

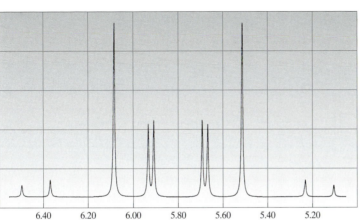

◄ FIGURE 4.1
Simulated 60 MHz ^{1}H NMR spectra
of CH_2F_2 (top) δ = 5.2 ppm
$^2J(HF)$ = 50.22 Hz, and $H_2C{=}CF_2$
(bottom) δ = 5.8 ppm,
$^2J(FF)$ = 30.72 Hz,
$^2J(HH)$ = −4.62 Hz,
$^3J(F_1H_1)$ = 0.61 Hz,
$^3J(F_1H_2)$ = 33.76 Hz, MacDonald,
C. J.; Schaefer, T. S. *Can. J. Chem.*
1967, *45*, 3157.

This must result from accidental chemical-shift equivalence. Magnetic equivalence can arise only if, by chance, the various coupling constants are equal. Such conditions give rise to "deceptively simple" spectra, such as that for furan, which should be compared with the spectrum of thiophene (Figure 4.2).

4.2 Spin-System Designation

It has become conventional to designate each NMR active nucleus in a molecule by a letter of the alphabet, using letters close together for nuclei whose chemical-shift differences are small relative to the coupling constant between them ($\delta \leq 6$ J), and letters at opposite ends of the alphabet for nuclei with large chemical-shift differences.* Thus, the molecule HF, containing two NMR active nuclei with quite different chemical shifts, would be designated as an

AX spin system, while the two protons of the molecule

$$\underset{Cl}{\overset{H}{\diagdown}}C{=}C\underset{Br}{\overset{H}{\diagup}}$$

would

be designated as an *AB* spin system. This spin-system designation is extended such that each anisochronous nucleus is assigned a different letter of the alphabet. When two or more nuclei are isochronous, subscripts are used to

---

*An excellent discussion may be found in Ault, A. *J. Chem. Educ.*, **1970**, *47*, 812.

Furan

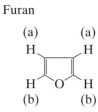

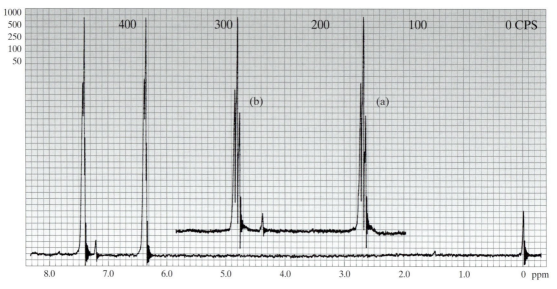

Thiophene

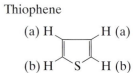

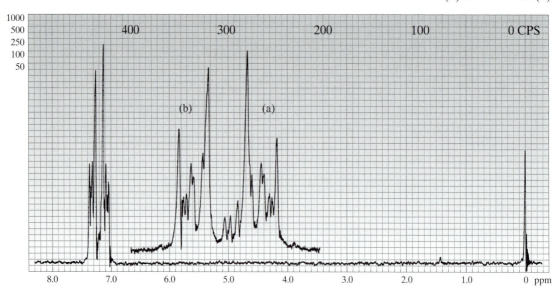

▲ FIGURE 4.2
60-MHz continuous wave ^{1}H NMR spectra of furan and thiophene.

denote their number. Thus, CHF_3 is designated as an AX_3 spin system. Iso-gamous nuclei, which are necessarily isochronous, are designated likewise such that CH_2F_2 is an $A_2 X_2$ spin system. In the past, anisogamous nuclei have been designated by primes, so that the molecule $H_2C=CF_2$ is an $AA'XX'$ spin system. This designation can become very cumbersome for molecules such as $[(C_6H_5CH_2)_2PPh]_2PdCl_2$, for which the methylene protons are designated as two sets of identical $AA'BB'XX'(A,B={}^1H,X={}^{31}P)$ spin systems. Even more cumbersome designations may result when more nuclei are involved. To sim-plify the designation of anisogamous nuclei, Haigh* has proposed an alter-native system that uses brackets to designate anisogamous nuclei. In his system, $AA'BB'XX'$ becomes $[ABX]_2$. Other similar correspondences are as follows:

$$AA'X \equiv [A]_2 X \qquad AA'XX' \equiv [AX]_2 \qquad A_2 A_2' B_2 B_2' XX' \equiv [A_2 B_2 X]_2$$
$$A_3 A_3' XX' \equiv [A_3 X]_2 \quad AA'MM'XX' \equiv [AMX]_2$$

It is necessary to become familiar with both systems, as the literature shows mixed usage. The following are some molecules whose NMR active nuclei exemplify some of the spin-system designations:

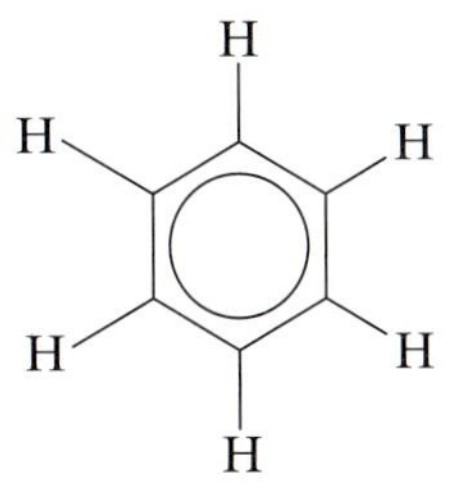

A_6 for the protons

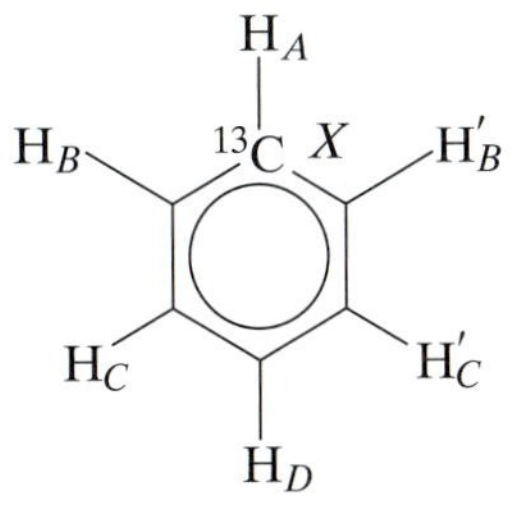

$ABB'CC'DX \equiv A[BC]_2DX$

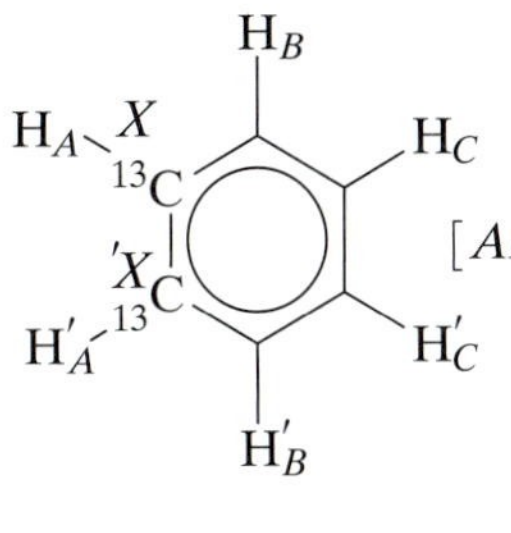

$[ABCX]_2$

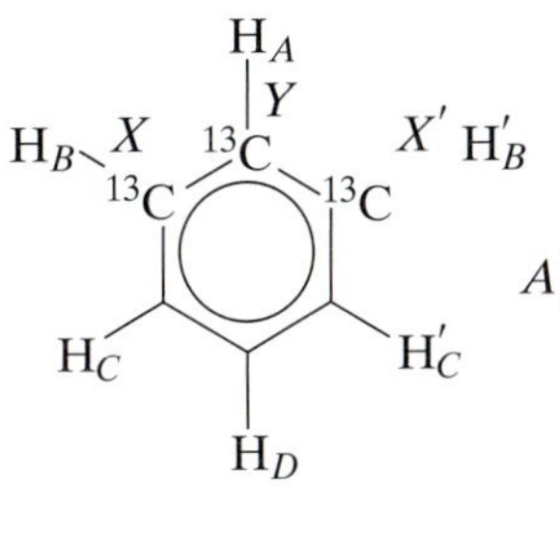

$A[BCX]_2DY$

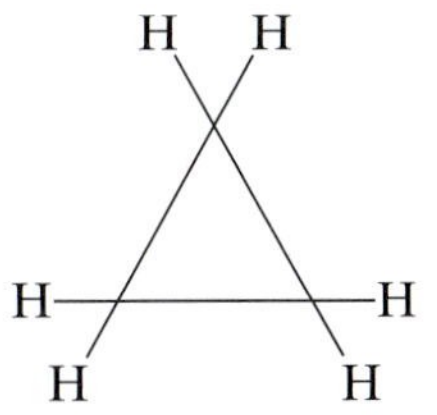

A_6

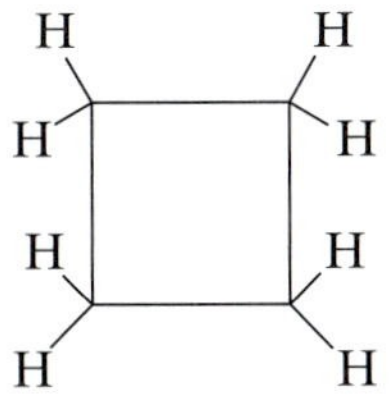

A_8

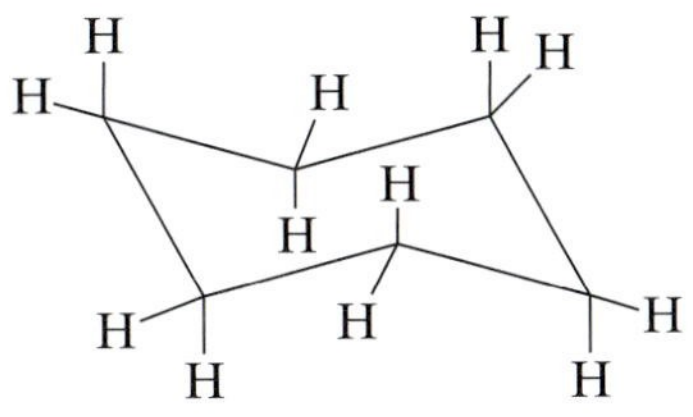

A_{12}

AB six times, not $[AB]_6$ with a rigid conformation, but A_{12} with rapidly interconverting conformations

---

*Haigh, C. W., *J. Chem. Soc. A*, **1970**, 1682.

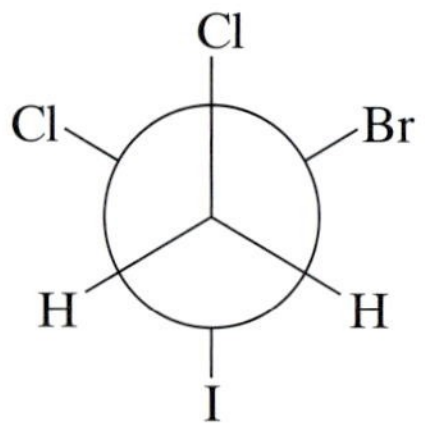

AB
or
AX

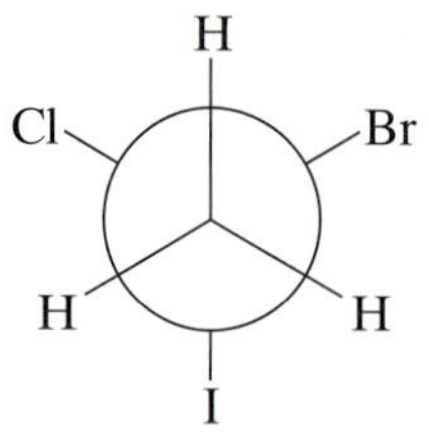

A_3 with rapid rotation about the C — C bond
AMX or ABC in a fixed conformer

Cl$_2$CHCHBr$_2$ AX Cl$_2$CHCH$_2$Cl AX_2 or A_2X Cl$_2$CHCH$_3$ AX_3 or A_3X

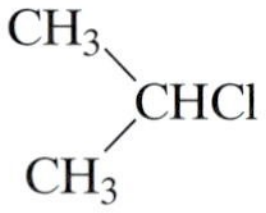

CH$_3$\
CHCl A_6X or AX_6 Cl$_2$CHCH$_2$CHCl$_2$ A_2X_2 CF$_3$-C≡C-H AX_3
CH$_3$

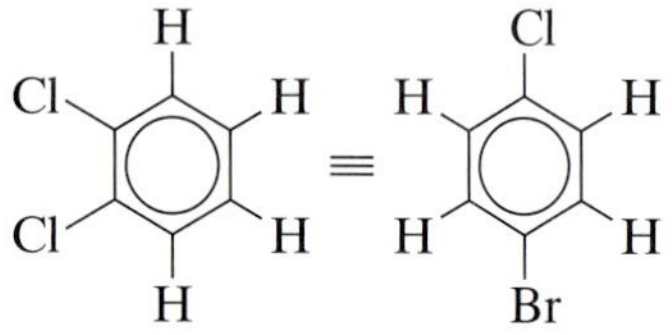

$AA'BB' = [AB]_2$

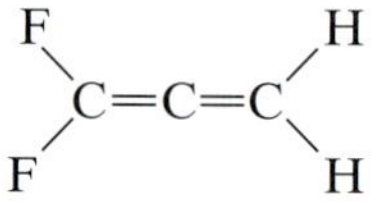

A_2X_2

(CH$_3$)$_2$PP(CH$_3$)$_2$ $X_6AA'X_6' = [AX_6]_2$

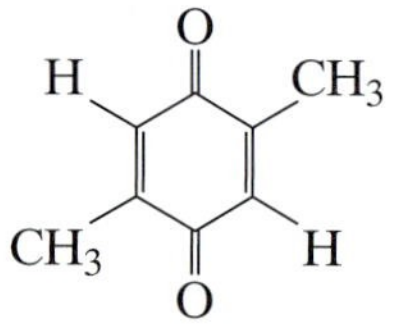

$X_3AA'X_3' = [AX_3]_2$

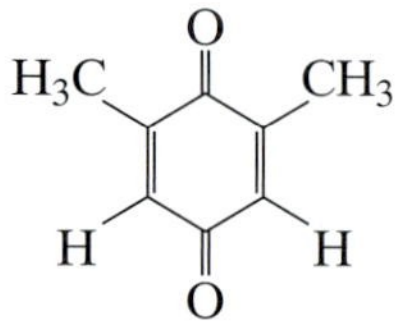

$X_3AA'X_3' = [AX_3]_2$

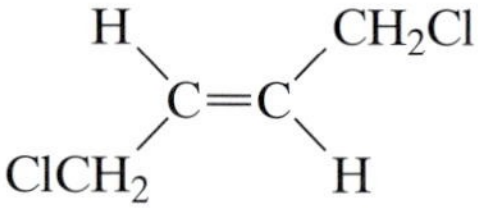

$X_2AA'X_2' = [AX_2]_2$

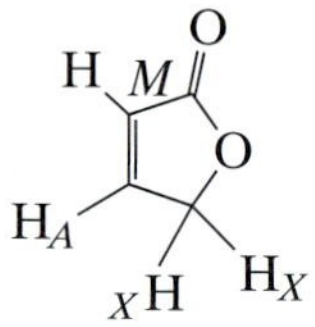

AMX_2

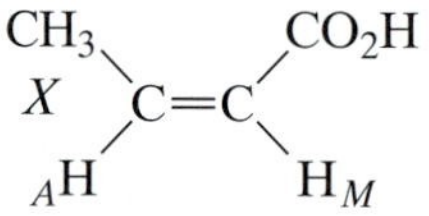

AMX_3

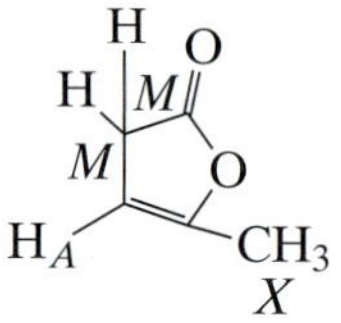

AM_2X_3

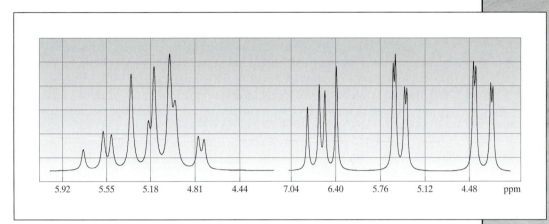

Transition from an ABC (left) to an AMX (right) spin system due to chemical shift differences.

Spin–Spin Coupling and NMR Spin Systems

5.1   General Considerations

When an NMR spectrum is obtained under high-resolution conditions, many of the resonances have fine structure. Unlike chemical shifts, the magnitudes of these splittings are independent of the field strength of the spectrometer. The splittings result from neighboring magnetic nuclei in the same molecule. Consider an ethanol molecule, CH_3CH_2OH. We will first give a simplified pictorial explanation and then a more complete and rigorous explanation of the ^1H NMR spectrum of this molecule. The influence of the external magnetic field $\mathbf{B_0}$, as stated in Chapter 1, separates the $I = +\frac{1}{2}$ and $I = -\frac{1}{2}$ spin states for each of the protons in the molecule. Now, each of the chemically and symmetrically equivalent protons resonates at the same field strength and radio frequency because they have the same shielding constant σ (Chapter 3). Hence, if we look at the field present at the protons of the CH_2 group, we see that they feel the influence of the external $\mathbf{B_0}$, and they also experience the much smaller field due to the neighboring CH_3 protons. As a group, the CH_3 protons have four different microfields (energy states), at $I = \frac{3}{2}, \frac{1}{2}, -\frac{1}{2}$, and $-\frac{3}{2}$. These arise in the following manner: Consider a vector representing the nuclear spin state $\left(\uparrow = \frac{1}{2}(\alpha)\right)$ and $\left(\downarrow = -\frac{1}{2}(\beta)\right)$; then, for the CH_3 group, we have

			Number of states
$\uparrow\uparrow\uparrow$	$(\alpha\alpha\alpha)$	$I = \frac{3}{2}$	1
$\uparrow\uparrow\downarrow, \uparrow\downarrow\uparrow, \downarrow\uparrow\uparrow$	$(\alpha\alpha\beta, \alpha\beta\alpha, \beta\alpha\alpha)$	$I = \frac{1}{2}$	3
$\downarrow\downarrow\uparrow, \downarrow\uparrow\downarrow, \uparrow\downarrow\downarrow$	$(\beta\beta\alpha, \beta\alpha\beta, \alpha\beta\beta)$	$I = -\frac{1}{2}$	3
$\downarrow\downarrow\downarrow$	$(\beta\beta\beta)$	$I = -\frac{3}{2}$	1

As a result, the CH_2 group "sees" four neighboring microfields, and these interact with the CH_2 protons to give the following result:

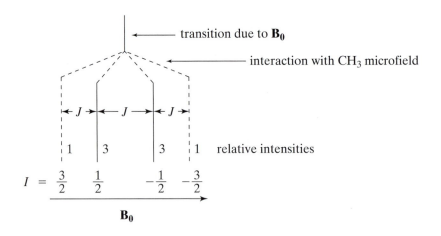

Thus, the CH_3 group splits the CH_2 group resonance into a four-line multiplet with $1:3:3:1$ relative intensities called a quartet. In a similar fashion, the CH_3 group "sees" three microfields ($I = 1, 0, -1$) due to the neighboring CH_2 protons, which diagrammatically are as follows:

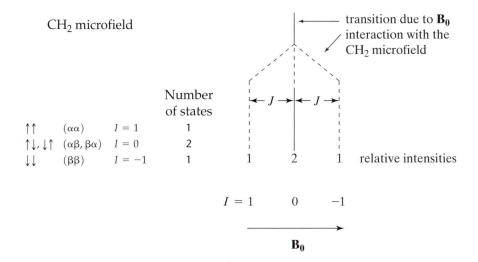

Consequently, the CH_2 group splits the CH_3 proton resonance into a triplet—a three-line multiplet with $1:2:1$ relative intensities. For the general $A_m X_n$ spin system, where both A and X have $I = \frac{1}{2}$, the spin function of the X nuclei has n factors, each of which can be either $\uparrow$, α, or $\downarrow$, β. Since n is fixed by the number of symmetry-equivalent X nuclei, the value of $\sum_{i=1}^{n} I_i$ is determined by the number of X nuclei having either $\uparrow$, α, or $\downarrow$, β, spin. Let this number be k. To determine the relative intensities within the multiplet resonance for the A nuclei, the number of different ways of assigning spin to k of the X nuclei must be found. In other words, we want the number of different ways of picking out k objects from a set of n numbered objects. This is by definition nC_k, the number of combinations of n things taken k at a time, and is given by the equation

$$nC_k = \frac{n!}{k!(n - k)!} \tag{5.1}$$

SAMPLE PROBLEM 5.1

How many ways can we assign three nuclei α spin?

Solution

This corresponds to $n = 3$ and $k = 3$ in Equation 5.1:

$$nC_k = \frac{3!}{3!(3 - 3)!} = \frac{3!}{3!0!} = 1 \text{ as } 0! \equiv 1.$$

SAMPLE PROBLEM 5.2

How many ways can we assign one of four equivalent nuclei α spin?

Solution

This corresponds to $n = 4, k = 1$ in Equation 5.1:

$$nC_k = \frac{4!}{1!(4-1)!} = \frac{4 \cdot 3 \cdot 2 \cdot 1}{1 \cdot 3 \cdot 2 \cdot 1} = 4.$$

For $I = \frac{1}{2}$ nuclei, these are also given by the coefficients of the binomial expansion

$$(a + b)^n = \sum_{k=0}^{n} nC_k a^{n-k} b_k$$

(e.g., for three nuclei, $n = 3$; $(a + b)^3 = a^3 + 3a^2b + 3b^2a + b^3$, and the coefficients are 1, 3, 3 and 1). These can be conveniently remembered by the mnemonic device of constructing Pascal's triangle:

n	$I = \frac{1}{2}$
0	1
1	1 1
2	1 2 1
3	1 3 3 1
4	1 4 6 4 1
5	1 5 10 10 5 1
6	1 6 15 20 15 6 1
7	1 7 21 35 35 21 7 1
8	1 8 28 56 70 56 28 8 1

In a similar manner, we may explain the ^{1}H NMR spectrum of $HD_2C(O)CD_3$, a common contaminant of $D_3CC(O)CD_3$. The proton in the former molecule "sees" two geminal deuterium nuclei, each with a spin $I = 1$ or a total spin of 2, which may have $2I + 1$ total spin states. These have M_I values of 2, 1, 0, −1, and −2. If we represent the individual spins as 1, 0, and −1, they may combine as follows:

	M_I	Number of states
$(1, 1)$	2	1
$(1, 0)(0, 1)$	1	2
$(1, -1)(0, 0)(-1, 1)$	0	3
$(-1, 0)(0, -1)$	−1	2
$(-1, -1)$	−2	1

As a consequence, this proton "sees" five microfields due to the neighboring deuterons and is split into a five-line multiplet with relative intensities of $1:2:3:2:1$. In general, the coupling of any number of A nuclei to n equivalent I spins gives a multiplet with $(2nI + 1)$ lines in the A spectrum. The following figure gives the relative intensities of these multiplets for $I = \frac{1}{2}, 1,$ and $\frac{3}{2}$ nuclei:*

n	$I = \frac{1}{2}$	$I = 1$	$I = \frac{3}{2}$
0	1	1	1
1	1 1	1 1 1	1 1 1 1
2	1 2 1	1 2 3 2 1	1 2 3 4 3 2 1
3	1 3 3 1	1 3 6 7 6 3 1	1 3 6 10 12 12 10 6 3 1
4	1 4 6 4 1	1 4 10 16 19 16 10 4 1	1 4 10 20 31 40 44 40 31 20 10 4 1

The relative intensities given by the coefficients of the binomial expansion or Equation 5.1 are those of what is termed a **first-order multiplet**. Conditions for a **first-order spectrum** are met whenever the chemical-shift difference of the interacting nuclei ($\Delta\delta$) is large relative to the coupling constant (J) between them. It can be shown that a rigorous first-order condition is $\Delta\delta_{12} \geq 12J_{12}$ for two interacting sets of nuclei. We will now consider in detail how the spectrum will appear when this condition is not met.

5.2   Quantum Mechanics and NMR Spectroscopy

A.   The *AB* Spin System

Consider a molecule such as *trans*-1-bromo-2-chloroethene, $\begin{smallmatrix} H^1 & & Cl \\ & C{=}C & \\ Br & & H^2 \end{smallmatrix}$,

which constitutes an *AB* spin system. This means that this molecule possesses two interacting nuclei, H^1 and H^2, with $I = \frac{1}{2}$ and no other nuclei with nonzero spin. The magnetic field due to the magnetic moment of the proton geminal to the chlorine, H^2, will interact with the magnetic moment of the proton geminal to the bromine, H^1, giving an additional term in the Hamiltonian describing the energy of the system classically:

$$-\mathbf{B_2} \cdot \hat{\mu}_1 = -g_N \beta_P \hbar^{-1} \mathbf{B_2} \cdot \hat{I}_1. \tag{5.2}$$

Here, $\mathbf{B_2}$ is the magnetic field at nucleus 1 produced by μ_2. Since $\mathbf{B_2}$ is proportional to μ_2 and hence to I_2, we can write Equation 5.2 as

$$\left(\frac{h}{\hbar^2}\right) J_{12} \hat{I}_1 \cdot \hat{I}_2, \tag{5.3}$$

where J_{12} is the proportionality constant called the nuclear spin–spin coupling constant. With the inclusion of the spin–spin interaction, the NMR

---

* See Orcutt, R. H. *J. Chem. Educ.* **1987**, *64*, 763.

Hamiltonian for a molecule with two magnetic nuclei in chemically different environments becomes

$$H = -\frac{g_N \beta_p \mathbf{B_0}}{\hbar}(1 - \sigma_1)\hat{I}_{Z_1} - \frac{g_N \beta_p \mathbf{B_0}}{\hbar}(1 - \sigma_2)\hat{I}_{Z_1} + \frac{hJ_{12}\hat{I}_1 \cdot \hat{I}_2}{\hbar^2}, \quad (5.4)$$

where the first and second terms represent the chemical shifts of nuclei 1 and 2, respectively, and the third term represents the interaction between these two nuclei. Because of the spin–spin coupling term, Equation 5.4 is not separable into the sum of Hamiltonians for the individual nuclei, and the corresponding Schrödinger equation is not separable. To deal with this problem, we shall use the method of expanding the unknown wave functions in terms of a set of known functions; that is, we will take linear combinations defined by the equation

$$\psi_i = \sum_n a_{in}\phi_1. \quad (5.5)$$

Equation 5.5 leads to the secular determinantal equation

$$\det\left[\int \phi_m H\phi_n - E\delta_{mn}\right] = 0 \quad (5.6)$$

and the secular equations for determining the coefficients a_{in}:

$$\sum_n\left[\int \phi_m H\phi_n - E\delta_{mn}\right]a_{in} = 0 \quad m = 1, 2 \ldots n. \quad (5.7)$$

Hence, we need a complete set of well-behaved spin functions ϕ_j. If the spin–spin coupling term is omitted from Equation 5.4 we obtain the Hamiltonian expressed by the equation

$$H = -g_N \beta_p \mathbf{B_0}\hbar^{-1}(1 - \sigma_1)\hat{I}_{Z_1} - g_N \beta_p \mathbf{B_0}\hbar^{-1}(1 - \sigma_2)\hat{I}_{Z_2}. \quad (5.8)$$

Now, the eigenfunctions for this Hamiltonian are

$$\psi_{MI_1}(1)\psi_{MI_2}(2); MI_1 = \pm\frac{1}{2}, MI_2 = \pm\frac{1}{2}. \quad (5.9)$$

Denoting $\psi_{\frac{1}{2}}$ by α and $\psi_{-\frac{1}{2}}$ by β and recognizing that Equation 5.8 is Hermitian, we find that the eigenfunctions so obtained are well behaved and form a complete set for two spin-$\frac{1}{2}$ nuclei. Moreover, since there are only a finite number of functions in this complete set, the secular determinant is of finite order and is easy to deal with. The complete set of eigenfunctions is

$$\phi_1 = \alpha(1)\alpha(2), \qquad \phi_2 = \alpha(1)\beta(2),$$

$$\phi_3 = \beta(1)\alpha(2), \quad \text{and} \quad \phi_4 = \beta(1)\beta(2),$$

where the α and β represent the spin and the (1) and (2) represent the nucleus with that spin, (i.e., $\phi_1 = \uparrow\uparrow$, $\phi_2 = \uparrow\downarrow$, $\phi_3 = \downarrow\uparrow$, and $\phi_4 = \downarrow\downarrow$). These functions are orthonormal, meaning that $\int \phi_i\phi_j = \delta_{ij}$, and they will serve as basis functions for our problem. To set up the secular determinant, we need to determine the effect of the Hamiltonian on our functions. We may write Equation 5.4 as

$$H = -C_1\hbar^{-1}\hat{I}_{Z_1} - C_2\hbar^{-1}\hat{I}_{Z_2} + hJ_{12}\hbar^{-2}\hat{I}_1 \cdot \hat{I}_2, \quad (5.10)$$

where $C_1 \equiv g_n \beta_p \mathbf{B}_0(1 - \sigma_1)$ and $C_2 \equiv g_n \beta_p \mathbf{B}_0(1 - \sigma_2)$. The effects of the first two terms in Equation 5.10 on the ϕ's can be determined from the formulas for the nuclear-shift operators, which are

$$\hat{I}_Z \alpha = \frac{1}{2} \hbar \alpha \quad \text{and} \quad \hat{I}_z \beta = -\frac{1}{2} \hbar \beta.$$

For example,

$$-C_1 \hbar^{-1} \hat{I}_{Z_1}[\alpha(1)\alpha(2)] = C_1 \hbar^{-1} \alpha(2) \hat{I}_{Z_1} \alpha(1)$$

$$= -\frac{1}{2} \hbar C_1 \hbar^{-1} \alpha(1)\alpha(2) = -\frac{1}{2} C_1 \alpha(1)\alpha(2).$$

To determine the effect of the third term in Equation 5.10, we express the dot product of the spin operators as

$$\hat{I}_1 \cdot \hat{I}_2 \equiv \hat{I}_{X_1} \hat{I}_{X_2} + \hat{I}_{Y_1} \hat{I}_{Y_2} + \hat{I}_{Z_1} \hat{I}_{Z_2}.$$

We then use the following rules, which can be derived from first principles in quantum mechanics:

$$\hat{I}_X \alpha = \frac{1}{2} \hbar \beta; \ \hat{I}_{X\beta} = \frac{1}{2} \hbar \alpha; \ \hat{I}_Y \alpha = \frac{1}{2} i \hbar \beta; \ \hat{I}_{Y\beta} = \frac{1}{2} i \hbar \alpha.$$

$\hat{I}_X$ and $\hat{I}_Y$ are the X and Y nuclear-shift operators. Then, for example,

$$hJ_{12} \hbar^{-2} \hat{I}_1 \cdot \hat{I}_2 \phi_1$$

$$= hJ_{12} \hbar^{-2} \left[\hat{I}_{X1} \hat{I}_{X2} + \hat{I}_{Y1} \hat{I}_{Y2} + \hat{I}_{Z1} \hat{I}_{Z2} \right] \phi_1$$

$$= \frac{4\pi^2}{h} J_{12} \left[\hat{I}_{X1} \hat{I}_{X2} + \hat{I}_{Y1} \hat{I}_{Y2} + \hat{I}_{Z1} \hat{I}_{Z2} \right] \alpha(1)\alpha(2)$$

$$= \frac{4\pi^2 J_{12}}{h} = \left[\frac{1}{2} \hbar \beta(1) \frac{1}{2} \hbar \beta(2) + \frac{1}{2} i \hbar \beta(1) \frac{1}{2} i \hbar \beta(2) + \frac{1}{2} \hbar \alpha(1) \frac{1}{2} \hbar \alpha(2) \right]$$

$$= \frac{4\pi^2 J_{12}}{h} \left[\frac{1}{4} \hbar^2 \beta(1)\beta(2) - \frac{1}{4} \hbar^2 \beta(1)\beta(2) + \frac{1}{4} \hbar^2 \alpha(1)\alpha(2) \right]$$

$$= \frac{1}{4} hJ_{12} \alpha(1)\alpha(2) = \frac{1}{4} hJ_{12} \phi_1.$$

Similarly, we find that

$$hJ_{12} \hbar^{-2} \hat{I}_1 \cdot \hat{I}_2 \phi_2 = \frac{1}{4} hJ_{12}(2\phi_3 - \phi_2),$$

$$hJ_{12} \hbar^{-2} \hat{I}_1 \cdot \hat{I}_2 \phi_3 = \frac{1}{4} hJ_{12}(2\phi_2 - \phi_3),$$

and

$$hJ_{12} \hbar^{-2} \hat{I}_1 \cdot \hat{I}_2 \phi_4 = \frac{1}{4} hJ_{12} \phi_4.$$

Combining all of the preceding results, we have

$$H\phi_1 = \left(-\frac{1}{2}C_1 - \frac{1}{2}C_2 + \frac{1}{4}hJ_{12}\right)\phi_1,$$

$$H\phi_2 = \left(-\frac{1}{2}C_1 + \frac{1}{2}C_2 - \frac{1}{4}hJ_{12}\right)\phi_2 + \frac{1}{2}hJ_{12}\phi_3,$$

$$H\phi_3 = \left(\frac{1}{2}C_1 - \frac{1}{2}C_2 - \frac{1}{4}hJ_{12}\right)\phi_3 + \frac{1}{2}hJ_{12}\phi_2,$$

and

$$H\phi_4 = \left(\frac{1}{2}C_1 + \frac{1}{2}C_2 + \frac{1}{4}hJ_{12}\right)\phi_4.$$

Hence, ϕ_1 and ϕ_4 are eigenfunctions that are already solutions of the Schrödinger equation. We wish to evaluate matrix elements of the secular determinant of the type

$$\int \phi_2 H\phi_2 \, d\tau = \left(-\frac{1}{2}C_1 + \frac{1}{2}C_2 - \frac{1}{4}hJ_{12}\right)\int \phi_1\phi_2 \, d\tau + \frac{1}{2}hJ_{12}\int \phi_2\phi_3 \, d\tau$$

$$= -\frac{1}{2}C_1 + \frac{1}{2}C_2 - \frac{1}{4}hJ_{12}.$$

Evaluating the remaining elements and substituting them into the secular determinant

$$\begin{vmatrix} H_{11}\text{-}E & H_{12} & H_{13} & H_{14} \\ H_{21} & H_{22}\text{-}E & H_{23} & H_{24} \\ H_{31} & H_{32} & H_{33}\text{-}E & H_{34} \\ H_{41} & H_{42} & H_{43} & H_{44}\text{-}E \end{vmatrix} = 0$$

gives

$$\begin{array}{c} \phi_1 \\ \phi_2 \\ \phi_3 \\ \phi_4 \end{array}\begin{vmatrix} \left[-\frac{1}{2}(C_1+C_2)+\frac{1}{4}hJ_{12}-E\right] & 0 & 0 & 0 \\ 0 & \left[\frac{1}{2}(C_1-C_2)-\frac{1}{4}hJ_{12}-E\right] & \frac{1}{2}hJ_{12} & 0 \\ 0 & \frac{1}{2}hJ_{12} & \left[\frac{1}{2}(C_1-C_2)-\frac{1}{4}hJ_{12}-E\right] & 0 \\ 0 & 0 & 0 & \left[\frac{1}{2}(C_1-C_2)+\frac{1}{4}hJ_{12}-E\right] \end{vmatrix} = 0.$$

with columns labeled ϕ_1, ϕ_2, ϕ_3, ϕ_4.

This secular determinant is block diagonalized, a direct result of using symmetry-adapted linear combinations of the ϕ_j as our basis functions. It is a fourth-order determinant in E and thus has four roots, two of which can be obtained immediately. These roots are

$$E_1 = -\frac{1}{2}(C_1 + C_2) + \frac{1}{4}hJ_{12} \quad \text{and} \quad E_4 = \frac{1}{2}(C_1 + C_2) + \frac{1}{4}hJ_{12}. \tag{5.11}$$

The remaining two roots are solutions of the equation

$$E_{2,3} + \frac{1}{2}hJ_{12}E - \frac{1}{4}(C_1 - C_2)^2 - \frac{3}{16}h^2J_{12}^2 = 0.$$

These roots are given by

$$E_{2,3} = -\frac{1}{4}hJ_{12} \pm [(C_1 - C_2)^2 + h^2J_{12}^2]^{\frac{1}{2}}. \qquad \textbf{(5.12)}$$

The four orthonormal wave functions ψ_1, ψ_2, ψ_3, and ψ_4 that correspond to these four energies are found by solving the secular equations and are

$$\psi_1 = \alpha(1)\alpha(2), \qquad\qquad \psi_2 = a\alpha(1)\beta(2) - b\beta(1)\alpha(2),$$

$$\psi_3 = b\alpha(1)\beta(2) + a\beta(1)\alpha(2), \quad \text{and} \quad \psi_4 = \beta(1)\beta(2),$$

where

$$a \equiv \frac{\{[h^2J_{12}^2 + (C_1 - C_2)^2]^{\frac{1}{2}} + (C_1 - C_2)\}^{\frac{1}{2}}}{2^{\frac{1}{2}}[h^2J_{12}^2 + (C_1 - C_2)^2]^{\frac{1}{4}}}$$

and

$$b \equiv \frac{\{[h^2J_{12}^2 + (C_1 - C_2)^2]^{\frac{1}{2}} - (C_1 - C_2)\}^{\frac{1}{2}}}{2^{\frac{1}{2}}[h^2J_{12}^2 + (C_1 - C_2)^2]^{\frac{1}{4}}}.$$

By using equations (5.10 to 5.12) we find that the energies of these wave functions are

$$E_1 = -g_N\beta_p\mathbf{B}_0\left(1 - \frac{1}{2}\sigma_1 - \frac{1}{2}\sigma_2\right), \qquad E_2 = -\frac{1}{2}g_N\beta_p\mathbf{B}_0(\sigma_2 - \sigma_1),$$

$$E_3 = \frac{1}{2}g_N\beta_p\mathbf{B}_0(\sigma_2 - \sigma_1), \qquad \text{and} \qquad E_4 = g_N\beta_p\mathbf{B}_0\left(1 - \frac{1}{2}\sigma_1 - \frac{1}{2}\sigma_2\right),$$

which are just the energies for two independent noninteracting spins. Now, in order to determine the spectral consequences of these results, the selection rules are needed. These are determined from integrals of the form

$$\int \psi_i|\hat{I}_{X1} + \hat{I}_{X2}|\psi_j\,d\tau,$$

where the magnetic field $\mathbf{B}_0$ is in the z direction and the radio frequency is in the x direction. The ψ's are linear combinations of the ϕ's, and we have seen that the selection rules for the ϕ's are such that only one spin changes at a time. The integral $\int \phi_k|\hat{I}_{X1} + \hat{I}_{X2}|\phi_\ell\,d\tau$ is nonzero for k and ℓ having the pairs of values 1,2; 1,3; 2,4; and 3,4 and vanishes for the k and ℓ values 1,4; 2,3; 2,2; and 3,3. These correspond to $\phi_1 = \alpha(1)\alpha(2)$, $\phi_2 = \alpha(1)\beta(2)$, $\phi_3 = \beta(1)\alpha(2)$, and $\phi_4 = \beta(1)\beta(2)$. Using these rules, we find that, for the two coupled spin-$\frac{1}{2}$ nuclei, $\psi_1 \leftrightarrow \psi_2$, $\psi_1 \leftrightarrow \psi_3$, $\psi_2 \leftrightarrow \psi_4$, and $\psi_3 \leftrightarrow \psi_4$ are allowed transitions and $\psi_1 \leftrightarrow \psi_4$ and $\psi_2 \leftrightarrow \psi_3$ are forbidden transitions. The relative intensities of the four allowed transitions are given by the proportionality

$$I \propto \left| \int \psi_i|\hat{I}_{X1} + \hat{I}_{X2}|\psi_j\,d\tau\right|^2,$$

which is the NMR transition-moment integral. We thus can construct the following table:

Line	Transition	Frequency	Relative intensity
1	$\psi_1 \rightarrow \psi_2$	$\nu_0\left(1 - \frac{1}{2}\sigma_1 - \frac{1}{2}\sigma_2\right) - \frac{1}{2}J_{12} - \frac{1}{2}[J_{12}^2 + \nu_0^2\delta_{12}^2]^{\frac{1}{2}}$	$(a - b)^2$
2	$\psi_3 \rightarrow \psi_4$	$\nu_0\left(1 - \frac{1}{2}\sigma_1 - \frac{1}{2}\sigma_2\right) + \frac{1}{2}J_{12} - \frac{1}{2}[J_{12}^2 + \nu_0^2\delta_{12}^2]^{\frac{1}{2}}$	$(a + b)^2$
3	$\psi_1 \rightarrow \psi_3$	$\nu_0\left(1 - \frac{1}{2}\sigma_1 - \frac{1}{2}\sigma_2\right) - \frac{1}{2}J_{12} + \frac{1}{2}[J_{12}^2 + \nu_0^2\delta_{12}^2]^{\frac{1}{2}}$	$(a + b)^2$
4	$\psi_2 \rightarrow \psi_4$	$\nu_0\left(1 - \frac{1}{2}\sigma_1 - \frac{1}{2}\sigma_2\right) + \frac{1}{2}J_{12} + \frac{1}{2}[J_{12}^2 + \nu_0^2\delta_{12}^2]^{\frac{1}{2}}$	$(a - b)^2$

In the table, the relative shift and the bare nucleus frequency are $\delta_{12} = \sigma_2 - \sigma_1$ and $\nu_0 = g_N\beta_p\mathbf{B}_0 h^{-1}$, respectively.

Thus, the spectrum of two coupled spin-$\frac{1}{2}$ nuclei consists of four lines symmetrically placed about the frequency:

$$\nu_0\left(1 - \frac{1}{2}\sigma_1 - \frac{1}{2}\sigma_2\right)$$

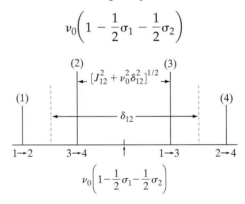

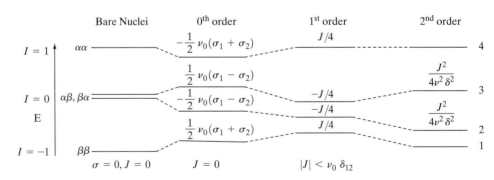

The labeling of the levels above ($\alpha\alpha$, $\beta\beta$, etc.) applies only to the bare nuclei and to the zeroth- and first-order cases. In the second-order perturbation case, the levels originally corresponding to the $\alpha\beta$ and $\beta\alpha$ functions are "mixed," and they should be represented by linear combinations of the type ($\alpha\beta \pm \beta\alpha$). (See ϕ_2 and ϕ_3 discussed earlier.) In the first-order case, where $|J| \geq \nu_0\delta$, the system is said to be weakly coupled and is usually called an *AX* rather than an *AB* spin system. The *AB* designation applies only when there is strong coupling between the nuclei (i.e., when $|J| \ll \nu_0 \delta$). Although the intensities and positions of individual lines within the AB multiplet change with changes in the magnitude of J, the total area under the AB multiplet is independent of J. If the two interacting nuclei have the same value of δ, and hence of $\nu_0 \delta_{AB}$, then only one line is observed that is independent of the magnitude of J. This is why the ^1H NMR spectra of CH_2Cl_2, C_6H_6, $H_2C{=}CH_2$, etc., each give only one resonance despite the fact that J_{HH} may not be zero in any of these cases. How will the spectrum of an AB system change as $\nu_0 \delta_{AB}$ and J change? We consider two extremes: (1) $\nu_0 \delta_{AB}$ is constant while J varies, and (2) J_{AB} is constant while $\nu_0 \delta_{AB}$ varies. The following figure illustrates these situations:

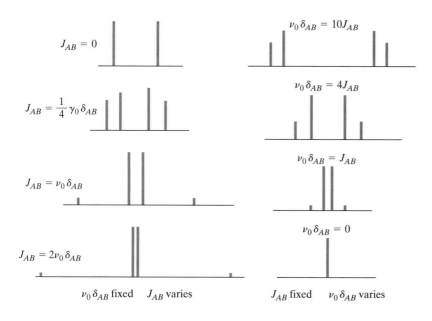

SAMPLE PROBLEM 5.3

Analyze the following AB spectrum to obtain ν_A, ν_B, and J_{AB}:

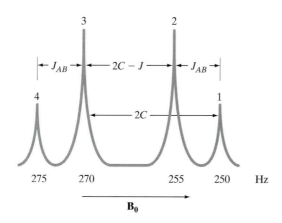

Solution

A. $J_{AB} = 5$ Hz, which is obtained directly from the separations of the lines 1 and 2, or 3 and 4,

B. $2C = 20$ Hz, the separation between lines 2 and 4 or 1 and 3.

C. From the definition of

$$C \equiv \frac{1}{2}[(\nu_A - \nu_B)^2 + J_{AB}^2]^{\frac{1}{2}},$$

we have

$$20 \text{ Hz} = [(\nu_A - \nu_B)^2 + 25 \text{ Hz}]^{\frac{1}{2}}, \quad \text{or} \quad 400 \text{ Hz} - 25 \text{ Hz} = (\nu_A - \nu_B)^2,$$

so that

$$375 \text{ Hz} = (\nu_A - \nu_B)^2, \quad \text{or} \quad 19.36 \text{ Hz} = \nu_A - \nu_B.$$

D. The center of the multiplet is

$$\frac{255 + 270}{2} = 262.5 \text{ Hz,}$$

whence ν_A or $\nu_B = 262.5 + \dfrac{19.36}{2} = 272.18$

or

$$\nu_B \quad \text{or} \quad \nu_A = 262.5 - \frac{19.36}{2} = 252.82.$$

If this is a 60-MHz spectrum, then

$$\delta_A = \frac{272.18}{60 \text{ Hz ppm}^{-1}} = 4.54 \text{ ppm}$$

and

$$\delta_B = \frac{252.82}{60 \text{ Hz ppm}^{-1}} = 4.21 \text{ ppm,}$$

E. $\sin 2\theta = \dfrac{1}{2}J_{AB}/C = \dfrac{1}{2}\left(\dfrac{5}{10}\right) = 0.25.$

Thus, the relative intensities are $0.75:1.25:1.25:0.75$, or, when normalized to the most intense line having unit intensity, $0.6:1:1:0.6$. Obviously, C may also be obtained from this relationship. It should be noted that the chemical shifts of A and B are generally not the averages of lines 1 and 2 or 3 and 4.

We may analyze an AB spectrum completely to obtain ν_A, ν_B, and J_{AB} whenever four lines are observed. In order to do so in a simple manner, we will construct a line position and intensity table in simpler terms by substituting $C = \frac{1}{2}[(\nu_A - \nu_B)^2 + J_{AB}^2]^{\frac{1}{2}}$ and $\sin 2\theta = J_{AB}/2C$ and relate the transitions to the center of the multiplet, whereupon we obtain Table 5.1.

Table 5.1   Transition Energies and Intensities for the *AB* Spin System

Line	Energy	Relative intensity
1	$\frac{1}{2}J + C$	$1 - \sin 2\theta$
2	$-\frac{1}{2}J + C$	$1 + \sin 2\theta$
3	$\frac{1}{2}J - C$	$1 + \sin 2\theta$
4	$-\frac{1}{2}J - C$	$1 - \sin 2\theta$

Some obvious conclusions from Table 5.1 are the following:

1. The *AB* spectrum is always symmetrical about the midpoint.
2. Lines 1 and 4 will always have the same intensity, and so will lines 2 and 3.
3. The separation between lines 2 and 3 is $2C - J$, and the separation between lines 1 and 2 or 3 and 4 is J.
4. The separation between lines 1 and 3 or lines 2 and 4 is $2C$.

B.   Systems with More than Two Interacting Nuclei

The spectra of systems with more than two nuclei become increasingly more complex, with more transitions involving more parameters. A system of N nuclei has N chemical-shift values and $\frac{1}{2}N(N - 1)$ coupling constants, which determine the line positions and intensities of the transitions. As an example, we will discuss the various three-spin spin systems.

In general, the three-spin spin system is denoted *ABC*, becoming *AMX* in the first-order limit and A_3 in the limit of all coincident chemical shifts or A_2B or A_2X if two chemical shifts are coincident. The following diagram is illustrative:

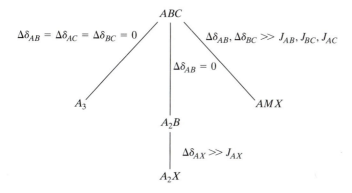

For the three-spin spin system, we must consider three chemical shifts (ν_A, ν_B, and ν_C) and three coupling constants (J_{AB}, J_{AC} and J_{BC}). Spectral analysis for *ABC* spin systems will thus be more difficult than for *AB* spin systems.

i. The *AMX* Case

For this spin system, which is first order, one should observe 12 lines of equal intensity, which arise as shown in the following stick diagram:

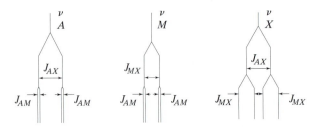

The 12 lines of equal intensity are grouped into three well-separated symmetrical doublets of doublets, centered about the chemical-shift positions of the individual nuclei. The spacings, or J's, are repeated twice each in the spectrum. Analysis of the spectrum reveals that it is first order and that the signs of the coupling constants cannot be determined.

A quantum mechanical treatment can be given as for the *AB* case. For N nuclei, there will be 2^N stationary states, or, in this case, $2^3 = 8$. Their wave functions are as follows:

ϕ	$\alpha\alpha\alpha$	$\alpha\alpha\beta$	$\alpha\beta\alpha$	$\beta\alpha\alpha$	$\beta\beta\alpha$	$\beta\alpha\beta$	$\alpha\beta\beta$	$\beta\beta\beta$
Serial No.	1	2	3	4	5	6	7	8
I	$\frac{3}{2}$		$\frac{1}{2}$			$-\frac{1}{2}$		$-\frac{3}{2}$

In each of the 12 allowed transitions between these states, the spin of only one nucleus is inverted, and all are subject to the selection rule $\Delta I = 1$. These criteria are not identical, as they are in the *AX* case. Thus, although the transition $\beta\beta\alpha \rightarrow \alpha\alpha\beta$ has $\Delta I = 1$, it is precluded because three spins are inverted. The secular determinant will be of the following form:

I \ ϕ	$\alpha\alpha\alpha$	$\alpha\alpha\beta$ $\alpha\beta\alpha$ $\beta\alpha\alpha$	$\beta\beta\alpha$ $\beta\alpha\beta$ $\alpha\beta\beta$	$\beta\beta\beta$
$\frac{3}{2}$	1×1	0	0	0
$\frac{1}{2}$	0	3×3	0	0
$-\frac{1}{2}$	0	0	3×3	0
$-\frac{3}{2}$	0	0	0	1×1

ii. The General *ABC* Case

There are three states, each of I_Z value $+\frac{1}{2}$ and $-\frac{1}{2}$. Consequently, if all the coupling constants have finite values, mixing of these states in sets of three will occur as shown in the secular determinant. Cubic equations must be solved, and simple analytical expressions, as in the *AX* case, do not result. On passing from the *AMX* to the *ABC* case, the intensity pattern is distorted, but the spacings of the four lines associated with each nucleus remain those of a symmetrical doublet of doublets. An additional complication results in the *ABC* case: There are three "combination transitions," which are allowed and can gain reasonable intensity; thus, while the A_2B, *AMX*, and *ABX* systems all have 12 possible transitions, the *ABC* system has 15.*

iii. Use of Symmetry to Simplify NMR Secular Equations; the AB_2 Spin System

For the systems considered thus far, the I_Z value has sufficed to tell us which of the first-order states will mix in non-first-order cases. This came from the first nonmixing rule: *States with different I_Z values cannot mix.* A further condition that also must be considered is the second nonmixing rule: *States of different symmetry cannot mix.*

The great simplification in the AB_2 case relative to the *ABC* case is due to the operation of this latter principle. The nuclear spin system (although not necessarily the whole molecule) has an important element of symmetry that was lacking in the *ABC* spin system, namely, a plane of symmetry as shown in the following figure:

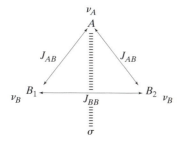

If we write the eight basic functions for the *AMX* case, then, in the AB_2 case, we must consider how each would be modified by "reflection" through the indicated plane:

Nucleus	AB_1B_2	AB_1B_2	AB_1B_2	AB_1B_2	AB_1B_2	AB_1B_2	AB_1B_2	AB_1B_2
ϕ	$\alpha\alpha\alpha$	$\alpha\alpha\beta$	$\alpha\beta\alpha$	$\beta\alpha\alpha$	$\beta\beta\alpha$	$\beta\alpha\beta$	$\alpha\beta\beta$	$\beta\beta\beta$
serial no.	1	2	3	4	5	6	7	8
symmetry	s	u	u	s	u	u	s	s
with respect to σ (s = symmetric, u = unsymmetric)								

Since *A* lies in the plane of symmetry, its spin remains unaltered. Interchanging B_1 and B_2, however, means that the spin of nucleus B_1 becomes that of B_2 and vice versa. Of the eight basic functions, four are symmetric with respect to σ. Thus, if we take the $I_Z = \frac{1}{2}$ block of the secular determinant, we have singled out $\beta\alpha\alpha$ and are left with $\alpha\alpha\beta$ and $\alpha\beta\alpha$. While these do not have any symmetry properties with respect to reflection, it is clear that their sum and difference do. Thus, if we replace $\alpha\alpha\beta$ and $\alpha\beta\alpha$ by $1/\sqrt{2}\,(\alpha\alpha\beta \pm \alpha\beta\alpha)$ or its

---

*The reader may consult Mathieson, *Nuclear Magnetic Resonance for Organic Chemists*, for exemplary spectra and their analysis. Also, Bovey, *Nuclear Magnetic Resonance Spectroscopy*, 2d ed., appendix D, gives the appearance of a very large number of spectra for three-spin spin systems.

equivalent, $1/\sqrt{2}\alpha(\alpha\beta \pm \beta\alpha)$, we find that $1/\sqrt{2}\alpha(\alpha\beta + \beta\alpha)$ is symmetric with respect to reflection and $1/\sqrt{2}\alpha(\alpha\beta - \beta\alpha)$ is antisymmetric with respect to reflection. The $I_Z = -\frac{1}{2}$ block can be treated similarly.

We have thus arrived at a new set of functions constructed from AMX functions to be symmetric (s) or antisymmetric (a) with respect to reflection:

I_z	Symmetric		Antisymmetric	
$\frac{3}{2}$	$\alpha\alpha\alpha$	$\equiv \psi_1$		
$\frac{1}{2}$	$1/\sqrt{2}\alpha(\alpha\beta + \beta\alpha)$	$\equiv \psi_2$	$1/\sqrt{2}\alpha(\alpha\beta - \beta\alpha) \equiv \psi_4$	
	$\beta\alpha\alpha$	$\equiv \psi_3$		
$-\frac{1}{2}$	$\alpha\beta\beta$	$\equiv \psi_6$		
	$1/\sqrt{2}\beta(\alpha\beta + \beta\alpha)$	$\equiv \psi_7$		
$-\frac{3}{2}$	$\beta\beta\beta$	$\equiv \psi_8$	$1/\sqrt{2}\beta(\alpha\beta - \beta\alpha) \equiv \psi_5$	

These are the correct wave functions for the AX_2 or AB_2 cases. The reduced number of transitions relative to the AMX case (8 versus 12) can now be explained most simply by the selection rule that *transitions are allowed only between states of the same symmetry*. Thus, the secular determinant, which was of the form

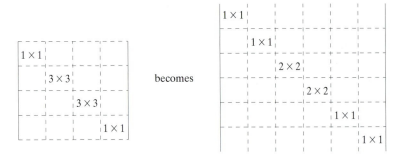

becomes

which is indeed much easier to solve. Analytical expressions (see Table 5.2) can be written for these transitions, with the following definitions:

$$C_+ = \frac{1}{2}\left(\Delta\nu^2 + \Delta\nu \cdot J + \frac{9}{4J^2}\right)^{\frac{1}{2}}; \qquad C_- = \frac{1}{2}\left(\Delta\nu^2 - \Delta\nu \cdot J + \frac{9}{4J^2}\right)^{\frac{1}{2}};$$

$$C_-\cos 2\phi_- = \frac{1}{2}\Delta\nu - \frac{1}{4}J; \qquad C_+\cos 2\phi_+ = \frac{1}{2}\Delta\nu - \frac{1}{4}J;$$

$$C_+\sin 2\phi_+ = C_-\sin 2\phi_- = \frac{J}{\sqrt{2}}.$$

We may write simple rules for obtaining the spectral parameters:

$$\nu_A = \nu_3; \nu_B = \frac{1}{2}(\nu_5 + \nu_7); J_{AB} = \frac{1}{3}(\nu_4 + \nu_8 - \nu_1 - \nu_6).$$

The appearance of the AB_2 spectrum changes drastically as the ratio $\Delta\nu_{AB}/J_{AB}$ changes. When this ratio is greater than approximately 18, the spectrum

Table 5.2 Transitions of an AB_2 Spin System

Line	Origin	Line position[a]	Relative intensity[b]
1	A	$\frac{1}{2}(\nu_A + \nu_B) + \frac{3}{4}J + C_+$	$(\sqrt{2}\sin\theta_+ - \cos\theta_+)^2$
2	A	$\nu_B + C_+ + C_-$	$[\sqrt{2}\sin(\theta_+ - \theta_-) + \cos\theta_+\cos\theta_-]^2$
3	A	ν_A	1
4	A	$\frac{1}{2}(\nu_A + \nu_B) + \frac{3}{4}J + C_-$	$(\sqrt{2}\sin\theta_- - \cos\theta_-)^2$
5	B	$\nu_B + C_+ - C_-$	$[\sqrt{2}\sin(\theta_+ - \theta_-) + \cos\theta_+\sin\theta_-]^2$
6	B	$\frac{1}{2}(\nu_A + \nu_B) + \frac{3}{4}J - C_+$	$(\sqrt{2}\cos\theta_+ + \sin\theta_+)^2$
7	B	$\nu_B - C_+ + C_-$	$[\sqrt{2}\cos(\theta_+ - \theta_-) - \sin\theta_+\cos\theta_-]^2$
8	B	$\frac{1}{2}(\nu_A + \nu_B) - \frac{3}{4}J - C_-$	$(\sqrt{2}\cos\theta - \sin\theta_-)^2$
9	Combination	$\nu_B - C_+ - C_-$	$[\sqrt{2}\sin(\theta_+ - \theta_-) + \sin\theta_+\sin\theta_-]^2$

[a] Since the AB_2 spectrum is not symmetrical, the line positions are referred to line 3, ν_A, rather than $\frac{1}{2}(\nu_A + \nu_B)$, as in Table 5.1. [b] Referred to line 3 as unity.

becomes a first-order AX_2 triplet and doublet. At intermediate values of the ratio, all eigth lines of the AB_2 system are easily seen. As the ratio becomes smaller still, some of the lines are very difficult to observe. These effects are illustrated in Figure 5.1.

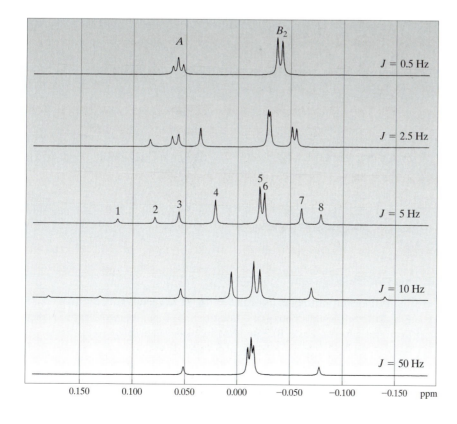

◄ FIGURE 5.1
Calculated AB_2 spectra for different values of J_{AB} with $\Delta\nu_{AB}$ fixed to 10 Hz.

SAMPLE PROBLEM 5.4

Analyze the 40.26-MHz ^{31}P{^{1}H} NMR Spectrum of *trans*-RuCl$_2$(CO)(PMe$_3$)$_3$ to obtain ν_a, ν_b, and J_{ab}.[a]

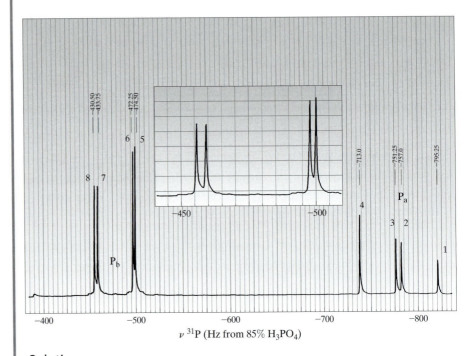

Solution

From the preceding rules, $\nu_a = \nu_3 = -751.25$; $\delta_a = -751.25/40.26 = -18.66$ ppm;

$\nu_b = (\nu_5 - \nu_7)/2 = (-474.5 - 433.75)/2 = -454.125$;

$\delta_b = -454.125/40.26 = -11.28$ ppm;

$J_{ab} = (\nu_4 + \nu_8 - \nu_1 - \nu_6)/3 = (-713 - 430.5 + 795.25 + 472.25)/3 = 41.33$ Hz.

iv. The *ABX* Spin System

For this spin system, the Hamiltonian of the *ABC* case is further broken down by the *X* approximation. Here, one nucleus (*X*) has a very different chemical shift from the others (*A* and *B*). Specifically, $\nu_X - \nu_A$ and $\nu_X - \nu_B \gg J_{AX}, J_{BX}$.

[a] Krassowaski, D. W., Nelson, J. H., Brower, K. R., Hauenstein, D., Jacobson, R. A., *Inorg. Chem.* **1988**, *27*, 4294.

When this occurs, it is possible to treat I_X as a good quantum number, separately from the A and B spins. Of the three $I_Z = \frac{1}{2}$ wave functions

$$\begin{vmatrix} ABX & ABX & ABX \\ & \text{and} & \\ \alpha\alpha\beta & \alpha\beta\alpha & \beta\alpha\alpha \end{vmatrix},$$

the first has $I_X = -\frac{1}{2}$ and therefore separates from the others, which both have $I_X = \frac{1}{2}$. The determinant now reduces from a 3×3 to a 1×1 and a 2×2 determinant, and the resultant quadratic equation may be solved explicitly.

This is a general result for X nuclei, and we can use these simple methods to handle any spin system, which, after removal of the X states, leaves only two strongly coupled nuclei. For example, the $ABMX$ system can be treated similarly, as both I_M and I_X are good quantum numbers.

For the ABX system, this procedure gives the transition energies and intensities shown in Table 5.3.

Table 5.3   Transition Energies and Intensities for the *ABX* Spin System

Transition	Origin	Energy[a]	Relative intensity
1	B	$\frac{1}{2}(\nu_A + \nu_B) - \frac{1}{2}(J_{AB} + N) - D_-$	$1 - \sin\phi_-$
2	B	$\frac{1}{2}(\nu_A + \nu_B) - \frac{1}{2}(J_{AB} - N) - D_+$	$1 - \sin\phi_+$
3	B	$\frac{1}{2}(\nu_A + \nu_B) + \frac{1}{2}(J_{AB} - N) - D_-$	$1 + \sin\phi_-$
4	B	$\frac{1}{2}(\nu_A + \nu_B) + \frac{1}{2}(J_{AB} + N) - D_+$	$1 + \sin\phi_+$
5	A	$\frac{1}{2}(\nu_A + \nu_B) - \frac{1}{2}(J_{AB} + N) + D_-$	$1 + \sin\phi_-$
6	A	$\frac{1}{2}(\nu_A + \nu_B) - \frac{1}{2}(J_{AB} + N) + D_+$	$1 + \sin\phi_+$
7	A	$\frac{1}{2}(\nu_A + \nu_B) + \frac{1}{2}(J_{AB} - N) + D_-$	$1 - \sin\phi_-$
8	A	$\frac{1}{2}(\nu_A + \nu_B) + \frac{1}{2}(J_{AB} + N) + D_+$	$1 - \sin\phi_+$
9	X	$\nu_X - N$	1
10	X	$\nu_X + D_+ - D_-$	$\frac{1}{2}[1 + \cos(\phi_+ - \phi_-)]$
11	X	$\nu_X - D_+ + D_-$	$\frac{1}{2}[1 + \cos(\phi_+ - \phi_-)]$
12	X	$\nu_X + N$	1
13	Comb.	$\nu_A + \nu_B - \nu_X$	0
14	Comb. (X)	$\nu_X - D_+ - D_-$	$\frac{1}{2}[1 - \cos(\phi_+ - \phi_-)]$
15	Comb. (X)	$\nu_X + D_+ + D_-$	$\frac{1}{2}[1 - \cos(\phi_+ - \phi_-)]$

[a] $D_\pm \cos\phi_\pm = \frac{1}{2}(\delta_{AB} \pm L)$; $D_\pm \sin\phi_\pm = \frac{1}{2}J_{AB}$. That is, $D_\pm = \frac{1}{2}[(\delta_{AB} \pm L)^2 + J_{AB}^2]^{\frac{1}{2}}$; $N = \frac{1}{2}(J_{AX} + J_{BX})$; $L = \frac{1}{2}(J_{AX} - J_{BX})$.

There are 14 allowed transitions, given in terms of v_A, v_B, v_X, J_{AB}, J_{AX}, and J_{BX}. We define $v_A > v_B$ such that A is the low-field nucleus, but the X nucleus can be either high or low field, as neither δv_{AX} nor δv_{BX} enters into the table.

This table is too complex to use directly to analyze the spin system. A more convenient method is to consider the given spectrum as an extension of the AB spectrum. Because the X spin may be considered independently of the A and B spins, we can consider the AB part of the spectrum as being made up of two ab subspectra, one for each orientation of the X spin.

Each of these two equally intense ab subspectra is a straightforward AB quartet and may be analyzed as such. In such an analysis, the value of J_{AB} determined is correct; however, the chemical shifts v_A and v_B are not the true chemical shifts, but rather effective chemical shifts v^* that are equal to the chemical shifts plus the effect of the X nucleus. For the $\alpha\left(+\frac{1}{2}\right)$ X orientation, the effective chemical shifts will be $v_A^* = v_A + \frac{1}{2}J_{AX}$ and $v_B^* = v_B + \frac{1}{2}J_{BX}$, and for the $\alpha\left(+\frac{1}{2}\right)$ X orientation, the effective chemical shifts will be $v_A^* = v_A - \frac{1}{2}J_{AX}$ and $v_B^* = v_B - \frac{1}{2}J_{BX}$. Substituting these quantities into the expressions for AB gives the following for the two ab subspectra:

$\alpha(x)$ orientation: $\quad \delta_{AB}^* = \delta_{AB} + \dfrac{1}{2}(J_{AX} + J_{BX}) = \delta_{AB} + L,$

with midpoint

$$\frac{1}{2}(v_A + v_B) + \frac{1}{4}(J_{AX} + J_{BX}) = \frac{1}{2}(v_A + v_B) + \frac{1}{2}N;$$

$\beta(x)$ orientation: $\quad \delta_{AB}^* = \delta_{AB} - \dfrac{1}{2}(J_{AX} - J_{BX}) = \delta_{AB} - L,$

with midpoint

$$\frac{1}{2}(v_A + v_B) - \frac{1}{4}(J_{AX} + J_{BX}) = \frac{1}{2}(v_A + v_B) - \frac{1}{2}N.$$

In order to analyze the ABX spin system, we need to identify the two ab subspectra. Examples of typical ABX spectra are given in Figure 5.2.

Note that in part (d) of the figure, lines 1 and 7 of one of the ab subspectra have zero intensity. Comparing Table 5.3 with Figure 5.2, we see that the lines in the two ab quartets are labeled 2, 4, 6, 8 for the $\alpha\left(+\frac{1}{2}\right)$ X spin and 1, 3, 5, 7 for the $\beta\left(-\frac{1}{2}\right)$ X spin. We may identify these two quartets by noting that J_{AB} must be repeated in the line spacings four times: $J_{AB} = v_3 - v_1 = v_4 - v_2 = v_7 - v_5 = v_8 - v_6$.

The X part of the spectrum is of considerable importance. It is always symmetric about its midpoint and usually consists of six lines. The absolute value of the sum of the X couplings, $|J_{AX} + J_{BX}|$ is obtained directly from the separation of the two major lines ($v_{12} - v_9$). The concepts of effective chemical shifts and subspectra are very useful concepts for many spin systems.

The ABX spectrum is the first spectrum thus far considered for which the *relative* signs of the coupling constants can be determined. For the AB and AB_2 spectra, changing the sign of J_{AB} merely alters the assignments of the transitions without affecting the appearance of the spectra, and this is also the case for all first-order spectra. For the ABX spectrum, changing the

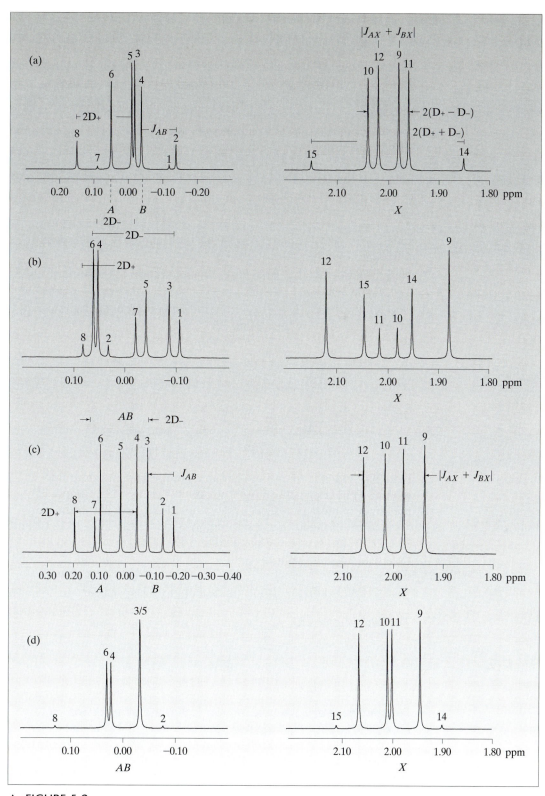

▲ FIGURE 5.2

Typical examples of *ABX* spectra at 100 MHz: (a) $\nu_A - \nu_B = 10$ Hz, $|J_{AB}| = 10$ Hz, $J_{AX} = 8$ Hz, $J_{BX} = -4$ Hz; (b) $\nu_A - \nu_B = 2$ Hz, $|J_{AB}| = 2$ Hz, $J_{AX} = 8$ Hz, $J_{BX} = 16$ Hz; (c) $\nu_A - \nu_B = 20$ Hz, $|J_{AB}| = 10$ Hz, $J_{AX} = 8$ Hz, $J_{BX} = 4$ Hz; (d) $\nu_A - \nu_B = 2$ Hz, $|J_{AB}| = 10$ Hz, $J_{AX} = 8$ Hz, $J_{BX} = 4$ Hz.

sign of J_{AB} also merely alters the assignments, so, for convenience, we may regard J_{AB} as positive. However, changing the relative signs of J_{AX} and J_{BX} does produce changes in the observed spectra, and thus, the relative signs of these couplings can be determined in principle. This is demonstrated as follows: In the first-order AMX limit, the spacings $\nu_2 - \nu_1$ and $\nu_4 - \nu_3$ would equal J_{BX}. Similarly, $\nu_8 - \nu_7$ and $\nu_6 - \nu_5$ would equal J_{AX}. These splittings are not equal to the actual couplings in the ABX case, but it is convenient for our discussion to call them J_{AX} and J_{BX}. If J_{AX} is negative, the A splittings are simply reversed, and the assignments in Figure 5.2a, for example, would become 7,8,5,6,4,3,2,1, from left to right. This reverses the assignment of the two ab subspectra, which must always be 1,3,5,7 and 2,4,6,8—hence the final analysis.

A general procedure for the analysis of ABX spectra is as follows:

1. Identify the two ab quartets on the basis of frequency and intensity relations. Note the value of J_{AB}.
2. Find the value of $\frac{1}{2}|J_{AX} + J_{BX}|$ from the separation of the centers of the two ab quartets.
3. Check the value of $|J_{AX} + J_{BX}|$ from the separation of the two strongest X lines, and identify lines 9 and 12.

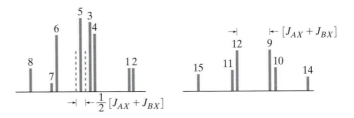

4. Find $2D_+$ and $2D_-$ from the separations of the first and third lines in the two ab quartets; $2D_+ = \nu_8 - \nu_4$ and $2D_- = \nu_7 - \nu_3$. Choose $2D_+$ as the larger of these two. Check the values of $2D_+$ and $2D_-$ from the separations of the lines in the X region, and identify lines 10,11,14, and 15. $2|D_+ + D_-| = \nu_{15} - \nu_{14}$ and $2|D_+ - D_-| = \nu_{10} - \nu_{11}$.
5. Calculate M and N, where

$$2M = (4D_+^2 - J_{AB}^2)^{\frac{1}{2}} \quad \text{and} \quad 2N = (4D_-^2 - J_{AB}^2)^{\frac{1}{2}}.$$

The two solutions for $\nu_A - \nu_B$ and $\frac{1}{2}(J_{AX} - J_{BX})$ are then

	1	2
$\nu_A - \nu_B$	$M + N$	$M - N$
$\frac{1}{2}(J_{AX} - J_{BX})$	$M - N$	$M + N$

6. Find the value of ϕ_+, where $0 < \phi_+ < 90°$, from the relation $\sin \phi_+ = J_{AB}/2D_-$. Find the two possible values of ϕ_-, where $0 < \phi_- < 180°$, from the relation $\sin \phi_- = J_{AB}/2D_-$. Calculate the two possible values of $\cos(\phi_+ - \phi_-)$, and from Table 5.3, compute the intensities of the X lines for each solution. If the smaller value of ϕ_- gives X intensities consistent with the observed spectrum while the larger value does not, then choose solution 1, point 5, as the correct solution. If the converse is true, choose solution 2.

7. Find $\frac{1}{2}(\nu_A + \nu_B)$, which is the average of the centers of the two *ab* quartets, or equivalently, the average of the frequencies of all eight *AB* lines. From this value and the correct value of $\nu_A - \nu_B$ determined in steps 5 and 6, calculate ν_A and ν_B.

8. Assign to the sum $\frac{1}{2}(J_{AX} + J_{BX})$, for which the absolute value was found in step 2, a positive sign if the $(ab)_+$ quartet is centered at a higher frequency than the $(ab)_-$ quartet or a negative sign if the reverse order is true. From this value and the correct value of $\frac{1}{2}(J_{AX} - J_{BX})$ determined in steps 5 and 6, calculate J_{AX} and J_{BX}.

SAMPLE PROBLEM 5.5

Analyze the following 60-MHz ^{1}H NMR spectrum of malic acid $(HO_2CCH_2CHOHCO_2H)$ in $NaOD/D_2O$:

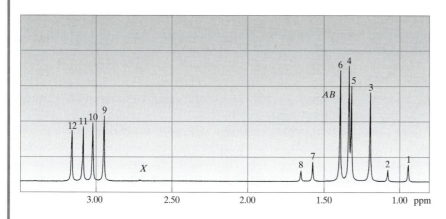

	Hz from TMS
1	56.4
2	64.7
3	72.0
4	80.4
5	79.3
6	83.7
7	94.6
8	98.9
9	176.9
10	181.1
11	185.1
12	189.7

Solution

The spectrum is due to the CH_2CH fragment of this molecule in which the CH_2 protons are not equivalent. (They are diastereotopic.) Since the CH resonance is well separated from those of the CH_2 protons, this is an *ABX* spectrum. We first identify the two *ab* subspectra: lines 8,6,4, and 2, and 7,5,3, and 1. Then $J_{AB} = 3 - 1 = 4 - 2 = 7 - 5 = 8 - 6 = 15.5$ Hz, and

$$\delta_{AB}^* = \sqrt{(\nu_8 - \nu_2)(\nu_6 - \nu_4)} = 10.62 \text{ Hz} = \delta_{AB} + \frac{1}{2}(J_{AX} - J_{BX}),$$

or

$$\delta_{AB}^* = \sqrt{(\nu_8 - \nu_1)(\nu_5 - \nu_3)} = 16.70 \text{ Hz} = \delta_{AB} - \frac{1}{2}(J_{AX} - J_{BX}), 10.62 - 16.70$$

$$= J_{AX} - J_{BX} = -6.08 \text{ Hz}.$$

Then, since $J_{AX} + J_{BX} = \nu_{12} - \nu_9 = 12.8$ Hz, it follows that $J_{AX} = 6.72$ Hz or $J_{AX} = 3.36$ Hz and $J_{BX} = 9.44$ Hz. The midpoint of lines 2,4,6,8 = $(\nu_A + \nu_B) + \frac{1}{2}N$, or $82.05 = \frac{1}{2}(\nu_A + \nu_B) + \frac{1}{4}(J_{AX} + J_{BX})$, or $\nu_A + \nu_B = 157.7$ Hz. Also, $2\delta_{AB} = 16.70 + 10.62$, or $\delta_{AB} = 13.66 = \nu_A - \nu_B$. Then $\nu_A = 85.68$ Hz and $\nu_B = 72.02$ Hz. ν_X is the midpoint of lines 10 and 11 = 183.1 Hz. Hence, the parameters for this spectrum are $\nu_X = 183.1$ Hz, $\nu_A = 85.68$ Hz, $\nu_B = 72.02$ Hz, $J_{AB} = 15.5$ Hz, $J_{AX} = 3.36$ Hz, and $J_{BX} = 9.44$ Hz.

v. Deceptively Simple Spectra

The previous analysis of *ABX* spectra assumes that all eight *AB* lines and either four or six *X* lines are experimentally observed. Frequently, however, some of the lines coincide, creating a spectrum with an appearance that is not of a typical *ABX* pattern. For example, Figure 5.3 shows calculated *ABX* spectra as a function of changing only one parameter, ν_B. The top spectrum is readily recognized as an *ABX* pattern. As ν_B approaches ν_A, the appearance of the spectrum changes drastically, and when $\nu_A = \nu_B$, the *AB* region appears to be simply a doublet, while the *X* region appears to be a 1:2:1 binomial triplet (bottom spectrum). Actually, there are other weak lines (indicated by the ↓), but these would normally be lost in the noise. This spectrum is an excellent example of what has been termed *deceptively simple spectra*. If one were confronted with such a spectrum and did not realize that it is a special case of *ABX* (actually *AA′X*), the doublet and triplet might be mistakenly interpreted as the components of a first-order A_2X spectrum, with J_{AX} equal to the line separations in the apparent doublet and triplet. In fact, the separation of lines 9 and 12 is $|J_{AX} + J_{BX}| = |J_{AX} + J_{A'X}|$. In general, these two coupling constants are not equal in magnitude and often are different in sign. The spectrum of 2,5-dichloronitrobenzene (Figure 5.4) is an example.

Deceptively simple spectra are widespread and are not limited to *ABX* systems. This spectrum is an instance of what has been termed *virtual coupling*. Nucleus *X*, although coupled to nuclei *A* and *B* by different magnitudes, appears to be equally coupled to both.

This phenomenon occurs widely in transition-metal phosphine complexes containing two identical phosphines—for example, *cis*- and *trans*-$(MePPh_2)_2Pd(N_3)_2$. The $^{13}C\{^1H\}$, ^1H, and $^{31}P\{^1H\}$ spectra of a mixture of these two compounds are shown in part in Figure 5.5.

▶ **FIGURE 5.3**
Calculated *ABX* spectra as a function of changing only one parameter, ν_B.

The bottom spectrum is an example of a deceptively simple *AA′X* spectrum. Note the very weak lines indicated by arrows.

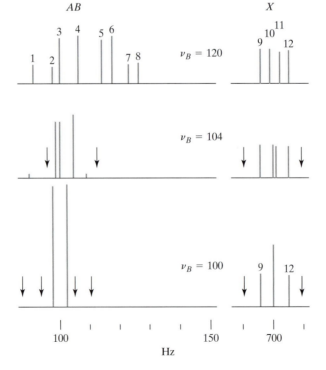

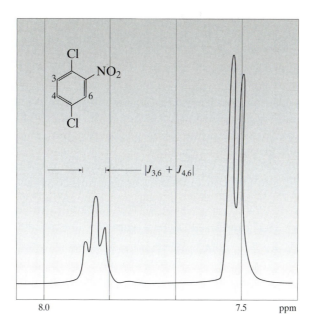

60-MHz ^{1}H NMR spectrum of 2,5-dichloronitrobenzene in $CDCl_3$.

The low-field *apparent* triplet is due to H_6, and the *apparent* doublet to H_3 and H_4, which are accidentally chemical shift equivalent. The observed splitting of 3.2 Hz is about what one would expect for $|^4J_{HH} + {}^5J_{HH}| = |3$ Hz (meta) + 0 Hz (para)$|$.

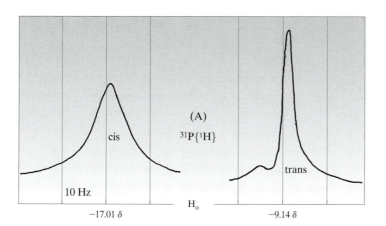

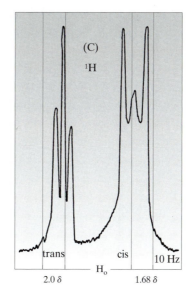

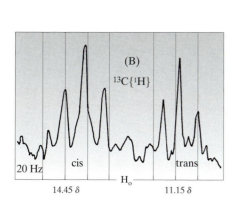

◄ FIGURE 5.5

NMR spectra at ambient temperature for $CDCl_3$ solutions of $[(C_6H_5)_2PCH_3]_2Pd(N_3)_2$:
(A) 40.5 MHz $^{31}P\{^1H\}$, H_3PO_4 external reference; (B) 25.2 MHz $^{13}C\{^1H\}$ in the methyl region, TMS internal reference; (C) 100-MHz ^{1}H in the methyl region, TMS internal reference.

The CH_3 carbons in these molecules form the A part of $A[X]_2$ spin systems; $X = {}^{31}P\left(I = \frac{1}{2}, 100\%\right)$ because of the low natural abundance of ${}^{13}C$. For this spin system, a triplet will be observed for the A resonance whenever $|J_{AX} - J_{A'X}|^2 < 8J_{XX'} \cdot \Delta\nu_{1/2}$, where $\Delta\nu_{1/2}$ is the resolving power of the instrument. Otherwise, one of the line shapes shown in Figure 5.6 will be observed. For *trans*-isomers, ${}^2J_{PP'} = {}^2J_{XX'}$ is usually greater than 300 Hz, and for *cis*-isomers, ${}^2J_{PP'}$ is usually less than 40 Hz. Hence, the line shape is often a good indicator of the geometry of the complex, with *trans*-isomers usually giving rise to apparent 1:2:1 triplets and *cis*-isomers other line shapes.

When five or six lines are observed in the A part of the $A[X]_2$ spectrum, all coupling constants may be determined with the use of the equations in Table 5.4; otherwise, only $|J_{AX} + J_{AX'}|$ may be determined.

vi. $[AB]_2$ and $[AX]_2$ Spin Systems

These four-spin spin systems are characterized by two chemical shifts and four coupling constants, $J_{AA'}$, $J_{BB'}$, J_{AB}, and $J_{A'B'}$. The last two are not equal, a situation that leads to magnetic inequivalence. Strict $[AX]_2$

▶ FIGURE 5.6

Computer simulation* of the $A[X]_2$ spin system, where $J_{AX} = 25$ Hz, $J_{AX'} = 0.5$ Hz, and J_{XX} is varied between 500 and 0.0 Hz. Similar spectra are obtained using differing values of J_{AX} and $J_{AX'}$, which exhibit the same general changes with changing $J_{XX'}$. The line width at half-height for these simulations is 0.8 Hz.

*Redfield, D. A., Cary, L. W., Nelson, J. H., *Inorg. Chem.* **1975**, *14*, 50.

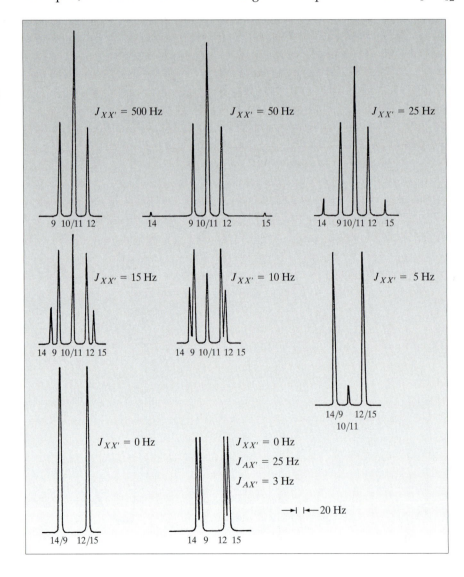

Table 5.4   The A Transitions of the $AXX' \equiv A[X]_2$ Spin System

Line	Origin	Line position	Relative intensity
9	A	$\nu_A - \dfrac{1}{2}(J_{AX} + J_{AX'})$	1
10	A	$\nu_A + D_+ - D_-$	$\cos^2(\phi_+ - \phi_-)$
11	A	$\nu_A - D_+ + D_-$	$\cos^2(\phi_+ - \phi_-)$
12	A	$\nu_A + \dfrac{1}{2}(J_{AX} + J_{AX'})$	1
13	Combination	$2\nu_X - \nu_A$	0
14	Combination	$\nu_A - D_+ - D_-$	$\sin^2(\phi_+ - \phi_-)$
15	Combination	$\nu_A + D_+ + D_-$	$\sin^2(\phi_+ - \phi_-)$

$$D_+ \cos 2\phi_+ = \frac{1}{4}(J_{AX} - J_{AX'}) \qquad D_- \cos 2\phi_- = \frac{1}{4}(J_{AX} - J_{AX'})$$

$$D_+ \sin 2\phi_+ = \frac{1}{2}(J_{XX'}) \qquad D_- \sin 2\phi_- = \frac{1}{2}(J_{XX'})$$

spectra are given by pairs of nuclei of different species as with the following species:

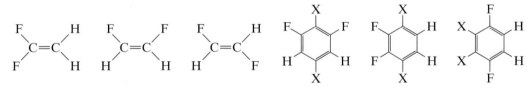

For convenience, we define

$$K = J_{AA'} + J_{XX'}, \quad L = J_{AX} - J_{AX'}, \quad M = J_{AA'} - J_{XX'},$$

and

$$N = J_{AX} + J_{AX'}$$

and, for the relative intensities,

$$\cos 2\phi_s : \sin 2\phi_s : 1 = K : L : (K^2 + L^2)^{\frac{1}{2}};$$
$$\cos 2\phi_a : \sin 2\phi_a : 1 = M : L : (M^2 + L^2)^{\frac{1}{2}}.$$

Figure 5.7 shows the A portion of a typical $AA'XX'$ spectrum. The X portion is identical in form. Another example (the ^{1}H NMR spectrum of $F_2C{=}CH_2$) is shown on page 93.

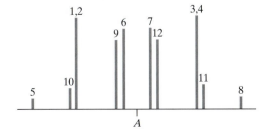

◄ FIGURE 5.7
The A portion of a calculated $AA'XX'$ spectrum with the following parameters:

$J_{AX} = 13.0$ Hz; $N = 16$; $J_{AA'} = 9.0$ Hz; $K = 12$;
$J_{AX'} = 3.0$ Hz; $L = 10$; $J_{XX'} = 3.0$ Hz; $M = 6$.

SAMPLE PROBLEM 5.6

Analyze the A part of the following $A[X]_2$ multiplet:

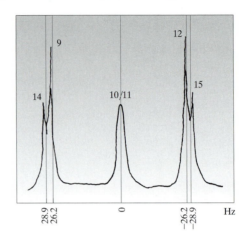

Solution

All line intensities are adjusted so that the intensity of line 9 equals that of line 12 (=1).

1. Line 14 − line 15 = $2|D_+ + D_-| = 4D$, since $|D_+| = |D_-|$ unless six lines are observed in this system. $4D = 57.9$, and $D = 14.45$.

2. The relative intensity of lines 14 and 15 is $\sin^2(\phi_+ - \phi_-) = 0.63$; thus, $\sin(\phi_+ - \phi_-) = 0.7937$, and $\phi_+ - \phi_- = 52.535°$.

3. In general, for this system, $2\phi_- = (180 - 2\phi_+)$, so $\phi_+ + \phi_- = 90°$. Hence, $2\phi_+ = 90° + 52.535°$, or $2\phi_+ = 142.535°$, and $\phi_+ = 71.268°$. Since $\phi_+ - \phi_- = 52.535$, we have $\phi_- = 18.733°$.

4. Since $D_+ \sin 2\phi_+ = D_- \sin 2\phi_- = \frac{1}{2}J_{XX'}$ and $D_+ = D_-$, it follows that $J_{XX'} = 2D \sin(2\phi_+) = 2(14.45)(0.6083) = 17.58$ Hz.

5. $|J_{AX} - J_{AX'}| = 4D \cos 2\phi_+ = -45.877$

$$= 4D \cos 2\phi_- = 45.877 \text{ can be either one.}$$

Now, $\nu_9 - \nu_{12} = 52.4$ Hz $= |J_{AX} + J_{AX'}|$. Combining these two results, $2J_{AX} = 6.52$, whereupon $J_{AX} = 3.26$ Hz and $J_{AX'} = 49.14$. In general, there are two indistinguishable solutions:

$J_{AX} = 49.14$ and $J_{AX'} = 3.26$ (the other solution in this case).

The entire $[AX]_2 \equiv AA'XX'$ spectrum contains 24 lines: 12 in the A part and an identical 12 in the X part. Each region is symmetric about its midpoint, as in Figure 5.7. Only 10 lines can be observed, as transitions 1 and 2 and 3 and 4 are always degenerate. Line positions and intensities are given in Table 5.5.

As indicated in Table 5.5, lines 1,2 and 3,4 together will always contain half the intensity of the A multiplet and, equivalently, of the X multiplet. Also, we note the following:

   a. The separation of lines 1,2 and 3,4 equals $N = J_{AX} + J_{AX'}$.

   b. The separation of lines 5 and 6 (and of 7 and 8) gives K.

Table 5.5 Transitions of $AA'XX' \equiv [AX]_2$ Spin System

Line	Line position[a]	Relative intensity
1	$\frac{1}{2}N$	1 $\Big\}$ 2
2	$\frac{1}{2}N$	1
3	$-\frac{1}{2}N$	1 $\Big\}$ 2
4	$-\frac{1}{2}N$	1
5	$\frac{1}{2}K + \frac{1}{2}(K^2 + L^2)^{\frac{1}{2}}$	$\sin^2 \theta_s$
6	$-\frac{1}{2}K + \frac{1}{2}(K^2 + L^2)^{\frac{1}{2}}$	$\cos^2 \theta_s$
7	$\frac{1}{2}K - \frac{1}{2}(K^2 + L^2)^{\frac{1}{2}}$	$\cos^2 \theta_s$
8	$-\frac{1}{2}K - \frac{1}{2}(K^2 + L^2)^{\frac{1}{2}}$	$\sin^2 \theta_s$
9	$\frac{1}{2}M + \frac{1}{2}(M^2 + L^2)^{\frac{1}{2}}$	$\sin^2 \theta_a$
10	$-\frac{1}{2}M + \frac{1}{2}(M^2 + L^2)^{\frac{1}{2}}$	$\cos^2 \theta_a$
11	$\frac{1}{2}M - \frac{1}{2}(M^2 + L^2)^{\frac{1}{2}}$	$\cos^2 \theta_a$
12	$-\frac{1}{2}M - \frac{1}{2}(M^2 + L^2)^{\frac{1}{2}}$	$\sin^2 \theta_a$

[a] Relative to ν_A.

    c. The separation of lines 9 and 10 (and of 11 and 12) gives M.

    d. Lines 5,6,7, and 8 form a quartet centered on ν_A.

    e. Lines 9,10,11, and 12 form a quartet, also centered on ν_A.

    f. The spacings of lines 5 and 7 (and of 6 and 8) give $(K^2 + L^2)$; the spacings of lines 9 and 11 (and of 10 and 12) give $(M^2 + L^2)^{\frac{1}{2}}$.

There is some ambiguity in $AA'XX'$ spectra, because, in the absence of prior knowledge, we do not know the relative signs of K and N and therefore cannot know how to assign the outer and inner lines of each quartet. (The inner lines will, however, always be more intense.) Thus, we can find K and M as a pair, but we cannot know a priori which is which, nor can we know their signs.

When the chemical-shift difference between the A and X resonances is comparable to the magnitude of the couplings, we have an $AA'BB'$ spin system. Some examples are o-disubstituted benzenes (see Figure 5.8), p-disubstituted benzenes with unlike substituents, 4-substituted pyridines, thiophene, furan (Figure 4.2), 1,2-disubstituted ethanes with unlike substituents, and appropriately substituted cyclopropanes.

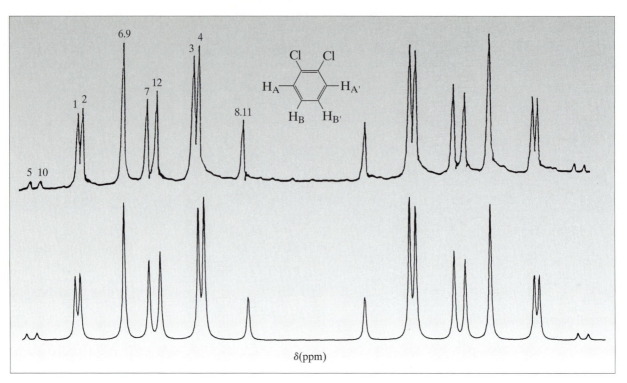

▲ FIGURE 5.8
**Observed (top) and calculated (bottom) spectra for neat o-dichlorobenzene
at 25°C (100 MHz).**

In treating such systems, we may again consider only the A portion, as the complete spectrum is centrosymmetric about $\frac{1}{2}(v_A + v_B)$. However, the A and B portions of the spectrum are not themselves symmetric, as they are in $AA'XX'$ spectra.

Transitions 1 and 2, and 3 and 4 are now not degenerate, so 24 total lines may be observed. Some of the line positions and intensities of the $AA'BB'$ spectrum can be expressed explicitly by algebraic expressions, and some cannot. (Those positions and intensities left blank in Table 5.6 cannot.) Note that none of the lines whose positions can be expressed depend upon K, making K difficult to determine. In addition to the parameters K, L, M, and N, we define the angles ϕ, ψ_+, and ψ_- by means of the following equations:

$$\cos 2\phi: \sin 2\phi: 1 = (v_A - v_B): N: [(v_A - v_B)^2 + N^2]^{\frac{1}{2}};$$
$$\cos 2\psi_+: \sin 2\psi_-: 1 = [(v_A - v_B) + M]: L: \{[(v_A - v_B) + M]^2 + L^2\}^{\frac{1}{2}};$$
$$\cos 2\psi_-: \sin 2\psi_-: 1 = [(v_A - v_B) - M]: L: \{[(v_A - v_B) - M]^2 - L^2\}^{\frac{1}{2}}.$$

The separation of lines 1 and 3 gives N, and from the sum of their spacings from the center of the spectrum, the chemical shifts v_A and v_B can be obtained. Analysis of the spectrum of o-dichlorobenzene gives the following results:

$$v_A - v_B = 25.0 \text{ Hz}; \qquad J_{AA'} = 0.3 \text{ Hz}; \qquad K = 7.9 \text{ Hz};$$
$$v_A = 7.23 \text{ ppm}; \qquad J_{BB'} = 7.5 \text{ Hz}; \qquad M = -7.2 \text{ Hz};$$
$$v_B = 6.98 \text{ ppm}; \qquad J_{AB} = 8.15 \text{ Hz}; \qquad N = 9.7 \text{ Hz};$$
$$J_{AB'} = 1.55 \text{ Hz}; \qquad L = 6.6 \text{ Hz}.$$

Table 5.6   Transitions of $AA'BB' = [AB]_2$ System

Line	Line position[a]	Relative intensity
1	$\dfrac{1}{2}N + \dfrac{1}{2}[(\nu_A - \nu_B)^2 + N^2]^{\frac{1}{2}}$	$1 - \sin 2\phi$
2	—	—
3	$-\dfrac{1}{2}N + \dfrac{1}{2}[(\nu_A - \nu_B)^2 + N^2]^{\frac{1}{2}}$	$1 + \sin 2\phi$
4	—	—
5	—	—
6	—	—
7	—	—
8	—	—
9	$\dfrac{1}{2}\left\{[(\nu_A - \nu_B) + M]^2 + \dfrac{1}{2}(M^2 + L^2)\right\}^{\frac{1}{2}}$	$\sin^2(\theta_a - \psi_+)$
10	$\dfrac{1}{2}\left\{[(\nu_A - \nu_B) - M]^2 + \dfrac{1}{2}(M^2 + L^2)\right\}^{\frac{1}{2}}$	$\cos^2(\theta_a + \psi_-)$
11	$\dfrac{1}{2}\left\{[(\nu_A - \nu_B) + M]^2 - \dfrac{1}{2}(M^2 + L^2)\right\}^{\frac{1}{2}}$	$\cos^2(\theta_a - \psi_+)$
12	$\dfrac{1}{2}\left\{[(\nu_A - \nu_B) - M]^2 - \dfrac{1}{2}(M^2 + L^2)\right\}^{\frac{1}{2}}$	$\sin^2(\theta_a - \psi_-)$

[a] Relative to $\frac{1}{2}(\nu_A + \nu_B)$, i.e., center of spectrum.

To obtain these values requires computer simulation. Further details of spin-system analysis may be found in R. J. Abraham, *The Analysis of High Resolution NMR Spectra*, Elsevier, Amsterdam, 1971.

Additional Reading

1. POPLE, J. A., SCHNEIDER, W. G., and BERNSTEIN, H. J., *High Resolution Nuclear Magnetic Resonance*, McGraw-Hill, New York, 1959.

2. ABRAHAM, R. J., *Analysis of High Resolution NMR Spectra*, Elsevier, Amsterdam, 1971.

3. CORIO, P. L., *Structure of High-Resolution NMR Spectra*, Academic, New York, 1966.

4. EMSLEY, J. W., FEENEY, J., and SUTCLIFFE, L. H., *High Resolution Nuclear Magnetic Resonance Spectroscopy*, Pegamon, Oxford, 1965.

5. Several computer programs for NMR spin-system analysis are available, and most commercial instruments have one as part of their software. (See DIEHL, P., KELLERHALS, H., and LUSTIG, E. "Computer Assistance in the Analysis of High-Resolution NMR Spectra," in DIEHL, P., FLUCK, E., and KOSFELD, R., eds., *NMR: Basic Principles and Progress*, Vol. 6, Springer-Verlag, Berlin, 1972.

$\emptyset \sim 60°$
$^3J_{HH}$ 1 to 3 Hz

$\emptyset \sim 0°$
$^3J_{HH}$ 6 to 12 Hz

Scalar couplings have a geometric dependence as exemplified by the vicinal $^3J_{HH}$ couplings.

Typical Magnitudes of Selected Coupling Constants

6.1   General Considerations

For a stationary molecule in the solid state, the magnetic moments of the nuclei interact directly by a dipolar interaction

$$D_{ij} = -\frac{\hbar \gamma_i \gamma_j}{4\pi} \left\langle \frac{(3\cos^2 \phi_{ij} - 1)}{r_{ij}^3} \right\rangle \tag{6.1}$$

that can be very large. For example, for two protons separated by 2 Å, D_{ij} can be as much as 30,000 Hz. Since D_{ij} is proportional to the product of the gyromagnetic ratios of the interacting nuclei, D_{ij} for all other pairs of interacting nuclei will be smaller than it is for two protons. (See Tables 1.1 and 1.2.) D_{ij} also decreases rapidly with increasing separation between the two interacting nuclei. This dipolar interaction gives rise to very large line widths for a proton spectrum of a solid and obscures any fine structure due to chemical shifts. In liquids and gases, rapid molecular motion averages the $3\cos^2 \phi_{ij} - 1$ term to zero, the dipolar coupling effectively vanishes, and, as a consequence, one observes very narrow resonances for solutions and gases.

The scalar couplings observed in high-resolution spectra of liquids and gases are transmitted via bonding electrons by one of three mechanisms: (1) the interaction of nuclear moments with the magnetic fields produced by the orbiting electrons, called an *orbital term*, (2) the dipolar interaction of the nuclear and electronic magnetic moments, called a *dipolar term*, and (3) the interaction of the nuclear spin with the electron spin of electrons in *s*-orbitals, called the *Fermi-contact interaction*.

The Fermi-contact interaction depends upon the gyromagnetic ratios of the interacting nuclei (γ_A and γ_B), the fractional *s* character of the two atoms participating in the AB bond (α_A^2 and α_B^2), the valence *s*-electron densities at the two nuclei ($S_A^2(0)$ and $S_B^2(0)$), and the mean triplet excitation energy ($^3\Delta E$). It can be rationalized qualitatively using the simplified equation

$$J_{AB} \propto \frac{\gamma_A \gamma_B S_A^2(0) S_B^2(0) \alpha_A^2 \alpha_B^2}{^3\Delta E}. \tag{6.2}$$

All other factors being constant, $^1J_{AB}$ should increase as the product $\gamma_A \gamma_B$ increases. As an example, for the homologous series of $M(CH_3)_4$ compounds, one observes that $|^1J_{MC}|$ generally increases as $|\gamma_M|$ increases: $^{29}Si(-50$ Hz$)$, $^{73}Ge(18.7$ Hz$)$, $^{119}Sn(340$ Hz$)$, and $^{207}Pb(250$ Hz$)$. As shown in Figure 6.1, the

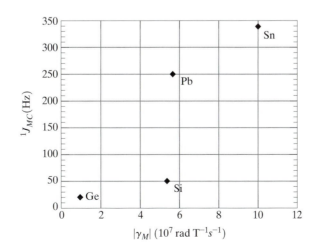

◀ FIGURE 6.1
Relation of $^1J_{MC}$ to γ_M for $M(CH_3)_4$ compounds.

relation is not linear, which suggests that other factors in addition to γ_M contribute to $^1J_{MC}$ for this series of compounds.

For all couplings involving hydrogen, the Fermi-contact term is dominant and the other terms may be neglected. For this reason, there is often considerable similarity between $H \cdots H$ and analogous $H \cdots X$ couplings. The magnitude of the contact term is proportional to, among other factors, the percentage s character at the coupling nucleus, and there are many correlations of $H—X$ couplings with the percentage s character in the HX bond.

For couplings between heavier elements, the other terms may increase in importance, and when this occurs, these couplings obey completely different rules from those obeyed by the corresponding $H \cdots H$ and $H \cdots X$ couplings. This occurs to some extent for $C \cdots C$, $C \cdots F$, and $F \cdots F$ couplings, for which the orbital and dipolar contributions can be as large as the contact contribution.

Coupling constants are proportional to the product of the gyromagnetic ratios of the interacting nuclei. For comparison purposes, when different nuclei are involved, it is convenient to define a reduced coupling constant K by the equation*

$$K_{AB} = \frac{(4\pi^2 J_{AB})}{h\gamma_A\gamma_B}. \tag{6.3}$$

The units of K are $NA^{-2}\,m^{-3}$(SI) or cm^{-3} (c.g.s.) and have typical values of 10^{20} to $10^{23}\,NA^{-2}\,m^{-3}$ ($1\,NA^{-2}\,m^{-3} = 10\,cm^{-3}$; 1 tesla $= 1\,NA^{-1}\,m^{-1}$; $1\,J = 1\,Nm$).

SAMPLE PROBLEM 6.1

If $^1J_{CH}$ for methane is 125 Hz, what is $^1K_{CH}$?

Solution

$^1K_{CH}$

$$= \frac{4(3.14\ \text{rad})^2(125\ s^{-1})}{(6.63\times10^{-34}\,\text{Nms})(6.7283\times10^7\,\text{rad AmN}^{-1}s^{-1})(26.7519\times10^7\,\text{rad AmN}^{-1}s^{-1})}$$

$$= 4.131 \times 10^{20}\,NA^{-2}\,m^{-3}$$

SAMPLE PROBLEM 6.2

If $^1J_{CH}$ in $CHCl_3$ is 209 Hz, predict the magnitude of $^1J_{CD}$ in $CDCl_3$.

Solution

From Equation 6.3 we have

$$J_{AB} = h\gamma_A\gamma_B\frac{K_{AB}}{4\pi^2}$$

and

$$^1J_{CD} = \frac{(\gamma_D K_{CD})}{(\gamma_H K_{CH})J_{CH}}.$$

---

*More thorough discussions of the theory of couplings may be found in Harris, R. K., Mann, B. E., eds., *NMR and the Periodic Table*, Academic Press, New York **1978**; and Mason, J., ed. *Multinuclear NMR*, Plenum, New York, **1987**.

If we assume that $K_{CH} = K_{CD}$, then

$$^1J_{CD} = \frac{^1J_{CH}\,\gamma_D}{\gamma_H} = (209 \text{ Hz})\left(\frac{4.1064}{25.7510}\right) = 32.08 \text{ Hz}.$$

The experimental value is 32.23 Hz.

The mechanism for Fermi-contact coupling can be visualized as follows:

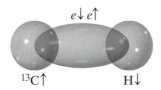

The ^{13}C nuclear moment interacts with the carbon 2s electrons through coupled orientation of nuclear and electron spins. The electron spins in the C—H bond interact in an antiparallel manner by virtue of the Pauli exclusion principle. The interaction transmits a preferred orientation to the hydrogen nucleus for its nuclear spin. This favored orientation of opposed carbon and proton spins results in a positive coupling constant.

The coupling constant is assigned a positive sign if its effect is to stabilize the state in which the nuclear spins are antiparallel, as in the case of $^1J_{CH}$. We therefore expect that, for nuclei with $\gamma > 0$, ^1J should be positive, and this has been experimentally confirmed in most cases.

For coupling through two bonds, there is an additional atom between the coupled nuclei, as in a CH_2 group. In this group, the bonding electrons responsible for the coupling belong to two different carbon orbitals as shown in the accompanying figure. According to Hund's rule, the energetically preferred state is that in which the electron spins of the two bonding electrons of carbon are parallel. In the figure, this gives parallel spins for the two hydrogens, as the lower energy state and $^2J_{HH}$ should be negative. That the two bond coupling is negative has been experimentally verified. Often, coupling constants alternate in sign as a function of the number of intervening bonds, with $^nJ_{XH}$ being positive for n odd and negative for n even. Because these couplings are transmitted via the bonding electrons, they also attenuate rapidly with an increasing number of bonds between the coupled nuclei, as is exemplified by the data in Table 6.1 for typical $^nJ_{HH}$.

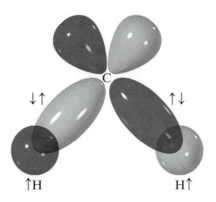

Table 6.1   Typical Proton–Proton Coupling Constants (Hz)

System	Coupling range	System	Coupling range

OPEN CHAIN

$^2J_{HH}$

$\overset{|}{\underset{|}{C}}\overset{H}{\underset{H}{}}$: −10 to 18

$-N=C\overset{H}{\underset{H}{}}$: 8 to 16

$^3J_{HH}$

$CH-CH$, $CH-CH=C$: 5 to 10

$CH_3.CH_2-$: 7 to 8

$C=C$ (H,H,H) : $^2J_{gem}$ −3 to +7 ; $^3J_{cis}$ 3 to 18 ; $^3J_{trans}$ 12 to 24

$CH-CHO$: 1 to 3

$C=CH-CHO$: 7 to 9

$CH-SH$: 6 to 8

$CH-OH$, $CH-NH-$: 4 to 8   } $^3J_{HH}$

$CH-C-CH$: 0 to 1

$CH-C=CH-$: 0 to −2

$CH-C\equiv CH$: −2 to −3   } $^4J_{HH}$

$CH-C=C-CH$: 0 to 2

$CH-C\equiv C-CH$: 2 to 3   } $^5J_{HH}$

CYCLIC

Arenes & alkenes	*ortho*	*meta*	*para*
Benzene derivatives	5–9	2–3	0–1

	J_{23}	J_{34}	J_{24}	J_{25}
Furan	1.8	3.5	0.8	1.6
Pyrrole	2.6	3.4	1.4	2.1
Thiophene	5.2	3.6	1.3	2.7
Cyclopentadiene	5.1	1.9	1.1	1.9

$\overset{H}{}\overset{H}{}C=C\ (CH_2)_{n-2}$

	$n=3$	4	5	6	7	8	9
	ca 1	2.7	5.1	8.8	10.8	10.3	10.7

Saturated	$^2J_{gem}$	$^3J_{cis}$	$^3J_{trans}$
Cyclopropane	−4.5	9.2	5.4
Ethylene oxide	+5.5	4.5	3.2
Ethylene imine	+2.0	6.3	3.8
Ethylene sulphide	ca 0	7.1	5.6
Cyclobutane derivatives	−11 to −15	6 to 11	3 to 9
Cyclopentane derivatives	−11 to −17	7 to 11	2 to 8
Cyclohexane derivatives	−12 to −15	J_{ae} 2 to 5	J_{aa} 8 to 13 ; J_{ee} 1 to 4

Couplings over more than three bonds are usually quite small (<3 Hz) and are often not resolved. The division into couplings over one, two, and three bonds (i.e., ^1J, ^2J, and ^3J couplings) is also a convenience for discussion.

6.2   Vicinal ($^3J_{HH}$) Couplings

The vicinal ($^3J_{HH}$) coupling is one of the more useful and best understood couplings. Some typical examples are given in Tables 6.1 and 6.2. Our understanding stems largely from the theoretical work of Karplus,* based upon the contact term mentioned previously. His theory led to three main conclusions:

1. The couplings are transmitted largely by sigma electrons and are always positive for both saturated and unsaturated systems.
2. *Trans* coupling is larger than *cis* coupling in alkenes.
3. For the saturated H—C—C—H fragment, the coupling is proportional to cos ϕ and cos 2ϕ, where ϕ is the dihedral angle between the two vicinal CH bonds.

We shall first consider alkene and arene couplings.

Substituents have a large effect on the actual magnitude of $^3J_{HH}$ in alkenes, but $^3J_{HH}$ (*trans*) is always greater than $^3J_{HH}$ (*cis*) in any given molecule. The substituent effect is simply related to the electronegativity (E_X) of the attached substituent by the equations

$$^3J_{HH}(trans) = 19.0 - 3.2\,(E_X - E_H),$$

$$^3J_{HH}(cis) = 11.7 - 4.1\,(E_X - E_H),$$

Table 6.2   Selected Values of Vicinal Couplings (Hz) in Alkene and Arene Systems

Molecule	Coupling constant		Molecule	Coupling constant
	J_{cis}	J_{trans}		
$CH_2{:}CH_2$	11.7	19.0	CH:CH	9.5
$CH_2{:}CHOR$	6.7	14.2	C_6H_6	7.6
cis- and *trans*-CHCl:CHCl	5.3	12.1		J_{12} 8.6 J_{23} 6.0
	J_{23}	5.5		10–12
	J_{34}	7.5		

---

*Karplus, M., *J. Chem. Phys.* **1959**, *30*, 11. Karplus, M., *J. Am. Chem. Soc.*, **1963**, *85*, 2870. Karplus, M., *J. Chem. Phys.*, **1960**, *33*, 1842.

and

$$^2J_{HH}(gem) = 8.7 - 2.9\,(E_X - E_H)$$

where E_H is the electronegativity of hydrogen (2.2) and the constants are those observed for ethylene. The couplings in a monosubstituted ethylene may be predicted from these equations and the E_X values (Table 6.3) quite well. By assuming an additivity relationship, the couplings in disubstituted alkenes may also be estimated, though, particularly for *cis* couplings, not as well.

Many of the preceding factors also influence the magnitude of vicinal $^3J_{HH}$ couplings in a saturated HCCH fragment, but, in addition, there is the very important dihedral-angle dependence. Further examples are given in Table 6.4.

For a freely rotating fragment (e.g., CH_3CHXY), a simple dependence on electronegativity is found. If $^3J_{HH}$ (avg.) is the average over all conformers of the $HCX^1X^2CHX^3X^4$ fragment, then

$$^3J_{HH}(\text{avg.}) = 8.0 - 0.80 \sum_{n=1}^{4}(E_X - E_H),$$

where the summation refers to the groups X^1 to X^4 and the constant is the value for ethane.

The dihedral-angle dependence is expressed by the equation

$$^3J_{HH} = 7.0 - 1.0 \cos\theta + 5.0 \cos 2\theta,$$

which is shown graphically in Figure 6.2.

The predictions of the Karplus relationship are best exemplified by the observed couplings in six-membered rings of fixed geometry. Thus, when $\phi = 180°$, as in a *trans*-diaxial moiety, $^3J_{aa}$ should be about 10 Hz. When $\phi = 60°$, as in an equatorial–equatorial or axial–equatorial moiety, $^3J_{ae} = {}^3J_{ee} = 2.5$ Hz. Note the important result that $^3J_{HH} = 0$ when $\phi = 90°$. These predictions are generally borne out in experiment, but are modified by substituent effects. Thus, $^3J_{aa}$ typically is found in the 8- to 12-Hz range, $^3J_{ae}$ from 2 to 4 Hz, and $^3J_{ee}$ from 1 to 3 Hz, depending upon the number of electronegative substituents present. Vicinal couplings have recently been more thoroughly discussed.[*]

Table 6.3 Selected Electronegativity Values			
X	**E_X**	**X**	**E_X**
SiMe$_3$	1.9	I	2.65
H	2.2	Br	2.95
Ph$_2$P	2.1		
RS	2.5		
Me		Cl	3.15
CN	2.6	OH	3.50
COCH$_3$			
NH$_2$	3.05	F	3.90

[*] Barfield, M., Smith, W. B., *J. Am. Chem. Soc.* **1992**, *114*, 1574. Sternhell, S., *Quart. Rev.,* **1969**, *23*, 236.

Table 6.4 Selected Examples of Vicinal Couplings across Saturated Bonds

Molecule	Coupling constant	Molecule	Coupling constant	
			J_{gauche}	J_{trans}
CH_3CH_2Li	8.4			
CH_3CH_3	8.0	$CH_3.CH\ Br.CH\ Br.CH_3$	2.9	10.3
$CH_3.CH_2.CH_3$	7.3	$CHCl_2.CHF_2$	1.6	8.4
CH_3CH_2OH	7.0	CH.CH:C	3.5	11.5
$CH_3CH:CH_2$	6.4	CH.CHO	0.1	8.3
$CH_3CH(OH)_2$	5.3			
$CH_3.CHO$	2.9			

(cyclohexyl OAc, J_{aa})	11.4	(cyclohexyl OAc, J_{ae})	2.7		
(J_{ae})	4.2	(J_{ee})	2.7		

(dioxane) J_{aa}	11.5
J_{ae}	2.7
J_{ee}	0.6

(dioxolane) J_{cis}	7.3
J_{trans}	6.0

	J_{aa}	13.2		
(H.OH) (H.OH)	J_{ae}	3.6		
	J_{ee}	3.1		

X	CH_2	CO	O	S
J_{cis}	7.4	7.2	10.7	10.0
J_{trans}	4.6	2.2	8.3	7.5

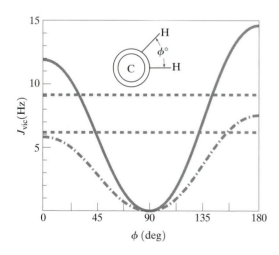

◀ FIGURE 6.2
Karplus curves relating the dihedral angle in an HC—CH fragment and the vicinal proton–proton coupling constant. Two curves are shown relating to differently substituted fragments and are differentiated by the heaviness of the line. The dotted lines indicate the typical range of values obtained when a group can rotate freely, giving rise to an averaged $^3J_{vic}$. The inset shows a view along the carbon–carbon bond. (After Jackman, L. M. and Sternhell, S., *Application of Nuclear Magnetic Resonance Spectroscopy in Organic Chemistry*, Pergamon, Oxford, U.K., 1969.)

6.3 Geminal ($^2J_{HH}$) Couplings

Selected values of $^2J_{HH}$ are given in Table 6.5.

Whereas $^3J_{HH}$ is normally positive, $^2J_{HH}$ may be either positive or negative and spans the range from -20 to $+40$ Hz. Molecular orbital theory predicts the following for $^2J_{HH}$:

a. Increasing the HCH angle increases the s character of the orbitals and makes the coupling more positive.

b. For both sp^2 and sp^3 CH_2 groups, substituting an electronegative atom on carbon makes $^2J_{HH}$ more positive (an inductive effect consistent with Bent's rules).*

c. Substituting an electronegative atom in a β position makes $^2J_{HH}$ more negative.

Table 6.5 Selected Examples of Geminal Coupling Constants $^2J_{HH}$ (Hz)

Molecule	Coupling constant	Molecule	Coupling constant
CH_4	−12.4	$CH_2\!:\!C\!:\!C\!\!<$	−9
$CH_3.CCl_3$	−13.0	$CH_2(CN)_2$	−20.3
$CH_3.CO.CH_3$	−14.9	$CH_2{=}O$	+41
$CH_3.C_6H_5$	−14.4	$CH_2\!:\!CH_2$	+2.5
CH_3Cl	−10.8	$CH_2\!:\!CHF$	−3.2
CH_3OH	−10.8	$CH_2\!:\!CHLi$	+7.1
CH_3F	−9.6	$BrCH_2.CH_2OH$	−10.4
		$BrCH_2.CH_2OH$	−12.2
		$BrCH_2.CH_2CN$	−17.5

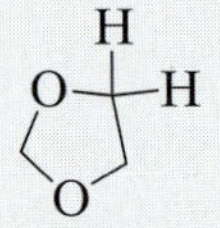

−8.3

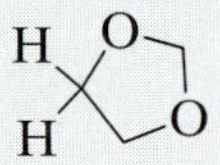

−0

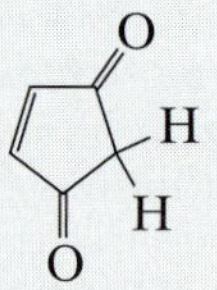

−21.5

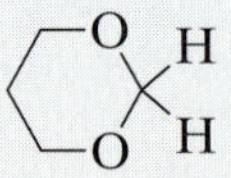

−5-6

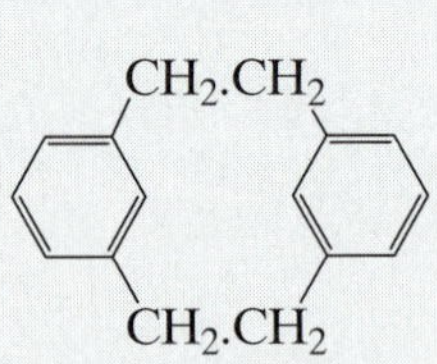

−12.0

Metacyclophane

*Bent, H. A., *J. Chem. Educ.*, **1960**, *37*, 616; *Chem. Rev.*, **1961**, *61*, 275.

d. Introducing a substituent that withdraws electrons from antibonding orbitals (i.e., hyperconjugative effects) makes $^2J_{HH}$ more negative.

The observed trends for $^2J_{HH}$ in alkanes, cycloalkanes, and alkenes are consistent with interaction (a). Thus, for CH_4($^2J_{HH}$ = −12.4 Hz), cyclopropane ($^2J_{HH}$ = −4.5 Hz), and $H_2C=CH_2$($^2J_{HH}$ = +2.5 Hz), one observes that $^2J_{HH}$ becomes more positive as the HCH angle increases.

Electronegative substituents affect $^2J_{HH}$ for both sp^2 and sp^3 carbons (e.g., C_2H_4, +2.5; $CH_2 = O$, +41; CH_4, −12.4; CH_3OH, −10.8; CH_3Cl, −10.8; CH_3F, −9.6) in a manner consistent with the predictions of point (b).*

6.4   Long-range H—H Couplings

Several different mechanisms are involved in H—H couplings over four or more bonds. Table 6.6 gives some selected values. One well-understood mechanism is the σ–π interaction in allylic (CH=C—CH) $^4J_{HH}$ and homoallylic (CH—C=C—C—H) $^5J_{HH}$ couplings. There is a certain amount of overlap, or mixing, of the π orbital of the double bond and the hydrogen 1s orbital in CH= and CHC= systems, as shown schematically in the following figure:

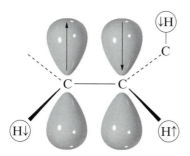

The olefinic-proton and adjacent π-electron spins are opposed, whereas the allylic and nearest π electron prefer parallel spins. As a consequence, this mechanism gives a positive contribution to $^3J_{CH=CH}$ couplings, a negative contribution to allylic $^4J_{HH}$, and a positive contribution to $^5J_{HH}$ homoallylic couplings. For freely rotating methyl groups, it happens that the interaction of the π electron and methyl group is about the same magnitude as, but of opposite sign to, the π-electron–olefinic-proton interaction. This results in the allylic CH_3—C=C—H and homoallylic CH_3—C=C—CH_3 couplings having similar magnitudes, but opposite signs.

Furthermore, since the interaction in an HC—C=C fragment is proportional to the H(1s), π-orbital overlap, the coupling is proportional to $\cos^2 \phi$, where ϕ is the dihedral angle between the CH bond and the π orbital. This results in a $\cos^2 \phi$ dependence for the allylic and a $\cos^2 \phi_1 \cdot \cos^2 \phi_2$ dependence for the homoallylic couplings, respectively, which implies that the contribution from this mechanism will be zero when $\phi = 90°$.

---

*Bothner-By, A. A. in *Advances in Magnetic Resonance*, Waugh, J. S. and Warren, W. S., eds., Academic, New York, Vol. 1, 1965.

Table 6.6 Selected Examples of Long-Range HH Couplings

Molecule	Coupling (Hz)	Molecule	Coupling (Hz)
CH_2:$CH.CH_3$	$^4J_{cis}$ −1.7 $^4J_{trans}$ −1.3		1–2
$CH_3.CH$:$C(CH_3).CO_2H$	$^5J_{cis}$ 1.2 $^5J_{trans}$ 1.5		1–1.5
HC:$C.CH_2.CH_3$ $CH_3.C$:$C.CH_2CH_3$ $CH_3.C$:$C.C$:$C.CH_3$	$^4J.$ −2.4 $^5J.$ +2.6 $^7J.$ 1.3		7–8
	$^4J_{(Me-H_6)}$ −0.63 $^5J_{(Me-H_3)}$ +0.40 $^6J_{(Me-H_4)}$ −0.58		$^4J_{(Me-H)_g}$ 0− −0.3 $^4J_{(Me-H)_t}$ + 0.4− +1.0
	$^4J_{(Me-H_3)}$ −1.6 $^5J_{(Me-H_4)}$ 2.7		$^5J_{(CHO-5)}.$ 0.7
	$^5J_{35}.$ +0.32 $^7J_{27}.$ +0.34 $^4J_{45}.$ −0.31 $^5J_{48}.$ +0.65		$^5J_{(CHO-3)}.$ 0.6
	$^5J_{48}.$ 0.8		

In saturated systems, $^4J_{HH}$ couplings are very small (approximately zero), except for specific orientations in which they may be appreciable.* The $^4J_{HH}$ coupling is largest (1 to 2 Hz) in nonstrained systems if the bonds form a W orientation, as, for example, the equatorial–equatorial coupling

* Sternhell, S., *Rev. Pure Appl. Chem.*, **1964**, *14*, 15. Barfield, M., Chakrabarti, B., *Chem. Rev.* **1969**, *69*, 757.

in cyclohexanes, as shown in the accompanying figure. In rigid bicyclic systems, this coupling can be quite large (7 to 8 Hz), as exemplified by the indicated coupling constants for the following two isomers:

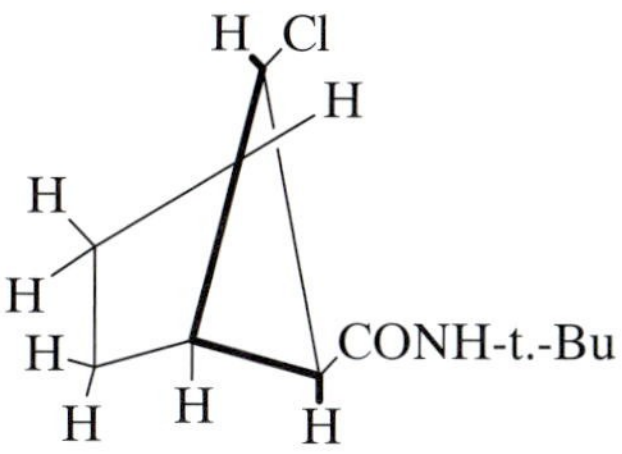

$^{4}J_{HH} = 7$ Hz $\qquad\qquad$ $^{4}J_{HH} = 0$

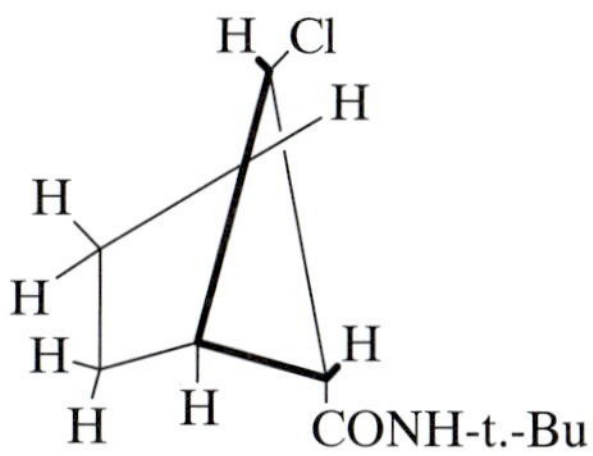

6.5 $^{1}J_{CH}$ Couplings

Directly bonded $^{1}J_{CH}$ couplings have been studied extensively and are of considerable theoretical interest.* For simple hydrocarbons, these couplings are directly proportional to the fractional carbon s character in the C—H bond (ρ) and range from 120 to 270 Hz:

$$^{1}J_{CH} = 500\ \rho. \tag{6.4}$$

This equation predicts $^{1}J_{CH}$'s for $CH_4(sp^3)$, $H_2C{=}CH_2(sp^2)$ and $H{-}C{\equiv}C{-}H$ (sp) of 125, 167, and 250 Hz, respectively. The observed couplings of 125, 156, and 248 Hz agree very well with these predictions and have led to the use of this coupling to investigate the hybridization of atoms in molecules for which the bonding is uncertain. It is strictly valid only in the absence of strongly perturbing substituents. For substituted methanes, $^{1}J_{CH}$ changes considerably with the introduction of substituents (Table 6.7).

The substituent effects are largely additive, and they may be used to estimate $^{1}J_{CH}$ in multiply substituted compounds. For example, the coupling in 1,1,2,2-tetrabromoethane may be estimated from the data given in Table 6.7 to be 179 Hz, in close agreement with the observed value of 181 Hz. The coupling constants are essentially the same for cycloaliphatic systems, such as cyclohexane, or heteroalicylics, such as dioxane, as they are for comparable open-chain compounds. Ring strain increases $^{1}J_{CH}$ considerably. For example, $^{1}J_{CH}$ is 161, 136, 131, and 127 Hz for cyclopropane, cyclobutane, cyclopentane, and cyclohexane, respectively. For disubstituted alkenes, $^{1}J_{CH}$ is larger for the *cis* isomer than for the *trans* isomer.

For halo- and pseudohalobenzenes, $^{1}J_{CH}$ decreases slightly with decreasing substituent electronegativity and distance from the substituent.

---

* See also Marshall, J. L., *Carbon–Carbon and Carbon–Proton NMR Couplings: Applications to Organic Stereochemistry and Conformational Analysis*, VCH, Deerfield Beach, FL, 1983.

Table 6.7 Some Characteristic $^1J_{CH}$ Couplings

Molecule	Coupling (Hz)	Molecule	Coupling (Hz)
C_2H_6	125	CH_4	125
Cyclohexane	123	$(CH_3)_4Si$	118
C_2H_4	156	$CH_3.CO.CH_3$	127
Cyclopropane	162	$CH_3.CO_2H$	130
Benzene	159	$C\underline{H}_3.C\vdots CH$	132
Acetylene	248	CH_3NH_2	133
$CH_3C\underline{H}O$	173	$CH_3.CN$	136
$CCl_3C\underline{H}O$	207	$CH_3.OH$	141
$H.C\underline{O}_2H$	222	$CH_3.\overset{\oplus}{N}H_3$	145
$Me_2N.C\underline{H}O$	191	$CH_3.NO_2$	147
$Me_2\overset{\oplus}{C}\underline{H}{}^{\ominus}SbF_5Cl$	168	CH_3F	149
$CH_2\vdots CHX$		CH_3Cl	150
		CH_3Br	152
		CH_3I	151
		$CH_2(OEt)_2$	161
		CH_2Cl_2	178
		$CHCl_3$	209

X	α	cis	trans
F	200	159	162
Cl	195	163	161
CHO	162	157	162
CN	177	163	165

	C_2	C_3	C_4
(pyridine)	170	163	152

	C_2	C_3
X=O	201	175
NH	184	170
S	185	167
CH_2	170	170

	C_2	
(pyranone)	C_2	200
	C_3	169

(fluorobenzene)	C_2	155
	C_3	163
	C_4	161

6.6 $^2J_{CH}$ and $^3J_{CH}$ Couplings

The patterns for $^2J_{CH}$ and $^3J_{CH}$ couplings are similar to those of the analogous $^2J_{HH}$ and $^3J_{HH}$ couplings. The ratio of the gyromagnetic ratios for carbon and hydrogen suggests that, for the same electronic transmission, $^nJ_{CH} \sim 0.25\,^nJ_{HH}$, but because of the increased electron density around the carbon nucleus, the observed couplings are somewhat larger than this. Table 6.8 gives some selected examples. A useful general rule is that $^nJ_{CH} \cong (0.6 \text{ to } 0.7)\,^nJ_{HH}$.

As with $^2J_{HH}$, $^2J_{CH}$ becomes more positive as the CCH angle increases, with values of -4 to -6 Hz for 109.5° angles and about $+3$ Hz for 120° angles.

Table 6.8 Selected Examples of $^2J_{CH}$ and $^3J_{CH}$ Couplings

	*CR=	CH₃	CH₂I	CCl₃	CHO	CO₂Me	PH	CN	C⫶CH
CH₃.*CR	^2J	−4.5	−5.0	5.9	−6.6	6.9	6.0	9.9	−10.6
(CH₃)₃C.*CR	^3J	4.65	5.99	–	4.60	4.11	–	5.38	–

CH₃.*CH₃	^2J	−4.5	
CH₂⫶*CH₂		−2.4	
CH⫶*CH		+49.3	
Cyclopropane		−2.55	
H*C⫶C.CH₃	^3J	4.8	
H₃*C.C⫶CH	^3J	3.6	

(benzene, * marked)

^2J	1.0
^3J	7.4
^4J	−1.1

(*CO₂H benzene)

^3J	+4.1
^4J	+1.1
^5J	+0.5

HCR₂.*CH₃	CR₂=	CH₂	CHCl	CCl₂	CO	C⫶CCl₂
	^2J	−4.5	−2.6	<1	+26.7	+3.2

(chlorinated bicyclic structure, *CO₂H)

^2J	−6.35
$^3J_{cis}$	+4.54
$^3J_{trans}$	+2.51

(alkene: $R, H_a, C_1=C_2, H_c, H_b$)

R	CO₂H	Cl	OAc
$^2J_{cis}(C_1H_a)$	−4.55	−8.3	−7.9
$^2J_{trans}(C_1H_b)$	+1.55	+7.1	+7.6
$^2J(C_2H_c)$	−0.6	+6.8	+9.7

(acrylic acid: HO₂*C, H, C₁=C, H, H)

^2J	4.1
$^3J_{cis}$	7.6
$^3J_{trans}$	14.1

(2-bromothiophene, positions 5 ¹S 2-Br, 4 3)

$^2J_{34}$	5.6
$^2J_{43}$	4.2
$^2J_{45}$	3.4
$^2J_{54}$	5.6
$^3J_{35}$	5.6
$^3J_{53}$	11.0

(sugar / pyranose structure)

$^3J_{15}.^3J_{64}.$ 2–3
$^3J_{13}.^3J_{31ax}.^3J_{51ax}$ < 1
$^3J_{3(5).eq},$ 5–6

$^{\ominus}O_2$*C.CH($\overset{\oplus}{N}H_3$).CH₂R
R=CO₂H.CH₂OH

^3J_g 0.4(±0.5)
^3J_t 11.9(±1.5)

(pyrrole, H, 5 N 2, 1, 4 3)

$^2J_{23}$	7.4
$^2J_{32}$	4.7
$^2J_{34}$	5.9
$^3J_{24}$	10.0
$^3J_{25}$	5.2
$^3J_{35}$	9.5

For disubstituted alkenes, $^2J_{CH}$ is larger for the *cis* isomer than for the *trans* isomer. For monosubstituted benzenes, $^2J_{CH}$ becomes more positive as the electronegativity of the substituent decreases. Particularly large values of $^2J_{CH}$ (24 to 27 Hz) are observed for aldehydes.

▶ FIGURE 6.3
Karplus–Conroy Relation for
$^3J_{CH}$.

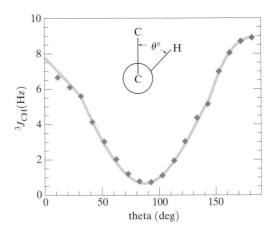

For aliphatic and cycloaliphatic compounds, $^3J_{CH}$ is related to the C—C, C—H dihedral angle θ and follows a Karplus–Conroy relation (Equation 6.4) analogous to the dihedral-angle dependence of $^3J_{HH}$:[*]

$$^3J_{CH} = 4.26 - \cos\theta + 3.56\cos 2\theta. \tag{6.5}$$

This relation is shown graphically in Figure 6.3.

Thus, from the relation $^3J_{CH} = (2\,^3J_{CH\,syn} + {}^3J_{CH\,anti})/3$, an average value of $^3J_{CH}$ of 4 to 4.5 Hz is expected for alkyl groups with free rotation. For alkyl-substituted alkenes, it is generally found that $^3J_{CH}$ is greater to *trans* than to *cis* alkyl substituents. The magnitude of $^3J_{CH}$ increases with increasing *s* character of the coupling carbon atom, as shown in the series consisting of propene, 3-methylenecyclohexene, and 2-methyl-1-buten-3-yne:

7.5 Hz	7.8 Hz	8.1 Hz
12.6 Hz	11.9 Hz	14.7 Hz
propene	3-methylenecyclohexene	2-methyl-1-buten-3-yne

Electronegative substituents within the coupling pathway decrease $^3J_{CH}$, while those at the coupling carbon atoms increase $^3J_{CH}$, with *cis* and *trans* protons for appropriately substituted alkenes. This effect is illustrated by the couplings observed for 2- and 3-bromopropene:

4.6 Hz	9.1 Hz
8.9 Hz	15.5 Hz
2-bromopropene	3-bromopropene

The ^{13}C NMR spectra of benzene derivatives have been studied extensively.[†] Two general trends are evident in the data shown in Table 6.9:

---

[*]Wasylishen, R., Schäfer, T., *Can. J. Chem.*, **1972**, *50*, 3686; **1973**, *51*, 961.

[†]Ernst, L., Wray, V., Chertkow, V. A., Sergeyew, N. M., *J. Magn. Reson.* **1977**, *25*, 123.

Table 6.9 $^nJ_{CH}$ for (a) Benzene, (b) Pyridine, (c) Pyrimidine, and (d) Monosubstituted Benzenes

(a)

$^1J_{CH} = 158-165$

$^2J_{CH_o} = 0-5$

$^3J_{CH_m} = 6-11$

$^4J_{CH_p} = 0.5-2$

(d)

(b)

$J_{C-2-2-H} = 172$

$J_{C-3-3-H} = 163$

$J_{C-4-4-H} = 160$

$^3J_{C-2-3-H} = 2$

$^3J_{C-2-4-H} = 7$

$^3J_{C-2-6-H} = 12$

$^2J_{C-3-2-H} = 7$

$^3J_{C-3-4-H} = 2$

$^3J_{C-3-5-H} = 6$

$^3J_{C-4-2-H} = 4$

(c)

$J_{C-2-2-H} = 203$

$J_{C-4-4-H} = 181.5$

$J_{C-5-5-H} = 168$

$^3J_{C-2-4-H} = 10.4$

$^2J_{C-4-5-H} = 3$

$^3J_{C-4-2-H} = 9$

$^3J_{C-4-6-H} = 5.3$

$^3J_{C-5-4-H} = 7.8$

$^4J_{C-5-2-H} = 1.6$

	C-1			C-2				
X	$^2J_{12}$	$^3J_{13}$	$^4J_{14}$	$^1J_{CH}$	$^2J_{23}$	$^3J_{24}$	$^3J_{26}$	$^4J_{25}$
F	−4.89	10.95	−1.73	162.55	1.10	8.29	4.11	−1.50
Cl	−3.38	10.93	−1.83	164.92	1.47	8.08	5.06	−1.36
Br	−3.39	11.20	−1.87	165.72	1.46	8.04	5.41	−1.32
I	−2.53	10.79	−1.88	165.62	1.55	7.97	6.07	−1.26
OH	−2.80	9.71	−1.56	158.35	1.20	8.09	4.68	−1.40
OCH$_3$	−2.79	9.22	−1.51	158.52	1.42	8.01	4.80	−1.42
NH$_2$	−0.88	8.62	−1.37	155.71	1.28	7.90	5.38	−1.36
CH$_3$	0.54	7.61	−1.40	155.89	1.19	7.78	6.59	−1.37
Si(CH$_3$)$_3$	4.19	6.34	−1.10	156.14	1.37	7.50	8.63	−1.15
H	1.15	7.62	−1.31	158.43	1.15	7.62	7.62	−1.31
NO$_2$	−3.57	9.67	−1.75	168.12	1.84	8.07	4.45	−1.33
CHO	0.29	7.19	−1.26	160.95	1.40	7.84	6.25	−1.29
CN	0.14	9.01	−1.44	165.45	1.75	7.90	6.13	−1.28

	C-3					C-4		
X	$^1J_{CH}$	$^2J_{32}$	$^2J_{35}$	$^3J_{35}$	$^4J_{36}$	$^1J_{CH}$	$^2J_{43}$	$^3J_{42}$
F	161.14	−0.57	1.74	9.02	−0.76	161.37	0.82	7.57
Cl	161.41	0.27	1.57	8.41	−0.87	161.35	0.90	7.51
Br	161.53	0.35	1.62	8.37	−1.05	161.27	0.89	7.51
I	161.26	0.56	1.52	8.20	−1.17	161.04	0.92	7.45
OH	158.99	−0.30	1.66	8.70	−0.74	160.84	0.77	7.42
OCH$_3$	158.35	−0.32	1.77	8.73	−0.75	160.43	0.85	7.52
NH$_2$	156.92	0.20	1.69	8.35	−0.82	160.39	0.82	7.41
CH$_3$	157.58	1.07	1.35	7.91	−1.06	158.83	1.07	7.54
Si(CH$_3$)$_3$	157.70	1.58	1.14	7.26	−1.35	158.10	1.24	7.45
H	158.43	1.15	1.15	7.62	−1.31	158.43	1.15	7.62
NO$_2$	165.12	−0.28	1.50	8.18	−0.74	162.75	1.25	7.65
CHO	161.92	0.76	1.28	7.58	−1.06	160.43	1.46	7.62
CN	163.84	0.43	1.27	7.55	−1.07	162.42	1.02	7.48

a. For substituted benzene carbon atoms, $^3J_{CH}$ decreases with decreasing electronegativity of the substituent.

b. When the electronegative substituent is located within the coupling pathway, $^3J_{CH}$ increases with decreasing electronegativity of the substituent.

6.7 $^nJ_{CC}$ Couplings

A. $^1J_{CC}$ Couplings

The trends for $^1J_{CC}$ coupling constants parallel those observed for $^1J_{CH}$ coupling constants. The former are generally smaller in magnitude than the latter, because of the smaller magnitude of carbon's magnetic moment. There is considerably less data for C—C couplings than for C—H couplings, since the observation of J_{CC} generally requires ^{13}C-enriched samples. However, inadequate two-dimensional spectroscopy, as described in Chapter 11, can provide values of $^1J_{CC}$ for natural-abundance samples. Some typical values of $^1J_{CC}$ are given in Table 6.10.

As might be expected, the magnitude of $^1J_{CC}$ increases with increasing bond multiplicity and hence increasing carbon s character in the carbon–carbon bond. The relationship is expressed quantitatively as

$$^1J_{CC} = 730\,\rho_{C1}\rho_{C2} - 17 \text{ Hz}, \qquad (6.6)$$

where ρ is the fractional s character in the carbon hybrid orbital.

SAMPLE PROBLEM 6.3

Predict the values of $^1J_{CC}$ for ethane, ethylene, benzene, and acetylene.

Solution

The predictions from Equation 6.6, together with the corresponding experimentally obtained values, are as follows:

Equation	**Experimental**
C_2H_6: $^1J_{CC} = 730\,(0.25)(0.25) - 17 = 28.6$ Hz	34.6 Hz
C_2H_4: $^1J_{CC} = 730\,(0.33)(0.33) - 17 = 62.5$ Hz	67.6 Hz
C_6H_6: $^1J_{CC} = 730\,(0.33)(0.33) - 17 = 62.5$ Hz	57.0 Hz
C_2H_2: $^1J_{CC} = 730\,(0.5)(0.5) - 17 = 165.5$ Hz	171.5 Hz

The agreement between the calculated and experimental results is good, with the exception of C_6H_6. For the four compounds, replacing −17 Hz with −11 Hz would give even better agreement.

Equation 6.6 and the observed ranges of values for $^1J_{CC}$ are compared graphically in Figure 6.4.

It is clear from Sample Problem 6.3 that effects other than hybridization are operative in determining the magnitude of $^1J_{CC}$. Nonetheless, Equation 6.6 is of diagnostic value.

Substituent effects on $^1J_{CC}$ are not large. For example, $^1J_{CC} = 34.6$ Hz for ethane and 37.7 Hz for ethanol. However, if the hybridization of one of the carbon atoms is affected by the substituent, then larger differences are

Table 6.10   Representative Values of $^1J_{CC}$

Class	Compound	Formula	$^1J_{CC}$(Hz)
sp^3–sp^3	Ethane	H_3C-CH_3	34.6
	2-Methylpropane	$H_3C-CH(CH_3)_2$	36.9
	Ethylbenzene	$H_3C-CH_2-C_6H_5$	34.0
	Propionitrile	H_3C-CH_2-CN	33.0
	1-Propanol	$H_3C-CH_2-CH_2OH$	34.2
		$H_3C-CH_2-CH_2OH$	37.8
	Ethanol	HC_3-CH_2OH	37.7
	2-Propanol	$(H_3C)_2CHOH$	38.4
	t-Butylamine	$(H_3C)_3CNH_2$	37.1
	t-Butylalcohol	$(H_3C)_3COH$	39.5
Cyclopropanes	Methylcyclopropane		44.0 (1–α)
	Dicyclopropylketone	$R=CH_3$	54.0 (1–α)
			10.2 (1–2)
	Cyclopropanecarboxylic acid	$R=COOH$	72.5 (1–α)
			10.0 (1–2)
	Cyclopropyl cyanide	$R=CN$	77.9 (1–α)
			10.9 (1–2)
	Cyclopropyl iodide	$R=I$	12.9 (1–2)
	Cyclopropyl bromide	$R=Br$	13.3 (1–2)
	Cyclopropyl chloride	$R=Cl$	13.9 (1–2)
sp^3–sp^2	2-Butanone	$H_3C-\overset{O}{\overset{\|}{C}}-C_2H_5$	38.4
	Acetaldehyde	$H_3C-\overset{O}{\overset{\|}{C}}-H$	39.4
	Acetone	$H_3C-\overset{O}{\overset{\|}{C}}-CH_3$	40.1
	3-Pentanone	$H_3C-CH_2-\overset{}{C}-CH_2-CH_3$ with O below	35.7
		$H_3C-CH_2-\overset{}{C}-CH_2-CH_3$ with O below	39.7
	Acetophenone	$H_3C-\overset{O}{\overset{\|}{C}}-C_6H_5$	43.3
	Acetate anion (aq.)	$H_3C-\overset{O}{\overset{\|}{C}}-O^{\ominus}$	51.6
	N,N-Dimethylacetamide	$H_3C-\overset{O}{\overset{\|}{C}}-N(CH_3)_2$	52.2

For the cyclopropanes, the structure is:

$$\begin{array}{c} H_2C_2 \\ | \quad \rangle CH-R \\ H_2C_1 \end{array}$$

with Dicyclopropylketone having $R = -\overset{O}{\overset{\|}{C}}-\triangleleft$

(continues on next page)

Table 6.10 (*cont.*)

Class	Compound	Formula	$^1J_{CC}$(Hz)
sp^3–sp^2 (*cont.*)	Acetic Acid	$H_3C-\overset{\overset{O}{\|\|}}{C}-OH$	56.7
	Ethyl acetate	$H_3C-\overset{\overset{O}{\|\|}}{C}-OC_2H_5$	58.8
sp^3–sp	*t*-Butyl cyanide	$(H_3C)_3C-C\equiv N$	52.0
	iso-Propyl cyanide	$(H_3C)_2CH-C\equiv N$	54.8
	Propionitrile	$H_3C-CH_2-C\equiv N$	55.2
	Acetonitrile	$H_3C-C\equiv N$	56.5
	Propyne	$H_3C-C\equiv C-H$	67.4
sp^2–sp^2 alkenic	Ethene	$H_2C=CH_2$	67.6
	Acrylic acid	$H_2C=CH-COOH$	70.4
	Acrylonitrile	$H_2C=CH-CN$	70.6
	Styrene	$H_2C=CH-C_6H_5$	70.0 ± 3
sp^2–sp^2 aromatic	Benzene	$X=H$	57.0
	Nitrobenzene	$X=NO_2$	55.4 (1–2)
			56.3 (2–3)
			55.8 (3–4)
	Iodobenzene	$X=I$	60.4 (1–2)
			53.4 (2–3)
			58.0 (3–4)
	Anisole	$X=OCH_3$	58.2 (2–3)
			56.0 (3–4)
	Aniline	$X=NH_2$	61.3 (1–2)
			58.1 (2–3)
			56.6 (3–4)
	Pyridine		53.8 (2–3)
			56.2 (3–4)
	Thiophene	$X=S$	64.2
	Pyrrole	$X=NH$	65.9
	Furan	$X=O$	69.1
sp^2–sp	Benzonitrile	$\langle\ \rangle C-C\equiv N$	80.3
	1,1-Dimethylallene	$(H_3C)_2C=C=CH_2$	99.5
sp–sp	Phenylethynyl cyanide	$C_6H_5-C\equiv C-C\equiv N$	155.8
	Ethyne	$H-C\equiv C-H$	171.5
	Phenylethyne	$C_6H_5-C\equiv C-H$	175.9

Note: Coupling carbons are printed in boldface.

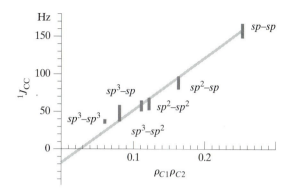

◀ **FIGURE 6.4**
**Equation 6.6 (line) and
observed ranges of $^1J_{CC}$ (Hz)
(bars).**

observed. For example, $^1J_{CC}$ = 34.6 Hz for ethane, 56.7 Hz for acetic acid, and 80.3 Hz in benzonitrile.

The correlation (Equation 6.7) between $^1J_{(C—CH_3)}$ and the corresponding CH couplings is

$$^1J_{(C—CH_3)} = 0.27 \; ^1J_{CH}. \qquad \textbf{(6.7)}$$

B. $^2J_{CC}$ and $^3J_{CC}$ Couplings

The magnitude of coupling constants often attenuates with increasing distance, but $^2J_{CC}$ is generally smaller than $^3J_{CC}$, as in both hexanoic and (E)-2-hexenoic acid:

2.9 Hz 52.2 Hz	6.9 Hz 71.6 Hz
∿∿*CO$_2$H	∿∿*CO$_2$H
1.0 Hz 1.7 Hz	0.7 Hz
hexanoic acid	(E)-2-hexenoic acid

1.5 Hz   123.0 Hz

$H_3C \; CH_2 \; CH_2 — C≡C — \overset{*}{C}O_2H$

19.3 Hz

2-hexynoic acid

$^2J_{CC}$ couplings of larger magnitude are observed if one or both of the coupled carbon atoms have a high *s* character (hexynoic acid, propyne, cyclobutane ring carbons) or if the coupling pathway includes an electron-deficient carbon atom such as a carbonyl moiety (Table 6.11).

There are simple relations of $^3J_{CC}$, $^3J_{CH}$, and $^3J_{HH}$ such that $^3J_{CC}$ = (0.6 to 0.8) $^3J_{CH}$ = (0.6 to 0.8) $^3J_{HH}$. These relations generally hold for couplings in aliphatic compounds (the couplings are averaged, due to free rotation) and for *trans* couplings in alkenes.

For aliphatic compounds, $^3J_{CC}$ follows a Karplus relation, being largest for dihedral angles of 0° and 180° (2 to 6 Hz) and zero at 90°.*

*More data and a discussion of carbon–carbon couplings may be found in Werli, F. W., Wirthlin, T., *Interpretation of Carbon–13 NMR Spectra*, Heyden, London 1976; Stothers, J. B. *Carbon–13 NMR Spectroscopy*, Academic, New York, N.Y., 1972; Breitmaier, E.; Voelter, W., *Carbon–13 NMR Spectroscopy*, 3d ed., VCH, Weinheim, 1990 and Kalinowski, H.-O., Berger, S., Braun, S. *Carbon–13 NMR Spectroscopy*, Wiley, New York, N.Y., 1988.

Table 6.11   Some Representative Values of $^2J_{CC}$ and $^3J_{CC}$

Compound	Formula		$^2J_{CC}$ (Hz)
Propyne	$H_3C—C\equiv C—H$		11.8
2-Butanone	$H_3C—C(O)—CH_2—CH_3$		15.2
Cyclobutanone	(cyclobutanone structure)		9.7
Cyclobutyl bromide	(cyclobutyl bromide structure)		9.0
			$^3J_{CC}$
Pyridine	(pyridine structure)	2–5	13.95
Aniline	(benzene ring with X) $X=NH_2$	2–5	7.9
Iodobenzene	$X=I$	2–5	8.6
Nitrobenzene	$X=NO_2$	2–5	7.6

6.8 $^nJ_{NH}$ and $^nJ_{NC}$ Couplings

A.   General Considerations

The magnetic moments of ^{14}N and ^{15}N are small compared with those of ^1H and ^{13}C ($\mu_H:\mu_C:\mu_{15N}:\mu_{14N} = 10:2.5:1:1.2$), and that for ^{15}N is negative. Thus, $^1H—^{15}N$ and $^{13}C—^{15}N$ coupling constants have small magnitudes, and their signs are, for the most part, reversed relative to the comparable $^nJ_{HH}$, $^nJ_{CH}$, and $^nJ_{CC}$ coupling constants. Due to the low natural abundance of both ^{13}C (1.1%) and ^{15}N (0.37%), the observation of $^nJ_{NC}$ generally requires enriched samples. Because ^{13}C is the more sensitive nucleus, convenient observation of $^nJ_{NC}$ is usually done by recording the ^{13}C spectra of ^{15}N enriched samples. Since ^{14}N has $I = 1$ and is therefore a quadrupolar nucleus, coupling to this nucleus is usually washed out by scalar relaxation of the second kind (Chapter 7), except in highly symmetric molecules. The magnitude of $^nJ^{15}_{NX}$ is related to the magnitude of $^nJ^{14}_{NX}$ by the ratio of the gyromagnetic ratios of the two nuclei, $|\gamma(^{15}N)/\gamma(^{14}N)| = 1.402$, such that $^nJ^{15}_{NX} = -1.402\,^nJ^{14}_{NX}$ is generally expected. Coupling to nitrogen is dominated by the Fermi-contact interaction and thus depends upon electron densities at each nucleus. Since only s orbitals have finite density at the nucleus, a dependence on atom hybridization is expected. Thus,

$$^nJ_{NX} \propto \frac{\rho_N \rho_X}{^3\Delta E},$$

where $^3 \Delta E$ is the mean triplet excitation energy and ρ_N and ρ_X are s-electron densities at the two nuclei.

B. $^n J_{NH}$ Couplings

Some representative values of $^n J_{NH}$ are listed in Table 6.12. These data show that, in accord with theory, all $^1 J^{15}_{NH}$ and $^3 J^{15}_{NH}$ coupling constants are negative. For $^2 J^{15}_{NH}$, the situation is rather intriguing. For saturated systems, $^2 J^{15}_{NH}$ is found to be quite small (0 to 2 Hz). However, the incorporation of a multiple bond to either the nitrogen or the carbon atom leads to a substantial increase in $^2 J^{15}_{NH}$. For example, in formaldoxime (shown to the right) the signs of $^2 J^{15}_{NH}$ are geometry dependent. But for both *cis* and *trans*-formanilide, both $^1 J^{15}_{NH}$ and $^2 J^{15}_{NH}$ are negative.

Table 6.12   Representative Values of $^n J_{NH}$

Compound	J(Hz)	Compound	J(Hz)
N—H Coupling		*N—C—H Coupling*	
pyrrole–^{15}N	−96.5	quinolinium–^{15}N-ethiodide	−1.6
2,5-di-t-butylpyrrole–^{15}N	−91.5	$HCO^{15}N(CH_3)_2$	+1.1[a], +1.2[b]
$HCO^{15}NH_2$	−91.3[a],	$CH_3{}^{15}NO_2$	+2.3
	−86.9[b]	$CH_3{}^{15}NH_2$	−1.0
$HCO^{15}NHCOCH_3$	−90.2	$CH_3{}^{14}N \equiv C$	−2.3
$(CF_3)_2 P^{15}NH_2$	−85.6	$[(CH_3)_3{}^{14}NCH=CH_2]^{\oplus}Br^{\ominus}$	+3.5[c]
$CH_3{}^{15}NH_2$	−65.0		
$^{14}NH_4^{\ominus}Br$	+52.8	*N—C—C—H Coupling*	
		pyrrole–^{15}N	−5.4
N—C—H Coupling		2,5-di-tert-butylpyrrole–^{15}N	−5.2
$CH_3CH={}^{15}N-OH$	−15.9[a],	quinoline–^{15}N oxide	−5.0
	+2.9[b]	^{15}N-ethylquinolinium iodide	−4.3
$HCO^{15}NH_2$	−14.5	$CH_3CH={}^{15}N-OH$	−4.2[a], −2.6[b]
$CH_2={}^{15}N-OH$	−13.9[a],	$(CH_3)_2C={}^{15}N-OH$	−4.0[a], −2.2[b]
	+2.7[b]	quinoline–^{15}N	−1.4
quinoline–^{15}N	−11.1	$CH_3C \equiv {}^{15}N$	−1.8
pyrrole–^{15}N	−4.5	$[(CH_3)_3{}^{14}NCH=CH_2]^{\oplus}Br^{\ominus}$	+5.6[d], +2.5[e]
		$CH_3CH_2{}^{14}N \equiv C$	+5.6[d], +2.5[e]
		$(CH_3)_3C^{14}N \equiv C$	+2.0
	−3.6($H_{(1)}$)	$[(CH_3CH_2)_4{}^{14}N]^{\oplus}OH^{\ominus}$	+1.8
	−1.7($H_{(3)}$)		
	−1.4($H_{(2)}$)		

[a] Coupling to proton *trans* to oxygen atom.
[b] Coupling to proton *cis* to oxygen atom.
[c] Geminal coupling to vinyl proton.
[d] *Trans* coupling.
[e] *Cis* coupling.

SAMPLE PROBLEM 6.4

Predict the magnitude of $^2J^{15}_{NH}$ in $CH_3N{\equiv}C$ from the magnitude of $^2J^{14}_{NH}$ given in Table 6.12.

Solution

$^2J^{15}_{NH}$ = 1.402 (2.3 Hz) = 3.2 Hz, compared with 3.2 Hz obtained by experiment.

Experimentally, $^2J_{NH}$ becomes progressively more positive, undergoing a sign inversion, as the hybridization of nitrogen changes from sp^3 to sp^2 to sp. Similar behavior has been noted for $^2J_{CH}$, which becomes progressively more negative with the same sequence of carbon hybridization.

SAMPLE PROBLEM 6.5

Given the following values of $^1J^{15}_{NH}$, show that $^1J^{15}_{NH}$ is linearly related to the nitrogen s character in the N—H bond:

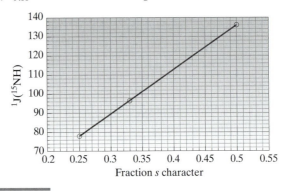

$-NH_3^+$ $^1J_{15NH} = 76.9$ Hz

$-C{\equiv}N^+-H$ $^1J_{15NH} = 136.0$ Hz

$^1J_{15NH} = 96.0$ Hz

Solution

The plot of $^1J^{15}_{NH}$ vs. the fractional nitrogen s character in the N—H bond is linear.

Extensive studies on sp^3 and sp^2 systems have led to the two relations

$$\% \, s = 0.43(^1J^{15}_{NH}) - 6$$

and

$$\% \, s = 0.34(^1J^{15}_{NH}).$$

Hybridization conclusions based upon these relations should be used with caution, because $^1J_{NH}$ is highly solvent dependent.

C. $^nJ_{NC}$ Couplings

Although the relation $^1J^{15}_{NC} = 125\rho^{15}N\rho^{13}C$ gives reasonable values for coupling constants and hybridization at the two nuclei, large deviations are noted, particularly for those compounds for which the nitrogen atoms are highly deshielded. The latter phenomenon seems to be associated with low-lying excited states and strongly indicates that contributions to the total coupling mechanism from the orbital terms are not negligible. As additional data summarized in Table 6.13 show, the results provide little hope for relating $^{1}J^{15}_{NC}$ to carbon and nitrogen s characters.*

Table 6.13   Representative Values of $^{1}J_{NC}$ and $^{2}J_{NC}$

Compound	$J(^{15}N-^{13}C)$ (Hz)	Compound	$J(^{15}N-^{13}C)$ (Hz)
One-Bond Coupling		$CH_3{}^{13}CO^{15}NHC_6H_5$	13.1
$^{13}CH_3{}^{15}N{=}CHC_6H_5$ (*trans*)	<3	$^{13}CH_3{}^{15}N{=}C{=}S$	13.4
$(CH_3{}^{13}CH_2)_4{}^{15}N$	4.0	$H^{13}CO^{15}N(CH_3)_2$	13.4
	3.1 (C_α),[a]5.5 (C_β)[a]	$CH_3{}^{13}CO^{15}NH_2$	<15
	4.9 (C_α),[b]5.9 (C_β)[b]	$^{13}CH_3{}^{15}NH(CS)NHC_6H_5$	<15
		$(CH_3)_3C^{13}C{\equiv}^{15}N$	15.0
		$(CH_3)_2CH^{13}C{\equiv}^{15}N$	15.4
		$CH_3CH_2{}^{13}C{\equiv}^{15}N$	16.4
		$CH_3{}^{13}C{\equiv}^{15}N$	17.5
$^{13}CH_3{}^{15}NH_2$	4.5, 7		
$CH_3{}^{13}CH(^{15}NH_3)^{\oplus}CO_2^{\ominus}$	5.6	*Two-Bond Coupling*	
$(^{13}CH_3)_4{}^{15}N^{\oplus}$	5.8	$^{13}CH_3C{\equiv}^{15}N$	3.0
$^{13}C{\equiv}^{15}N^{\ominus}$	5.9	$CH_3{}^{13}CH_2CO^{15}NH_2$	6.6
$C_6H_5{}^{13}CH{=}^{15}NCH^3$ (*trans*)	7.1	$^{13}CH_3CO^{15}NHC_6H_5$	9.3
$(CH_3)_3C^{15}N{\equiv}^{13}C$	7.2	$^{13}CH_3CO^{15}NH_2$	9.5
$CH_3CH_2{}^{15}N{\equiv}^{13}C$	7.3		
$^{13}CH_3{}^{15}NH_3^{\oplus}$	<8		
$CH_3{}^{15}N{\equiv}^{13}C$	9.1		
$(CH_3)_3{}^{13}C^{15}N{\equiv}C$	9.4		
$CH_3{}^{13}CH_2{}^{15}N{\equiv}C$	10.2		11.4
$^{13}CH_3{}^{15}N{\equiv}C$	10.6		
	9.9 $^{15}N{-}^{13}CH_2$	$(^{13}CH_3)_2N^{15}NH_2$	<1
	12.9 $^{15}N{-}^{13}CO$	$(^{13}CH_3)_2N^{15}NO$	7.5 (*anti*), 1.4 (*syn*)

[a] *trans*-isomer, [b] *cis*-isomer

---

*More data concerning coupling to nitrogen may be found in Witanowski, M., Webb, G. A., *Nitrogen NMR*, Plenum, New York, N.Y., 1973. Levy, G. C., Lichter R. L., "Nitrogen-15 Nuclear Magnetic Resonance Spectroscopy," Wiley-Interscience, New York, N.Y., 1979, and Martin, G. J., Martin, M. L., Gouesnard, J.-P., "*^{15}N-NMR Spectroscopy*," Springer-Verlag, Berlin, 1981.

6.9 $^{n}J_{FX}$ Couplings

A. $^{n}J_{FF}$ Couplings

Because $^{n}J_{FF}$ and $^{n}J_{HF}$ couplings are transmitted to an observable degree through a greater number of intervening bonds than are $^{n}J_{HH}$ couplings, ^{19}F resonances are frequently quite complex multiplets and may appear poorly resolved and broad. $^{n}J_{FF}$ couplings display unexpected behavior, particularly in highly fluorinated compounds. $^{3}J_{FF}$ couplings in perfluoroalkyl chains are often nearly zero, while $^{4}J_{FF}$ and $^{5}J_{FF}$ may be on the order of 5 to 15 Hz. In several instances, these larger, long-range couplings make $^{3}J_{FF}$ indeterminate because of "virtual-coupling" effects (Chapter 5). In these cases, only the sums of $^{3}J_{FF}$ and the long-range coupling may be directly observed.

Substituting other atoms for fluorine markedly affects $^{3}J_{FF}$ couplings. For example, for $CF_3 CF_3$, $^{3}J_{FF} = 3.5$ Hz; for C_2F_5I, $^{3}J_{FF} = 4.6$ Hz; for $CF_3 CF_2 H$, $^{3}J_{FF} = 2.8$ Hz; for CF_3CFH_2, $^{3}J_{FF} = 15.5$ Hz; and for CF_3CFCl_2, $^{3}J_{FF} = 6.1$ Hz. Thus, $^{3}J_{FF}$ is not a simple function of either the electronegativity or the steric bulk of the substituent.

The geometrical dependence of $^{n}J_{FF}$ is not as well understood as is that of $^{n}J_{HH}$, but it is clear that very strong effects may occur. For substituted ethanes, $^{3}J_{FF}$ (*gauche*) is about the same magnitude as $^{3}J_{FF}$ (*trans*) (3 to 20 Hz) and is much smaller than $^{2}J_{FF}$ (*gem*) (150 to 290 Hz).

$^{3}J_{FF}$ in alkenes is much larger than $^{3}J_{FF}$ in aliphatic compounds and is reminescent of $^{3}J_{HH}$ in alkenes. One usually finds that $^{3}J_{FF}$ (*trans*) (115 to 145 Hz) is larger than $^{3}J_{FF}$ (*cis*) (8–58 Hz). However, $^{2}J_{FF}$ (*gem*) (150–290 Hz) is much larger than $^{2}J_{HH}$ (*gem*) (0–5 Hz) and may exceed $^{3}J_{FF}$ (*cis*) in magnitude. For perfluoropropene, the *cis*-like coupling ($^{4}J_{FF} = 22$ Hz) is larger than the *trans*-like coupling ($^{4}J_{FF} = 8$ Hz), similar to what is observed for the $^{4}J_{HH}$ couplings for propene.

$^{3}J_{FF}$ (about 20 Hz), $^{4}J_{FF}$ (2 to 13 Hz), and particularly $^{5}J_{FF}$ (9 to 18 Hz) in fluorobenzenes are much larger than comparable $^{n}J_{HH}$ values and are smaller than $^{3}J_{FF}$ and $^{4}J_{FF}$ in alkenes.

$^{2}J_{FF}$ values in alicyclic compounds are typically large and increase with increasing ring size. Some representative values of $^{n}J_{FF}$ are given in Table 6.14.

Very large values of $^{n}J_{FF}$ with $n > 4$ may be observed when the coupling nuclei are proximate in space. For example,

$^{5}J_{FF} = 174$ Hz $^{6}J_{FF} = 43.2$ Hz $^{5}J_{F^1F^2} = 68.4$ Hz

These are among the earliest examples of through-space coupling, which has since also been observed for other nuclei as well.*

---

*Hilton, J., Sutcliffe, L. H., *Prog. in Nucl. Magn. Reson. Spectros*, **1975**, *10*, 27: For a recent theoretical discussion of through-space fluorine-fluorine coupling, see Arnold, W. D.; Mao, J.; Sun, H.; Oldfield, E., *J. Am. Chem. Soc.* **2000**, *122*, 12164.

Table 6.14 Representative $^nJ_{FF}$ Couplings

Compound	J (Hz)	Compound	J (Hz)
$CF_{3(2)}-CF_{2(1)}$ $\quad\quad\quad\quad N-CF_{3(2)}$ $CF_{3(2)}-CF_{2(1)}$	$J_{12}+J_{12}$: 10.2 J_{13}: 15.8 J_{23}: 6.8 J_{11}: ca.9	(cyclopropane with CH_3, H, H, H, F, F)	J_{gem}: 157
$[CF_{3(2)}CF_{2(1)}]_2NCF_{2(1')}CF_{2(2')}$ $CF_{3(2)}CF_{2(1)}N[CF_{3(3)}]_2$	$J_{12}+J_{12}$: 13.6 J_{12}: ≤1 J_{13}: 16 J_{23}: 6	(cyclobutane with Cl Cl, Cl, Cl, H, F, H, H, F)	J_{gem}: 187
$CF_{3(2)}CF_{2(1)}OCF_{2(1)}CF_{3(2)}$	$J_{12}+J_{12'}$: 3.4	(cyclobutanone with C_6H_5, CH_3, O, H, H, F)	J_{gem}: 249
$[CF_{3(2)}]_3CF_{(1)}$	J_{12}: 1.4		
$CF_{3(2)}CF_{2(1)}H$	J_{12}: 2.8		
$CF_{3(2)}CF_{(1)}H_2$	J_{12}: 15.5		
$CF_{3(3)}CF_{2(2)}CHF_{2(1)}$	J_{12}: 4.5 J_{13}: 7.3	(cyclobutene with C_6H_3, H, C_6H_3, Cl, F, F)	J_{gem}: 192
$CF_{3(3)}CF_{2(2)}CH_2F_{(1)}$	J_{12}: 15.2 J_{13}: 7.9		
$CF_{2(3)}Cl-CF_{2(2)}\cdot CH_2F_{(1)}$	J_{12}: 15.1 J_{13}: 7.7 J_{23}: 3.9	(cyclobutane: F 3, H, F 4, H, F 2, ClH$_2$C 1, F H)	$J_{2\text{-gem}}$: 230 $J_{3\text{-gem}}$: 224
$CF_{2(3)}Br\cdot CF_{2(2)}CH_2F_{(1)}$	J_{12}: 15.5 J_{13}: 7.7 J_{23}: 3.9		
$CF_{2\underline{(2a,2b)}}Br\cdot CHF_{(1)}Br$	$J_{1.2a}, J_{1.2b}$: 21, 24 $J_{2a,2b}$: 174	(perfluorocyclohexane structure)	J_{gem}: 284
$CF_{2\underline{(2a,2b)}}Br\cdot CHF_{(1)}Cl$	$J_{1.2a}, J_{1.2b}$: 18, 18 $J_{2a,2b}$: 177		
$CF_{2\underline{(2a,2b)}}Br\cdot CF_{(1)}ClBr$	$J_{1.2a}, J_{1.2b}$: 13, 14 J_{2ab}: 159	(benzene ring with NO_2, Br, F, positions 1 2 3 4 5 6, Br, F)	J_{23}: 20.2
$CF_{2\underline{(2a,2b)}}(SiCl_3)CF_{(1)}ClH$	$J_{1.2a}, J_{1.2b}$: 16.8, 16.8 $J_{2a,2b}$: 343		
$CF_3CF_{2\underline{(2a,2b)}}CF_{(1)}ICl$	$J_{2a,2b}$: 270.4	(benzene ring with F_1, Cl, C_1, Cl, positions 6 2 5 3 4, F, F, F)	J_{35}: 1.9
$F_{(2)}\quad Cl$ $\quad C=C$ $F_{(1)}\quad F_{(3)}$	J_{12}: 76 J_{13}: 56 J_{23}: 116		
$F\quad F$ $\;C=C$ $Cl\quad Cl$	J_{cis}: 37.5		

(continues on next page)

Table 6.14 (*cont.*)

Compound	J (Hz)	Compound	J (Hz)
	J_{trans}: 129.57		J_{35}: 3.1
	J_{12}: 87 J_{13}: 33 J_{23}: 119		J_{14}: 12.0
	J_{12}: 57 J_{13}: 39 J_{23}: 116 J_{14}: 8 J_{24}: 22 J_{34}: 13		J_{36}: 14.4
	$J_{\alpha\text{-gem}}$: 185 $J_{\beta\text{-gem}}$: 278 $J_{\gamma\text{-gem}}$: 284		

B. $^{n}J_{HF}$ Couplings

The behavior of $^{n}J_{HF}$ is more "normal" than that of $^{n}J_{FF}$ in that $^{n}J_{HF}$ generally decreases monotonically as n increases. Also, $^{n}J_{HF}$ is strongly geometrically dependent. $^{2}J_{HF}$ falls in the 45- to 80-Hz range and increases with an increase in the electronegativity of X for CHF_2X compounds. $^{3}J_{HF}$ couplings are usually less than 0.5 $^{2}J_{HF}$. For fluoroethanes of the type CH_3CF_2X, $^{3}J_{HF}$ decreases slightly as the electronegativity of X increases.

Vicinal $^{3}J_{FH}$ couplings do not follow a clear Karplus-type dependence on dihedral angle. But there is evidence that there are maxima at 0° and 180° and a minimum near 90°. Near 0°, $^{3}J_{FH}$ ranges from 10 to 30 Hz; at 60°, $^{3}J_{FH}$ ranges from 3 to 6 Hz. Representative values of $^{n}J_{FH}$ are given in Table 6.15.

For fluorobenzenes, $^{3}J_{FH}$ is usually slightly larger than $^{4}J_{FH}$, and both are three to five times $^{5}J_{FH}$. $^{3}J_{FH}$ increases with an increasing number of fluorines in the ring, while $^{4}J_{FH}$ and $^{5}J_{FH}$ change little.

C. $^{n}J_{CF}$ Couplings

Some representative $^{n}J_{CF}$ couplings are listed in Table 6.16. $^{1}J_{CF}$ is experimentally found to be negative. $^{1}J_{CF}$ decreases with an increase in the number of fluorines for fluoromethanes, but an increasing number of hydrogens has an even greater effect. There is no direct relationship between carbon s character and $^{1}J_{CF}$, as has been found for $^{1}J_{CH}$.

Table 6.15   Representative Values of $^nJ_{FH}$

Compound		J (Hz)	Compound		J (Hz)
$CH_{3(2)}CH_{2(1)}F$	$J_{H(1)-F}$:	47.5		$J_{H-F(cis)}$:	ca. 1
	$J_{H(2)-F}$:	25.7		$J_{H-F(trans)}$:	34
$CH_{3(2)}CH_{(1)}F_2$	$J_{H(1)-F}$:	57.2			
	$J_{H(2)-F}$:	20.8		$J_{H-F(trans)}$:	20.4
CH_3CF_3	J_{H-F}:	12.8		$J_{H-F(gem)}$:	72.7
$CF_{2(2)}H_{(2)}CF_{2(1)}H_{(1)}$	$J_{H(1)-F(1)}$:	52.1		$J_{H-F(cis)}$:	4.4
	$J_{H(1)-F(2)}$:	4.8		$J_{H-F(gem)}$:	74.3
$CF_{3(2)}CF_{(1)}H_{2(1)}$	$J_{H(1)-F(1)}$:	45.5			
	$J_{H(1)-F(2)}$:	8.0		$J_{H-F(cis)}$:	<3
$CF_{3(2)}CF_{2(1)}H_{(1)}$	$J_{H(1)-F(1)}$:	52.6		$J_{H-F(trans)}$:	12
	$J_{H(1)-F(2)}$:	2.6		$J_{H-F(gem)}$:	72
CF_3CH_2Cl	J_{H-F}:	8.5			
CF_3CH_2Br	J_{H-F}:	9.0			
$CF_2CH_{2(1)}OH$	$J_{H(1)-F}$:	8.9		$J_{H-F(cis)}$:	20.1
CH_3CF_2Cl	J_{H-F}:	15.0		$J_{H-F(trans)}$:	52.4
CH_3CF_2Br	J_{H-F}:	15.9		$J_{H-F(gem)}$:	84.7
$CH_2Cl \cdot CF_2Cl$	J_{H-F}:	6, 16 (*AA'XX'*)			
$CH_2Cl \cdot CF_2Br$	J_{H-F}:	3, 21 (*AA'XX'*)		$J_{H-F(cis)}$:	8
$CH_2Br \cdot CF_2Br$	J_{H-F}:	4, 22 (*AA'XX'*)		$J_{H-F(trans)}$:	40
$CF_{2(2a,2b)}Br \cdot CH_{(1)}F_{(1)}Br$	$J_{H(1)-F(1)}$:	48			
	$J_{H(1)-F(2a)}$:			$J_{H-F(cis)}$:	<3
	$J_{H(1)-F(2b)}$:	3.2, 9.1		$J_{H-F(trans)}$:	13
$CF_{2(2a,2b)}Br \cdot CH_{(1)}F_{(1)}Cl$	$J_{H(1)-F(1)}$:	48			
	$J_{H(1)-F(2a)}$:				
	$J_{H(1)-F(2b)}$:	3.5, 6.3		$J_{F-H(1)}$:	89.9
$CF_{3(3)}CF_{2(2)}CH_{2(1)}F_{(1)}$	$J_{H(1)-F(1)}$:	46.0		$J_{F-H(2)}$:	41.8
	$J_{H(1)-F(2)}$:	11.7		$J_{F-H(3)}$:	2.6
$CF_{3(3)}CF_{2(2)}CH_{(1)}F_{2(1)}$	$J_{H(1)-F(1)}$:	52.1			
	$J_{H(1)-F(2)}$:	4.5			
$CF_{2(2)}Cl \cdot CF_{2(2)}CH_{2(1)}F_{(1)}$	$J_{H(1)-F(1)}$:	45.9		$J_{F-H(1)}$:	84.8
	$J_{H(1)-F(2)}$:	11.8		$J_{F-H(2)}$:	19.9
	$J_{H(1)-F(3)}$:	1.0		$J_{F-H(3)}$:	3.3
$CF_{2(3)}Br \cdot CF_{2(2)}CH_{2(1)}F_{(1)}$	$J_{H(1)-F(1)}$:	46.0			
	$J_{H(1)-F(2)}$:	11.7			
	$J_{H(1)-F(3)}$:	0.9			
	J_{H-F}:	81	$F-C \equiv C-H$	J_{F-H}:	21

(continues on next page)

Table 6.15 (cont.)

Compound	J (Hz)		Compound	J (Hz)	
C₆H₅, C₆H₇ / H(b)—Cl / H(a) F	$J_{F-H_{(a)}}$: 17.7 $J_{F-H_{(b)}}$: 6.3		NO₂ / Br, F / Br F (1,2,3,4,5,6)	J_{24}: 6.8 J_{34}: 8.7	
Cl C₆H₅ / Cl—CH₃ / F(b)—H(a) / F(a) H(b)	$J_{F_{(a)}-H_{(a)}}$: 8.79 $J_{F_{(a)}-H_{(b)}}$: 0.91 $J_{F_{(b)}-H_{(a)}}$: 21.02 $J_{F_{(b)}-H_{(b)}}$: 13.93		NO₂ / NH₂ / F, F / Cl (1,2,3,4,5,6)	J_{36}: −2.2 J_{56}: 9.4	
C₆H₅ H(a) / H₁C—H(b) / H(a)—F(a) / H(b) F(b)	$J_{F_{(a)}-CH_3}$: ca. 0 $J_{F_{(b)}-CH_3}$: 2.0 $J_{F_{(a)}-H_{(a)}}$: 8.90 $J_{F_{(a)}-H_{(b)}}$: 10.60 $J_{F_{(b)}-H_{(a)}}$: 12.52 $J_{F_{(b)}-H_{(b)}}$: 12.45		F / Cl / Cl, Cl / F (1,2,3,4,5,6)	J_{16}: 8.4 J_{46}: 6.3	
CH₃ O / C₆H₅ / H(a)—F(a) / H(b) F(b)	$J_{F_{(a)}-H_{(a)}}$: 10.97 $J_{F_{(a)}-H_{(b)}}$: 15.09 $J_{F_{(b)}-H_{(a)}}$: 16.17 $J_{F_{(b)}-H_{(b)}}$: 10.28		NO₂ / F, Cl / F / Cl (1,2,3,4,5,6)	J_{35}: 6.3 J_{56}: 8.2	
H₂ F(ax) / H₂ H(eq) / H₂ H(eq) / H₂ H(ax) (5,4,3,6,1,2) ⇄ H₂ H(ax) / H₂ / H₂ F(eq) / H₂ H(eq) H(ax) (4,3,5,2,6,1)	$J_{H-F(gem)}$: 49 $J_{H_{(2ax)}-F_{(ax)}}$: 43.5 $J_{H_{(2eq)}-F_{(ax)}}$: <3 $J_{H-F(gem)}$: 49 $J_{H_{(2ax)}-F_{(eq)}}$: <3 $J_{H_{(2eq)}-F_{(eq)}}$: <3		CH₃ / / F (1,2,3,4,5,6)	$J_{24}(J_{46})$: 7.0 $J_{34}(J_{45})$: 8.6	
			F / O=C—N—CH₃(a) / CH₃(b) (1,2,3,4,5,6)	$J_{F-CH_{3(a)}}$: 1.2 $J_{F-CH_{3(b)}}$: ca. 0	

Table 6.16   Representative $^nJ_{CF}$ Couplings

Compound	^1J (Hz)	^2J (Hz)	Compound	^1J (Hz)	^2J (Hz)
$CFBr_3$	372			−299.67	+26.94
CF_2Br_2	358				
CF_3Br	324				
$CFCl_3$	337				
CF_2Cl_2	325				
CF_3Cl	299			252	17.5
CF_4	257				
CF_3H	272				
CF_2H_2	232				
CH_3F	−157.5				
	−291.0	+54.5		260	
$CF_2{=}CD_2$	287				
CF_3COOH	283				
CF_3CF_3	281.3	46.0		242.0	$F_1{-}C_2: 24.3$
	298	−25.8			$F_1{-}C_4: 8.5$
					$F_1{-}C_5: 3.8$

$^2J_{CF}$ is typically 0.1 to 0.2 times $^1J_{CF}$ and may be of opposite sign even in closely related molecules.*

6.10 $^nJ_{PX}$ Couplings

A. $^nJ_{PH}$ Couplings

$^1J_{PH}$ coupling constants always appear to be positive and vary from about 120 to 1,180 Hz. The presence of one or more lone pairs decreases $^1J_{PH}$. For example, couplings are observed for PH_4^+ (547 Hz), PH_3 (189 Hz), and PH_2^- (139 Hz). Although there is a general increase in $^1J_{PH}$ with increasing phosphorus oxidation state or upon coordination of the phosphine to a transition metal, the ranges for the various oxidation states overlap considerably. Typical values for $^1J_{PH}$ in complexes are $L = PH_3$; $(OC)_4Cr(PBu_3)L$ (307 Hz), $Me_{3-n}BH_nL$, $n = 0,1,2$ (366 to 372 Hz); PF_2H (+181.7 Hz) in BH_3L (467 Hz); and B_4H_8L (651 Hz).

Electronegative substituents on phosphorus increase $^1J_{PH}$ in phosphonium ions; $^1J_{PH}$ ranges from 457 Hz to 870 Hz as the alkyl groups of HPR_3^+ are replaced by Ph, OR, and OPh groups. Similar effects are observed for $HPF_nCl_{3-n}^+$ and $HPF_nBr_{3-n}^+$ ($n = 0{-}3$) ions, for which a dramatic, nearly linear rise (810–1,191 Hz) in $^1J_{PH}$ occurs as the sum of the substituent

---

*An extensive compilation of fluorine coupling constants may be found in Emsley, J. W., Phillips, L., Wray, V., *Prog. in Nucl. Magn. Res: Spectrosc.* **1977**, *10*, 85.

Table 6.17   Representative Values of $^2J_{PH}$ Couplings

Structural Class (or structure)	$^2J_{PH}$ (Hz)	Structural Class (or structure)	$^2J_{PH}$ (Hz)
P(III)		**P(IV)**	
⟩PCH	0 to 18	—P(O)CH	7 to 30
$P(CH_3)_3$	+2.7	$P(O)(CH_3)_3$	−13.3
$P(C_2H_5)_3$	−0.5	$P(O)(CH_3)F_2$	19.5
PCl_2CH_3	17.6	—$\overset{+}{P}$CH	12 to 18
⟩PNH	13 to 28		
		$P(CH_3)_4{}^+$	−14.6
P(V)		$PH_3CH_3{}^+$	−17.6
⟩PCH	10 to 18	—P(S)CH	11 to 15
$P(CH_3)_3F_2$	17.2	$P(S)(CH_3)_3$	−13.0
$P(CH_3)F_3H$	17.8	$P(S)(CH_3)_2H$	14.4

electronegativities increases. In the series of $HPR_{3-n}H_n^+$, $HPPhR_{2-n}H_n^+$, and $HPPh_2R_{1-n}H_n^+$ cations, for which electronegativity changes are expected to be small, $^1J_{PH}$ increases with decreasing substituent steric bulk.

Some representative values of $^2J_{PH}$ are given in Table 6.17. There is a Karplus-like geometric dependence on $^2J_{PH}$ in the moiety $H{-}C{-}P{-}X$ (X = lone pair or heteroatom), as illustrated in Figure 6.5.

Representative values of $^3J_{PH}$ are listed in Table 6.18. As might be expected from the dihedral-angle dependence of vicinal $^3J_{HH}$ coupling constants, similar Karplus-like curves have been established for vicinal $^3J_{PH}$ in the PCCH, POCH, PNCH, and PSCH moieties. Caution is recommended in attempting to apply these relationships universally, since separate correlations seem to exist for each structural class of compound. In all cases, a minimum value of

▶ FIGURE 6.5
Dihedral-angular dependence of $^2J_{PH}$ in phosphines.[a]

[a] Albrand, J. P., Gagnaire, D., Robert, J. B., *J. Chem. Soc. Chem. Commun.*, **1968**, 1469.

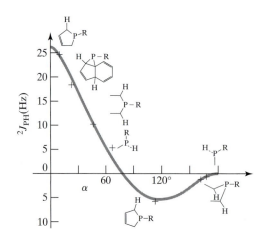

Table 6.18 Representative Values of $^3J_{PH}$ Couplings

P(III)

\>POCH	0.15	\>PSCH	2.20
$P(OCH_3)_3$	10.8–11.8	$P(SCH_3)_3$	9.8
$P(OC_2H_3)_3$	7.3–8.0	$P(SCH_2)_3CCH_3$	2.1
\>PCCH	10–16	\>PNCH	3–14
$P(C_2H_5)_3$	+13.7	$P[N(CH_3)_2]_3$	8.8–9.0
$P(C_2H_3)(C_6H_5)_2$	16.5	$P(Cl)_2N(CH_3)_2$	12.9–13.0
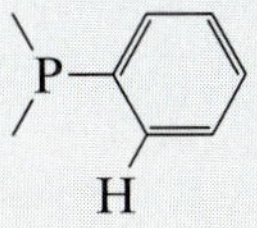 P–(C$_6$H$_4$)–H (ortho)	12.5		

P(IV)

$-\overset{+}{P}OCH$	7–11	$-\overset{+}{P}SCH$	16–20
$\overset{+}{P}(OCH_3)_4$	11.2	$\overset{+}{P}(SCH_3)(NH_2)_3$	16.2
		$P(O)(SCH_3)_3$	15.1
$-\overset{+}{P}CCH$	15–22	$-\overset{+}{P}NCH$	9–17
$P(C_2H_3)_4$	+18.1	$P[N(CH_3)_2]_3Cl$	13.0
$P(C_2H_5)_3H$	+20.0	$P(O)[N(CH_3)_2]_3$	9.4–9.5
		$P(S)[N(CH)_3)_2]_3$	11.0–11.3
\>P(O)OCH	0–13	\>P(S)OCH	10–14
$P(O)(OCH_3)_3$	10.2–11.4	$P(S)(OCH_3)_3$	13.4
\>P(M)CCH	14–25		
(M=O, S)			
$P(O)(C_2H_5)_3$	+18.0		
$P(S)(t\text{-}C_4H_9)_3$	+14.0		

P(V)

\>PCCH	20–27	\>POCH	12–17
$P(C_2H_5)F_3H$	26.4	$P(OCH_3)_5$	12
\>PSCH	13–25	\>PNCH	2–15
$P(SCH_3)F_4$	25.2	$P[N(CH_3)_2]_3F_2$	10.6

$$^3J_{(PCCH)} = 18 \cos^2 \phi \, PH \quad 0° \le \phi \le 90°$$
$$^3J_{(PCCH)} = 41 \cos^2 \phi \, PH \quad 90° \le \phi \le 180°$$

$$^3J_{(POCH)} = 18.1 \cos^2 \phi \, PH - 4.8 \cos \phi \, PH \quad 0° \le \phi \le 90°$$
$$^3J_{(POCH)} = 15.3 \cos^2 \phi \, PH - 6.1 \cos \phi \, PH + 1.6 \quad 90° \le \phi \le 180°$$

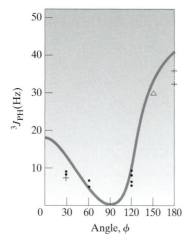

▲ FIGURE 6.6
Dihedral-angular dependence of
$^3J_{(PCCH)}$ **in phosphonates.**

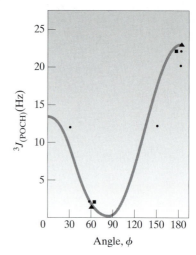

▲ FIGURE 6.7
Dihedral-angular dependence of
$^3J_{(POCH)}$ **in phosphates.**

$^3J_{PH}$ is found around a 90° dihedral angle, and maxima are observed at 0° and 180°. Typical relations are shown in Figures 6.6 and 6.7.

The data in Table 6.19 reveal several trends in $^nJ_{PH}$ for vinyl phosphorus derivatives. In general, $^3J_{PH}$ (*trans*) is larger than $^3J_{PH}$ (*cis*), as is found for $^3J_{HH}$ in alkenes. However, $^2J_{PH}$ (*gem*) can be larger than $^3J_{PH}$ (*trans*), as is found in the behavior of $^nJ_{FH}$ in fluoroalkenes. Electronegative substituents on phosphorus and coordination to a transition metal have little effect on $^nJ_{HH}$ and small, but noticeable, effects on $^3J_{PH}$ (*trans*). Coordination to a transition metal substantially increases $^2J_{PH}$ (*gem*), and $^3J_{PH}$ (*cis*) and electronegative substituents have a variable effect on $^3J_{PH}$ (*cis*).

As with $^4J_{HH}$, $^4J_{PH}$ is maximized when the phosphorus and hydrogen adopt a "W" relationship:

For example, in the system shown, ($X{=}O$), $^4J_{PHeq}$ is 2 to 3 Hz, whereas $^4J_{PHax}$ is only 0 to 1 Hz. There seems to be little influence of the phosphorus oxidation state on $^4J_{PH}$. For the system in the next figure,

Table 6.19 Representative Values of $^{n}J_{HH}$ and $^{n}J_{PH}$ for Vinyl Phosphorus Derivatives

Compound	$^{n}J_{AB}$	$^{n}J_{AC}$	$^{n}J_{BC}$	$^{n}J_{PA}$	$^{n}J_{PB}$	$^{n}J_{PC}$
(A)H, H(B) / Ph₂P, H(C) C=C	12.44	18.06	2.20	11.47	31.74	14.16
(A)H, H(B) / PhP, H(C) ‒₂ C=C	11.70	17.90	2.30	11.70	31.30	14.70
(A)H, Ph / Cl₂P, H(C) C=C	—	18.0	—	15.0	—	15.0
(A)H, Ph / (Me₂N)₂P, H(C) C=C	—	17.5	—	20.4	—	15.4
(A)H, Ph(B) / (EtO)₂P, H(C) C=C	—	18.0	—	14.0	—	12.0
(A)H, Ph / Ph₂P, H(C) C=C	—	18.0	—	11.2	—	11.2
H₃C, H(B) / (EtO)₂P, H(C) C=C	—	—	2.3	—	28.0	11.0
Ph, H(B) / (EtO)₂P, H(C) C=C	—	—	2.3	—	11.0	10.0
Ph, H(B) / Ph₂P, H(C) C=C	—	—	2.3	—	14.0	7.0
A(H), H(B) / PhP, H(C) ‒₂ / Mo(CO)₅ C=C	11.72	17.58	1.46	20.76	36.13	17.60
(A)H₃C, H(B) / Ph₂P, H(C) C=C	−1.7	−1.1	1.7	9.2	25.3	10.5

(continues on next page)

Table 6.19   (*cont.*)

Compound	$^nJ_{AB}$	$^nJ_{AC}$	$^nJ_{BC}$	$^nJ_{PA}$	$^nJ_{PB}$	$^nJ_{PC}$
(A)H, Ph$_2$P / C=C / CH$_3$(B), CH$_3$(C)	−2.0	−1.5	0	4.0	0.3	0.5
(A)H, Ph$_2$P / C=C / H(B), CH$_3$(C)	8.5	−0.8	6.0	2.7	20.0	0
(A)H, Ph$_2$P / C=C / CH$_3$(B), H(C)	1.5	13	4.9	7.0	0.3	10.0
(A)H$_3$C, Ph$_2$P / C=C / CH$_3$(B), H(C)	0	−1.1	6.0	8.5	0.3	10.0
(A)H$_3$C, Ph$_2$P / C=C / H(B), CH$_3$(C)	1.4	1.6	7.0	3.0	23.7	0.5
cis-PdCl$_2$ ((A)H, Ph$_2$P / C=C / H(B), H(C))$_2$	11.89	18.15	1.50	23.48	40.69	20.35
trans-PdCl$_2$ ((A)H, Ph$_2$P / C=C / H(B), H(C))$_2$	12.52	18.78	1.57	24.73	38.19	18.76

when the CH$_3$ group is axial, $^4J_{PH}$ is less than 1 Hz. However, when the CH$_3$ group is equatorial, $^4J_{PH}$ is 1.4 to 3.0 Hz. Examples of ^4J(POCCH) values of 1.5 to 2.0 Hz are reported for acyclic phosphate esters when C—C and C—O torsional angles have values appropriate for the "W" relation.

B. $^nJ_{PC}$ Couplings

Some representative $^nJ_{PC}$ values are listed in Tables 6.20 through 6.29.

The data in Table 6.20 show that, for trialkylphosphines, $^2J_{PC}$ is usually larger than either $^1J_{PC}$ or $^3J_{PC}$, which are of similar magnitude. Upon coordination to Ni(0), $^1J_{PC}$ increases substantially, $^2J_{PC}$ decreases substantially, and $^3J_{PC}$ increases slightly. For methyl phosphorus derivatives (Table 6.21) $^1J_{PC}$ is usually negative for P(III) compounds and substantially larger and positive for P(IV) and P(V) compounds.

Also, the magnitude of $^1J_{PC}$ increases with increasing electronegativity of the substituents on phosphorus.

Table 6.20 ^{13}C NMR Chemical Shifts and $J(^{13}C^{31}P)$ Coupling Constants for Trialkylphosphines and Their $LNi(CO)_3$ Complexes[*]

Phosphine		C(1)	C(2)	C(3)	C(4)	C(5)	C(6)	C(7)	C(8)
$P(ethyl)_3$									
	ligand	18.0	9.0						
		(11.5)	(12.2)						
	complex	20.18	7.89						
		(22.3)	(–)						
$P(propyl)_3$									
	ligand	29.71	19.27	15.95					
		(11.8)	(13.2)	(11.2)					
	complex	31.11	18.05	15.94					
		(19.6)	(3.0)	(13.7)					
$P(butyl)_3$									
	ligand	26.80	27.94	24.34	13.60				
		(12.0)	(12.3)	(10.8)					
	complex	28.09	26.27	42.25	13.55				
		(19.5)	(2.0)	(12.9)					
$P(pentyl)_3$									
	ligand	26.80	25.24	33.38	22.07	13.57			
		(11.2)	(12.2)	(10.3)					
	complex	28.33	23.81	33.36	22.03	13.50			
		(19.0)	(2.5)	(12.7)					
$P(hexyl)_3$									
	ligand	27.11	25.60	30.85	31.30	22.24	13.66		
		(12.7)	(13.2)	(10.8)					
$P(octyl)_3$									
	ligand	27.10	25.71	31.31	29.16	29.03	31.67	22.42	13.80
		(12.2)	(12.7)	(10.7)					
	complex	28.45	24.20	31.33	29.06	29.06	31.77	22.55	13.83
		(18.6)	(2.4)	(13.2)					

[*]Bodner, G. M., Bauer, L., *J. Organomet. Chem.* **1982**, *226*, 85.

Table 6.21   Representative $^1 J_{PC}$ Values for Methyl Phosphorus Compounds

Compound	$^1 J_{PC}$	Compound	$^1 J_{PC}$
CH_3PH_2	9.3	$(CH_3)_2P(S)SeMe$	+50
CH_3PCl_2	−45	$CH_3P(O)(OMe)_2$	+142.2
$CH_3P(S)Cl_2$	81	$(CH_3)_3P$	−13.6
$CH_3P(O)Cl_2$	104	$(CH_3)_4P^+I^-$	+56
$CH_3P(O)F_2$	147	$(CH_3)_2PC_6H_5$	−14
$(CH_3)_2PH$	11.6	$(CH_3)_2P(C_6H_5)H^+Br^-$	56
$(CH_3)_2PSeMe$	−25	$(C_6H_5)_3PCH_3^+I^-$	51.7 (CH_3)
			88.4(C_i)

Table 6.22   Representative $^2J_{POC}$ and $^3J_{POCC}$ Couplings for Alkoxy-Substituted Phosphorus Compounds

Compound	$^2J_{PC}$	$^3J_{PC}$	Compound	$^2J_{PC}$	$^3J_{PC}$
$(CH_3O)_3P$	+10.05		$(CH_3O)_3PS$	−5.6	
$(CH_3CH_2O)_3P$	11.0	4.9	$(CH_3CH_2O)_3PS$	−5.2	+6.4
$(C_6H_5O)_3P$	3.0	4.85	$(CH_3O)_2P(O)H$	−6.0	
$(CH_3O)_3PMo(CO)_5$	−2.3		$(CH_3CH_2O)_2P(O)H$	−4.5	+6.1
$(CH_3O)_3PO$	−5.8		$(CH_3O)_2P(O)Me$	−6.0	
$(CH_3CH_2O)_3PO$	−5.8	+6.8	$CH_3OP(O)Cl_2$	−10.5	
$(CH_3CH_2CH_2CH_2O)_2PO(OH)$	−5.6	6.0	$CH_3CH_2OP(O)Cl_2$	−8.9	+8.8
$(CH_3CH_2CH_2CH_2O)_3PO$	−5.9	6.5	$(CH_3CH_2O)_2P(O)Cl$	−7.1	+7.5

Table 6.23 $^nJ_{PC}$ Values for Aminophosphines[a]

Compound	Phenyl				Benzyl			Amine		
	C–1	o	m	p	C–1	o	m	p	CH$_2$	CH$_3$
$Ph_2PN(CH_2Ph)_2$	14.4	20.4	3.9	0	1.7	0.5	0	0	14.4	
$PhP[N(CH_2Ph)_2]_2$	f(T)	f(T)	5.7	1.1	1.9	1.2	0	0	15.0	
$PhP[N(CH_2CH_3)_2]_2$	f(T)	f(T)	3.1	2.0					16.9	3.2
$Cl_2PN(CH_3)_2$										(+)21.3
$P[N(CH_3)_2]_3$										+19.15
$P[N(CH_2Ph)_2]_3$					0	0	0	0	13.4	

$Ph—P$ (cyclic diamine with two N–CH_3 groups)

| | 42.1 | 19.1 | 5.1 | 0 | | | | | 8.4 | 23.5 |

$(CH_3)_2N—P$ (cyclic diamine with two N–CH_3 groups)

$$\left[-N(CH_3)_2 \quad \begin{matrix} 17.4 & at & \sim 30°C \\ \sim 43° & at & -138°C \end{matrix} \right]$$

| | | | | | | | | | 9.1 | 23.8 |

$(CH_3CH_2)_2N—P$ (cyclic diamine with two N–CH_3 groups)

$$\left[-N(CH_2CH_3)_2 \quad \begin{matrix} & CH_2 & 18.9 & CH_3 & 1.8 & \text{at ambient temp.} \\ syn & CH_2 & 47.6 & CH_3 & 0 \\ anti & CH_2 & -6.1 & CH_3 & 0 \end{matrix} \right\} \text{at } -128°C \right]$$

| | | | | | | | | | — | — |

[a] Gray, G. A., Nelson, J. H., *Org. Magn. Reson.*, **1980**, *14*, 8.   *(continues on next page)*

Table 6.23 (cont.)

Compound	Phenyl				Benzyl				Amine	
	C–1	o	m	p	C–1	o	m	p	CH$_2$	CH$_3$

For the first compound $(i-\text{Pr})_2\text{N}-\text{P}$ (with N–CH$_3$ ring):

$$\begin{bmatrix} -\text{N}(i\text{-Pr})_2 & \text{CH} & 9.3 & \text{CH}_3 & 8.0 & \text{at ambient temp.} \\ syn & \text{CH} & 26.0 & \text{CH}_3 & \sim 0 \\ anti & \text{CH} & -8.8 & \text{CH}_3 & 12.6 \end{bmatrix} \text{at } -116°\text{C}$$

Amine: CH$_2$ = 8.8, CH$_3$ = 23.6

For the second compound MeEtCHO–P (with N–CH$_3$ ring):

$$\begin{bmatrix} \text{CH} & 12.8 & \text{CH}_2 & 3.0 \\ | & & | & \\ \text{CH}_3 & 2.7 & \text{CH}_3 & 0 \end{bmatrix}$$

Amine: CH$_2$ = 10.2, CH$_3$ = 23.8

The magnitude of $^n J_{PC}$ is highly dependent on structural parameters. Not only do the phosphorus coordination number, oxidation state, and value of n play major roles, but the stereochemical dispositions of the coupled nuclei can also have a profound influence.

In tetracovalent phosphorus compounds, $^1 J_{PC}$ has a positive sign and correlates in many cases with the s character in the P—C bond. Tertiary phosphine oxides and sulfides have $^1 J_{PC}$ near 55 Hz, while phosphonic acids and their derivatives have larger values (130 to 220 Hz) and exhibit significant sensitivity to α-substituent effects.

Table 6.24   Representative $^nJ_{PC}$ to sp^3, sp^2, and sp Carbons

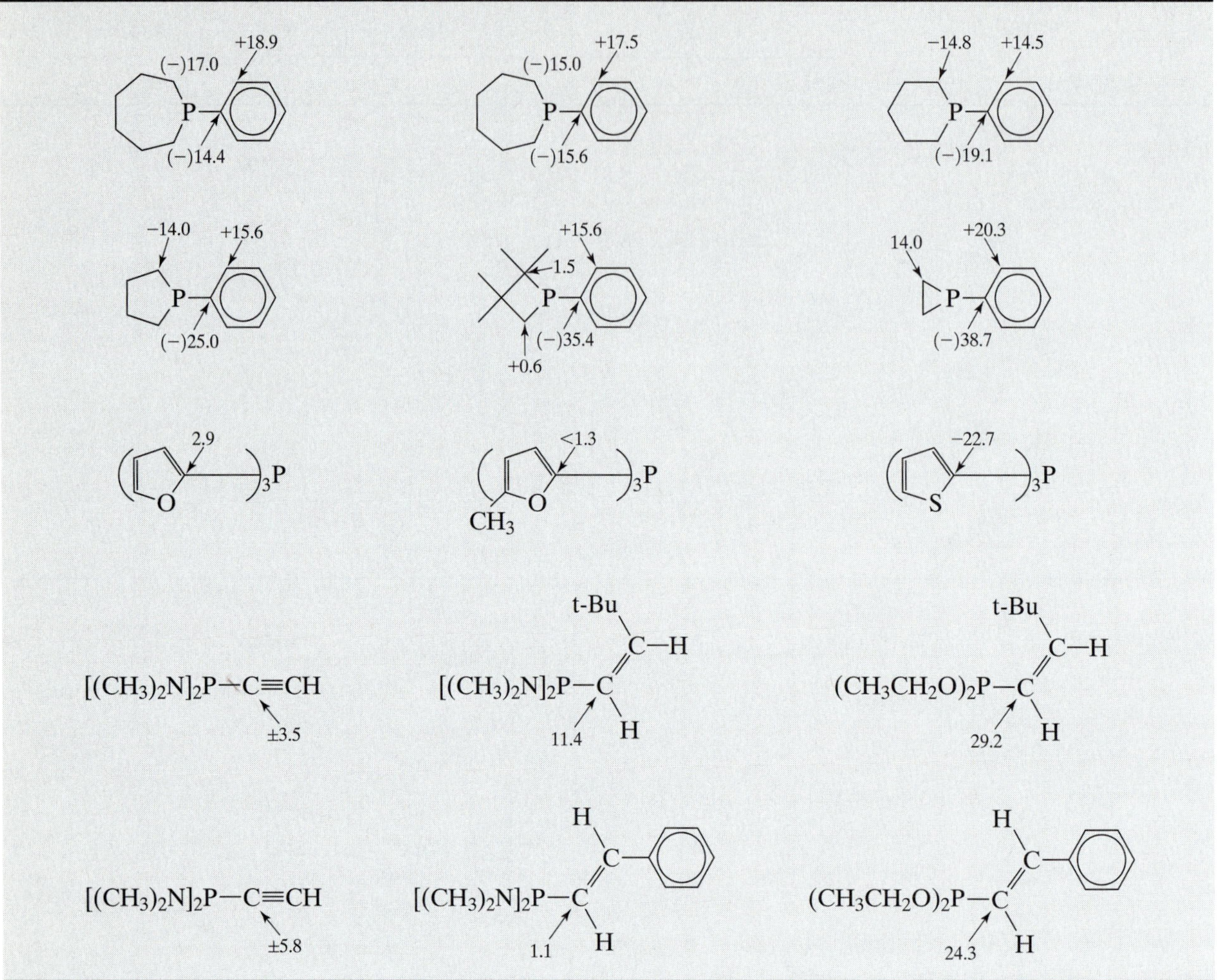

For alicyclic rings, a severe modification of bond angles or unusual modification of substituents may bring about large changes in $^1J_{PC}$, as, for example, in the following figure:

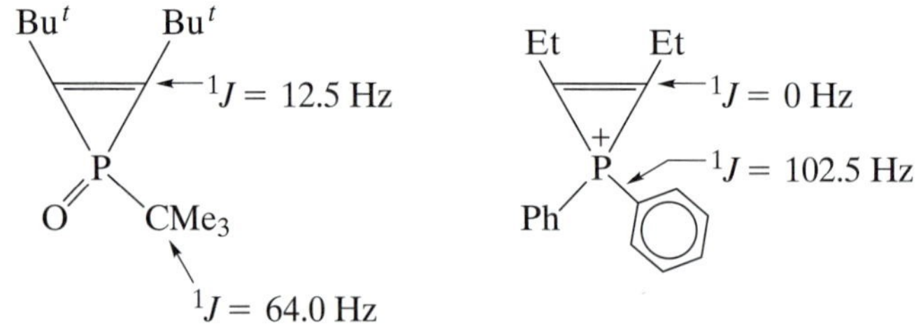

The diversion of s character into the exocyclic bond causes an increase in its $^1J_{PC}$, while the value to the internal carbon diminishes. Rings with four or more members (but not larger rings) show some reduction of the value of $^nJ_{PC}$

Table 6.25 ^{13}C Chemical Shifts and $^n J_{PC}$ Spin Couplings for Benzylphenylphosphines, Ph_3P, Ph_4P^+, and $Ph_3P = O$

	$(PhCH_2)_3P$	$(PhCH_2)_2PhP$	$PhCH_2(Ph)_2P$	Ph_3P	Ph_4P^+	$Ph_3P=O$
			δ			
CH_2	34.69	35.70	36.30			
C—1 ($PhCH_2$)	138.26	138.05	137.80			
o	129.52	129.48	129.65			
m	128.56	128.41	128.40			
p	125.98	125.96	126.04			
C—1 (Ph)		137.87	139.03	137.8		136.6
o		133.10	133.21	134.0		132.3
m		128.30	128.49	128.8		128.8
p		129.13	128.68	128.5		132.3
			$J(CP)$			
CH_2	20.8	18.3	16.7			
C—1($PhCH_2$)	5.7	6.5	8.3			
o	5.6	5.7	7.1			
m	1.0	1.6	1.5			
p	2.0	2.4	2.5			
C—1(Ph)		20.6	16.2	−12.5	88.4	104.4
o		19.8	18.9	+19.6	10.9	9.8
m		7.0	6.4	+6.8	12.8	12.1
p		0.6	0.2	0.3	2.9	2.8

to the internal carbons with tetracovalent phosphorus functions, but tricoordinate P heterocycles give more noticeable effects (Table 6.24).

An increase in the steric bulk of substituents on the carbon attached to phosphorus causes a reduction in $^1 J_{PC}$ as in the series $(Ph_3PCH_3)^+$, 57.1; $(Ph_3PCH_2Me)^+$, 51.6; $(Ph_3PCHMe_2)^+$, 47.0; $(Ph_3PCMe_3)^+$, 42.5 Hz; α substitution by alkyl groups on phosphonates or an increase in the size of the n substituent in a phosphinamide causes a reduction in $^1 J_{PC}$. The value of 51.27 Hz for $Bu^t_3P=O$ seems unexceptional.

An effect noted with steric congestion in acylic phosphines also finds an explanation in s-character changes. For Me_3P, $^1 J_{PC}$ is −13.6 Hz, while for Bu^t_3P, $^1 J_{PC}$ is −33.9 Hz; the large bond angle in the latter (109.9°; Me_3P has 98.8°) increases s character in the P—C bonds. The very large $^1 J_{CP}$ of 71.7 Hz found for the sterically encumbered $Ph_2P C(SiMe_3)_3$ presumably has a similar origin.

In the trigonal bipyramidal geometry adopted by pentacoordinate phosphorus compounds the bonds to axial groups (p) have little or no s character, while bonds to equatorial groups (sp^2) have appreciably greater s character. Consequently, axial carbon substituents should have much smaller $^1 J_{CP}$ than equatorial carbon substituents. The normal Berry-pseudorotation of pentacoordinate phosphorus compounds must be arrested by rigid structural geometry

Table 6.26 ^{1}H and ^{13}C NMR Data for Phenyldibenzophosphole (PhDBP) and Derivatives (δ in ppm relative to TMS, J in Hz)[a]

Compound	Position	δ (^{1}H)	δ (^{13}C)	^nJ (PH)	^nJ (PC)	^nJ (HH)
PhDBP	1	7.88	121.36	1.25	0.5	1.2 (7.25); 1.3 (1.25);
	2	7.39	128.61	0	0	1.4 (−1.25); 2.3 (7.25);
	3	7.25	127.55	2.75	7.8	2.4 (1.25); 3.4 (7.25)
	4	7.63	130.27	5.38	21.9	
	i	—	136.16	—	19.1	
	o	7.23	132.63	8.8	20.1	
	m	7.16	128.61	1.3	7.5	
	p	7.19	129.22	1.7	1.1	
	α	—	142.58	—	2.3	
	β	—	143.60	—	2.9	
PhDBP=O	1	7.74	121.21	3.6	10.2	1.2 (7.9); 1.3 (1.0);
	2	7.49	133.38	1.9	2.3	1.4 (−0.85); 2.3 (7.9);
	3	7.28	129.41	4.0	11.2	2.4 (1.54); 3.4 (7.92);
	4	7.64	129.76	10.1	9.6	
	i	—	130.85	—	103.3	
	o	7.59	130.94	13.2	10.9	
	m	7.32	128.65	3.4	12.4	
	p	7.41	132.15	2.1	2.9	
	α	—	132.74	—	107.0	
	β	—	141.69	—	21.8	
PhDBP=S	1	7.82	121.37	3.2	9.6	1.2 (7.95); 1.3 (1.25);
	2	7.54	132.74	2.0	3.0	1.4 (−0.75); 2.3 (7.38);
	3	7.37	129.70	4.2	12.6	2.4 (1.5); 3.4 (7.5)
	4	7.68	129.40	11.7	11.1	
	i	—	131.02	—	81.6	
	o	7.69	130.83	14.6	12.1	
	m	7.33	128.57	3.4	13.6	
	p	7.42	131.81	2.6	4.0	
	α	—	135.74	—	90.6	
	β	—	141.28	—	15.6	
PhDBP=Se	1	7.78	121.53	3.1	9.2	1.2 (8.10); 1.3 (0.80);
	2	7.49	132.64	1.5	2.0	1.4 (−1.0); 2.3 (7.3);
	3	7.33	129.70	3.9	9.02	2.4 (1.5); 3.4 (7.4)
	4	7.68	129.61	10.9	7.88	
	i	—	130.67	—	93.53	
	o	7.72	131.19	15.0	11.30	
	m	7.29	128.54	3.3	12.28	
	p	7.37	131.84	2.2	4.0	
	α	—	135.15	—	82.66	
	β	—	141.16	—	16.03	
PhDBP—W(CO)$_5$	1	7.93	121.80	2.37	5.6	1.2 (8.13); 1.3 (1.25);
	2	7.55	131.00	1.40	2.5	1.4 (−1.13); 2.3 (8.13);
	3	7.46	128.77	3.80	9.6	2.4 (1.36); 3.4 (8.13)
	4	7.78	130.42	8.75	14.6	
	i	—	133.64	—	36.8	
	o	7.45	131.13	12.4	13.1	
	m	7.32	128.88	2.3	11.1	
	p	7.33	130.56	0	1.5	
	α	—	139.87	—	48.3	
	β	—	141.49	—	8.5	
	CO[b]	—	198.42	—	20.67	
	CO[c]	—	196.17	—	7.54	

PhDBP

[a] Nelson, J. H., Affandi, S., Gray, G. A., Alyea, E. C., *Magn. Reson. Chem*, **1987**, *25*, 774. [b] CO *trans* to PhDBP. [c] CO *cis* to DBP.

Table 6.27 $\delta\,^{13}C$ and J (PC) for Phospholes and Analogs[a]

Carbon	^{13}C chemical shifts (ppm)								
2	136.95	129.31	135.11	128.56	128.52	128.79	131.47	126.69	129.18
3	128.10	142.93	136.71	147.38	148.75	148.62	147.72	148.67	147.82
4		34.21		140.07					
5		25.90		136.00					
6				18.35	17.57	17.35	17.67	17.69	17.48
α						31.66	7.21	32.68	29.64
β								29.17	23.78
γ									23.92
δ									13.50
i	137.31	140.64	129.59	130.62	132.42	138.07			
o	132.32	131.84	133.31	132.98	133.37	128.16			
m	128.48	128.23	128.33	127.99	128.78	127.67			
p	128.56	128.35	128.95	128.09	129.88	125.30			
	$^{13}C—^{31}P$ nuclear spin coupling constants (Hz)								
2	−13.91	−16.03	−5.18	1.2	7.3	5.9	3.7	6.1	3.7
3	+23.16	+1.87	+8.17	8.8	8.5	6.6	7.3	6.1	7.3
4		5.37		7.6					
5		−10.24		5.3					
6				3.7	3.7	2.4	3.7	2.4	3.7
α						18.3	17.01	8.6	7.3
β								11.0	22.0
γ									0
δ									0
i	9.17	25.87	−8.63	10.5	12.2	2.2			
o	+18.70	+18.54	+19.09	19.0	18.3	4.9			
m	+6.74	+6.32	+8.29	8.0	6.1	1.7			
p	0.48	0.61	1.43	0	0	2.2			

[a] Gray, G. A., Nelson, J. H., *Org. Magn. Reson.*, **1980**, *14*, 14.

Table 6.28 $\delta^{13}C$ in ppm and ^nJ (PC) (in parentheses in Hz) for $(RPPh_2)Mo(CO)_5$ Complexes[a]

R	C_i	C_o	C_m	C_p	C_1	C_2	C_3	C_4	CO_{eq}	CO_{ax}
H	132.44 (86)	132.29 (11.9)	129.03 (9.7)	130.29 (0)					205.12 (8.9)	209.58 (23.2)
CH_3	137.43 (36.1)	131.16 (12.1)	128.71 (9.4)	129.95 (1.9)	19.72 (24.5)				205.67 (9.1)	210.17 (23.2)
C_2H_5	136.37 (34.6)	131.88 (11.8)	127.62 (8.5)	129.88 (1.3)	25.50 (22.7)	8.27 (0)			205.69 (8.8)	210.12 (22.6)
$n-C_3H_7$	136.70 (34.9)	131.84 (11.5)	128.65 (9.1)	129.88 (1.2)	34.56 (21.9)	17.86 (2.8)	15.77 (15.6)		205.73 (9.5)	210.22 (22.3)
$i-C_3H_7$	134.28 (32.1)	132.80 (10.8)	124.38 (8.8)	129.80 (1.9)	28.55 (19.4)	18.15 (1.9)			205.79 (8.8)	210.40 (22.9)
$n-C_4H_9$	136.76 (34.2)	131.82 (11.5)	128.62 (9.3)	129.85 (0)	31.99 (22.0)	26.11 (0)	24.06 (18.4)	13.58 (0)	205.72 (9.1)	210.19 (22.5)
$sec-C_4H_9$	134.62 (31.9)	132.74 (10.7)	128.39 (8.7)	129.78 (1.6)	35.25 (18.9)	25.16 (3.1)	12.72 (12.8)		205.80 (8.6)	210.42 (22.9)
	134.49 (31.9)	132.68 (10.8)	128.38 (8.8)	129.74 (1.9)		14.29 (2.1)				
$-CH_2C{\equiv}CH$	135.34 (34.2)	131.86 (11.0)	128.67 (8.6)	130.36 (2.4)	24.84 (19.5)	77.18 (6.1)	73.72 (7.3)		205.51 (9.8)	209.81 (24.4)
$-C{\equiv}CH-CH_3$	135.60 (42.9)	131.16 (14.3)	128.65 (10.4)	130.11 (2.0)	73.80 (80.3)	109.68 (14.5)	5.25 (1.6)		205.31 (9.1)	210.32 (23.5)
$-CH_2-CH{=}CH_2$	136.09 (33.8)	131.99 (11.4)	128.60 (9.1)	130.00 (1.8)	38.23 (20.1)	120.85 (11.2)	129.11 (2.5)		205.60 (9.1)	210.98 (23.1)
$t-CH_2CH{=}CH(CH_3)$	136.45 (33.3)	132.05 (12.0)	128.54 (8.9)	129.91 (0)	36.99 (20.5)	121.54 (2.0)	131.83 (3.3)	17.97 (1.5)	205.71 (9.1)	210.21 (23.3)
$-CH_2C_6H_5$	136.67 (33.3)	132.38 (11.4)	128.58 (9.1)	128.17 (2.6)	40.09 (16.2)				205.35 (8.9)	210.10 (24.0)
	126.98[b] (3.4)	130.07[b] (5.0)	130.05[b] (8.1)	133.04[b] (1.9)						

[a] Maitra, K., Wilson, W. L., Jemin, M. M., Yeung, C., Rader, W. S., Redwine, K. D., Striplin, D. P., Catalano, V. J., Nelson, J. H., *Syn. React. Inorg. Met. – Org. Chem.*, **1996**, *26*, 967.
[b] $CH_2C_6H_5$ resonances.

or by lowering the temperature, in order for such effects to be observed. For the compounds

$^1J_{PC} = 7.3$ Hz

$^1J_{PC} = 116.0$ Hz

and

$^1J_{PC} = 17$ Hz

$^1J_{PC} = 258$ Hz

at −90°C and −108°C, respectively, the $^1J_{PC}$ values are as indicated.*

---

*Additional data on phosphoranes may be found in Cavell, R. G., Gibson, J. A.; Kwat, I. T. *J. Am. Chem. Soc.* **1977**, *99*, 7841.

Table 6.29 $\delta^{13}C$ in ppm and ^nJ (PC) (in parentheses in Hz) for (RDBP)Mo(CO)$_5$ Complexes[a]

R	C_α	C_β	C_1	C_2	C_3	C_4	C'_1	C'_2	C'_3	C'_4	CO_{eq}	CO_{ax}
H	134.33 (43.2)	143.47 (7.2)	121.84 (5.0)	130.92 (0)	128.58 (10.6)	130.63 (15.9)					204.63 (9.5)	208.60 (22.5)
CH$_3$	140.89 (41.9)	141.32 (7.9)	121.74 (5.0)	130.66 (0)	128.51 (10.3)	129.37 (15.7)	20.37 (20.0)				205.15 (9.3)	209.24 (21.2)
C$_2$H$_5$	138.76 (40.1)	142.25 (6.5)	121.65 (4.2)	130.66 (0)	128.42 (10.1)	129.84 (15.2)	27.55 (18.7)	8.42 (0)			205.24 (8.1)	209.20 (22.0)
n—C$_3$H$_7$	139.20 (39.8)	142.08 (6.4)	121.58 (4.6)	130.54 (0)	128.37 (10.1)	129.76 (15.3)	36.24 (17.8)	17.97 (3.2)	15.29 (13.8)		205.26 (8.5)	209.24 (21.1)
i—C$_3$H$_7$	138.84 (37.8)	142.23 (5.9)	121.51 (4.5)	130.48 (1.4)	128.19 (9.8)	130.32 (14.7)	32.07 (17.4)	18.04 (2.1)			205.38 (8.8)	209.05 (21.4)
n—C$_4$H$_9$	139.29 (40.4)	142.01 (6.9)	121.59 (4.5)	130.53 (0)	128.36 (10.0)	129.75 (15.1)	33.74 (17.8)	26.32 (3.0)	23.68 (12.9)	13.42 (0)	205.26 (8.3)	209.24 (21.0)
sec—C$_4$H$_9$	139.09 (37.6)	142.29 (5.9)	121.50 (4.7)	130.43 (2.0)	128.19 (9.6)	130.42 (14.5)	38.34 (17.0)	24.76 (3.0)	12.55 (13.6)		205.29 (8.8)	209.12 (21.37)
	138.67 (37.3)	142.13 (5.9)	121.46 (4.5)	130.42 (1.6)	128.15 (9.6)	130.29 (14.7)		13.86 (1.5)				
—CH$_2$C≡CH	138.45 (39.9)	141.97 (7.7)	121.65 (4.8)	131.19 (0)	128.55 (10.1)	120.06 (15.3)	25.77 (15.9)	76.72 (6.1)	73.10 (8.2)		204.86 (9.8)	208.64 (24.0)
—CH$_2$CH=CH$_2$	138.67 (40.1)	142.11 (6.7)	121.61 (4.8)	130.70 (0)	128.31 (10.1)	129.93 (15.2)	39.48 (15.7)	119.94 (9.6)	129.00 (7.2)		205.14 (9.1)	209.04 (22.4)
1—CH$_2$CH=CH(CH$_3$)	139.28 (39.3)	141.91 (6.2)	121.56 (4.6)	130.59 (0)	128.22 (9.9)	129.98 (14.9)	38.44 (16.6)	121.73 (6.3)	131.44 (11.0)	17.87 (2.3)	205.24 (9.1)	209.16 (22.9)[b]
CH$_2$C$_6$H$_5$[c]	138.34 (40.0)	142.30 (6.2)	121.39 (4.5)	130.68 (1.6)	128.16 (10.2)	130.29 (15.5)	42.03 (12.4)				205.08 (9.1)	209.03 (22.6)

[a] Maitra, K., Wilson, W. L., Jemin, M. M., Yeung, C., Rader, W. S., Redwine, K. D., Striplin, D. P., Catalano, V. J., Nelson, J. H., *Syn. React. Inorg. Met. - Org. Chem.*, **1996**, 26, 967.

[b] C$_1$—C$_4$. C$_\alpha$ and C$_\beta$ are phosphole carbons; C'$_1$ —C'$_4$ are R carbons.
[c] C 132.91(7.7); C$_{ax}$ 129.14(4.3); C$_{eq}$ 127.66(2.6); C$_p$ 126.55(3.3).

Extensive data support the idea that $^2J_{PC}$ in trivalent phosphorus compounds is dominated by the proximity of the phosphorus lone pair to the coupled carbon atom. In general, coupling is large and positive when the lone pair is close, and it is quite small, and even negative, when the lone pair is remote. This effect is particularly well documented among aminophosphines (Table 6.23), for which structural parameters are available for a number of N-methyl derivatives from X-ray or electron diffraction studies. The nitrogen atom is planar (or nearly so) in these compounds, as shown in the following Newman projection for an aminophosphine of the type $Me_2NP(X)(Y)$:

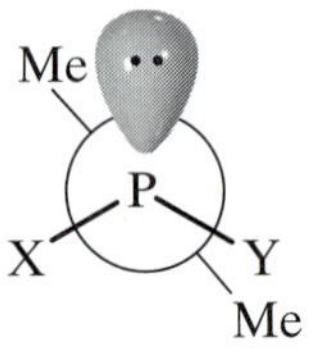

If the lone pair is assumed to lie in the plane that bisects the X—P—Y angle along the P—N bond, then one can express a dihedral angle (θ) relating the orientation of the lone pair to the coupled N-methyl groups. In any fixed conformation, the two methyl groups are diasteriotopic, and two resonances should be observed for them. If the molecule is stereochemically nonrigid, then a single resonance should be observed, but $^2J_{PC}$ may then be temperature dependent. $^1J_{PC}$ may be temperature dependent in aminophosphines for the same reason. $^2J_{PC}$ through oxygen or silicon is subject to similar stereochemical control. For phosphines, this effect is of greatest value in cyclic systems, in which rotation about P—C bonds is restricted and the lone pair is held in a definite location. For example, in the compound

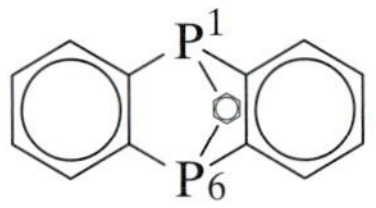

$^2J_{PC}$ is +44.1 Hz and $^2J_{P6C_{1a}}$ is −3.9 Hz. Replacing P_6 with a CH group gives a compound with similar properties ($^2J_{PC2} = 36.5$ Hz; $^2J_{PC5a}$ is about zero), as seen in Table 6.30.

A considerable amount of data is available for group-six metal carbonyl complexes of the type $(R_3P)M(CO)_5$ (M=Cr, Mo, W)(cf. Tables 6.28 and 6.29). There are two types of chemical-shift inequivalent CO moieties in these compounds: CO(ax) *cis* to R_3P and CO(eq) *trans* to R_3P. It is generally found that $^2J_{PC}$ (*trans*) is a little greater than twice $^2J_{PC}$ (*cis*) in these compounds. Also, comparing the data in Tables 6.20 and 6.25 with the data in Table 6.30, one observes that, upon coordination to a metal carbonyl, $^1J_{PC}$ increases substantially, $^2J_{PC}$ decreases substantially for aliphatic carbons, but only moderately for aromatic carbons, and $^3J_{PC}$ and $^4J_{PC}$ change little for either type of carbon.

The control of $^3J_{PC}$, which is positive, in tetracovalent phosphorus compounds by the dihedral angle (θ) relating the coupled nuclei is very useful for stereochemical analysis. As with vicinal couplings of other nuclei, this Karplus* effect allows maximum values at $\Theta = 0°$ and $180°$; little or no coupling is observed at $\Theta = 90°$. Karplus relations are found for $^3J_{POCC}$ in

---

*Minch, M. J., *Concepts Magn. Reson.*, **1994**, *6*, 41 discusses the Karplus relationships in general.

Table 6.30 δ^{13}C and J(PC) for Dibenzophospholes and Analogs[a]

Carbon						
^{13}C chemical shifts (ppm)						
10a, b	142.68	143.72	143.6	118.96	141.1	145.32
9a, b	143.61	144.97	144.7	155.64	149.6	134.42
1	128.61	128.76	129.1	118.20	129.2	128.24
2	121.37	121.35	122.1	131.25	125.2	
3	127.51	127.14	127.8	124.09	125.2	
4	131.01	129.49	130.6	135.25	132.8	
α		13.97	22.5			
β			9.5			
i	135.24		133.96	140.86		
o	132.06		133.05	132.16		
m	129.62		128.43	128.32		
p	129.22		128.52	128.91		
^{13}C—^{31}P nuclear spin coupling constants (Hz)						
10a, b	3.0	2.2	6.7	4.4	8.7	−7.6
9a, b	2.9	3.9	1.5	0	0	−3.9
1	0	0	0	0	0	+1.8
2	0.4	0	0	0.5	0	−0.1
3	7.3	7.6	7.4	11.4	13.9	+12.5
4	20.1	22.0	21.3	34.8	36.5	+44.1
α		19.5	19.0			
β			6.6			
i	19.0			13.2		
o	21.9			18.5		
m	7.8			7.6		
p	1.0			0		

[a]Gray, G. A., Nelson, J. H., *Org. Magn. Reson.*, **1980**, *14*, 14.

nucleotides and for $^3J_{PCCC}$ in α hydroxyphosphonates and phosphonates, phosphine oxides ($Me_2P(O)R$), phosphine sulfides ($Me_2P(S)R$), and phosphonium iodides (Me_3PR^+). Quin[†] discusses many applications for both freely rotating and cyclic systems.

C. $^nJ_{PF}$ Couplings

Representative values of $^nJ_{PF}$ are listed in Table 6.31. $^1J_{PF}$ is apparently always negative.

For tricovalent phosphorus compounds, $^1J_{PF}$ seems to increase with increasing electronegativity of the phosphorus substituents.

When a fluorophosphine is coordinated to BH_3, $^1J_{PF}$ decreases slightly. (The range is 1,114 to 1,288 Hz.) When $PF_{3-n}Cl_n$ ($n = 0$ to 2) are protonated, $^1J_{PF}$ decreases by 100 Hz. Coordination to transition metals frequently causes

Table 6.31   Representative Values of $^nJ_{PF}$ Couplings

$^1J_{PF}$

Compound class (or structure)	$^nJ_{PF}$ (Hz)	Compound class (or structure)	$^nJ_{PF}$ (Hz)
P(III)		P(IV)	
\>P—F	820 to 1,450	\>P(E)F	1,000 to 1,400
		E=(O,S)	
PF_3	−1,441	$P(O)F_3$	1,063
P_2F_4	−1,194	$P(S)F_3$	1,177
		$F_2P(O)O(O)PF_2$	−1,063
P(V)		P(V)	
\>P—F	530 to 1100		
PF_5	938	PF_6^-	708–714[*]
$P(CF_3)_3F_2$	541		

$^2J_{PF}$

Compound class (or structure)	$^nJ_{PF}$ (Hz)	Compound class (or structure)	$^nJ_{PF}$ (Hz)
P(III)		P(IV)	
\>P—C—F	40 to 149	P(O)CF	100 to 130
$P(CF_3)_3$	85.5	$P(O)(CF_3)_3$	113.4
P(V)		P(V)	
\>PCF	124 to 193	—PCF	130 to 160
$P(CF_3)Cl_4$	154	$P(CF_3)F_5^-$	145

[*]Solvent dependent.

[†]Quin, L. D. in *Phosphorus-31 NMR Spectroscopy in Stereochemical Analysis*, Verkade, J. G.; Quin, L. D. eds., VCH, Deerfield Beach, FL., 1987.

a decrease in $^1 J_{PF}$ of 50 to 100 Hz, the range of values being about 1,070 to 1,555 Hz. For $Hg_2PF_3^{2+}$ (1,555 Hz) and $Ag(PF_3)^+$ (1,487 Hz), however, the coupling is somewhat greater than for free PF_3 ($-1,441$ Hz). For $V(PF_3)_6^+$ (1,200 Hz) and $Nb(PF_3)_6^+$ (1,250 Hz), $^1 J_{PF}$ is 16% less than for free PF_3. For the series of compounds $M(n-C_3H_7OPF_2)_6$, $^1 J_{PF}$ increases in the sequence $1,138 < 1,150 < 1,187$ Hz for W < Mo < Cr.

Tetracovalent fluorophosphorus compounds generally have $^1 J_{PF}$ from -760 Hz to $-1,406$ Hz. For trimeric cyclophosphazenes, $^1 J_{PF}$ is found in the range from 760 to 1,056 Hz, with covalently bound anions of the type FPO_2Z^- (Z=O, S) and $F_2P(Z)Z^-$ (Z=O, S) at the lower end of this range. Increasing n in $(NPF_2)_n$ from 3 to 9 causes a general increase in $^1 J_{PF}$. The coupling in the $F_2PS_2^-$ ion (1,150 Hz) increases when coordination through sulfur occurs (about 1,200 to 1,300 Hz). The couplings in the cations $Me_2PF_2^+$ (1,145 Hz) and $RPF(NR_2)_2^+$ (1,032 to 1,051 Hz) are somewhat larger than those for $[RPF(N=PR_3)_2]^+$ (948 to 961 Hz).

For pentacovalent phosphorus compounds, which are generally trigonal bipyramidal, $^1 J_{PF}$ is usually, but not always, larger to equatorial fluorines than to axial fluorines. The same is true of $^2 J_{PF}$ in CF_3-substituted compounds.*

D. $^n J_{PP}$ Couplings

Representative values of $^n J_{PP}$ are listed in Table 6.32 for organophosphorus compounds.

For transition-metal complexes containing two or more phosphine ligands, $^2 J_{PP}$ may be directly observed if the phosphorus nuclei are anisochronous. If the

Table 6.32 Representative Values of $^n J_{PP}$ for Organophosphorus Compounds

Structural class (or structure)	$^1 J_{PP}$ (Hz)	Structural class (or structure)	$^2 J_{PP}$ (Hz)
P(III)			
>P—P<	100 to 400	PEP	70 to 90
H_2PPH_2	-108	(E=S, C) E=N	100–600
F_2PPF_2	-227.4		
P(III)–P(V)	$^2 J_{PP}$	P(V)	$^2 J_{PP}$
>POP< (O double bond)	0 to 23	C-triphosphate	19.8
$F_2P(O)OP(O)F_2$	0 to 4	C-diphosphate	23.1
$Me_2PP(S)Me_2$	-220	P(V)–P(V)	
$Y^1Y^2PP(O)Y^3Y^4$	157 to 392	$[O_2FPPFO_2]^{2-}$	$+766$
Me_3PPF_5	$+715$	$[O_2FPPO_3]^{3-}$	$+650$
		$[O_2HPPO_3]^{3-}$	$+465.5$

*Cavell, R. G., in "*Phosphorus-31 NMR Spectroscopy in Stereochemical Analysis*" Verkade, J. G., Quin, L. D., eds. VCH, Deerfield Beach, FL, 1987.

two phosphorus nuclei are isochronous in a bis-phosphine complex, they are usually anisogamous, and $^2J_{PP}$ may often be calculated from the second-order ^1H or, more commonly, $^{13}C\{^1H\}$ NMR spectra of these complexes.[*]

In *cis*-coordinated bis (monodentate phosphine) complexes, the coupling between the two phosphorus atoms occurs predominantly through the metal and may be described as a $^2J(PMP)$. Incorporating the two phosphorus donor atoms into a chelate ring provides an additional pathway for $^2J_{PP}$ coupling through the sigma bonds of the chelate backbone. For example, in $(Ph_2PCH_2CH_2PPh_2)M(CO)_4$, the values of $^2J_{PP}$ are lower than might be expected. This is because the two pathways for P—P coupling—through the metal (^2J) and via the ligand backbone (^3J)—often have similar magnitudes and opposite signs.[†] Similar observations have been made for a variety of metal complexes containing chelating diphosphines.

For most transition-metal bis phosphine complexes, the magnitude of $^2J_{PP}$ is dependent upon whether the two phosphorus nuclei are mutually *cis* or *trans* to one another. The same phenomenon also seems to hold for $^2J_{PMX}$, where $X = {^1H}, {^{13}C}, {^{15}N}, {^{107,109}Sn}$, and ^{199}Hg. Perhaps it is a rather general phenomenon. Usually, $^2J_{PX}$ (*trans*) is larger than $^2J_{PX}$ (*cis*), often by an order of magnitude.

For $(R_3P)_2M(CO)_4$ ($M = Cr$, Mo, W) complexes, $^2J_{PP}$ (*cis*) decreases along the series Cr > Mo > W, while $^2J_{PP}$ (*trans*) shows somewhat opposite behavior. Some illustrative data are as follows:

	cis complexes				*trans* complexes		
R_3P	Cr	Mo	W	R_3P	Cr	Mo	W
PH_3	−26.2	18.9	13.4	PMe_3	−28.5		
PMe_3	−36	−29.7	−25.0	$P(NMe_2)_3$	−17	101	81
PF_3	78	55	38	PF_3	34	312	315

For $(R_3P)_2MX_2$ ($M = Ni$, Pd, Pt) complexes, $^2J_{PP}$ (*cis*) decreases along the series Ni > Pd > Pt and often decreases as the atomic weight of the halide increases. The following data are illustrative:

cis complexes				*trans* complexes	
Compound	$^2J_{PP}$	Compound	$^2J_{PP}$	Compound	$^2J_{PP}$
$PdCl_2(PMe_3)_2$	−8	$PtCl_2(PMe_2H)_2$	0	$PdCl_2(PMe_3)_2$	+610
$PdBr_2(PMe_3)_2$	−1	$PtBr_2(PMe_2H)_2$	0	$PdBr_2(PMe_3)_2$	+612
$PdCl_2(PMe_2H)_2$	0	$PtCl_2(PMe_3)_2$	−18.9	$PdI_2(PMe_3)_2$	+604
$PdBr_2(PMe_2H)_2$	2	$PtBr_2(PMe_3)_2$	−16.2	$PtCl_2(PMe_3)_2$	+510
$PdCl_2[P(OMe)_3]_2$	+79.9	$PtI_2(PMe_3)_2$	−14.0	$PtBr_2(PMe_3)_2$	+514
$PdBr_2[P(OMe)_3]_2$	84.8			$PtI_2(PMe_3)_2$	+498
$PdI_2[P(OMe)_3]_2$	61.9				

[*]A large amount of data is available in Pregosin, P. S., Kunz, R. W., *^{31}P and ^{13}C MNR of Transition Metal Phosphine Complexes*, Springer-Verlag, Berlin, 1979.

[†]Grim, S. O., Briggs, W. L., Barth, R. C., Tolman, C. A., Jesson, J. P., *Inorg. Chem.*, **1974**, 13, 1095.

For $RhX(PPh_3)_3$, $^2J_{PP}$ is −38, −37, and −36 Hz for $X = Cl$, Br, and I, respectively. For *trans*-$RhX(CO)(PR_3)_2$, $^2J_{PP}$ ranges from 306 to 538 Hz, and for $RhX(CO)(PP)$, where PP is a chelating diphosphine, $^2J_{PP}$ (*cis*) is about 34 Hz. For $(PR_3)_3IrX_3$ complexes, $^2J_{PP}$ (*cis*) is 16 to 20 Hz, while $^2J_{PP}$ (*trans*) is 320 to 450 Hz. As can be seen from the data presented, the magnitude of $^2J_{PP}$ in transition-metal complexes may often be used to ascertain whether two phosphorus nuclei are mutually *cis* or *trans* to one another.

E. $^1 J_{MP}$ Couplings

$^1J_{PM}$ has been measured for 21 metals. For quadrupolar metals $\left(I > \frac{1}{2}\right)$, measurements have usually involved CP/MAS ^{31}P NMR spectroscopy of solids (Chapter 14). For dipolar metals $\left(I = \frac{1}{2}\right)$, measurements have come from observations of the ^{31}P or M spectrum, or both, for solutions.

$^1J_{PM}$ is believed to increase as the *s* character in the M—P bond increases. This is consistent with the larger values of $^1J_{PM}$ found for equatorial P ligands than for axial P ligands in trigonal bipyramidal Rh(I) or Pt(II) complexes. It is also consistent with a general decrease in $^1J_{PM}$ as the oxidation state increases for platinum complexes. For *cis*-$Cl_2Pt(PR_3)_2$ complexes, $^1J_{PtP}$ is larger than for the *trans* isomers. This behavior has been explained in terms of the weaker *trans* influence of a Cl ligand than of a phosphine ligand. For a given type of metal complex, changing from a phosphine to a phosphite ligand induces an increase in $^1J_{PM}$ of between 50 and 100%. The remainder of this section will discuss $^1J_{PM}$ by the element *M*, according to its position in the periodic table.

i. $^1 J P^{51}V$

For an extensive series of $(\eta^5—C_5H_5)V(CO)_3PR_3$ complexes, $^1J P^{51}V$ ranges from 110 to 427 Hz and increases with increasing phosphorus-substituent electronegativity. For a series of $[V(CO)_{6-n}(PR_3)_n]^-$ and $(\eta^5—C_5H_5)V(CO)_{4-n}(PR_3)_n$ complexes ($n = 1, 2$), the $^1J P^{51}V$ correlate nearly linearly with the sums of the Taft τ_1 values for the ligand substituents.[*] $^1J P^{51}V$ increases by a factor of 1.2 from *cis*-$(\eta^5—C_5H_5)V(CO)_2(PR_3)_2$ to the *trans* isomer, a rise that has been attributed to the greater *trans* influence of CO than of PR_3.[†]

ii. $^1 J P^{93}Nb$

The significantly larger $^1J P^{93}Nb$ coupling in $[Nb(PF_3)_6]^-$ (1050 Hz) compared with $^1J P^{51}V$ for $[V(PF_3)_6]^-$ (510 Hz) has been ascribed to greater *s* character in the Nb—P bond and enhanced π bonding.

iii. $^1 J P^{95}Mo$

For $Mo(CO)_5(PR_3)$ complexes, $^1J P^{95}Mo$ ranges from 90 to 284 Hz and increases quite linearly with increasing phosphorus-substituent electronegativity. Couplings in $Mo(CO)_{6-n}(PR_3)_n$ ($n = 2$ to 6) for a variety of isomers are apparently less sensitive to a stereochemical environment than is $^1J P^{183}W$ in

---

[*] Rehder, D., Dorn, W. L., *Transition Metal Chem.*, **1977**, *1*, 233. Rehder, D., Schmidt, J., *Transition Metal Chem.*, **1977**, *2*, 41.

[†] Appleton, T. G., Clark, H. C., Manzer, L. F., *Coord. Chem. Rev.*, **1973**, *10*, 335.

analogous complexes. However, the uncertainty surrounding $^1JP^{95}Mo$ is about ± 5 Hz, due to the broad line widths of ^{95}Mo resonances. Some representative data are as follows:*

PR_3	$Mo(CO)_5(PR_3)$	cis-$Mo(CO)_4(PR_3)_2$	fac-$Mo(CO)_3(PR_3)_3$
PBu_3	123	123	—
PPh_3	137	140	120
PPh_2Cl	165	166	154
$PPhCl_2$	197	194	198
PCl_3	250	250	251
PBr_3	242	235	—
PF_3	284	281	290

iv. $^1JP^{183}W$

The range of $^1JP^{183}W$ is about 112 to 526 Hz. For $W(CO)_5PR_3$ complexes, the range is 143 to 485 Hz, and $^1JP^{183}W$ increases linearly with increases in the electronegativities of the phosphorus substituents.

As expected on the basis of the *trans* influence, cis-$W(CO)_4(PR_3)_2$ complexes exhibit smaller couplings than those found for the analogous *trans* isomers. For complexes of the type $W(CO)_4(PR_3)(PR_3')$, $^1JP^{183}W$ correlates linearly with $^2JPP'$. For $(\eta^5-C_5H_5)W(CO)_2(Me)(PR_3)$ systems, the coupling for the *cis* isomer is greater than that for the *trans* isomer, because the *trans* influence of the Me group is greater than that of CO.

v. $^1JP^{55}Mn$, $^1JP^{199}Tc$, $^1JP^{57}Fe$, $^1JP^{99}Ru$, and $^1JP^{187}Os$

Very little data are available for these five nuclei. For $Mn_2(CO)_9PPh_3$ and $Mn_2(CO)_8(PPh_3)_2$, $^1JP^{55}Mn$ is 297 and 335 Hz, respectively. For ruthenium carbonyl phosphine complexes of various types, $^1JP^{99}Ru$ ranges from 100 to 174 Hz. The only report of $^1JP^{199}Tc$ is for $Tc[P(OMe)_3]_6^+$ (909 Hz). $^1JP^{57}Fe$ of 26 to 27 Hz have been observed for the $Fe(CO)_4$ $(PEt_{3-n}Ph_n)$ series, and for $Fe(CO)_2(CS_2)(PMe_3)$, $^1JP^{57}Fe = 31$ Hz. For osmium carbonyl phosphine complexes of various types, $^1JP^{187}Os$ ranges from 71 to 317 Hz.

vi. $^1JP^{59}Co$

For $[Co(PF_3)_4]^-$, $^1JP^{59}Co$ is 1,222 Hz. For complexes of the type $[Co(PR_3)_6]^{3+}$, $^1JP^{59}Co$ ranges from 375 to 453 Hz and decreases linearly with increasing ligand field strength of PR_3. This decrease is opposite to trends in $^1J_{PM}$ discussed previously. It is likely that $^1JP^{59}Co$ in these complexes is dominated by the $^3\Delta E$ term, which is a measure of the ligand field strength. For compounds of the type $(PR_3)Co(DH)_2X$, where DH is dimethylglyoxime, $^1JP^{59}Co$ ranges from 226 to 615 Hz, increasing as the *trans* influence of X decreases.

vii. $^1JP^{103}Rh$

A large number of $^1JP^{103}Rh$ couplings (ranging from 21 to 374 Hz) have been reported. The magnitude of the coupling is a function of the rhodium oxidation state, coordination geometry, and other ligands coordinated to

*Alyea, E. C., Song, S., *Inorg. Chem.*, **1995**, *34*, 3864.

rhodium. Some of the data for four-coordinate and five-coordinate Rh(I) compounds and six-coordinate Rh(III) compounds is discussed in the next four sections.

a. Monomeric Four-Coordinate Rh(I) Compounds   For Rh(CO)(5-Me-8-oxyquinolate)PR$_3$ compounds, $^1JP^{103}$Rh increases linearly (155 to 170 Hz) with decreasing phosphine basicity. For complexes of the type *trans*-ZRh(X)(PR$_3$)$_2$, where Z is CO, N$_2$, CS, H$_2$C=CH$_2$, or RC≡CR, $^1JP^{103}$Rh (104 to 217 Hz) increases with increasing electronegativities of PR$_3$ and Z. For complexes of the type (PR$_3$)$_3$RhX, $^1JP^{103}$Rh for PR$_3$ *trans* to X (189 to 195 Hz) is greater than that for the mutually *trans* PR$_3$ ligands (139 to 145 Hz), and both change very little with a change in the halide X.

b. Dimeric Four-Coordinate Rh(I) Compounds   For [(PR$_3$)$_2$Rh(μ—Cl)]$_2$ compounds, $^1JP^{103}$Rh ranges from 195 to 294 Hz and, for [(PR$_3$)$_2$Rh(μ—Cl)$_2$ Rh(cyclooctadiene)] complexes, from 141 to 311 Hz. For the latter compounds, $^1JP^{103}$Rh is linearly related to δ^{103}Rh and to the basicity of PR$_3$. For bridging diphosphine complexes of the type

the ^{31}P NMR spectra are second-order $[A]_2[X]_4$ ($A=^{103}$Rh, $X=^{31}$P) spin systems for which only the sum $|^1J_{RhP} + {}^3J_{RhP}| = 114$ Hz can be determined when $n = 1$ and $X=$Cl. The values when $n = 3(125$ Hz) and $n = 4(121$ Hz) suggest that $^3J^{103}$RhP is quite small.

c. Five-Coordinate Complexes   For trigonal bipyramidal complexes of the type *trans*-YRh(PR$_3$)$_2$X$_2$ ($Y=$H, (O)CCH$_2$CH$_2$Ph), $^1JP^{103}$Rh is 106 to 108 Hz. For the stereochemically nonrigid [Rh(SnX$_3$) (diolefin) (PR$_3$)$_2$] complexes, $^1JP^{103}$Rh is 128 to 135 Hz. For complexes of the type [HRh(CO)(PR$_3$)$_3$], $^1JP^{103}$Rh spans the range 123 to 154 Hz, with little difference in the magnitude of coupling to either axial or equatorial phosphines. For Rh[P(OR)$_3$]$_5^+$ complexes, pseudorotation is sufficiently slow at low temperature to differentiate the coupling to axial (about 140 Hz) and equatorial (about 205 Hz) ligands.

d. Six-Coordinate Rh(III) Complexes   For complexes of the type [(η^5—C$_5$H$_5$)Rh(PR$_3$)Cl$_2$], $^1JP^{103}$Rh is 133 to 150 Hz. For *fac*-[RhCl$_3$(PR$_3$)$_3$], $^1JP^{103}$Rh is 110 to 122 Hz, and for the *mer* isomers, coupling to the phosphine *trans* to chloride (103 to 115 Hz) is larger than that to the mutually *trans* phosphines (82 to 85 Hz), consistent with the greater *trans* influence of a phosphine than of Cl$^-$.

viii. $^1JP^{61}$Ni

For complexes of the type (PR$_3$)Ni(CO)$_3$ and (PR$_3$)$_4$Ni, $^1JP^{61}$Ni is 161 to 232 Hz and 398 to 482 Hz, respectively. For the latter, the magnitude of the coupling increases with increasing electronegativity of the phosphorus substituents.

ix. $^1JP^{105}$Pd

For complexes of the type *cis*- and *trans*-(PR$_3$)$_2$ PdX$_2$, $^1JP^{105}$Pd ranges from 160 to 420 Hz, with the coupling usually being larger for the *cis* isomers.

x. $^1JP^{195}Pt$

The literature concerning these couplings is extensive.* For the most part, trends in $^1JP^{195}Pt$ parallel those of $^1JP^{103}Rh$, and, as expected, the metal oxidation state, geometry, and *trans* influence are all important in determining the magnitude of $^1JP^{195}Pt$. The magnitude of the coupling generally decreases with increasing platinum oxidation state, and the subject will be considered on this basis.

a. Platinum (0) Complexes For complexes of the type $(PR_3)_nPt$, $^1JP^{195}Pt$ generally increases with decreasing n and hence increasing Pt s character, as illustrated by the following data:

	PR$_3$	(PR$_3$)$_2$Pt	(PR$_3$)$_3$Pt	(PR$_3$)$_4$Pt
$R=$	*i*-Pr	4,155		
	t-Bu	4,399		
	Cy	4,159	4,236	
	Et		4,220	3,740
	Ph		4,370	3,780
	Me			3,835

For $[P(OR)_3]_4Pt$ complexes, $^1JP^{195}Pt$ ranges from 5,371 to 5,798 Hz and increases with increasing steric bulk of R.

b. Platinum (II) Complexes Platinum (II) complexes are generally square planar when four coordinate and either square pyramidal or trigonal bipyramidal when five coordinate. For these types of complexes, $^1JP^{195}Pt$ ranges from 1,400 to more than 5,000 Hz. For square planar complexes of the type $(R_3P)_2PtX_2(X=Cl^-, Br^-, I^-)$, $^1JP^{195}Pt$ is near 2,400 Hz for the *trans* isomers and near 3,000 Hz for the *cis* isomers. This criterion for distinguishing isomers should be applied with caution, as exceptions are known and it is ambiguous for $X=NO^-_2$ and R_2S. Couplings often increase with an increase in the inverse cube of decreasing PtP bond lengths. $^1JP^{195}Pt$ increases slightly with increasing phosphine basicity for complexes of the type *cis*-PtCl$_2$[P(C$_6$H$_4$—p—Z)$_3$]$_2$ (Z=Cl, 3,652 Hz; F, 3,666 Hz; H, 3,676 Hz; CH$_3$, 3,691 Hz; OCH$_3$, 3,708 Hz). The magnitude of $^1JP^{195}Pt$ is a function of the *trans* influence of the *trans* ligand, as illustrated by the data in Tables 6.33 and 6.34. As the *trans* influence of the group *trans* to phosphorus increases, $^1JP^{195}Pt$ decreases. As can be seen from the data in Table 6.34, similar effects have been noted for the relation of the *trans* influence to ^1JPtH, ^1JPtC, and ^2JPtH.

For *sym-trans*–[(R$_3$P)PtX$_2$]$_2$ complexes, analysis of second-order [AX]$_2$ ^{31}P{^{1}H} NMR spectra provides values of ^1JPtP, ^3JPtP, ^4JPP, and ^2JPtP. Typical values are given in Table 6.35 and again show effects that may be rationalized on the basis of the *trans* influence.

---

*Pregosin, P. S., Kunz, R. W. "^{31}P and ^{13}C NMR of Transition Metal Complexes," Springer-Verlag: New York, 1979. Pregosin, P. S., *Coord. Chem. Rev.* **1982**, *44*, 247. Pregosin, P. S., *Chimia* **1981**, *35*, 43.

Table 6.33   Typical $^{1}JP^{195}Pt$ Couplings as a Function of the Anionic Ligand in Square Planar Pt(II) Complexes

X	trans–[PtX$_2$(Bu$_3$P)$_2$]	[PtX(Et$_3$P)$_3$]ClO$_4$		cis–[PtX$_2$(Bu$_3$P)$_2$]
		P trans to P	P trans to X	
O—NO$_2$	2506	2380	3497	3795
N$_3$		2312	3124	3409
NCO	2322			3396
NCS	2208			3344
NO$_2$	2607	2412	2700	2998
CN	2158	2100	2530	
Cl	2400	2233	3499	3504
Br	2327	2255	3490	3477
I	2265	2247	3313	3345

Table 6.34   Representative Values of $^{1}J_{PtP}$, $^{1}J_{PtH}$, $^{1}J_{PtC}$, and $^{2}J_{PtH}$

trans–H PtX(PEt$_3$)$_2$			trans–CH$_3$PtX(PEt$_3$)$_2$			
X	$^{1}J_{PtP}$	$^{1}J_{PtH}$	X	$^{1}J_{PtP}$	$^{1}J_{PtC}$[b]	$^{2}J_{PtH}$
Cl	2,723	1,275	Cl	2,821	673	84.2
Br	2,685	1,345	Br	2,797		83.0
I	2,632	1,369	I	2,753	664	82.4
CN	2,685	778	NO$_3$	2,925		86.0
NCO	2,715	1,080	N$_3$	2,836		78.6
NCS	2,700	1,086	NCO	2,800		79.5
SCN	2,618	1,233	NCS	2,763	632	78.6
NO$_2$	2,768	1,003	NO$_2$	2,868	564	70.9
NO$_3$	2,824	1,322	CN	2,595	495	60.2
N$_3$	2,730					
CH$_3$CN[a]	2,633					
CO[a]	2,443	967				
PPh$_3$[a]	2,480	890				
PEt$_3$	2,515	790				

[a] as BF$_4$ salts,
[b] trans-CH$_3$PtX(PMe$_2$Ph)$_2$.

For square-based pyramidal complexes of the type (R$_3$P)$_3$PtX$_2$, $^{1}JPtP$ is larger to the axial phosphine (3,154 to 3,379 Hz) than to the basal phosphines (2,141 to 2,251 Hz).

c. Platinum (IV) Complexes   For mer-[(R$_3$P)$_3$PtX$_3$]$^+$ complexes $^{1}J_{PtP}$ trans to P (1,454 to 1,461 Hz) is smaller than $^{1}J_{PtP}$ trans to X (1,975 to 2,054 Hz).

Table 6.35 Representative Values of Coupling Constants for *sym-trans*–[(R₃P)PtX₂]₂ Complexes

$$
\begin{array}{c}
X \quad\quad X \quad\quad PR_3 \\
\diagdown \quad / \diagdown \quad / \\
Pt \quad\quad Pt \\
/ \quad \diagdown \quad / \diagdown \\
P_3R \quad\quad X \quad\quad X
\end{array}
$$

R	X	$^1J_{PtP}$	$^3J_{PtP}$	$^4J_{PP}$	$^2J_{PtPt}$
CH₃	Cl	3,834			
	Br	3,862			
	I	3,840			
C₂H₅	Cl	3,849	23.6	3.0	193
n—C₃H₇	Cl	3,816	−23		200
	Br	3,669	−24		239
	I	3,489	−24		391
n—C₄H₉	Cl	3,822	23.5	2.8	199
	I	3,528	26.0	5.1	380

For *cis*-(R₃P)₂PtCl₄, $^1J_{PtP}$ is larger than for the *trans* isomers (about 2,070 Hz vs. about 1,460 Hz). For complexes of the type *fac*-[PtXMe₃(R₃P)₂], $^1J_{PtP}$ ranges from 1,080 to 1,230 Hz and generally increases with increasing phosphine basicity.

xi. $^1JP^{63,65}$ Cu

For copper (I) complexes of the type (R₃P)ₙCuX, the ratio $^1J^{63}CuP/^1J^{65}CuP$ is generally equal to $\gamma^{63}Cu/\gamma^{65}Cu = 0.933$ when resolvable. It is also found that $^1J_{CuP}$ increases with an increase in $1/(dCuP)^3$ (Figure 6.8) and increases as *n* decreases as follows: *n* = 4 (750 to 1,030 Hz); 3 (879 to 980 Hz); 2 (903 to 1,400 Hz); 1 (1,486 to 2,160 Hz) (Table 6.36).

▶ **FIGURE 6.8**

Correlation of $1/r^3$ (*r*—Cu—P bond length) and the scalar $a(^{63.65}$Cu—P) coupling constant (as approximated by Δv_2, $1/(r/Å^3) = (1.582 \times 10^{-5})$ $a/(h/Hz) + 6.441 \times 10^{-2}$ ($r^2 = 0.90$). The numbers on the points correspond to the compound numbers in Table 6.36.

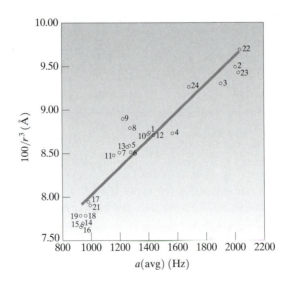

Table 6.36   Scalar Coupling Constants and Copper–Phosphorus Bond Distances for a Variety of Complexes[a]

no.	Compound Formula	$d(CuP)(avg)$, Å	$100/r^3$	$a(avg)$, Hz
1	$[(DMPP)CuI]_4$	2.2535	8.738	1409
2	$[(Ph_3P)CuCl]_4$	2.192	9.495	2010
3	$[(Ph_3P)CuBr]_4$	2.207	9.302	1910
4	$[(Ph_3P)CuI]_4$	2.254	8.732	1570
5	$[(Ph_3P)_2CuCl] \cdot 0.5C_6H_6$	2.266	8.594	1270
6	$[(Ph_3P)_2CuBr] \cdot 0.5C_6H_6$	2.2725	8.521	1280
7	$[(Ph_3P)_2CuI]$	2.273	8.515	1200
8	$[(T_aPh_2P)_2CuCl]$	2.249	8.791	1275
9	$[(T_aPh_2P)_2CuBr]$	2.240	8.897	1225
10	$[(T_aPh_2P)_2CuI]$	2.255	8.721	1400
11	$[(Ph_3P)_2CuBH_4]$	2.276	8.482	1160
12	$[(Ph_3P)_2CuNO_3]$	2.256	8.709	1440
13	$[(Ph_3P)_2Cu(CH_3CN)_2]ClO_4$	2.2675	8.577	1255
14	$[(Ph_3P)_3CuCl]$	2.351	7.696	940
15	$[(Ph_3P)_3CuBr]$	2.353	7.676	935
16	$[(Ph_3P)_3CuI]$	2.355	7.656	930
17	$[(Ph_3P)_3CuCl] \cdot 2Me_2CO$	2.326	7.946	980
18	$[(Ph_3P)_2CuBr] \cdot 2Me_2CO$	2.342	7.785	960
19	$[(Ph_3P)_3CuI] \cdot C_9$	2.342	7.785	925
20	$[(Ph_3P)_4Cu]^+ClO_4^-$	2.5645	5.929	1030
21	$[(Ph_3P)_3CuCH_3CN]^+ClO_4^-$	2.330	7.906	993
22	$[P(2,4,6)_3CuCl]$	2.177	9.692	2040
23	$[P(2,4,6)_3CuBr]$	2.197	9.430	2030
24	$[PPh_3Me][Ph_3PCuBr_2]$	2.210	9.265	1690

[a] Attar, S., Bowmaker, G. A., Alcock, N. W., Frye, J. S., Bearden, W. H., Nelson, J. H., *Inorg. Chem.*, **1991**, *30*, 4743.

xii. $^1JP^{107,109}Ag$

The naturally occurring isotopes of silver (^{107}Ag, 51.82% natural abundance; ^{109}Ag, 48.18%) both have nuclear spin $I = \frac{1}{2}$, and their gyromagnetic ratios are similar in magnitude ($\gamma^{109}Ag/\gamma^{107}Ag = 1.15$). Thus, the ^{31}P NMR spectra of complexes of silver (I) with phosphorus donor ligands at reduced temperatures, where ligand exchange is slowed, often contain two doublets in which the separate couplings of ^{31}P to the ^{107}Ag and ^{109}Ag nuclei are resolved. Typical values of $^1J^{107}AgP$ are listed in Table 6.37. For complexes of the type $(R_3P)_nAgX$, $^1J^{107}AgP$ decreases as n increases (Figure 6.9) and as the donor ability of X increases (Figure 6.10). As for the analogous $(R_3P)_nCuX$ complexes, $^1J^{107}Ag$ increases as $1/r^3$ ($r = dAgP$) increases (Figure 6.11).

Table 6.37 ^{31}P NMR Chemical Shifts (δ/ppm) and One-Bond ^{107}Ag—^{31}P Spin–Spin Coupling Constants (^1J/Hz) for Selected L_nAgX Complexes[a]

L	X	$\delta(^{31}\text{P})/{}^1J^{107}_{\text{Ag}-\text{P}}$			
		$n = 1$	$n = 2$	$n = 3$	$n = 4$
Ph$_3$P	Cl	3.3/282	3.0/277	3.0/277	
	I				b/262
	PF$_6$	15.8/755	15.3/507	11.5/319	5.6/224
	BF$_4$			b/318	7.5/222
	NO$_3$			b/309	
	OH				5.9/221
	SnCl$_3$			8.3/300^c	
				5.9/223^c	
Ph DBP	Cl			−4.7/262	−1.4/222
	Br			−6.9/259	−2.2/222
	I			10.4/247	−2.1/222
	PF$_6$	3.0.730	4.9/510	2.7/316	1.9/223
	NO$_3$				b/225
DMPP	Cl	4.3/544		2.5/249	5.2/215
	Br	2.2/505		1.3/247	5.3/217
	I	2.6.460		−0.5/232	5.5/215
	BF$_4$				4.6/215
Et$_3$P	BF$_4$	10.3/712	8.9/470	1.9/304	7.7/218
Me$_2$PhP	BF$_4$				−31.6/212
MePh$_2$P	BF$_4$				−16.0/230
(n-Bu)$_3$P^d	BF$_4$	b/759	32.9/470	26.6/304	15.5/219
	CH$_3$	−6.8/279			
(t-Bu)$_3$P^d	Cl	82.5/593			
	Br	73.8/561			
	I	74.5/544			
	PF$_6$		79.9/437		
(t-Bu)$_3$P^d	BF$_4$		80.0/444		
	NO$_3$	86.3/683	79.4/442		
	ClO$_4$		79.7/442		
	CN	79.7/498			
	SCN	83.3/635			
	OAc		−28.4/513		
(MeS)$_3$P^e	PF$_6$		−28.4/513	2.5.273	5.3/224
(p-Tol)$_3$P^f	F		8.1/450	2.7/280	5.3/225
	Cl		1.4/378	1.7/278	5.5/230
	Br			−1.2/266	

(continues on next page)

Table 6.37   (*cont.*)

| L | X | $\delta(^{31}P)/^1J^{107}_{Ag-P}$ | | | |
		$n = 1$	$n = 2$	$n = 3$	$n = 4$
	I			10.6/321	5.6/225
	PF_6		13.4/496	7.3/312	5.5/224
	NO_3		9.8/470		5.6/225
	ClO_4		13.1/503	4.2/230	
	CN		3.8/278	5.0/270	
	SCN			7.3/310	5.7/230
	F_3OAc^g		9.5/451		
$(Me_2N)_3P$	Cl		121.2/535		
	I		119.3/597		
	BF_4	121.5/910			
	BPh_4		115.4/610	122.1/393	
	NO_3		118.6/592		
	CN		124.3/437		

[a] For $^1J^{109}Ag-P = 1.15\,(^1J^{107}Ag-P)$, the values have not been reported or are not listed here. The $\delta(^{31}P)$ and $^1JAg-P$ values have been obtained in common NMR solvents (CH_2Cl_2, CD_2Cl_2, $CDCl_3$) at temperatures ranging from -80 to $-100°C$, unless specified otherwise.
[b] $\delta(^{31}P)$ not reported.
[c] The major species is the ionic $[(Ph_3P)_3Ag]^+SnCl_3^-$ ($J = 223$ Hz).
[d] All parameters obtained at ambient temperature.
[e] $(Mes)_3P$ = trimesitylphosphine, where mesityl = 2,4,6-trimethylphenyl.
[f] $(p\text{-}Tol)_3P$ = tris (p-tolyl) phosphine = $(p\text{-}CH_3C_6H_4)_3P$.
[g] $F_3OAc = F_3CCO_2^-$. Attar, S., Alcock, N. W., Bowmaker, G. A., Frye, J. S., Bearden, W. H., Nelson, J. H., *Inorg. Chem.*, **1991**, *30*, 4166.

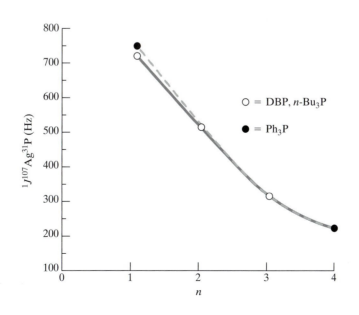

◀ FIGURE 6.9
Plot of $^1J^{107}Ag^{31}P$ (Hz) versus n for L_nAgPF_6 complexes. $L = (PhDBP, Ph_3P, n\text{-}Bu_3P$; $n = 1–4)$; data from Table 6.37; for $n = 2, 3$, the values are essentially equal.

○ = DBP, n-Bu$_3$P

● = Ph$_3$P

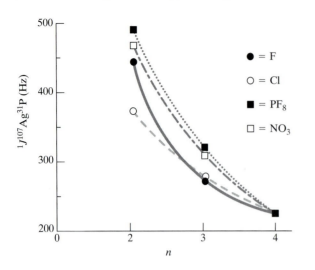

▲ FIGURE 6.10
Plot of $^1J^{107}Ag^{31}P$ (Hz) versus n for {(P-
to1)$_3$P}$_n$AgX complexes. Data from Table 6.37.
When $n = 4$, the values are independent of X.

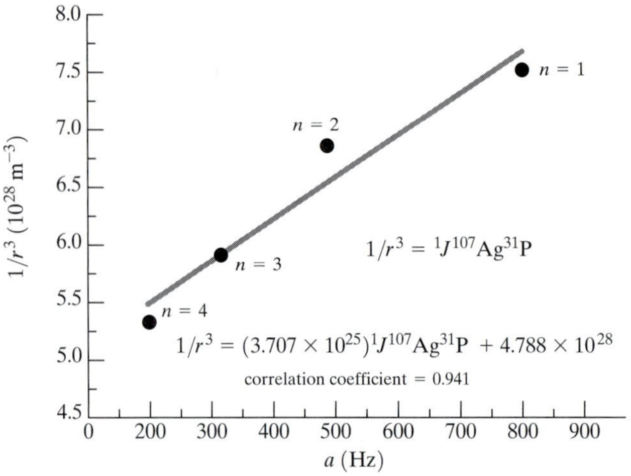

▲ FIGURE 6.11
Plot of $1/r^3$ (10^{28} m^{-3}) versus $^1J^{107}Ag^{31}P$ (Hz) for the
(Ph$_3$P)$_n$AgNO$_3$ ($n = 1$–4) complexes.

xiii. $^1JP^{111,\,113}Cd$

As with silver, the naturally occurring isotopes of cadmium (^{111}Cd, 12.75% natural abundance; ^{113}Cd, 12.26%) both have nuclear spin $I = \frac{1}{2}$, and their
gyromagnetic ratios are similar in magnitude (γ^{113}Cd$/\gamma^{111}$Cd $= 1.046$). $^1JP^{111}$Cd
ranges from 1,000 to 2,700 Hz. The $^1JP^{111}$Cd coupling decreases in the se-
ries (Bu$_3$P)$_2$Cd^{2+} (2,320 Hz), (Bu$_3$P)$_3$Cd^{2+} (1,570 Hz), (Bu$_3$P)$_3$Cd$_2$I$_4$ (1,455 Hz),
(Bu$_3$P)$_2$CdI$_2$ (1,323 Hz). For CdX$_2$[P(C$_6$H$_4$—p—Y)$_3$]$_2$ complexes, the coupling
rises linearly over a range of about 200 Hz with increasing electronegativity
of Y, and for a given X, it increases by about the same amount with increasing
phosphine pK_a.

Dimeric complexes of the type (R$_3$P)XCd(μ—X)$_2$CdX(R$_3$P) have couplings in the 1,620–2,500-Hz range for X = halide, while for X = CH$_3$CO$^-_2$ or
NO$^-_3$, the range is from 2,340 to 2,700 Hz. Some typical coupling constants are
listed in Table 6.38.

xiv. $^1JP^{199}Hg$

The values of $^1JP^{199}$Hg extend over more than two orders of magnitude, namely, from 143 to 17,528 Hz. For linear [(PMe$_3$)HgX]$^{n+}$ ions, the coupling is very
sensitive to the *trans* ligand, ranging from 7,852 Hz for X = Cl, $n = 1$, to
5,173 Hz for X = PMe$_3$, $n = 2$, to 1,745 Hz for X = CH$_3$, $n = 1$. For a series of
{Hg[P(C$_6$H$_4$—p—Z)$_3$]$_2$}(ClO$_4$)$_2$ compounds, $^1JP^{199}$Hg decreases with increasing ligand basicity (5,585 to 3,218 Hz). For Hg[P(OR)$_3$]$_n$$^{2+}$ species, $^1JP^{199}$Hg decreases with increasing n as follows: $n = 1$ (about 17,000 Hz); $n = 2$ (about
11,000 Hz); $n = 3$ (about 6,700 Hz); and $n = 4$ (about 4,300 Hz). For analogous
[Hg(R$_3$P)$_n$]$^{2+}$ species, $^1JP^{199}$Hg also decreases with increasing n as follows:
$n = 2$ (3,730 to 5,550 Hz); $n = 3$ (3,050 to 3,255 Hz); $n = 4$ (1,980 to 2,090 Hz).

As expected, coordinating anions lower $^1JP^{199}$Hg because of their influence on Hg s character in the Hg—P bond. Using a series of substituted ac-
etates as anions (X) in (R$_3$P)$_n$HgX$_2$ reveals that the anions remain bidentate
from $n = 1$ (about 8,200 to 10,440 Hz) to $n = 2$ (about 5,500 to 6,300 Hz), and
$^1JP^{199}$Hg increases with increasing electronegativity of X. Anions such as

Table 6.38 Representative $^{1}J_{PCd}$ Couplings for $(R_3P)_2CdX_2$, $(R_3P)_3Cd_2X_4$ and $(R_3P)_2Cd_2X_4{}^{a}$ Complexes

	$(R_3P)_2CdX_2$				$(R_3P)_3Cd_2X_4$		
T(K)	R_3P	x	$^{1}JP^{111}Cd$	$^{1}JP^{113}Cd$	R_3P	X	$^{1}JP^{113}Cd$
183	Et$_3$P	Cl	1,603	1,677	Bu$_3$P	Br	1,669
183	Et$_3$P	Br	1,479	1,547	Bu$_3$P	I	1,502
183	Et$_3$P	I	1,301	1,367			
203	Bu$_3$P	Cl	1,626	1,697		$(R_3P)_2Cd_2X_4$	
183	Bu$_3$P	Br	1,490	1,558	Et$_3$P	Cl	2,504
183	Bu$_3$P	I	1,323	1,384	Et$_3$P	Br	2,355
213	Ph$_3$P	Br	1,265	1,323	Et$_3$P	I	1,768
243	Ph$_3$P	I	1,077	1,125			

[a] Kessler, J. M., Reeder, J., Vac, R., Yeung, C., Nelson, J. H., Frye, J. S., Alcock, N. W., *Magn. Reson. Chem.*, **1991**, *29*, S94.

halides are monodentate and more strongly coordinating, and they generally give rise to monomeric $(R_3P)_2HgX_2$ or dimeric $(R_3P)HgX(\mu{-}X)_2HgX(R_3P)$ complexes. For the monomers, $^{1}JP^{199}Hg$ ranges from 2,600 to 5,900 Hz for a variety of phosphines and from 6,240 to 8,100 Hz for $P(OEt)_3$. For a given ligand, these couplings increase with increasing electronegativity of the halide. They also increase with increasing $\Delta\delta^{31}P$, with increasing phosphine basicity, and with decreasing temperature (owing to slowing of ligand exchange). Some typical data are listed in Table 6.39.

Table 6.39 Representative ^{31}P {^{1}H} NMR Data for $(R_3P)_2HgX_2$ and $(R_3P)_2Hg_2X_4$ Complexes^{a}

			$(R_3P)_2HgX_2$						$(R_3P)_2Hg_2X_4$		
Γ	PR$_3$	X	$\delta^{31}P$ (ppm)	$\Delta\delta^{31}P$ (ppm)	$^{1}J(^{199}Hg^{31}P)$ (Hz)	Γ	PR$_3$	X	$\delta^{31}P$ (ppm)	$\Delta\delta^{31}P$ (ppm)	$^{1}J(^{199}Hg^{31}P)$ (Hz)
230	PEt$_3$	Cl	37.90	57.90	5,109.9	230	PEt$_3$	Cl	80.35	100.35	7,451.2
230	PEt$_3$	Br	32.27	52.27	4,829.1	230	PEt$_3$	Br	74.36	94.36	6,718.7
230	PEt$_3$	I	17.92	37.92	4,153.0	230	PEt$_3$	I	52.80	72.80	5,273.5
230	PBu$_3$	Cl	28.96	61.46	5,129.4	233	PBu$_3$	Cl	33.12	65.62	7,460.9
213	PPb$_3$	Cl	27.71	33.71	4,766.1	233	PBu$_3$	Br	27.55	60.05	6,743.2
213	PPb$_3$	Br	21.29	27.29	4,240.3	223	PBu$_3$	I	6.27	38.77	5,342.7
243	PPb$_3$	I	6.10	12.10	3,055.3	213	PPh$_3$	Cl	33.63	39.63	7,670.2
213	DMPP	Cl	26.23	28.73	4,465.4	181	PPh$_3$	Br	26.64	32.64	6,450.0
213	DMPP	Br	21.47	23.97	3,886.1	213	PPh$_3$	I	6.42	12.42	4,774.4
208	DMPP	I	8.32	10.82	2,509.5	233	DMPP	Cl	28.52	31.02	7,180.2
213	PhDBP	Cl	16.58	27.58	4,081.0	181	DMPP	Br	24.89	27.39	6,352.6
213	PhDBP	Br	8.65	19.65	3,394.6	185	DMPP	I	9.56	12.06	unresolved
213	PhDBP	I	−8.62	2.38	2,234.3	213	PhDBP	Cl	22.06	32.06	7,424.5
						185	PhDBP	I	−2.36	13.36	unresolved

[a] Bowmaker, G. A., Clase, H. J., Alcock, N. W., Kessler, J. M., Nelson, J. H., Frye, J. S., *Inorg. Chim. Acta.*, **1993**, *210*, 107.

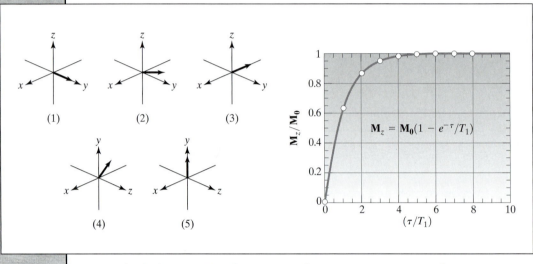

CHAPTER 7

Longitudinal relaxation (T_1). The recovery of a magnetization vector (shown on resonance in the rotating frame, left). As the (x–y) magnetization diminishes, the z magnetization, $\mathbf{M}_z$, is restored to its equilibrium value (right).

Nuclear Spin Relaxation

7.1   Introduction

C hemical shifts and coupling constants are mainly related to the structural properties of molecules, but relaxation phenomena are related to the microdynamical behavior of the spin systems. Were it not for relaxation, the NMR experiment would be impossible, since relaxation allows the spins to return to their thermal equilibrium. (See Chapter 2.)

Relaxation involves exchange of energy between spins or between spins and the surrounding environment, called the *lattice*. This energy transfer implies magnetic interactions with the magnetic dipoles associated with the spins. For nuclei with $I > \frac{1}{2}$, the associated electric quadrupole moment may also interact with the electric field gradient, leading to a mechanical torque on the nucleus. Relaxation results from a fluctuation in these interactions, which arise from the Brownian motion of the molecules.

Two models will be employed to describe relaxation phenomena. At a microscopic level, individual spins will be considered and quantum mechanics will be applied. At a macroscopic level, the vector sum of all the microscopic magnetic moments of the nuclei contributing to the macroscopic magnetization will be used.

7.2   Different Types of Relaxation

A.   Longitudinal Relaxational

If the macroscopic magnetization $\mathbf{M_0}$ is tilted by an angle α from its equilibrium position along the main magnetic field, and the system is allowed to relax, it will return to its equilibrium position. The behavior of the components $\mathbf{M}_z$ and $\mathbf{M}_x$ or $\mathbf{M}_y$ will not be the same. $\mathbf{M}_z$ will return to its equilibrium value $\mathbf{M_0}$ according to a first-order rate process characterized by the time constant T_1 in the equation

$$\frac{d\mathbf{M}_z}{dt} = -\frac{(\mathbf{M}_z - \mathbf{M_0})}{T_1}. \tag{7.1}$$

T_1 is called the *longitudinal relation time* or the *spin–lattice relaxation time*. (See also Sections 1.4 through 1.6).

This effect involves the z component of the magnetization. At the microscopic level, that component is related to the Zeeman energy of each spin. The relaxation mechanism governs both the transition between states and the populations of the different states. For independent spins, this evolution is given by $dn/dt = n/T_1$, where n is the excess of the spin population compared with the population at thermal equilibrium. The effect is then an enthalpic phenomenon characterizing the exchange of energy between the spin system and the surroundings.

B.   Transverse Relaxation

The $\mathbf{M}_x$ and $\mathbf{M}_y$ components of $\mathbf{M}$ also return to their equilibrium state, which is zero. This return corresponds to the dephasing of the spins. Dephasing arises from an exchange of energy between spins. On a microscopic level, when a spin exchanges energy with its neighbors, it acquires or loses energy and its precession frequency changes. The resulting component in the xy-plane, which is the sum of all individual moments, will then decrease. As in the case of longitudinal relaxation, this phenomenon is a first-order process characterized by the time constant T_2, called the *spin–spin relaxation time* or the *transverse relaxation time*. T_2 is given by the equation

$$\frac{d\mathbf{M}_x}{dt} = -\frac{\mathbf{M}_x}{T_2} \quad \text{or} \quad \frac{d\mathbf{M}_y}{dt} = -\frac{\mathbf{M}_y}{T_2}. \tag{7.2}$$

This dephasing process disorganizes the spin system and is thus an entropic phenomenon. The effect, which changes the Larmor frequencies of the spins, will lead to a distribution of frequencies around the unperturbed resonance frequency and is thus the origin of the line width, which is related, under line-narrowing conditions, to T_2 by the equation

$$\Delta\nu_{1/2} = \frac{1}{\pi T_2}. \tag{7.3}$$

Other phenomena outside the spin system may also contribute to the decrease of $\mathbf{M}_x$ and $\mathbf{M}_y$. Magnetic field inhomogeneities are an example. When a spin in a molecule diffuses in the sample, it is acted upon by slightly different polarizing fields. This changes its Larmor precession frequency, resulting in an extra contribution to the fanning out of the spins and to the rate constant $k = 1/T_2$. If we call T_2' the time constant describing those external effects, we have

$$\frac{1}{T_2^*} = \frac{1}{T_2} + \frac{1}{T_2'}, \tag{7.4}$$

where T_2^* is the experimental value measured from the line width. T_2 must be smaller than or equal to T_1. It is usually the case that T_2 is smaller than T_1, except for liquids, for which $T_1 = T_2$. In the solid state, T_2 is shorter than T_1.

C.   Relaxation in the Rotating Frame

If we consider the rotating frame precessing at an angular velocity ω_e, and if a continuous RF field is applied, a constant magnetic field $\mathbf{B}_1$ will be applied along the x-axis in the rotating frame. If we turn the magnetization along the x-axis, in the rotating frame, with, for instance, a $-90°y$ pulse, then the magnetization $\mathbf{M}$ will be continuously aligned along $\mathbf{B}_1$. $\mathbf{M}$ will return to its equilibrium value along the main field $\mathbf{B}_0$ such that the component of $\mathbf{M}$ along $\mathbf{B}_1$ will decrease according to a first-order rate process characterized by a new time constant $T_{1\rho}$, which is different from either T_1 or T_2. This is the *relaxation time in the rotating frame.*

If $\mathbf{M}$ is not along $\mathbf{B}_1$, the process is applied to the component of $\mathbf{M}$ parallel to $\mathbf{B}_1$. In the liquid state, $T_{1\rho} = T_2$.

D.   Order of Magnitude of Relaxation Times

Relaxation times vary over a very large range. T_1 may have values that are as long as some months in the solid state at low temperatures or as short as 10^{-10} s. When the value of T_2 (also, for liquids) is very short, the NMR signal will not be observed, because, as Equation 7.3 shows, the line width will be very broad. Usually, for most nuclei in systems of interest to chemists, T_1 ranges from minutes to 10^{-4} or 10^{-5} s. Typical values for representative nuclei are given in Appendix.

7.3   Microscopic Origin of Relaxation

A.   General Phenomena

We must describe the microscopic effects of the motions of the fluctuating surroundings on the spins in order to obtain, by summation, the macroscopic evolution of the magnetization, which is experimentally observed to satisfy the equation

$$\overline{\mathbf{M}} = \sum_{i=1}^{N} \overline{\mathbf{M}}_i$$

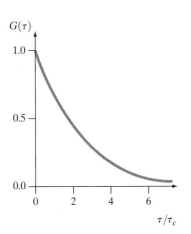

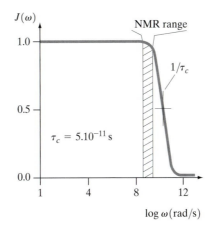

◀ FIGURE 7.1
General form of an autocorrelation
function and of a spectral density
function.

where N is on the order of Avogadro's number. This cannot be done, spin by spin, and a statistical analysis is necessary. All such treatments are based upon the notion of an *autocorrelation function*.

If we consider a variable y, such as a distance or an angle, that is a function of time $y(t)$, and if $y(t)$ varies randomly with time, as is the case for Brownian motion, the value of y at time t depends upon a probability $p(y, t)$. The mean value of y observed at time t is obtained from the sum of all possible values of y, weighted by their probabilities: that is,

$$\overline{y(t)} = \int p(y, t)\, dy, \tag{7.5}$$

where the bar indicates an *ensemble average* on the system.

Suppose now that we measure a parameter f, say, a magnetization, that depends on the variable y. Then the observed value is given by

$$\overline{f(t)} = \int p(y, t) f(y)\, dy. \tag{7.6}$$

If we measure $f(t)$ at two different times, t_1 and t_2, the result is not necessarily random, but a correlation may occur. The system has some "memory" of the past. One can write the function $P(y_1, t_1, y_2, t_2)$ indicating that the system takes the value y_2 at t_2 when it has the value y_1 at t_1. If the system is stationary, the value of this new function is independent of the value of t_1, and we can set $t_1 = 0$ to obtain $P(y^0, y, \tau)$. We call this an *autocorrelation function*, and it is given by

$$G(\tau) = \overline{f(0)f(t)} = \iint P(y^0)f(y^0)P(y^0, y, \tau)f(y)\, dy^0\, dy. \tag{7.7}$$

$G(\tau)$ describes the evolution of the memory of the system from its initial state. It is a decreasing function which may be characterized by a *correlation time* τ_c that corresponds to the time at which a certain percentage of the memory remains as it appears, for instance, in the classical function $G(\tau) = \exp - (\tau/\tau_c)$. The Fourier transform of $G(\tau)$ is $J(\omega) = \int G(\tau) \exp(-i\omega\tau)\, d\tau$, where $J(\omega)$ is called the *spectral density*. The behavior of $G(\tau)$ and $J(\omega)$ are illustrated in Figure 7.1. The decrease in $J(\omega)$ is characterized by τ_c.

B.   Basic Equations for Relaxation

In order to modify the spins over a period of time, the surrounding environment must induce transitions between the energy levels of the spin system, as illustrated in Figure 7.2.

▶ FIGURE 7.2
**Transition probabilities
among energy levels.**

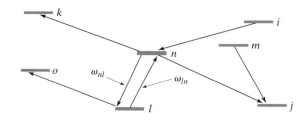

If we consider the level n, populated with P_n spins, and if ω_{nl} is the transition probability between levels n and l, then the evolution of P_n is given by

$$\frac{dP_n}{dt} = \sum_l P_l \omega_{ln} - P_n \sum_l \omega_{nl}.$$

Since, by symmetry, $\omega_{ln} = \omega_{nl}$, this equation becomes

$$\frac{dP_n}{dt} = \sum_l \omega_{nl}(P_l - P_n).$$

Solving these differential equations, applied to the macroscopic magnetization, gives

$$\mathbf{M}_z(t) = \gamma\hbar \sum_n M_n P_n(t), \tag{7.8}$$

where M_n is the eigenvalue of the z component of the total momentum of the spin system of state n. Usually, in a relaxation-time measurement, we measure the difference between the actual magnetization $\mathbf{M}_z(t)$ and its equilibrium value $\mathbf{M}_z^0$, as given in Equation 7.1. Then, we have

$$\mathbf{M}_z(t) - \mathbf{M}_z^0 = \gamma\hbar \sum_n M_n(P_n(t) - P_n^0).$$

If we take $P_n(t) = P_n(t) - P_n^0$, which represents the deviation of the population of level n from its equilibrium value, we then obtain the following equations:

$$\mathbf{M}_z(t) - \mathbf{M}_z^0 = \gamma\hbar \sum_n M_n P_n(t); \tag{7.9}$$

$$\frac{dP_n(t)}{dt} = \sum_l \omega_{nl}(P_l - P_n). \tag{7.10}$$

The solution of Equation 7.8 is a sum of exponential terms, which is not necessarily an exponential, but often is in practice. The only remaining problem is to calculate the transition probabilities ω_{mn}, which are given by the Schrödinger equation of the second type:

$$\left(-\frac{\hbar^2}{2m}\right)\nabla(\phi) + V(\phi) = H(\phi) = \left(-\frac{\hbar}{i}\right)\frac{\partial y}{\partial t}.$$

Applying a second-order perturbation treatment to this equation leads to

$$\omega_{nl} = \int_0^t (n|H(t)|m)(m|H(t')|l)\,\exp[i\omega_{nl}(t - t')]\,dt' + \text{complex conjugate,}$$

where $H(t)$ is the fluctuating Hamiltonian describing the interaction of each spin with the surroundings. This operator varies as a function of the different possible mechanisms, as we shall see later. But if we take as an example the most common interaction, namely, the dipole–dipole interaction, the energy of this interaction, given in all classical textbooks on magnetism, is

$$E_{ij} = \left(\frac{\mu_0}{4\pi}\right)\left[\left(\frac{\overline{\mu}_i \overline{\mu}_j}{r^3}\right) - 3\frac{(\overline{\mu}_i \overline{r})(\overline{\mu}_j \overline{r})}{r^5}\right],$$

where r is the distance between the two magnetic dipoles μ_i and μ_j. Since μ_i corresponds to $\gamma \hbar I$ and r fluctuates with time, this gives

$$H_1(t) = \gamma_i \gamma_j \hbar^2 \left(\frac{\mu_0}{4\pi}\right)[I_{iz} I_{jz}]\left[\left(\frac{3\cos^2 \Theta - 1}{r^3}\right)\right],$$

where Θ (which also varies with time) is the angle between r and the main magnetic field $\mathbf{B_0}$. We see that $H_1(t)$ consists of three parts: some constants, a spin operator part, and a time-dependent part, which describes the random motions of the geometrical factors. This form is similar for all types of interaction. We then have $H(t) \sim S(I) \cdot F(t)$, where $S(I)$ operates on the spin functions and $F(t)$ on the space parameters. This leads to the equation

$$\omega_{nl} = |<n|S|1>|^2 J(\omega)_{nl} \tag{7.11}$$

with

$$J(\omega)_{nl} = \int_{-\infty}^{+\infty} G(\tau) \exp(i\omega_{nl}\tau)\, d\tau, G(\tau)$$

being the autocorrelation function of the spatial part of the Hamiltonian: $G(\tau) = F(0)F(\tau)$. The result depends now on the interaction and on the model used to describe the fluctuation of the geometrical parameters. Starting with Equation 7.8, we have

$$\frac{d\mathbf{M}_z}{dt} = \gamma\hbar \sum_n m_n \frac{dP_n(t)}{dt},$$

and substituting Equation 7.1 into this equation gives

$$\frac{d\mathbf{M}_z(t)}{dt} = \gamma\hbar \sum_n \sum_l \omega_{nl}(P_l - P_n)M_n = -\frac{(\mathbf{M}_z - \mathbf{M_0})}{T_l},$$

where P_l and P_n are given by the Boltzmann law with the approximation that the perturbation of the populations by the NMR excitation is sufficiently low to be considered negligible. As we have ω_{nl}, we can compare the form of our result with that of Equation 7.11 to obtain $1/T_l$. Similar equations are obtained for $\mathbf{M}_x$ and $\mathbf{M}_y$. As an example, we give the results for the dipole–dipole relaxation mechanism between two different spin $I = \frac{1}{2}$ nuclei with an autocorrelation function $G(\tau) = \exp(-\tau/\tau_c)$:

$$\frac{1}{T_{1A}} = \left(\frac{\mu_0}{4\pi}\right)^2 \frac{\gamma_A^2 \gamma_X^2 \hbar^2}{r^6}\left(\frac{\tau_c}{10}\right)$$

$$\left\{\frac{1}{1 + (\omega_X - \omega_A)^2\tau_c^2} + \frac{3}{1 + \omega_A^2\tau_c^2} + \frac{6}{1 + (\omega_X + \omega_A)^2\tau_c^2}\right\}.$$

Notice that, under the extreme narrowing condition $\omega_A \tau_c \ll 1$ or $(\omega_X \pm \omega_A)\tau_c \ll 1$, the term in braces reduces to 10.

7.4 Different Types of Relaxation Mechanisms

A. Dipole–Dipole (DD) Relaxation

The origin of the fluctuating magnetic field on the observed nucleus A comes from another spin X on the same molecule, which moves under the influence of the collisions. The general form is given by

$$\frac{1}{T_{1DD}} = \left(\frac{\mu_0}{4\pi}\right)^2 \left(\frac{\gamma_A^2 \gamma_X^2 \hbar^2}{r^6}\right) I_X(I_X + 1) f_{DD}(\nu_Z, \nu_X, \tau_c),$$

where $f_{DD}(\nu_A, \nu_X, \tau_c)$ is the term arising from the calculation of the autocorrelation function. Previously, we saw that $f_{DD} = 2\tau_c$ under extreme narrowing conditions. If A and X are of the same type, say, two protons, and in the same molecule, then $f_{DD} = 4/3\tau_c$.

Under extreme narrowing conditions, the expressions of T_{1DD} and T_{2DD} are the same. This is not true in other cases, and T_1, T_2, and $T_{1\rho}$ have different behavior, especially in the solid state. The dipolar relaxation time is frequency dependent in the general case, with minima occurring near the conditions $(\omega_A + \omega_X)\tau_c = 1$. The frequency dependence vanishes under extreme narrowing conditions, as illustrated in Figures 7.3 (a) and (b).

B. Electron–Nuclear Relaxation (ED,EC)

This interaction requires the presence of unpaired electrons near the spin system. In that case, the large magnetic moment related to the spin $\frac{1}{2}$ momentum of the electron acts as an active nucleus by dipole–dipole interaction. This usually leads to very efficient relaxation, which is described by two different types of interaction.

i. Dipolar Interation

A dipole–dipole interaction between the observed nuclei and the magnetic moment μ of the unpaired electrons in the paramagnetic species takes the same form as the equation for T_{1DD}, but with μ_0 replaced by μ. When the un-

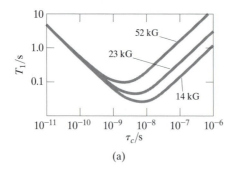
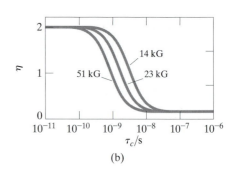

(a)   (b)

▲ FIGURE 7.3
(a) Log–log plot of the dipolar relaxation time T_1 versus the rotational correlation τ_c.
(b) Logarithmic plot of the nuclear Overhauser enhancement factor η against the rotational correlation time τ_c.

paired electron is not in the same molecule as the observed nucleus—as, for instance, in paramagnetic impurities or shift reagents—the expression

$$\frac{1}{T_{1ED}} = \left(\frac{\mu_0}{4\pi}\right)\left(\frac{\gamma_A^2\langle\mu^2\rangle}{r^6}\right)f_{ED}(\nu_A\nu_e\tau_c^A)$$

is used, where $\langle\mu^2\rangle$ is the mean value of the square of μ. Generally f_{ED} is different for T_1 and T_2, except in the extreme-narrowing case.

ii. Contact Interaction

A contact interaction arises from the scalar coupling constant A, called the *hyperfine coupling constant*, between unpaired s electrons and the nucleus. If S is the total electron spin momentum of the paramagnetic center (e.g., $S = 1$ for Ni^{2+}, and $S = \frac{3}{2}$ for Co^{2+} in most geometries), the relaxation time is expressed by

$$\frac{1}{T_{1EC}} = 4\pi^2 A^2 S(S + 1)f_{EC}(\nu_A, \nu_e, \tau_c^A).$$

Under extreme narrowing conditions, $f_{EC} = 2T_{1E}$, and for T_{2EC}, $f_{EC} = \frac{1}{3}(T_{1E} + T_{2E})$, where T_{1E} and T_{2E} are the two relaxation times of the electron.

C. Chemical-Shift Anisotropy (CSA)

If the surrounding environment of the spin is not spherically symmetrical, the screening of the nucleus, a tensor, is not isotropic, but depends upon the orientation of the molecule in the magnetic field. Tumbling of the molecule creates a fluctuating field, leading to a relaxation process. The expression for T_{1CSA} for a molecule with cylindrical symmetry is

$$\frac{1}{T_{1CSA}} = \frac{2}{15}\gamma_A^2\mathbf{B}_0^2(\sigma_\parallel - \sigma_\perp)^2 f_A(\tau_{CA}),$$

where $\sigma_\parallel$ and $\sigma_\perp$ respectively represent the screening tensor components parallel to and perpendicular to the principal axis of the external magnetic field. For T_2, the same expression applies, except that the factor $\frac{2}{15}$ is replaced by $\frac{7}{45}$. Note that T_{1CSA} is highly field dependent, as the relaxation rate is directly proportional to the square of the field strength, $\mathbf{B}_0^2$.

D. Scalar Relaxation (SC)

i. Scalar Relaxation of the First Kind

Sometimes it may happen that the observed nucleus A undergoes exchange between different sites, either intra- or intermolecularly. If, in some sites, A is J-coupled to some other nuclei, the exchange process destroys and regenerates the coupling, randomly creating a fluctuating field at nucleus A. A relaxation mechanism occurs, which is expressed in the extreme narrowing case by

$$\frac{1}{T_{1SCl}} = \frac{\left(\frac{8\pi^2}{3}\right)J_{AX}^2 I_X(I_X + 1)\tau_{eX}}{1 + 4\pi^2(\nu_A - \nu_X)^2\tau_{eX}^2},$$

where τ_{eX} is the time constant related to the exchange rate and

$$\frac{1}{T_{2SCl}} = \frac{1}{2T_{1SCl}} + \frac{4\pi^2}{3}J_{AX}{}^2 I_X(I_X + 1)\tau_{eX}.$$

Note that, even under extreme narrowing conditions, T_1 is different from T_2.

ii. Scalar Relaxation of the Second Kind

It is well known that if nucleus A is scalar coupled to a rapidly relaxing nucleus X, such as a quadrupolar nucleus like ^{14}N, ^{35}Cl, ^{59}Co, or ^{81}Br, and the AX bond is not broken, then no coupling is observed for A, but J_{AX} always exists and fluctuates rapidly. These random fluctuations lead to a relaxation mechanism characterized by the same expression as that for scalar relaxation of the first kind, but with τ_{eX} replaced by the spin–spin relaxation time of nucleus X, T_{2X}. This is true, however, only when $1/T_{2X} \gg J_{AX}$.

E. Spin-Rotation Relaxation (SR)

For small molecules rotating about their own axes, their electronic cloud moves, creating fluctuating magnetic fields at their nuclei. For a molecule possessing cylindrical symmetry,

$$\frac{1}{T_{1SR}} = \left(\frac{8\pi^2 kT}{3h^2} \right)(2I_\perp C_\perp^2 \tau_{SR\perp} + I_\parallel C_\parallel \tau_{SR\parallel})$$

where $I_\perp$ and $I_\parallel$ are the components of the inertial moment perpendicular and parallel, respectively, to the axis of symmetry, $C_\perp$ and $C_\parallel$ are, respectively, the perpendicular and parallel components of the spin-rotation coupling tensor expressed in hertz, and $\tau_{SR\perp}$ and $\tau_{SR\parallel}$ are the associated correlation times. Under extreme narrowing conditions, T_{1SR} equals T_{2SR}. Since the temperature appears explicitly in the formula, the relaxation rate increases with increasing temperature.

F. Quadrupole Relaxation (Q)

Nuclei with $I > \frac{1}{2}$ have an electric quadrupole moment Q that is assumed to represent a nonspherical distribution of the electric charges in the nucleus. This distribution interacts with the electric field gradient created by the surrounding environment. As it fluctuates with molecular tumbling, a relaxation process that is of an electrical nature occurs. Under extreme narrowing conditions, we have

$$\frac{1}{T_{1Q}} = \frac{1}{T_{2Q}} = \left(\frac{3\pi^2}{10} \right)\left[\frac{(2I + 3)}{I^2(2I - 1)} \right]\left[1 + \frac{\eta^2}{3} \right]\left[\frac{e^2 qQ}{h} \right]^2 \tau_C,$$

where τ_C is the reorientational correlation time of the A–X bond, $\eta = (E_{XX} - E_{YY})/E_{ZZ}$ is the asymmetry parameter, and q is the electric field gradient along the A–X bond taken as the principal axis of the tensor. $e^2 Qq/h$ is called the quadrupole coupling constant, expressed in Hz, and is often larger than 1 MHz. The quadrupole coupling constant may often be obtained experimentally by nuclear quadrupole resonance (NQR) spectroscopy.

This relaxation mechanism is very efficient, and it is for this reason that most nuclei with $I > \frac{1}{2}$ give broad resonances and can be observed only with difficulty. Note that the $(2I + 3)/[I^2(2I - 1)]$ factor decreases as I increases.

7.5 Experimental Behavior of Nuclear Relaxation

A. Simultaneous Mechanisms

Usually, all of the preceding mechanisms are active for a nucleus. The exceptions are the quadrupolar mechanism when only spin $\frac{1}{2}$ nuclei are present and the electron interactions in the absence of paramagnetic centers. The measured relaxation time is then

$$\frac{1}{T_{1\text{exp}}} = \sum_i \frac{1}{T_{1i}},$$

with similar expressions for T_2 and $T_{1\rho}$. In most cases, one of the mechanisms will be dominant. For instance, with $I = \frac{1}{2}$ nuclei, the electron interaction, when present, is dominant. This implies that all magnetic impurities (for example, O_2) must be removed from samples. For $I > \frac{1}{2}$ nuclei, the quadrupolar relaxation is dominant, except in highly symmetrical environments where the electric field gradient is theoretically zero.

B.   Effects of Different Parameters

i.   Temperature

As the temperature increases, Brownian motion increases and the correlation time τ_c decreases. Then the relaxation times increase, as they are approximately proportional to $1/\tau_c$. The only exception is for the spin-rotation interaction as previously seen.

ii. Effect of Correlation Time τ_c

The correlation time τ_c affects T_1, T_2, and $T_{1\rho}$ differently. Figure 7.4 shows the evolution of these parameters as a function of τ_c.

In the region where τ_c is below 10^{-10} s, $T_1 = T_2 = T_{1\rho}$. This corresponds to the extreme narrowing condition, which usually involves a rapidly tumbling lattice. In other regions, the three relaxation times have a different behavior. As illustrated in Figure 7.3a, this behavior is field dependent for T_1.

iii. Experimental Distinction between Mechanisms

In order to find the origin of the relaxation of a given nucleus, it is necessary to distinguish among the different relaxation mechanisms. We suppose that the necessary experimental precautions have been undertaken to remove paramagnetic impurities (such as degassing, filtering, etc.), since these species provide the dominant mechanism, except at very low concentrations. For dipolar nuclei, we suppose that the quadrupole mechanism is not dominant. We then note the following:

1. The $T_{1\text{DD}}$ contribution is easily obtained from nuclear Overhauser experiments (NOEs, to be defined in Chapter 8), since NOE is related to $T_{1\text{DD}}$. If $T_{1\text{DD}}$ is the most important mechanism, the NOE effect is maximum, but if other mechanisms are present, they quench the

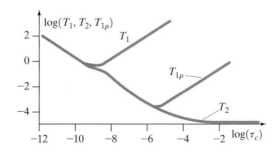

◄ FIGURE 7.4
Evolution of T_1, T_2, and $T_{1\rho}$ according to τ_c.

▶ FIGURE 7.5
Evolution of T_1 of the methyl carbon in acetone as a function of temperature.*

*H. W. Spiess, "Rotation of Molecules and Nuclear Spin Relaxation, in Dynamic NMR Spectroscopy," *NMR Basic Principles and Progress*, P. Diehl, E. Fluck, R. Kostfield, eds., **1978**, *15*, 55.

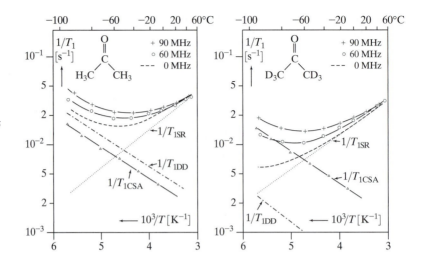

magnetization transfer, which gives rise to the NOE effect. Then we have

$$T_{1DD} = \frac{2\eta(\text{measured})T_1(\text{measured})\gamma(\text{observed})}{\gamma(\text{decoupled})}.$$

Subtracting $1/T_{1DD}$ from $1/T_1$ observed experimentally gives the contribution of the other mechanisms.

2. The next most easily determined mechanism is that arising from the chemical-shift anisotropy, since it is field dependent, as previously seen. Measuring T_1 at two distinct magnetic fields gives the contribution of T_{1CSA}.

3. It is possible to deduce the presence of a spin-rotation mechanism by ascertaining the temperature effect on T_1. Figure 7.5 shows a typical example.

   At low temperatures, all mechanisms except T_{1SR} play a part. At higher temperatures, the situation is reversed. Simulation of the experimental curves with the superimposition of two straight lines, as in Figure 7.5, gives the two contributions.

4. If some other contributions remain, these may be due to a scalar mechanism, some unknown paramagnetic species for a spin $\frac{1}{2}$ nucleus, or the quadrupolar contribution for quadrupolar nuclei. Previously, we noted that when scalar relaxation occurs, T_1 is different from T_2.

7.6   Experimental Measurement of Relaxation Times

A.   Sample Preparation

As we have noted, at low concentrations, paramagnetic impurities have an effect on the value of T_1. Thus, it is necessary to remove all traces of paramagnetic materials from a sample. One of the most important contributors is oxygen from the air. Two techniques are useful for removing oxygen from liquid samples. The first is to apply three or four freeze–thaw cycles under high vacuum. In the second, a noble gas (argon) is allowed to flow through the sample for 10 to 15 minutes. In both methods, the concentration of the sample will increase slightly, due to evaporation of some solvent. After oxygen removal, the sample is sealed.

Degassing also applies to solid-state samples. When possible, degassing may be done on the melt, as for liquids; for powder samples, the rotor must be filled with argon. Care must be taken to avoid incorporating solid paramagnetic impurities into the sample, such as catalysts used in the preparation of the compound.

B.   Measurements of T_1

i. Inversion Recovery*

The best method for measuring T_1 is the *inversion-recovery Fourier transform (IRFT)*. The pulse sequence $(180°–\tau–90°$ acquisition–$t)_n$ is used, where τ is a variable delay and t is a waiting time that allows the spin system to return to equilibrium. The duration of the sequence must be greater than $5T_1$. The intensity for each τ value is related to T_1 by

$$I_\tau = I_{eq}\left[1 - 2_{exp}\left(-\frac{\tau}{T_1}\right)\right].$$

To determine T_1, one plots $\ln(I_{eq} - I_\tau)$ vs. τ as

$$\ln(I_{eq} - I_\tau) = \ln 2 + \ln I_{eq} - \frac{\tau}{T_1}.$$

The slope of this plot is $-1/T_1$. Typical data are shown in Figure 7.6.

SAMPLE PROBLEM 7.1

Since relaxation is a first-order process, T_1 can be *estimated* from null times. As I_τ passes through zero, $T_1 = \tau_{null}/\ln 2 = \tau_{null}/0.693$. Use this relation to estimate T_1 from Figure 7.6.

Solution

The null time in Figure 7.6 is about 100 ms. Hence, $T_1 \cong 100$ ms$/0.693 = 144$ ms. This result is within 3% of the result obtained from a least-squares fit of the data (140 ms).

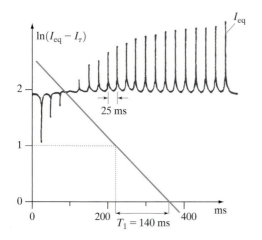

◀ FIGURE 7.6
Typical data obtained by the inversion recovery method for a single resonance.

---

*Vold, R. L., Waugh, J. S., Klein, M. P., Phelps, D. E., *J. Chem. Phys.*, **1968**, *48*, 3831.
Freeman, R., Hill, H. D. W., *J. Chem. Phys.*, **1969**, *51*, 3140.

The evolution of T_1 is sensitive to experimental conditions, and what is usually observed is an evolution given by

$$I_\tau = I_{eq}\left[1 - a\exp\left(-\frac{\tau}{T_1} \right) \right].$$

One can then use an optimization procedure with two parameters, a and T_1, but this frequently gives erroneous results. The best approach is to perform a three-parameter least-squares fit with at least 10 points in which the unknowns are I_{eq}, a, and T_1. With this approach, it is not necessary to measure I_{eq}, which is the most time-consuming experiment. Computer programs are available in the software supplied by most instrument manufacturers to do this automatically.

ii. Other Methods

   a. Saturation-Recovery Fourier Transform (SRFT)*   In this experiment, the first 180° pulse of the IRFT sequence is replaced by a 90° pulse, and we have

$$I_\tau = I_{eq}\left[1 - \exp\left(-\frac{\tau}{T_1} \right) \right].$$

   b. Progressive-Saturation Fourier Transform (PSFT)†   This is a very simple sequence: $[90°\text{–Acquisition–}\tau]_n$. The first 5 to 10 scans are deleted in order to reach the steady state. The evolution of the intensity is given by

$$I_\tau = I_{eq}\left\{ \frac{1 - \exp[(-\text{Aqt} + \tau)]}{T_1} \right\}.$$

This method is limited to systems with long T_1s, since the shortest value of the exponential factor is the acquisition time (Aqt). Figure 7.7 illustrates comparative data for the IRFT, SRFT, and PSFT methods applied to the same molecule. Note that in the IRFT method I_τ changes phase and passes through zero (the null), whereas in the other two techniques I_τ is always positive.

   For each of the techniques, it is necessary to measure intensities that correspond to the areas under the resonances. Since the line widths are constant in all these experiments, the peak height is a good measure of relative intensity and is usually employed. But some nuclei have a large temperature effect on their chemical shift. This leads to broadening and sometimes to distortion of the line during the experiment, and the peak height is no longer valid as a measure of the intensity of the signal.

C.   Measurement of T_2

T_2 is more difficult to measure than T_1, except when the line width is sufficiently large. Then, external effects may be neglected, and the line width gives T_2 directly from the relation

$$T_2 = \frac{1}{(\pi\Delta\nu_{1/2})}.$$

When T_2 exceeds about 200 to 300 ms, other methods are necessary.

---

* Markley, J. L., Horsley, W. H., Klein, M. P., *J. Chem. Phys.*, **1971**, *55*, 3604. McDonald, G. C., Leigh, J. S., Jr., *J. Magn. Reson.*, **1973**, *9*, 358.
† Freeman, R., Hill, H. D. W., *J. Chem. Phys.*, **1971**, *54*, 3367.

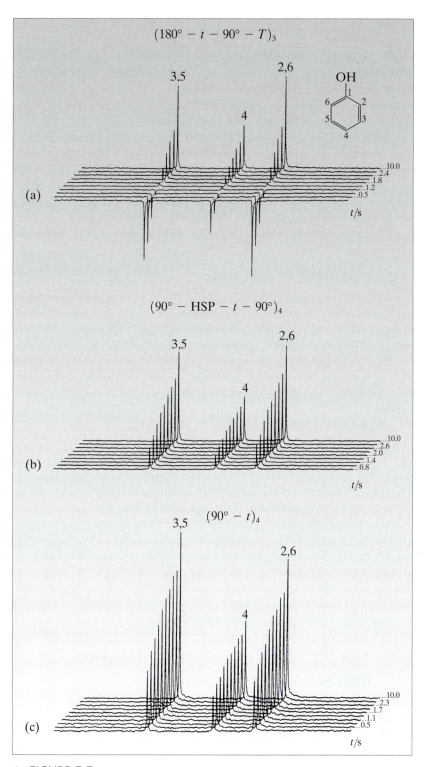

▲ FIGURE 7.7
[13]C Relaxation data for phenol. (a) IRFT, (b) SRFT, and (c) PSFT.

i. The Echo Method

The first sequence used was the Carr–Purcell sequence*

$$[90°-\tau-180°-\tau-(\text{Aqt} + t)-180°-\tau-(\text{Aqt} + t)\ldots T]_n$$

where $(\text{Aqt} + t) = \tau$ and T is a waiting time that allows the system to return to equilibrium. The result is then $I_\tau = I_{eq}\exp(-\tau/T_2)$, from which T_2 is obtained. Satisfactory results are difficult to obtain in practice, and a modification by Meiboom and Gill** improves the results. These authors introduce a 180° pulse, which is phase shifted by 90° with respect to the initial pulse.

The Carr–Purcell sequence is not easy to implement on older standard Fourier–transform specrometers, because they are not able to acquire the successive signals. In that case, after the initial $90°-\tau-180°$ burst, a certain number P of 180° pulses are applied, separated by 2τ, and the free-induction decay is acquired at a time τ after the last 180° pulse. Measurements are performed for several values of P, and after Fourier transformation one measures $I_{\tau\rho} = I_{eq}(-\tau\rho/T_2)$, where $\tau\rho = (P + 2)\tau$. The best result seems to be obtained for τ around 100 ms.

ii. Problems with T_2 Measurements

T_2 measurements are subject to the same errors as T_1 measurements, as well as other difficulties. During such experiments, the use of an external lock must be avoided and the sample must not be spun. When heteronuclear decoupling is used, the Hartmann–Hahn[†] conditions, $\gamma_A^{obs}\mathbf{B_1} = \gamma_X^{dec}\mathbf{B_2}$, must also be avoided.

If scalar coupling J exists between some nuclei, the echo will be modulated by J. This effect can be reduced by using a short τ value such that $|2\tau J| \ll 1$ or by selectively exciting each line one at a time.

D.   Measurement of $T_{1\rho}$

The proposed sequence to measure $T_{1\rho}$ is $[90°_X, (\mathbf{B_1}\gamma)\tau, \text{Aqt} - t]_n$, where $(\mathbf{B_1}\gamma)\tau$ means that the RF field is applied along y in the rotating frame for a time τ; t is the equilibrium recovery time. We then have $I_\tau = I_{eq}\exp(-\tau/T_{1\rho})$. If broadband heteronuclear decoupling is simultaneously in use, care must be taken to avoid the Hartmann–Hahn conditions. This method is not sensitive to pulse misadjustment and does not suffer from any offset effect or J modulation, but it requires special equipment and probes, which allow a continuous high RF field to be applied over several milliseconds. The technique may be used to measure T_2 in the liquid state under extreme narrowing conditions, since $T_2 = T_{1\rho}$ under such conditions. As $T_{1\rho}$ is sensitive to chemical exchange, it may be used just as T_2 may be used to study such exchange processes that occur in the 10^3- to 10^5-s range.

7.7   The ^{13}C Nucleus and Its Dominant Relaxation Mechanisms

For the ^{13}C nucleus in a system that contains neither quadrupolar nuclei nor unpaired electrons, we expect that the dipole–dipole, spin-rotation, scalar-coupling and chemical-shift anisotropy mechanisms may all contribute to

---

* Carr, H. Y., Purcell, E. M., *Phys. Rev.*, **1954**, *94*, 630.
** Meiboom S., Gill, D., *Rev. Sci. Instrum.*, **1958**, *29*, 688.
[†] Hartmann, S. R., Hahn, E. L., *Phys. Rev.*, **1962**, *128*, 2042.

[13]C relaxation times. Thus, the observed T_1 may be a combination of these individual effects such that

$$\frac{1}{T_{1obs}} = \frac{1}{T_{1DD}} + \frac{1}{T_{1SR}} + \frac{1}{T_{1SC}} + \frac{1}{T_{1CSA}}.$$

For the large majority of ^{13}C nuclei with directly attached protons, the dipolar mechanism is dominant and is often the only significant relaxation mechanism. Long-range CH interactions are generally much less important, because of the $1/r^6$ dependence. As a result, for relatively small molecules, the intensities of quaternary carbon resonances are generally low under normal acquisition conditions. This effect is often helpful in terms of chemical-shift assignment.

The spin-rotation mechanism is normally most important for small, highly symmetrical moieties such as short aliphatic side chains or freely rotating methyl groups. In these cases, segmental motion results in an increase in spin-rotation relaxation and, normally, a decrease in the effectiveness of the dominant dipolar relaxation. Consequently, when segmental motion occurs, those groups undergoing it have longer relaxation times than those groups that do not undergo segmental motion. As an example, the methyl and ipso carbon resonances of toluene have approximately equal intensities under normal observation conditions.

If one examines the [13]C T_1 values (given in seconds above the nuclei) for the molecules illustrated in the following figure, the effects of segmental motion become apparent:

Additional Reading

1. BERGER, S., BRAUN, S., and KALINOWSKI, H.-O., *NMR Spectroscopy of the Non-Metallic Elements*, Wiley, New York, 1997.

2. BREITMAIER, E. and VOELTER, W., *Carbon-13 NMR Spectroscopy*, 3d ed., VCH Publishers, Weinheim, Germany, 1990.

3. HARRIS, R. K. and MANN, B. E., *NMR and the Periodic Table*, Academic, London, 1978.

4. MARTIN, M. L., DELPUECH, J.-J., and MARTIN, G. L., *Practical NMR Spectroscopy*, Heyden, London, 1980.

5. LEVY, G. C., PERT, I. R., *J. Mag. Res., 1975*, **18**, 500. Gives a detailed discussion of the various experimental techniques, including a discussion of the sources of error.

6. WEHRLI, F. W. and WIRTHLIN, T., *Interpretation of Carbon-13 NMR Spectra*, Heyden, London, 1976.

7. Typical values of relaxation times for nearly all nuclei may be found in BREVARD, C., GRANGER, P., *Handbook of High Resolution Multinuclear NMR*, Wiley, New York, **1981**.

$$\frac{\mathbf{M}_C(^1H_{irrad})}{\mathbf{M}_C^0} = 1 + \left[\frac{6J_2(\omega_H + \omega_C) - J_0(\omega_H - \omega_C)}{J_0(\omega_H - \omega_C) + 3J_1\omega_C + 6J_2(\omega_C + \omega_H)}\right] \cdot \frac{\gamma_H}{\gamma_C} = 1 + \eta_{CH}$$

$$\frac{\mathbf{M}_C(^1H_{irrad})}{\mathbf{M}_C^0} = 1 + 0.5\frac{\gamma_H}{\gamma_C} = 2.988 \quad \text{or} \quad \eta_{CH} = 1.988$$

$$\frac{1}{T_1(\text{other})} = \frac{1}{T_1(\text{obs})}\left[1 - \frac{\eta_{obs}}{\eta_0}\right]$$

The Nuclear Overhauser Effect

8.1   Introduction

T he nuclear Overhauser effect has many important ramifications. It affects signal intensities and hence the time required to obtain spectra with a given signal-to-noise ratio. It can be used to ascertain whether nuclei are proximate or distant in space and, as a result, can provide structural information. The first effect is encountered in heteronuclear decoupling experiments, most commonly in decoupling proton spins from those of other nuclei—for example, from ^{13}C. For this case, let us symbolize a ^{13}C spin as I and a ^{1}H spin as S. If the protons are irradiated so that the spin S is saturated and $\mathbf{M}_S^t = 0$, then it can be shown that under steady-state conditions,

$$\frac{\mathbf{M}_I(S_{\text{irrad}})}{\mathbf{M}_I^0} = 1 + \frac{\sigma}{\rho}\left(\frac{\mathbf{M}_S^0}{\mathbf{M}_I^0}\right), \tag{8.1}$$

or, since equilibrium-spin populations and magnetizations are proportional to magnetogyric ratios, as

$$\frac{\mathbf{M}_I(S_{\text{irrad}})}{\mathbf{M}_I^0} = 1 + \frac{\sigma}{\rho}\left(\frac{\gamma_S}{\gamma_I}\right).$$

In these two equations, σ and ρ are transition probabilities. For carbon-13 enhancement under proton irradiation, we obtain,

$$\frac{\mathbf{M}_C(^1\text{H}_{\text{irrad}})}{\mathbf{M}_C^0} = 1 + \left[\frac{6J_2(\omega_H + \omega_C) - J_0(\omega_H - \omega_C)}{J_0(\omega_H - \omega_C) + 3J_1\omega_C + 6J_2(\omega_C + \omega_H)}\right] \cdot \frac{\gamma_H}{\gamma_C}, \tag{8.2}$$

$$= 1 + \eta_{CH}$$

where η is termed the ^{13}C—{^{1}H} NOE enhancement factor. The ratio $\gamma_H/\gamma_C = 3.976$. Under extreme motional narrowing, Equation 8.2 becomes

$$\frac{\mathbf{M}_C(^1\text{H}_{\text{irrad}})}{\mathbf{M}_C^0} = 1 + 0.5\frac{\gamma_H}{\gamma_C} = 2.988 \quad \text{or} \quad \eta_{CH} = 1.988. \tag{8.3}$$

Hence, if a ^{13}C{^{1}H} NMR spectrum were obtained with maximum NOE and compared with a ^{13}C NMR spectrum with no proton decoupling, the signal intensity of the former would be almost three times that of the latter, as illustrated in Figure 8.1. This means that a ^{13}C{^{1}H} experiment gives the same signal-to-noise ratio as a ^{13}C experiment in about one-ninth the time, since signal to noise is proportional to time squared. (See Section 2.15.)

The nuclear Overhauser effect depends upon dipole–dipole relaxation. Other competing relaxation pathways detract from it, and since such pathways are common, especially for smaller molecules, the carbon-proton NOE is often less than maximal even under fast-motion conditions. This may afford insight into these other mechanisms of relaxation, since

$$\frac{1}{T_1(\text{other})} = \frac{1}{T_1(\text{obs})}\left[1 - \frac{\eta_{\text{obs}}}{\eta_0}\right], \tag{8.4}$$

where η_0 is the maximal NOE, assuming fast motion. (See also Chapter 7, Section 7.5Biii.) The NOE will be less than maximal as well if the motional

Proton-coupled and decoupled ^{13}C spectra of 23% enriched formic acid exhibiting a 2.98 nuclear Overhauser enhancement $(\eta + 1)$ in integrated intensity.

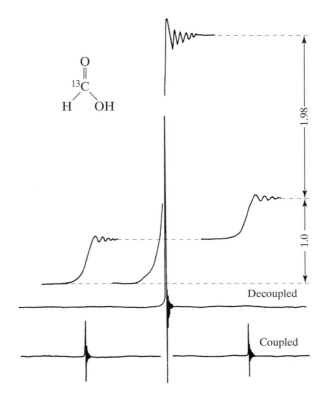

narrowing limit does not apply. Figure 8.2 shows plots of Equation 8.2 for the $^{13}C-\{^1H\}$ Overhauser enhancement at carbon-13 frequencies of 50.3 MHz and 125 MHz. The inflection points of the NOE plots are at the minimum value for T_1, which occurs for a correlation time τ_c of about 10^{-9} s.

For quantitative ^{13}C NMR measurements, where η may vary among carbons, it is sometimes desirable to quench all NOEs by means of a relaxation agent such as chromium acetylacetonate, $[Cr(AcAc)_3]$ or by employing a gated decoupling scheme as described in Section 2.14. The latter is usually preferable, because the sample may be recovered without the need to separate it from the relaxation agent.

The proton–proton Overhauser effect was the first of the Overhauser effects to be used for making resonance assignments and has much potential for dynamic and structural measurements. Since the magnetogyric term is now unity, Equation 8.2 may be written as

$$\frac{M_H(irrad)}{M_H^0} = 1 + \left[\frac{6J_2(2\omega_H) - J_0(0)}{J_0(0) + 3J_1(\omega_H) + 6J_2(2\omega_H)} \right] \tag{8.5}$$

$$= 1 + \eta_{HH}.$$

It must, of course, be assumed that the protons to be observed and those to be irradiated differ sufficiently in chemical shift, as in double-resonance experiments, to make the experiment feasible. A plot of Equation 8.5 is shown in Figure 8.2a for magnetic fields corresponding to proton frequencies

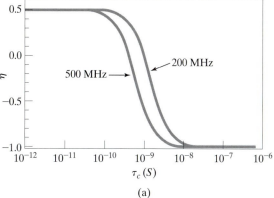

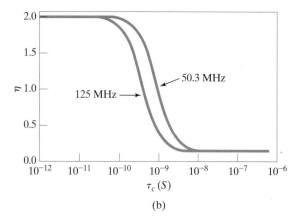

◀ FIGURE 8.2
(a) Nuclear Overhauser enhancement factor η for a proton, produced by irradiation of one or more neighboring protons as a function of the correlation time τ_c, calculated for proton frequencies of 200 MHz and 500 MHz [Eq. 8.5]. (b) Nuclear Overhauser enhancement factor η for a carbon-13 nucleus, produced by irradiation of one or more neighboring protons as a function of the correlation time τ_c, calculated for carbon-13 frequencies of 50.3 MHz and 125 MHz [Eq. 8.2].

$\nu_0 = \omega_0/2\pi$ of 200 MHz and 500 MHz. The maximum enhancement is only 0.5, and it is further notable that η passes through zero when $\tau_c \omega_0 \approx 1$ and becomes negative with an asymptote of -1.

SAMPLE PROBLEM 8.1

What is the maximum NOE for a $^{15}N\{^1H\}$ experiment?

Solution

From Equations 8.2 and 8.4,

$$\frac{\mathbf{M}_z(\text{irrad})}{\mathbf{M}_z^0} = 1 + 0.5\left(\frac{\gamma_H}{\gamma_N}\right)\left(\frac{T_1}{T_{1DD}}\right) = 1 + \eta_{NH}.$$

When $T_1 = T_{1DD}$, $\eta_{NH} = 0.5(-9.86) = -4.93$.

Note in this case that, since $\gamma^{15}N$ is negative, decoupling decreases the intensity of the signal. Figure 8.3 shows an example.

The observation that two of the resonances in Figure 8.3 show appreciable NOE can be used as an aid in the spectral assignment. It indicates that the nitrogen nuclei giving rise to these resonances either are coupled to nearby protons or are proximate in space to nearby protons (or both).

▶ FIGURE 8.3
30.46 MHz $^{15}N\{^1H\}$ NMR spectra of a mixture of 1- and 2-benzyl-5-phenyltetrazoles in $CDCl_3$ (a) with NOE and (b) without NOE.

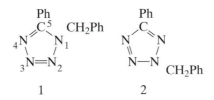

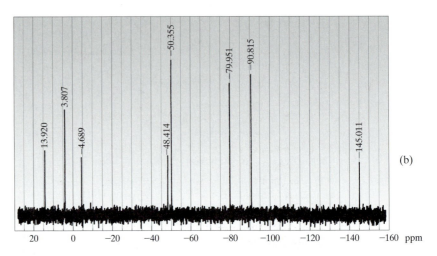

(b)

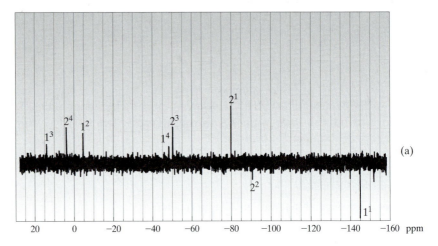

(a)

8.2 Experimental Measurement

The simplest method for measuring $X-\{^1H\}$ nuclear Overhauser enhancements might appear to be to obtain the integrated X spectra with and without broadband proton decoupling. In practice, this is found to be undesirable, because of the frequent overlapping of multiplets in the coupled spectrum, particularly for ^{13}C measurements. This makes it difficult to measure enhancement ratios for individual resonances. A better method is to gate the proton decoupler on only during data acquisition. The multiplets collapse instantaneously, but the NOE requires a time on the order of T_1 to build up. By gated decoupling and complete decoupling experiments, the decoupled spectra are obtained, respectively, without and with NOEs.

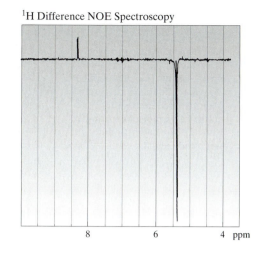

^{1}H Difference NOE Spectroscopy

(R)$_c$(S)$_{Ru}$

8   6   4 ppm

▲ **FIGURE 8.4**

A 300-MHz ^{1}H NOE difference spectrum of the molecule shown. Proton irradiation of the η^6—C$_6$H$_6$ resonance (δ = 5.3 ppm) gives rise to a positive NOE for the H$_a$ resonance (δ = 8.2 ppm).

For ^{1}H—^{1}H NOE measurements, one obtains the normal ^{1}H spectrum and stores it on the instruments' computer. Then, a single-frequency ^{1}H decoupled spectrum is obtained, and the normal spectrum is subtracted from it. What remains is the NOE. This technique is called NOE-difference spectroscopy. An example is shown in Figure 8.4. In this example, irradiation of the η^6—C$_6$H$_6$ resonance allows for an unequivocal assignment of the H$_a$ resonance, as the respective nuclei are proximate in space.[*] The weak signals near 7 ppm are subtraction artifacts, and these may often, but not always, be distinguished from genuine NOE signals by their phase behavior. Genuine NOE signals either have the opposite phase (if the NOE is positive) with respect to the irradiated resonance, or the same phase (if the NOE is negative). For this relatively small molecule, the NOE is positive. For molecules with molar masses that exceed about 1,000 atomic mass units, the NOE is often negative, and as Figure 8.2a illustrates, the NOE passes through zero for intermediate-sized molecules. In this case, the effect becomes unobservable. Since the NOE is field dependent (Equation 8.5), it is sometimes possible to observe it at one field strength, but not at a higher field strength. This is a case in which a higher field spectrometer could in fact be disadvantageous.

Shaka and coworkers[†] have devised an experiment that employs the use of pulsed-field gradients to eliminate subtraction artifacts from ^{1}H NOE difference spectra. A comparison of the results obtained from a normal ^{1}H NOE difference spectrum with those from the pulsed-field-gradient experiment is shown in Figure 8.5. The pulsed-field-gradient experiment is very sensitive, but it requires specialized equipment.

An example of the use of the pulsed-field gradient ^{1}H NOE experiment for the determination of molecular geometry is shown in Figure 8.6.[‡]

[*] Attar, S., Nelson, J. H., Fischer, J., De Cian, A., Pfeffer, M., *Organometallics*, **1995**, *14*, 4559. Attar, S., Catalano, V. J., Nelson, J. H., *Organometallics*, **1996**, *15*, 2932.

[†] Stott, K., Stonehouse, J., Keeler, J., Hwang, T.-L., Shaka, A. J., *J. Am. Chem. Soc.*, **1995**, *117*, 4199.

[‡] Maitra, K.; Catalano, V. J., Nelson, J. H., *J. Am. Chem. Soc.*, **1997**, *119*, 12560.

▶ FIGURE 8.5
(a) The conventional ¹H NMR spectrum of 1. (b) Conventional ¹H NOE difference spectrum with irradiation of the H_2 resonance of the pyrrolidine ring (δ = 3.76). Note the subtraction artifacts (indicated by asterisks). (c) The pulsed-field-gradient experiment. Note the elimination of the subtraction artifacts.

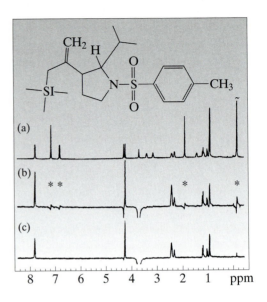

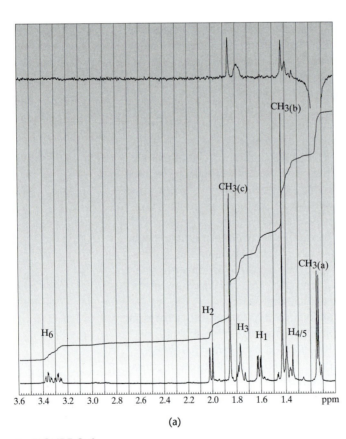

(a)

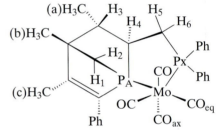

▲ FIGURE 8.6
Expansions of the normal 500-MHz ¹H NMR spectra of the molecule shown in CDCl₃ solutions (lower) with selective excitations (upper) of the (a) CH₃(a) resonance (δ = 1.14) and (b) predominantly the CH₃(b) resonance (δ = 1.43). The NOE at 3.33 ppm results from partial excitation of H_5 (δ ≈ 1.49). (c) H_1 resonance (δ = 1.60). Note that H_1 and H_2 are scalar coupled. (d) The H_2 resonance (δ = 2.03). Note that the magnitudes of the NOE from H_2 to H_1 and from H_1 to H_2 are not equal. *(continues on next page)*

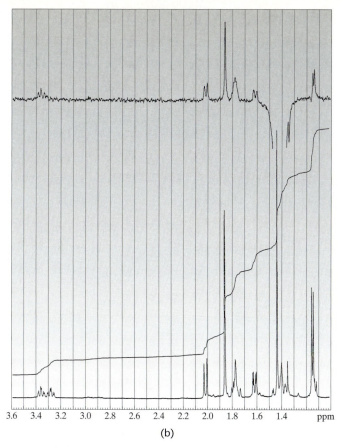

(b)

(c)

◀ FIGURE 8.6 (continues on next page)

211

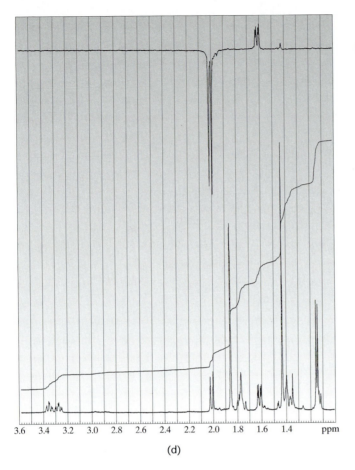

(d)

◀ FIGURE 8.6 *(cont.)*

Additional Reading

1. An excellent book on the nuclear Overhauser effect is *The Nuclear Overhauser Effect in Structural and Conformational Analysis* by NEUHAUS, D., WILLIAMSON, M., VCH Publishers, New York, 1989.

2. ATTA-UR-RAHMAN and CHOUDHARY, M. I., *Solving Problems with NMR Spectroscopy*, Academic, San Diego, 1996.

3. DUDDECK, H. and DIETRICH, W., *Structure Illucidation by Modern NMR*, Springer-Verlag, New York, 1989.

4. SANDERS, J. K. M. and HUNTER, B. K., *Modern NMR Spectroscopy*, 2d ed., Oxford University Press, Oxford, 1993.

5. a) BRAUN, S., KALINOWSKI, H.-O., and BERGER, S., *100 and More Basic NMR Experiments*, VCH Publishers, Weinheim, 1996; b) BERGER, S., *150 and More Basic NMR Experiments*, Wiley-VCH, Weinheim, 1998.

6. BERGER, S., BRAUN, S., and KALINOWSKI, H.-O., *NMR Spectroscopy of the Non-Metallic Elements*, Wiley, New York, 1997.

CHAPTER 9

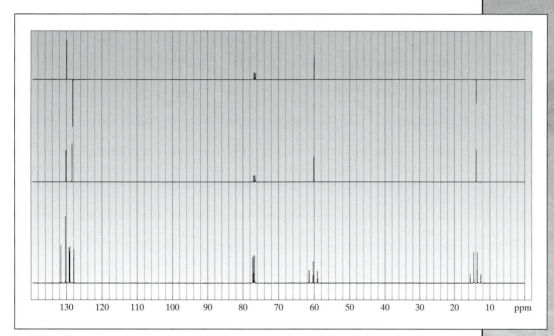

From top to bottom: APT, ^{13}C{^{1}H} gated and ^{13}C{^{1}H} inverse gated spectra of ethylacrylate

CH$_3$CH$_2$O

Editing ^{13}C NMR Spectra

9.1   Introduction

In ^{13}C{^{1}H} NMR spectra, because of proton decoupling, information from $^1J_{CH}$ is lost. On the other hand, proton-coupled spectra may be complicated and difficult to interpret, in addition to suffering from lower signal to noise due to the lack of an NOE. Thus, it may not be easy to assign the resonances in the decoupled spectra to methyl, methylene, methine, or quaternary carbons. In this chapter, two experiments, the **APT** and the **DEPT**, are described that have been developed for the purpose of simplifying ^{13}C shift assignments.

9.2   Attached-Proton Test (**APT**)*

The attached-proton test uses the pulse sequence illustrated in Figure 9.1. In this sequence, the initial 90_X° ^{13}C pulse generates ^{13}C transverse magnetization, which evolves under the influence of proton–carbon J *modulation* during the period τ. A 180_X° ^{13}C pulse then refocuses all effects other than J modulation, such as field inhomogeneities. Simultaneously, the ^{1}H broadband decoupler is turned on in order to observe a proton-decoupled FID signal with nuclear Overhauser enhancement during the detection period.

J modulation may be explained by Figure 9.2. After the proton decoupler is switched off, the multiplet components of transverse magnetization dephase with different velocities due to proton–carbon coupling. For CH doublets, the two components rotate with frequencies $\pm J/2$ relative to the rotating frame. In the CH_2 triplet case, the central component remains aligned along y', while the peripheral components dephase with frequencies $\pm J$. For CH_3 quartets, two slower, but stronger, vectors rotate with $\pm J/2$, while two faster, but weaker, ones dephase with $\pm 3J/2$. The resultant magnetization of each multiplet depends upon how long the proton decoupler is switched off, as shown in the figure. At time $\tau = 0.5\,J^{-1}$, the resultant is zero for all CH_n multiplets; at time $\tau = J^{-1}$, the vector sum is negative for CH_3 and CH, but positive for CH_2 moieties. The transverse magnetization of nonprotonated carbons, however, remains constant and positive. Neglecting attenuations due to relaxation, the resultant transverse magnetization and the responding signal intensities follow cosine functions, as shown in Figure 9.3.

In sum, J_{CH} modulation converts J_{CH} coupling and multiplicity into intensity and phase information, which is observed in the decoupled

▶ FIGURE 9.1
Pulse sequence for attached-proton test.

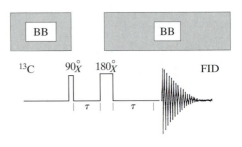

---

*Patt, S., Shoolery, J. N., *J. Magn. Reson.*, **1982**, *46*, 535.

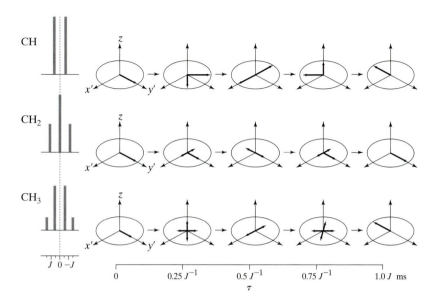

◀ FIGURE 9.2

J modulation: Components of transverse CH, CH$_2$, and CH$_3$ magnetization as a function of τ for the pulse sequence of Figure 9.1.

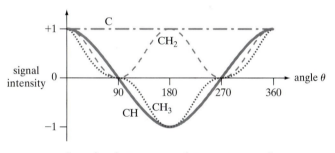

◀ FIGURE 9.3

J modulation: Signal intensities J_{CH_n} of C, CH, CH$_2$, and CH$_3$ carbon nuclei as a function of τ.

— · — Intensity of quaternary carbon resonance = I_0

——— Intensity of methine carbon resonance $I_{CH} = I_0 \cos(\pi\tau J)$

– – – Intensity of methylene carbon resonance $I_{CH_2} = I_0 [1 + \cos(2\pi\tau J)]$

·········· Intensity of methyl carbon resonance $I_{CH_3} = I_0 [3\cos(\pi\tau J) + \cos(3\pi\tau J)]$

spectrum. Figure 9.3 illustrates how the switch-off delay of the proton decoupler has to be adjusted in order to sort CH$_n$ multiplets. For $\tau = (2J)^{-1}$, only quaternary carbon atoms give rise to intense positive signals. For $\tau = J^{-1}$, nonprontonated and CH$_2$ carbon nuclei appear positive, while CH and CH$_3$ signals appear negative, as $\pi(J^{-1})(J) = \pi$. An additional experiment with $\tau = 3(J^{-1})/4$ may help to distinguish weaker methyl from stronger methine signals. Figure 9.4 shows an **APT** spectrum of camphor, which clearly allows assignment of the various types of carbon.

The optimum switch-off delays have to be adapted to the individual CH coupling constants, mostly ranging from 125 Hz for alkyl groups ($\tau = J^{-1} = 8$ ms) to 155 Hz for alkene and aromatic CH bonds ($\tau = J^{-1} = 6$ ms). If the pulse sequence is modified by using an initial ^{13}C pulse that is less than 90°, then if one adds an additional refocusing sequence Δ, 180°_X, Δ, where $\Delta < \tau$ (typically ms) and τ is set to 7.5 ms, then these modified APT experiments generally give good results, as illustrated in Figure 9.4.

▶ FIGURE 9.4
125-MHz ^{13}C **APT** spectrum
of camphor in C$_6$D$_6$ ($\tau = J^{-1}$
avg = 7.5 ms).

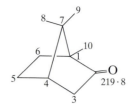

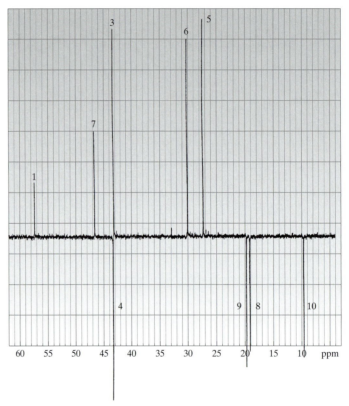

9.3   Distortionless Enhancement by Polarization Transfer (**DEPT**)*

A very useful experiment known as *distortionless enhancement by polarization transfer* (**DEPT**) permits the cleanest multiplicity selection known so far, with full **NOE** enhancement and low sensitivity to individual CH coupling constants. In addition, fully enhanced and undistorted coupled spectra can be recorded if desired. Finally, subspectra for CH, CH$_2$, and CH$_3$ groups can be generated.

Figure 9.5 illustrates the influence of the DEPT sequence on a ^{13}C—^{1}H two-spin system. The proton magnetization created by a 90°_X ^{1}H pulse (a) is modulated by CH coupling until, at time $\tau = \frac{1}{2} J_{CH}$, the CH doublet vectors have attained a 180° phase difference (b).

Then, a 90°_X ^{13}C pulse tips the carbon vectors onto the xy-plane (c), while a 180°_X ^{1}H pulse causes refocusing of inhomogeneity effects (b). Now,

---

*Bendall, M. R., Doddrell, D. M., Pegg, D. T., *J. Am. Chem. Soc.*, **1981**, *103*, 4603.

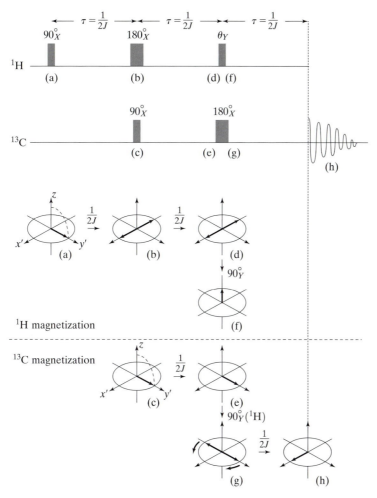

The **DEPT** pulse sequence and the associated motion of the proton and carbon magnetization vectors for a CH spin pair.

z magnetization is zero for both protons and carbons; in fact, the two nuclei are decoupled. Therefore, the ^{1}H and ^{13}C magnetization vectors remain standing in the subsequent period $\tau = \dfrac{1}{2J_{CH}}$ (d and e). At the end of this second τ period, a 90°_Y ^{1}H pulse rotates both doublet components of proton magnetization to the $\pm z$-axis (f). This is a polarization transfer, because the polarization of protons induced by the 90°_Y ^{1}H pulse also controls the ^{13}C population due to ^{13}C—^{1}H coupling. Subsequently, the enhanced ^{13}C—^{1}H doublet vectors rotate about the $\pm z$-axis due to the coupling protons and refocus along the x'-axis after the third period of $\tau = \dfrac{1}{2J_{CH}}$ (h). A 180°_X ^{13}C pulse at the beginning of the third τ period (g) cancels ^{13}C inhomogeneity effects, after which the FID signal is finally recorded and Fourier transformed.

The extent of polarization transfer is determined by the Θ_Y ^{1}H pulse angle for CH doublets, as illustrated in Figure 9.5. Under the same conditions, the resultant x' magnetization is zero for CH$_2$ and CH$_3$ carbon signals (Figure 9.6). Further, the dependence of the intensities on the polarization transfer angle Θ_Y ^{1}H varies in a sinusoidal fashion for each CH$_n$ signal.

▶ FIGURE 9.6
Sinusoidal dependence of CH$_n$ signal intensities in a **DEPT** experiment as a function of Θ_Y ^{1}H.

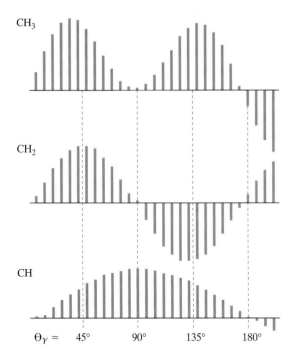

The **DEPT** sequence provides a clean method of CH$_n$-multiplet editing with sensitivity enhancement by polarization transfer. One experiment with Θ_Y ^{1}H of 90° generates the CH subspectrum. An additional one with Θ_Y ^{1}H of 135° gives negative CH$_2$, but positive CH and CH$_3$, signals. Those signals observed in the normal ^{13}C{^{1}H} spectrum, but not in the **DEPT** spectra, are due to quaternary carbons.

To edit a ^{13}C spectrum, four **DEPT** experiments with polarization transfer angles of Θ_Y = 45°, 90°, 90°, and 135° are recorded. These experiments are commonly called $s1$, $s2$, $s3$, and $s4$, respectively. The spectra illustrated in Figure 9.6 arise from linear combinations of the four experiments as follows:

1. All protonated carbons $0.23\,(s1 + s2) + 2s1$
2. CH: $s2 + s3$
3. CH$_2$: $s1 - s4$
4. CH$_3$: $-0.707\,(s1 + s3) + s1 + s4$

These calculations are normally done automatically in the instruments' computer and are provided in the software supplied by the instrument manufacturer. Note that quaternary carbons are suppressed in the **DEPT** spectrum.

The **APT** experiment (Figure 9.4) depends much less critically upon the setting of the 90° pulse widths and the actual values of $^1J_{CH}$ than does the **DEPT** experiment (Figures 9.6 and 9.7). Further, the **DEPT** experiment usually requires about four times as much instrument time as does the **APT** experiment, but assignments are less ambiguous in the **DEPT** experiment. Also, the signals in the **APT** experiment may be phased in two opposite ways, which may sometimes lead to confusion.

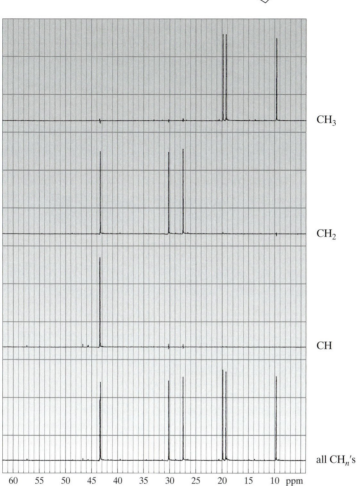

◀ FIGURE 9.7
125-MHz DEPT spectrum of camphor.

CH$_3$

CH$_2$

CH

all CH$_n$'s

60 55 50 45 40 35 30 25 20 15 10 ppm

Additional Reading

1. Braun, S., Kalinowski, H.-O., and Berger, S., *100 and More Basic NMR Experiments*, VCH Publishers, Weinheim, 1996.

2. Braun, S., Kalinowski, H.-O., and Berger, S., *150 and More Basic NMR Experiments*, VCH Publishers, Weinheim, 1999.

3. Breitmaier, E. and Voelter, W., *Carbon-13 NMR Spectroscopy*, 3d ed., VCH Publishers, Weinheim, 1990.

4. Derome, A. E., *Modern NMR Techniques for Chemistry Research*, Pergamon, Oxford, 1987.

5. Macomber, R. S., *A Complete Introduction to Modern NMR Spectroscopy*, Wiley, New York, 1998.

6. Sanders, J. K. M. and Hunter, B. K., *Modern NMR Spectroscopy*, 2d ed., Oxford University Press, Oxford, 1993.

CHAPTER 1

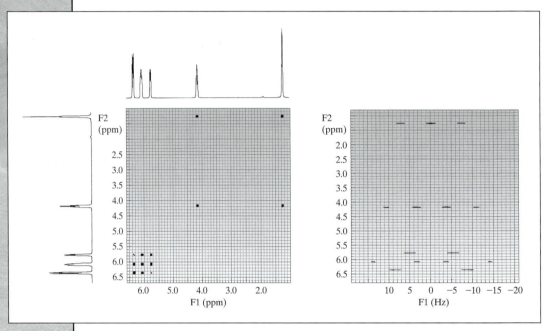

COSY (left) and **HOM2DJ** (right) spectra of ethylacrylate

Two-Dimensional NMR Spectroscopy

10.1   Introduction

One-dimensional NMR spectra have an ordinate that gives the intensity of the resonances and an abscissa that gives their frequencies. In a two-dimensional (2D) NMR spectrum, both the ordinate and the abscissa are frequency axes, and the signal intensity constitutes a third dimension.

There are two common types of 2D spectra. Those where chemical shifts are plotted along one axis and coupling constants are plotted along the other axis are called *two-dimensional J-resolved spectra*. Those where both axes are chemical shifts are called *two-dimensional (shift)-correlated spectra*. For either of these two types of 2D spectra, the nuclei may be of the same element (e.g., both ^1H) or of different elements (e.g., ^1H and ^{13}C). Thus, for *J*-resolved spectra, the acronyms **2DHOJ** or **HOM2DJ** and **2DHETJ** or **HET2DJ** have been used to indicate that the *J*-resolved spectra are homonuclear and heteronuclear, respectively. The most commonly encountered situations are ^1H chemical shifts vs. $^nJ_{HH}$ in the **2DHOJ** case and ^{13}C chemical shifts vs. $^nJ_{CH}$ in the **2DHETJ** case. For correlated spectroscopy, the acronyms **COSY** and **HETCOR** are used to indicate homonuclear and heteronuclear chemical-shift correlations, respectively.* The most commonly encountered situations here are $^1H/^1H$ and $^1H/^{13}C$ chemical-shift correlations, although other heteronuclear correlations are becoming more common.

Two-dimensional NMR spectroscopy methods are based upon the couplings between nuclear dipoles. These need not be scalar couplings, but, as in the nuclear Overhauser effect, may also be of a dipolar type through space. The latter enables the determination of the geometrical structure of molecules. Another kind of two-dimensional NMR spectroscopy is based upon the transfer of magnetization by chemical exchange, and it can be used to study dynamic processes.

We consider first the basic idea underlying two-dimensional NMR spectroscopy. First proposed by Jeener in 1971,[†] this idea became a practical reality when Ernst et al.[‡] and Freeman et al.[§] carried out their pioneering experiments. We will not consider the exact mathematical details. [The interested reader may consult Ernst, R. R., Bodenhausen, G., Wokaun, A., *Principles of Nuclear Magnetic Resonance in One and Two Dimensions* (Oxford: Clarendon Press, New York), 1987.]

10.2   The Basic Two-Dimensional NMR Experiment

A.   Preparation, Evolution and Mixing, Data Acquisition

In a normal Fourier-transform NMR experiment, the RF excitation pulse is followed immediately by the *data acquisition* (detection) *phase*, during which the free-induction decay signal (FID) is recorded and stored in the instrument's computer. When complex pulse sequences are employed, the spin system undergoes a *preparation phase* before data acquisition. In two-dimensional

---

*One also finds the acronyms **HOMCOR** for **COSY** and **H, C-COSY** for **HETCOR** used in the literature.

[†] Jeener, Ampére International Summer School, Basko Polje, Yugoslavia, 1971.

[‡] Ernst, R. R., *Chimia* **1975**, *29*, 179.

[§] Freeman, R., Morris, G. A., *Bull. Magn. Reson.* **1979**, *1*, 5.

▶ FIGURE 10.1

The basic two-dimensional NMR pulse sequence. The variable t_1, the mixing or evolution time, can be on the order of milliseconds to seconds.

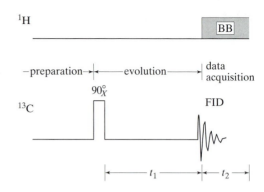

NMR experiments, the preparation and data acquisition phases are separated by an *evolution and mixing phase*. (See Figure 10.1.)

The principle may be illustrated as follows: Consider a two-spin AX spin system with $A = {}^1H$ and $X = {}^{13}C$, as in the ${}^{13}C$ NMR spectrum of $CHCl_3$. In the experiment, a variable delay time t_1 is inserted between the 90° ${}^{13}C$ pulse and the recording of the FID, and the ${}^1H$ broadband (BB) decoupler is switched on only during the data acquisition period. If n experiments, each with a different value of t_1, starting at $t_1 = 0$, were performed and the resulting n FIDs were Fourier transformed with respect to t_2, then n frequency-domain spectra F_2 would be obtained. If the decoupler were turned on during the entire experiment, all n of these spectra would contain a single resonance, but each spectrum would have a different intensity, due to differing amounts of relaxation during t_1. However, since the decoupler is turned on only during data acquisition, the ${}^{13}C$ and ${}^1H$ nuclei are coupled during the evolution phase. The vector diagrams in Figure 10.2 show how the C—H coupling affects the single peaks in the final spectra.

After the $90^\circ_{X'}$ pulse ($t_1 = 0$, Figure 10.2a), the macroscopic magnetization $\mathbf{M}_C$ of the ${}^{13}C$ nuclei lies along the y' direction in the rotating frame. $\mathbf{M}_C$ can be resolved into two components, $\mathbf{M}_C^{H\alpha}$ and $\mathbf{M}_C^{H\beta}$, corresponding to ${}^{13}C$ nuclei in $CHCl_3$ molecules containing protons in the α and β spin states, respectively.

The two vectors $\mathbf{M}_C^{H\alpha}$ and $\mathbf{M}_C^{H\beta}$ rotate with different Larmor frequencies according to the equations

$$\nu \mathbf{M}_C^\alpha = \nu_C - \frac{1}{2}J_{CH} \qquad (10.1)$$

and

$$\nu \mathbf{M}_C^\beta = \nu_C + \frac{1}{2}J_{CH}.$$

If the rotating frame coordinates x' and y' rotate with a frequency of ν_C that is the average of the two Larmor frequencies, then $\mathbf{M}_C^{H\beta}$ rotates faster than the frame by $\frac{1}{2}J_{CH}$, and $\mathbf{M}_C^{H\alpha}$ rotates slower than the frame by $\frac{1}{2}J_{CH}$, because ${}^1J_{CH}$ is positive. (See Chapter 6.) During the time t_1, the magnetization vectors $\mathbf{M}_C^{H\alpha}$ and $\mathbf{M}_C^{H\beta}$ respectively rotate through the angles

$$\phi_\alpha = 2\pi\left(\nu_C - \frac{1}{2}J_{CH}\right)t_1 \qquad (10.2)$$

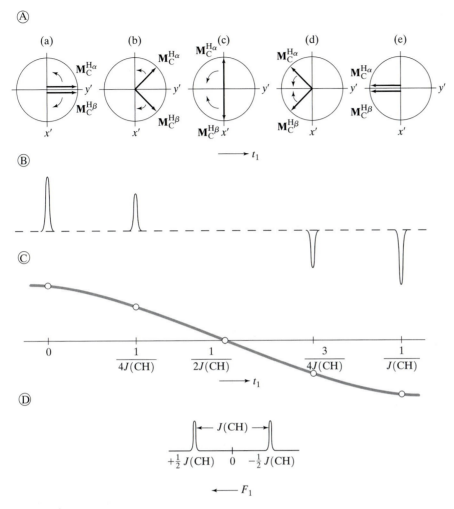

▲ FIGURE 10.2
(A) Vector diagrams of the evolution of the ^{13}C magnetization vectors for an AX ($A = {}^1H$, $X = {}^{13}C$) spin system at different values of t_1. The frequency of rotation of the rotating frame is ν_C. (a) $t_1 = 0$; (b) $t_1 = [4J_{CH}]^{-1}$; (c) $t_1 = [2J_{CH}]^{-1}$; (d) $t_1 = 3[4J_{CH}]^{-1}$; (e) $t_1 = [J_{CH}]^{-1}$.
(B) Fourier transformation with respect to t_2 gives ^{13}C NMR spectra composed of single resonances whose amplitudes depend upon t_1. (C) The curve shows that the amplitude modulation is related to J_{CH}^{-1}. (D) Fourier transformation with respect to t_1 yields two resonances separated by J_{CH}.

and

$$\phi_\beta = 2\pi\left(\nu_C - \frac{1}{2}J_{CH}\right)t_1$$

in the directions indicated by the small arrows. The phase difference Θ between $\mathbf{M}_C^{H\alpha}$ and $\mathbf{M}_C^{H\beta}$ is given by the equation

$$\Theta = \phi_\alpha - \phi_\beta = 2\pi(J_{CH})t_1. \tag{10.3}$$

When $t_1 = [4J_{CH}]^{-1}$ $\Theta = 90°$, when $t_1 = [2J_{CH}]^{-1}$ $\Theta = 180°$, and when $t_1 = [J_{CH}]^{-1}$, the two magnetization vectors are again in phase, but now lie along the $-y'$ direction (Figure 10.2A).

▶ FIGURE 10.3
Amplitude and phase modulation
of the spectra that result from the
first Fourier transformation of the
FIDs in Figure 10.2.

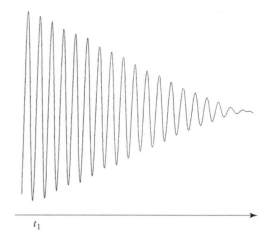

t_1

The information about the evolution of the spin system during the evolution time t_1 is encoded both in the phase and in the amplitude of the FID detected, and the modulation is passed through the first Fourier transformation to the spectra that exhibit combined phase and amplitude modulation as shown in Figures 10.2C and 10.3.

The amplitude modulation carries information on the motion of the components or the splitting of the multiplet (in this case, $^1J_{CH}$), and the phase modulation carries information about the chemical shift of the multiplet. A second Fourier transformation with respect to t_1 produces a 2D spectrum (Figure 10.4) that has a doublet in the direction of the F_1-axis, as is found in the ^{13}C NMR spectrum without ^1H decoupling. The center of this represents the ^{13}C chemical shift.

(a) Stacked plot   (b) Contour plot

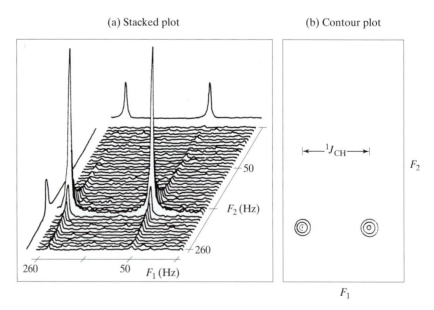

▲ FIGURE 10.4
2DHETJ spectrum of CHCl$_3$; δ^{13}C = 77.1 ppm, $^1J_{CH}$ = 209 Hz. The spectra plotted along the F_2 and F_1 axes were obtained in separate experiments, respectively with and without ^1H broadband decoupling.

10.3   Graphical Presentation of 2D NMR Spectra

The 2D NMR spectra or expansions thereof are customarily presented graphically in one of two forms: *stacked-trace plots* and *contour plots*. A stacked-trace plot is obtained by drawing a series of spectra, one behind the other, in order of increasing frequency, as in Figure 10.4a. The spectra are arranged in the figure in such a way that a three-dimensional impression is created. This impression is generally enhanced by "whitewashing," that is, by leaving out those parts of the traces that have a lower intensity than previous traces and therefore would not be seen in space.

The three-dimensional impression is further enhanced by a skewed presentation, with each trace starting not only a few millimeters above the preceding one, but also shifted slightly to the left or right. One common danger when using whitewashed presentation methods is that peaks which are "hidden" behind peaks in preceding traces might escape our attention, as would any peak with a negative intensity. The whitewashed, skewed stacked-trace plots can give an attractive appearance, but finding the exact location of a peak (i.e., its F_1 and F_2 coordinates) is difficult, as is a measurement of the distance between two peaks in such graphs. These types of plots were commonly used in the early days of the development of 2D NMR spectroscopy, but now are seldom used.

Contour plots depict the intensity distribution in the 2D NMR spectrum in the same way that contour lines indicate altitudes in topographical maps. The contour lines connect points (F_1, F_2) that have the same signal intensity (Figure 10.4b). Plotting is usually much faster than in the case of stacked-trace plots, but problems arise when both low- and high-intensity peaks are present in the spectrum. If the minimum contour level is chosen to show the presence of low-intensity peaks, the high-intensity peaks will then extend over a large area, and contours of spurious signals or noise will crowd the spectrum. The precision in determining the position of a peak depends upon the diameter of the highest contour level surrounding the peak.

10.4   Two-Dimensional *J*-Resolved Spectroscopy

A.   Heteronuclear (**HET2DJ**)

The most common application of **HET2DJ** NMR spectroscopy is *J*-resolved ^{13}C NMR. The mechanism described previously (Section 10.2) for the AX spin system of $CHCl_3$ can easily be extended to the A_2X and A_3X spin systems of CH_2 and CH_3 groups in organic compounds. In these systems, the total magnetization $\mathbf{M_C}$ of the ^{13}C nuclei is composed of a greater number of components. In the A_2X system, the three components $\mathbf{M_C^+}$, $\mathbf{M_C^0}$, and $\mathbf{M_C^-}$ must be considered. These components rotate during the evolution period with the frequencies of triplet components (i.e., with the frequencies $\nu_C + J_{CH}$, ν_C, and $\nu_C - J_{CH}$, respectively) in the rotating frame of reference. Similarly, the $\mathbf{M_C}$ magnetization in the A_3X system consists of four components, $M_C^{+3/2}$, $M_C^{+1/2}$, $M_C^{-1/2}$, and $M_C^{-3/2}$, which rotate with frequencies of $\nu_C + J_{CH} \Sigma_{ml}$, where Σ_{ml} is the sum of the magnetic quantum numbers of the ^{1}H spins in the particular state of the whole system. The sums are indicated as the index on the corresponding magnetization components. All these components rotate during the evolution period, and their vector sum would produce a complicated FID if it were detected. After the decoupler is switched on, the motion of the components is simplified. All components continue to rotate with the same frequency ν_C, and an FID corresponding to a decoupled spectrum is detected. Its phase and amplitude, however, depend again

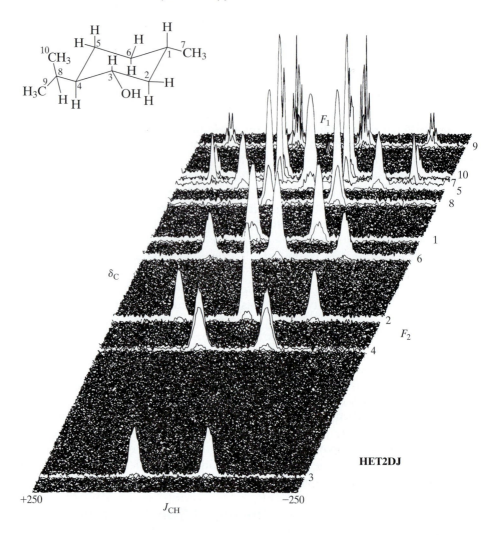

▶ FIGURE 10.5
75-MHz **HET2DJ** spectrum of menthol in $CDCl_3$. Note the long-range coupling to C_9 in this spectrum.

on the time t_1 and the frequencies that the components have had during that time. An example of an **HET2DJ** spectrum of menthol is shown in Figure 10.5.

Obviously, the **HET2DJ** spectrum spreads the multiplets of a linear ^{13}C NMR one-dimensional spectrum into the two-dimensional plane, and this enhances spectral resolution. Without such spreading, the 1D spectra of large molecules measured without decoupling are often difficult to analyze because of severe multiplet overlap.

The experiment that we have just described was first proposed by L. Müller, A. Kumar, and R. R. Ernst in 1975 and is called the basic **MKE** experiment.*

The simplicity of the basic **MKE** experiment (Figure 10.6a) and the suitability of the semiclassical model for providing an adequate description of this experiment provide a variety of opportunities for using the experiment to test the basic understanding of the model, the methods, and the concepts of 2D NMR spectroscopy.

A simple mathematical operation rotates the **MKE** spectrum by 90° (transposition). In a rotated spectrum, the stacked-trace plot presents cross sections through the spectrum that are parallel to the F_2-axis. The cross sections that show decoupled spectra are plotted in a left-to-right direction. 2D spectra with the

---

*Müller, L., Kumar, A., Ernst, R. R., *J. Chem. Phys.* **1975**, *63*, 5490.

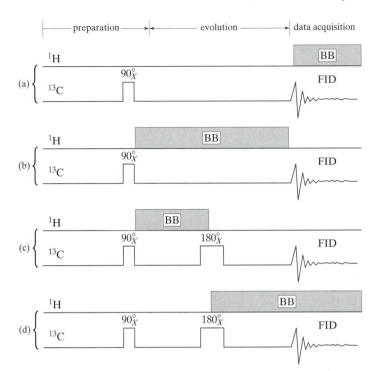

◀ FIGURE 10.6
Pulse sequences for measurement
of **HET2DJ** NMR spectra by a gated
decoupling method:
(a) **MKE** experiment;
(b) inverted **MKE** experiment;
(c) decoupling during defocusing;
(d) decoupling during refocusing.

same appearance can be obtained directly by using the pulse sequence shown in
Figure 10.6b. With this pulse sequence, which employs continuous decoupling
only during the evolution period, the spin system evolves during the evolution
periods solely under the influence of chemical shifts. During the data-acquisition
period, undecoupled spectra are measured—that is, spectra which show both
chemical-shift and coupling effects. As a result, 2DJ resolved spectra are obtained.

Both of the experiments just described place large demands on computer
memory. Since the spectra are spread in both directions by chemical shifts, the
two axes must cover large spectral widths ($\Delta\delta$ X Hz/ppm ~ 14,000 Hz for a
typical 75-MHz ^{13}C observation). At the same time, the spectra must be well resolved, which requires several data points per Hz in order to obtain good digital resolution. Memory requirements are considerably reduced if a $180°(^{13}C)$
pulse is placed in the middle of the evolution period (i.e., at $t_1/2$). This pulse
eliminates the effects of chemical shifts during that period, refocusing the magnetization components due to different chemical shifts so that at time t_1 the components are in the same position that they were immediately after the first pulse at
$t_1 = 0$. The introduction of the $180°$ ^{13}C pulse would also eliminate (refocus) the
effects of CH coupling, which one wants to retain in the spectrum. There are two
ways to prevent the refocusing of coupling interactions. One method is to apply
a $180°$ ^{1}H decoupler pulse simultaneously with the $180°$ ^{13}C pulse. The other
method is to use continuous decoupling during either the first or the second half
of the evolution period. The first method, two simultaneous $180°$ pulses, forms
the basis of the methods with proton (or spin) flip. Continuous decoupling, during either the defocusing time (Figure 10.6c) or the refocusing time (Figure 10.6d),
forms the basis of gated decoupling. With these modifications, the spin system
at the end of the evolution period is in the same state that it would have reached
if it had developed only under the influence of spin–spin coupling. The chemical-
shift effects are eliminated, significantly reducing the spectral width of the F_1-axis
(500 Hz would normally be sufficient) and also reducing the requirements on
memory data storage and measuring time.

B. Homonuclear (**HOM2DJ**)

The most common application of **HOM2DJ** NMR spectroscopy is J-resolved ^{1}H NMR, which is useful for the resolution of overlapping multiplets that are often encountered in large molecules. This experiment will be discussed by considering a two-spin AX spin system. A spin-echo pulse sequence (Figure 10.7) with the variable t_1 is used.

The $90^\circ_{X'}$ pulse rotates the macroscopic magnetization vectors $\mathbf{M}_A$ and $\mathbf{M}_X$ of the A and X protons into the y' direction. We will consider only $\mathbf{M}_A$, which we will separate into two components, $\mathbf{M}_A^{X\alpha}$, with the X protons in the α spin state, and $\mathbf{M}_A^{X\beta}$, with the X protons in the β spin state (Figure 10.7).

Like the individual spins, these two vectors rotate about the z-axis with the frequencies

$$\nu_A^{X\alpha} = \nu_A - \frac{1}{2}J_{AX} \quad \text{and} \quad \nu_A^{X\beta} = \nu_A + \frac{1}{2}J_{AX}.$$

Here, ν_A is the Larmor frequency of the A protons in the absence of coupling to the X spins. The frame is rotating at a frequency of ν_A. The sign of J_{AX} determines which of the two vectors rotates faster than the frame. Assuming that J_{AX} is positive, $\mathbf{M}_A^{X\beta}$ rotates faster, and $\mathbf{M}_A^{X\alpha}$ rotates slower, than the frame. After a time $t_1/2$, a phase difference has developed between these two vectors (Figure 10.7Bb). Each vector has also fanned out because of field inhomogeneities. The plus and minus signs indicate the faster and slower moving spins, respectively.

The $180^\circ_{X'}$ pulse reflects the spins through the x'-axis. If this were a heteronuclear spin system in which X were, for example, ^{13}C, then the directions of rotation relative to the frame would remain unchanged. However, in the homonuclear case, the $180^\circ_{X'}$ pulse also acts upon the X protons.

We will consider a single X nucleus to understand this step, rather than the macroscopic magnetization vector $\mathbf{M}_X$. Each individual nucleus always has some magnetic-moment component μ_z along the z direction. The $180^\circ_{X'}$ pulse reverses this z component so that an X nucleus in the α state becomes one in the β state and vice versa. For the macroscopic sample, this means that the $180^\circ_{X'}$ pulse changes $\mathbf{M}_A^{X\alpha}$ into $\mathbf{M}_A^{X\beta}$ and $\mathbf{M}_A^{X\beta}$ into $\mathbf{M}_A^{X\alpha}$. From vector diagram 10.7Bc, we see that after the $180^\circ_{X'}$ pulse, the two vectors

▶ **FIGURE 10.7**
HOM2DJ NMR spectroscopy.
(A) Pulse sequence. (B) Evolution of the magnetization vectors in the rotating frame for an AX spin system ($A, X = {}^1$H). Only the magnetization vectors $\mathbf{M}_A^{X\alpha}$ and $\mathbf{M}_A^{X\beta}$ are shown. These vectors fan out as a result of field inhomogeneities; the faster spins are indicated by the '+' sign and the slower ones by the '−' sign. The small arrows indicate the direction of rotation of the spin packet as a whole. The vector diagrams correspond to the points a–d shown in the pulse-sequence diagram.

$$90^\circ_X - t_1/2 - 180^\circ_X - t_1/2 - \text{FID}(t_2)$$

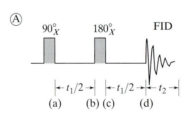

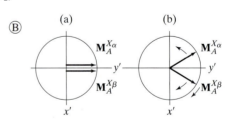

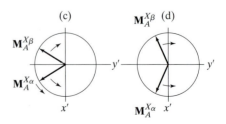

continue to move apart. After a further period $t_1/2$, the phase difference has doubled (Figure 10.7Bd).

The $180^\circ_{X'}$ pulse also has another effect. During the second $t_1/2$ time interval the fanning-out caused by field inhomogeneities is reversed. Although the $180^\circ_{X'}$ pulse reverses the direction of rotation of the magnetization vectors relative to the coordinate system, within each of the fans the faster spins are still the fastest, as their higher frequency arises from the external field and not from the coupling J_{AX}. The actual phase difference between the two magnetization vectors depends upon the value of t_1 chosen for the experiment and, more importantly, upon the magnitude of J_{AX}.

In practice, one records FIDs for a range of different t_1 values at constant increments of a few milliseconds. The Fourier transformation of an individual FID gives a frequency spectrum F_2 that contains information on both chemical shifts and coupling constants. A second Fourier transformation with respect to t_1 yields another frequency spectrum, F_1, and as in the heteronuclear case, this contains only the multiplets produced by the couplings and thus gives the coupling constants. However, the multiplets appear on lines that are tilted at an angle of 45° with respect to the F_2-axis if both dimensions have the same scale (Figure 10.8).

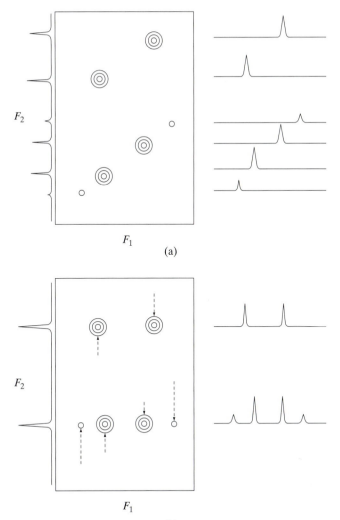

◀ FIGURE 10.8
A schematic representation of a typical homonuclear 2DJ spectrum in the contour format (a) before tilting and (b) after tilting.

F_2

F_1

(a)

F_2

F_1

(b)

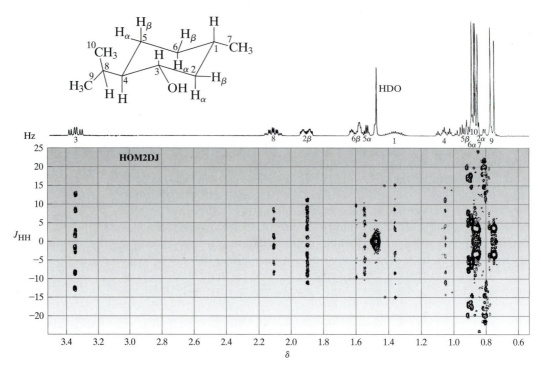

▲ FIGURE 10.9

300-MHz ^{1}H **HOM2DJ** spectrum of menthol with a high-resolution normal ^{1}H NMR spectrum plotted on the top.

By tilting the spectrum by 45° with respect to this axis, one can make all of a given multiplet's signals belonging to a single chemical shift δ appear on a line perpendicular to the F_2-axis. Projection of this spectrum onto that axis gives only single lines corresponding to the ^{1}H chemical shifts. The resulting projection then corresponds to a broadband ^{1}H decoupled ^{1}H NMR spectrum.

A **HOM2DJ** spectrum of menthol is shown in Figure 10.9. At the top of this figure a high-resolution ^{1}H NMR spectrum is also plotted, with the assignments indicated. Along the direction of the J_{HH} axis in the contour diagram, we recognize the same multiplet patterns that are seen in the normal one-dimensional spectrum.

The **HOM2DJ** spectroscopic technique is very useful when there is another nucleus, such as ^{31}P, in the molecule that also exhibits coupling with the proton spins. An example of a **HOM2DJ** spectrum of 1-phenyldibenzophosphole is given in Figure 10.10.

For situations like this, the **HOM2DJ** spectrum clearly separates the splittings due to homonuclear and heteronuclear spin–spin couplings. A projection of the spectrum in Figure 10.10 along the F_2-axis would provide a broadband ^{1}H decoupled ^{1}H NMR spectrum that would clearly evidence the PH couplings. For example, the H_4 resonance in this projection would be a doublet.

The **HOM2DJ** spectral technique is generally not very useful for strongly coupled second-order spin systems such as those illustrated by the H_p and H_m multiplets in Figure 10.10. The multiplets obtained for these types of spin systems are difficult to interpret in such spectra.

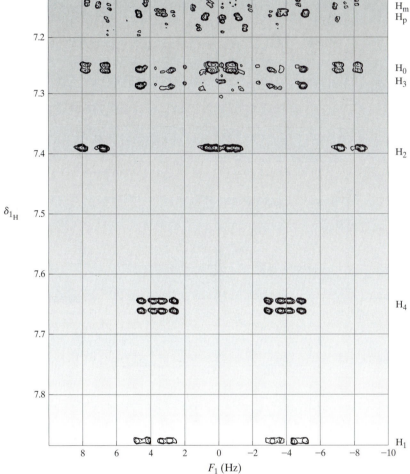

F_2 (ppm)

J_{HH}

◀ FIGURE 10.10
300-MHz ^{1}H homonuclear
2DJ spectrum of
1-phenyldibenzophosphole
(22 mg) in 0.5 ml $CDCl_3$. Spacings
in F_1 arise from H—H couplings,
while the apparently smaller
splittings in F_2 arise from PH
couplings.*

*Nelson, J. H., Alfandi, S., Gray,
G. A., Alyea, E. C., *Magn. Reson.
Chem.*, **1987**, *25*, 774.

H_m
H_p

7.2

H_0
H_3

7.3

7.4 H_2

$\delta_{^1H}$ 7.5

7.6

H_4

7.7

7.8

H_1

8   6   4   2   0   −2   −4   −6   −8   −10

F_1 (Hz)

10.5   Two-Dimensional Correlated Spectroscopy

Many problems in the assignment of NMR spectra can be greatly simplified by the aid of two-dimensional (*shift-*) *correlated* NMR spectroscopy. The basic principles will be illustrated by considering a simple hypothetical experiment and then seeing how the experiment can be modified for practical use. The heteronuclear C, H case will be treated first, followed by the homonuclear H, H case. For simplicity, the effects of relaxation and field inhomogeneity will be neglected.

A. Two-Dimensional Heteronuclear Correlated (**HETCOR**) Spectroscopy

Consider the AX spin system of $CHCl_3$ to which the pulse sequence shown in Figure 10.11A is applied. The vector diagrams of Figure 10.2B show how the applied pulse sequence $90^\circ_{X'}-t_1-90^\circ_{X'}$ affects the ^1H macroscopic magnetization vectors $\mathbf{M}_H^{C_\alpha}$ and $\mathbf{M}_H^{C_\beta}$. These vectors represent the magnetization vectors for $CHCl_3$ molecules whose ^{13}C nuclei are in the α and β spin states, respectively.

The first $90^\circ_{X'}$ pulse tips both magnetization vectors from the z direction onto the y' direction (Figure 10.11B). During the evolution time t_1, both vectors rotate with their Larmor frequencies according to the equations

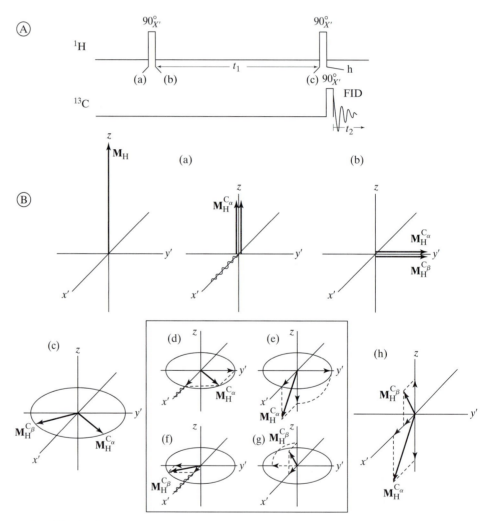

▲ FIGURE 10.11

Two-dimensional HETCOR NMR spectroscopy.

(A) Pulse sequence. (B) Diagrams showing the evolution of the ^1H magnetization vectors $\mathbf{M}_H^{C_\alpha}$ and $\mathbf{M}_H^{C_\beta}$ for an AX spin system ($A = {}^1H$, $X = {}^{13}C$) in the rotating frame. The vector diagrams (a) to (c) correspond to the times marked in the pulse sequence. Diagrams (d) and (e) (for $\mathbf{M}_H^{C_\alpha}$) and (f) and (g) (for $\mathbf{M}_H^{C_\beta}$) show how these vectors are affected by the second $90^\circ_{X'}$, ^1H pulse. Diagram (h) shows the situation immediately preceding the $90^\circ_{X'}$, ^{13}C pulse.

$$\nu \mathbf{M}_H^{C\alpha} = \nu_H - \frac{1}{2}J_{CH} \quad \text{and} \quad \nu \mathbf{M}_H^{C\beta} = \nu_H + \frac{1}{2}J_{CH}, \qquad \textbf{(10.4)}$$

where ν_H is the Larmor frequency in the absence of CH coupling.

The angles ϕ_α and ϕ_β through which the vectors $\mathbf{M}_H^{C\alpha}$ and $\mathbf{M}_H^{C\beta}$ travel during the time t_1 are given by

$$\phi_\alpha = 2\pi\left(\nu_H - \frac{1}{2}J_{CH}\right)t_1 \qquad \textbf{(10.5)}$$

and

$$\phi_\beta = 2\pi\left(\nu_H + \frac{1}{2}J_{CH}\right)t_1.$$

The phase difference between ϕ_α and ϕ_β depends only on t_1 and J_{CH} and is given by

$$\Theta = \phi_\beta - \phi_\alpha = 2\pi(J_{CH})t_1. \qquad \textbf{(10.6)}$$

Thus, when $t_1 = [4J_{CH}]^{-1}$, $\Theta = 90°$, and when $t_1 = [2J_{CH}]^{-1}$, $\Theta = 180°$. Equations 10.4 to 10.6 correspond exactly to Equations 10.1 to 10.3, except that the ^1H instead of the ^{13}C magnetization vectors are involved.

Figure 10.11Bc shows the situation after an arbitrary evolution time t_1, which is assumed to be several milliseconds. Because the precession frequencies of $\mathbf{M}_H^{C\alpha}$ and $\mathbf{M}_H^{C\beta}$ do not coincide with the frequency of the rotating frame, these two vectors are now inclined with respect to the y'-axis at angles that depend upon the frequency differences. Consider first the $\mathbf{M}_H^{C\alpha}$ vector. It has components in the $(+y')$ and $(+x')$ directions (Figure 10.11Bd). The second ^1H $90°_{X'}$ pulse tips the y' component onto the -z direction, while the x' component is unaffected. The new direction of $\mathbf{M}_H^{C\alpha}$ is the vector sum of these new components. In the example shown in the figure, $\mathbf{M}_H^{C\alpha}$ is in the lower front quadrant (Figure 10.11Be).

The same thing happens to the y' component of $\mathbf{M}_H^{C\beta}$, except that it is tipped onto the $(+z)$ direction, so that after the ^1H $90°_{X'}$ pulse, $\mathbf{M}_H^{C\beta}$ is in the front upper quadrant of the $x'z$-plane (Figures 10.11Bf and g). For the rest of the discussion, we are not concerned with the $\mathbf{M}_H^{C\alpha}$ and $\mathbf{M}_H^{C\beta}$ vectors themselves, but only with their z components (Figure 10.11Bh). The magnitudes of these longitudinal magnetization components are proportional to the population differences between energy levels ① and ③ (for $\mathbf{M}_H^{C\alpha}$) and between ② and ④ (for $\mathbf{M}_H^{C\beta}$). From this, two conclusions can be drawn:

Energy-level diagram for an *AX* spin system.

1. The pulse sequence $90°_{X'}$–t_1–$90°_{X'}$ has altered the level populations from those at the start. At the instant chosen in Figure 10.11, the population of level ③ is even greater than that of the lowest-lying level ①.

2. The state reached by the spin system depends on the time t_1—that is, on the angles covered by the vectors (Equation 10.5). These in turn are determined by the Larmor frequency ν_H and the coupling constant J_{CH}.

What effect does the ^1H pulse sequence have on the ^{13}C NMR spectrum? Consider again Figure 10.11Bh, which can be described by an energy-level scheme like the one described previously. The intensity of the ^{13}C NMR

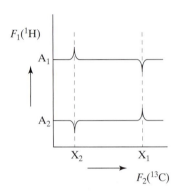

$F_1({}^1\text{H})$

A_1

A_2

X_2 X_1

$F_2({}^{13}\text{C})$

▲ **FIGURE 10.12**
Schematic two-dimensional
HETCOR spectrum of an *AX* spin
system using the pulse sequence
shown in Figure 10.11. The two
signals along the F_2 direction
correspond to the one-dimensional
^{13}C NMR spectrum without
decoupling, except that the
signals have opposite signs. The
carbon-coupled ^{1}H spectrum
appears along the F_1 direction, also
with opposite signal amplitudes.

signal is determined by the level populations after the second $90^\circ_{X'}$, pulse, which means that a transfer of polarization or magnetization from ^{1}H to ^{13}C occurs. The ^{13}C NMR signal is modulated as a function of t_1 by the Larmor frequencies of the protons, which cannot be illustrated diagrammatically in a clear way. Mathematical analysis gives the result that the magnetization vectors $\mathbf{M}_H^{C\alpha}$ and $\mathbf{M}_H^{C\beta}$, and therefore also $\mathbf{M}_C^{H\alpha}$ and $\mathbf{M}_H^{C\beta}$, which are connected to them as a consequence of the aforementioned energy-level scheme, are always changed by the same amount, but with opposite signs.

The $90^\circ_{X'}$, ^{13}C pulse tips these two longitudinal vectors onto the $(+y')$ and $(-y')$ directions, respectively. During the data-acquisition period t_2, they precess at the frequencies corresponding to the X_2 and X_1 transitions, giving the FID signal in the receiver coil. Fourier transformation with respect to t_2 gives two ^{13}C signals along the F_2 axis, whose modulation depends upon t_1, and after a second Fourier transformation with respect to t_1 is performed, a two-dimensional spectrum with four signals is obtained. In this 2D spectrum, two of the signals have negative amplitudes (Figure 10.12), the ^{1}H resonances occur along the F_1-axis, and the ^{13}C resonances occur along the F_2-axis.

The spectrum parallel to the F_2-axis is the one-dimensional proton-coupled ^{13}C spectrum, and the spectrum parallel to the F_1-axis is the one-dimensional carbon-coupled ^{1}H spectrum. In a molecule as simple as $CHCl_3$, which contains only two coupled nuclei, this gives spectra that are easy to interpret. For larger molecules, the interpretation becomes more difficult or even impossible. The experiment needs to be modified for practical use.

The experiment is modified as shown by the pulse sequence in Figure 10.13.

An attempt to explain the **HETCOR** experiment by means of vector diagrams in the rotating frame (Figure 10.13) is presented. The first $90^\circ_{X'}$, ^{1}H pulse tips both of the ^{1}H magnetization vectors $\mathbf{M}_H^{C\alpha}$ and $\mathbf{M}_H^{C\beta}$ onto the y' direction (Figure 10.13Ba). Because they have different Larmor frequencies, $\nu_H - \frac{1}{2}J_{CH}$ and $\nu_H + \frac{1}{2}J_{CH}$, these two vectors move apart. After a time $t_1/2$, their phase difference is $\Theta = \pi(J_{CH})t_1$. The 180° ^{13}C pulse reverses the identities of the ^{13}C nuclei in the α and β spin states, so that $\mathbf{M}_H^{C\alpha}$ becomes $\mathbf{M}_H^{C\beta}$ and $\mathbf{M}_H^{C\beta}$ becomes $\mathbf{M}_H^{C\alpha}$. The faster rotating vector (indicated by the thicker of the two arrows outside the circle) now follows the slower one. After a further time interval equal to $t_1/2$, $\mathbf{M}_H^{C\alpha}$ and $\mathbf{M}_H^{C\beta}$ are again in phase (Figure 10.13Bd). The total angles through which the vectors travel, ϕ, depends only upon the Larmor frequency for the ^{13}C nuclei in the absence of J_{CH}.

A $90^\circ_{X'}$ ^{1}H pulse immediately following t_1 would result in no polarization transfer and no modulation of the ^{13}C signal. But the insertion of an additional delay time Δ_1 before the second $90^\circ_{X'}$, ^{1}H pulse causes the ^{1}H magnetization vectors to again move apart. If $\Delta_1 = [2J_{CH}]^{-1}$, the phase difference of the two ^{1}H vectors will be 180°. The $90^\circ_{X'}$, ^{1}H pulse that follows now tips the y' components of these two vectors onto the $(+z)$ and $(-z)$ directions, respectively (Figures 10.13Be and f). This process gives rise to the polarization transfer, the magnitude of which depends only upon the angle ϕ. When the vectors are along the y' direction before the pulse, the polarization is maximum, and when the vectors are along the x' direction, the polarization is zero. The angle ϕ covered by the vectors during t_1 is a function of the Larmor frequency of the ^{13}C decoupled protons. The system continues to evolve during the time Δ_1, but this contribution is constant for all spectra recorded at different t_1 values. Thus, the polarization on which the ^{13}C signal intensities depend is determined solely by the proton Larmor frequency.

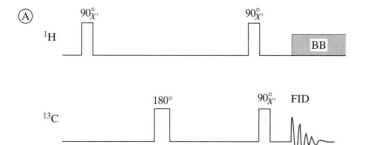

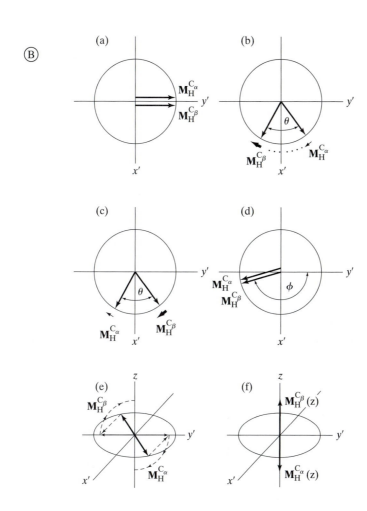

◀ **FIGURE 10.13**
(A) Pulse sequence for a **HETCOR** experiment. (B) The vector diagrams for an *AX* spin system. Diagrams (a) to (f) show the positions of the ^{1}H magnetization vectors $\mathbf{M}_H^{C_\alpha}$ and $\mathbf{M}_H^{C_\beta}$ or their *z* components (f) at the times indicated in *A*; in diagrams (a) to (d), only the *x′y′*-plane is shown.

The $90_{X'}^{\circ}$ ^{13}C pulse tips the ^{13}C magnetization vectors onto the $(+y')$ and $(-y')$ directions, and after a further delay time $\Delta_2 = [2J_{CH}]^{-1}$, the ^{13}C vectors, $\mathbf{M}_C^{H_\alpha}$ and $\mathbf{M}_C^{H_\beta}$, are in phase. Switching on the ^{1}H *BB* decoupling channel at this instant then eliminates CH coupling during data acquisition. A Fourier transformation with respect to t_2 gives a signal at ν_C. If one records a series of spectra with $\Delta_1 = \Delta_2 = [2J_{CH}]^{-1}$ (which is about 3.5 ms) and different t_1 values, the signal intensities are modified by the frequency ν_H. A second Fourier transformation with respect to t_1 then gives a two-dimensional spectrum (F_1, F_2) that contains only one signal with coordinates (ν_H, ν_C). Theoretically, the signal intensity is maximized for $\Delta_1 = [2J_{CH}]^{-1}$. It is also amplified by magnetization transfer from the sensitive ^1H nucleus to the much less sensitive ^{13}C nucleus. The same experiment can also be applied to multiple spin–spin systems.

Figure 10.14 shows the $^{13}C/^1H$ **HETCOR** spectrum of menthol. Along the top of this spectrum is plotted a **DEPT** ^{13}C spectrum obtained in another experiment. Along the right-hand side a high resolution ^1H spectrum, obtained in still another experiment, is plotted. One should note that quaternary carbons will not give rise to peaks in **HETCOR** spectra.

The cross peaks in the HETCOR spectrum connect proton chemical shifts with carbon chemical shifts. Since Δ_1 and Δ_2 delays were chosen that correspond to one-bond couplings, the cross peaks connect the carbon resonance with the multiplet of the directly bonded protons. For this reason, spectra such as the one shown in Figure 10.14 have been called *chemical-shift correlation maps*. If the assignment of the resonances in one of the two one-dimensional correlated spectra is known, even if only partially, assignment of the resonances in the other spectrum through the interconnecting cross peaks is possible. The procedure is illustrated in Figure 10.11 by the dashed line connecting the H_8 and C_8 resonances. The H_8 resonance may be identified by its apparent septet structure, due to equivalent $^3J_{HH}$ coupling to the methyl groups $CH_3(9)$ and $CH_3(10)$.

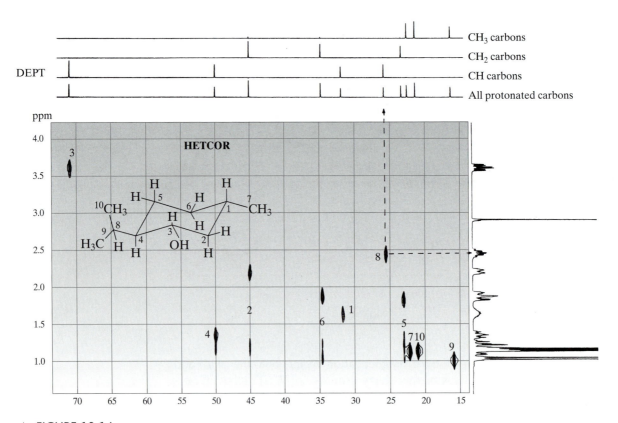

▲ FIGURE 10.14
75-MHz $^{13}C/^1H$ **HETCOR** spectrum of menthol. ^{13}C **DEPT** spectra are plotted along the top, and a high-resolution ^1H spectrum is plotted on the right.

Often, some of the resonances in the ^{1}H spectra can be assigned immediately on the basis of chemical shifts or multiplicities. The H$_3$ resonance is another example, as one expects a triplet of doublets, as observed, for this resonance. In this case, it is anticipated, on the basis of substituent effects, that the C$_3$ resonance would be the most downfield carbon resonance. The ^{1}H/^{13}C cross peak for these two resonances unequivocally affirms these expectations. The resonances due to the other two CH moieties, CH(1) and CH(4), may then be assigned by way of their cross peaks. For each of the CH$_2$ groups CH$_2$(2), CH$_2$(5), and CH$_2$(6), two proton cross peaks are observed for each ^{13}C chemical shift. This is because the geminal protons in each of these groups are diastereotopic and hence anisochronous. Although the methyl resonances for CH$_3$(7), CH$_3$(9), and CH$_3$(10) are clearly methyl resonances, they cannot be unambiguously assigned solely on the basis of this experiment.

^{13}C/^{1}H **HETCOR** spectra of molecules containing ^{31}P nuclei may be used to determine the relative signs of J_{PH} and J_{PC}. Consider Figure 10.15, which shows a ^{13}C/^{1}H **HETCOR** spectrum of 1-phenyldibenzophosphole, whose structure is as follows:

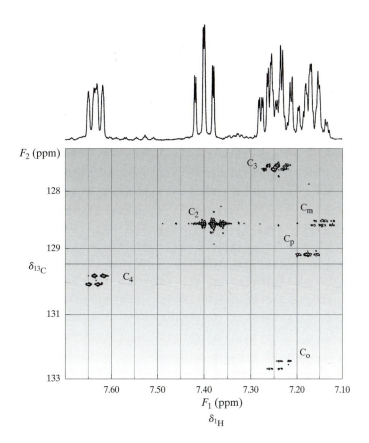

◀ **FIGURE 10.15**
100-MHz ^{13}C/^{1}H **HETCOR** spectrum of 1-phenyldibenzophosphole (50 mg) in 0.5 mL CDCl$_3$. This expansion shows all protonated carbons except C–1, which does exhibit a correlation (doublet in F_1 no observed splitting in F_2) at 121.4 ppm for ^{13}C and 7.8 ppm for ^{1}H. A high-resolution ^{1}H spectrum for the same region is plotted along the top. PC couplings are responsible for the splittings in F_2, while homonuclear HH and heteronuclear PH splittings are observed along F_1.*

*Nelson, J. H., Affandi, S., Gray, G. A., Alyea, E. C., *Magn. Reson. Chem.*, **1987**, *25*, 774.

One of the more noticeable aspects of the $^{13}C/^1H$ **HETCOR** spectra of organophosphorus compounds is the "tilt" of the correlation patterns. In these cases, correlation peaks should occur at $\left[\nu_H + \frac{1}{2}J_{PH}, \nu_C + \frac{1}{2}J_{PC} \right.$ and $\left. \nu_H - \frac{1}{2}J_{PH}, \nu_C - \frac{1}{2}J_{PC} \right]$. All of the correlations in the **HETCOR** spectrum in Figure 10.15 that exhibit PC splittings, have the low-frequency (high-field) proton shift matched with the low-frequency carbon-shift value. This indicates that $^nJ_{CiP}$ and $^nJ_{HiP}$ are of like sign. Since the J_{PC} couplings are known to be positive (from other experiments), the $^nJ_{PH}$'s for this molecule are likely also positive.

B. Two-Dimensional Homonuclear (H,H)-Correlated NMR Spectroscopy (**COSY**)

The two-dimensional homonuclear (H,H)-correlated NMR experiment gives NMR spectra in which ^1H chemical shifts along both frequency axes are correlated with each other. This technique is known as **COSY** (*correlated spectroscopy*) or sometimes as **HOMCOR**. It is based upon the pulse sequence $90°_{X'}-t_1-\Theta_{X'}$ shown in Figure 10.16.

Consider the **COSY** experiment with $\Theta_{X'} = 90°$ applied to an AX spin system, where both A and X are ^1H. There is an important difference from the heteronuclear AX case because the first $90°_{X'}$ pulse tips both magnetization vectors, $\mathbf{M}_A$ and $\mathbf{M}_X$, onto the y'-axis. Due to the coupling J_{AX}, the A nuclei have two macroscopic magnetization vectors, $\mathbf{M}_H^{X\alpha}$ and $\mathbf{M}_H^{X\beta}$, depending upon whether the X nucleus is in the α or β spin state, respectively. Similarly, there are two $\mathbf{M}_X$ vectors, $\mathbf{M}_X^{A\alpha}$ and $\mathbf{M}_X^{A\beta}$. These four vectors all rotate in the $x'y'$-plane around the z-axis with the frequencies $\nu_A \pm \frac{1}{2}J_{AX}$ and $\nu_X \pm \frac{1}{2}J_{AX}$.

During the time t_1, the four magnetization vectors fan out in the $x'y'$-plane as a consequence of their frequency differences. At the instant t_1, each vector has components in the x' and y' directions. The second $90°_{X'}$ pulse tips each of the y' components onto the $(+z)$ or $(-z)$ direction. This causes a transfer of polarization. The amount of magnetization that is transferred depends on the situation in the spin system at the instant t_1 and, therefore, on the Larmor frequencies of the A and X spins as well as on J_{AX}.

The x' components of the magnetization vectors continue to rotate in the $x'y'$-plane and give an FID, which, after Fourier transformation with respect to t_2, yields a four-line AX-type spectrum with the frequencies

$$\nu_A + \frac{1}{2}J_{AX}(A_1), \nu_A - \frac{1}{2}J_{AX}(A_2)$$

and

$$\nu_X + \frac{1}{2}J_{AX}(X_1), \nu_X - \frac{1}{2}J_{AX}(X_2).$$

These frequencies correspond to the transitions labeled A_1, A_2, X_1, and X_2 in Figure 10.17. The signals as functions of t_1 are modulated with those four frequencies. The second Fourier transformation with respect to t_1 then gives a two-dimensional spectrum with four groups, each of which contains four signals. Two of these groups are centered around the positions (ν_A, ν_A) and

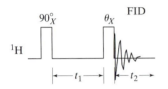

▲ FIGURE 10.16

Pulse sequence for the **COSY** NMR experiment. The variable is t_1. The pulse angle Θ is usually 90°, 45°, or, occasionally, 60°, called **COSY**-90, **COSY**-45, and **COSY**-60, respectively.

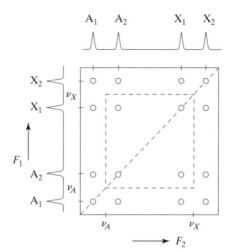

Schematic representation of a **COSY**-90 experiment for an *AX* spin system. The signal amplitudes are shown in the absolute-value mode. In an actual spectrum, the peaks on the diagonal are dispersion signals, while the correlations (cross peaks) are absorption signals with alternating signs. The diagonal peaks of a mutually coupled pair of nuclei and their cross peaks form the apices of a square.

(ν_X, ν_X) and are called the *diagonal peaks*, while the other two groups are centered around (ν_A, ν_X) and (ν_X, ν_A) and are called the *correlation peaks* or *cross peaks*. The diagonal peaks and the cross peaks are located at the apices of a square. Cross peaks always occur when two nuclei, in this case *A* and *X*, are *scalar coupled* to each other.

Within a group, the separation in each dimension (F_1 and F_2) between two adjacent signals is exactly equal to J_{AX}. The projection of such a **COSY** spectrum onto the F_1-or F_2-axis corresponds to the one-dimensional ^{1}H NMR spectrum.

Figure 10.17 shows a spectrum of this type for the *AX* spin system as a contour plot in the absolute-value mode. This form of presentation is usually chosen, since, in the **COSY**-90 experiment, the phases of the diagonal and cross peaks always differ by 90°.

If we look only at the cross peaks, we find that, in both the horizontal and vertical directions, we have absorption signals with alternating positive and negative amplitudes. For multispin spin systems, these phase relationships between the cross peaks provide an aid in determining coupling constants. It can also happen that superimposed signals with opposite phases may cancel each other. Because in most cases we are interested only in correlations, we usually dispense with phase-sensitive detection and present the data in the absolute-value mode.

If a proton is coupled to more than one neighboring proton, the diagonal peak will appear at the corner of more than one square. This makes it possible to identify the positions of the resonances of the coupled nuclei, even in complicated spectra. The **COSY** experiment is far superior to homonuclear-decoupling experiments as an aid in assigning ^{1}H resonances, because it gives connectivity information on *all* the coupled nuclei in a single experiment. As a result of the high proton sensitivity, a **COSY** experiment can routinely be performed in less than one hour of instrument time. An example of a **COSY**-90 spectrum of 2-butanol is shown in Figure 10.18.

The diagonal (dashed line) runs from the upper left-hand corner to the lower right-hand corner in this contour map. We begin with the group of

▶ **FIGURE 10.18**
COSY-90 2D NMR spectrum of 2-butanol in $CDCl_3$; preparation period 3 s, 512 increments of t_1, FID and spectrum matrices 1024 × 1024, spectral width 680 Hz in both dimensions, four transients for each value of t_1, total measuring time 2.5 h. The ^1H spectrum plotted on the top and right-hand side was obtained in a separate experiment.

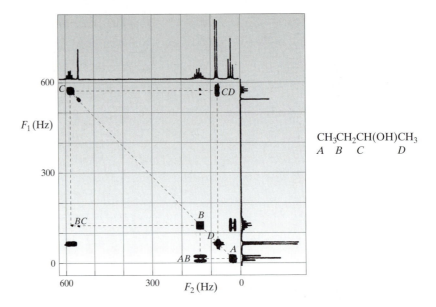

$$CH_3CH_2CH(OH)CH_3$$
$$A \quad B \quad C \qquad D$$

peaks near the diagonal with the lowest frequencies, which are due to the triplet in the one-dimensional spectrum. The diagonal peaks labeled "A" have the same F_1 coordinates as the AB cross peak. (The other AB cross peak is located symmetrically across from the diagonal.) When we proceed along the dashed line, we move from the AB cross peak to the second group of diagonal peaks labeled "B," which belong to the coupled protons of the neighboring CH_2 group (diastereotopic protons). Group B has another, much weaker group of cross peaks "BC," which brings us to the diagonal group C. The C peaks correspond to the one proton of the CH group. This proton is coupled (cross peaks CD) with the protons of the terminal methyl group (diagonal peaks D). The remaining diagonal peak, which is not accompanied by any cross peak, is due to the singlet of the OH proton, which does not show any coupling interaction in the solution being measured. The same procedure is used in analyzing the spectra of more complex compounds. In the procedure described, it is crucial not to overlook the weak cross peaks labeled "BC." These peaks are of very low intensity, since they correlate broad multiplets of low intensity.

COSY-45 spectra allow spectral simplification around the diagonal (for reasons that we will not discuss here) and are usually preferred to **COSY**-90 experiments in the study of large molecules. They are obtained in the same manner, the only difference being that the second pulse is a 45°, rather than a 90°, pulse. Both **COSY** spectra are interpreted in the same way.

10.6 Two-Dimensional Exchange Spectroscopy (**NOESY** and **EXSY**)

For all of the two-dimensional NMR techniques discussed thus far, the magnetization transfer occurs between scalar-coupled nuclei. There are two additional mechanisms for magnetization transfer. Magnetization may be transferred through space by dipole–dipole interactions, as we encountered when the nuclear Overhauser effect (**NOE**) was discussed in Chapter 8. Magnetization may also be transferred as a consequence of chemical-exchange processes (Chapter 11). For example, consider an exchange from site A to site X governed by a rate constant k. In this case, if the A nuclei are polarized, there is transfer of polarization from A to X with the rate constant k.

In the homonuclear case (A and X are both protons, for example), polarization transfer by either of these two mechanisms can be detected using the pulse sequence $90^\circ_{X'} - t_1 - 90^\circ_{X'} - \Delta - 90^\circ_{X'} - \text{FID}(t_2)$. The first part of this sequence, $90^\circ_{X'} - t_1 - 90^\circ_{X'}$ corresponds to the **COSY**-90 experiment. As discussed earlier, in this experiment the A nuclei are in a polarized condition immediately after the second $90^\circ_{X'}$ pulse. During the following time interval Δ, polarization is transferred from A to X either by NOE or by chemical exchange. Applying a further $90^\circ_{X'}$ pulse after the mixing or evolution time Δ then allows the FID to be recorded. The process is repeated for a series of different values of t_1. By performing a Fourier transformation with respect to t_2, a series of signals for the X nuclei is obtained whose t_1 dependence is modulated with the Larmor frequencies of the A nuclei. The second Fourier transformation with respect to t_1 then yields, as in the **COSY** experiment, cross peaks at the positions (ν_A, ν_X) and (ν_X, ν_A).

In order for such an experiment to be successful, correct setting of the evolution time Δ is essential. In the **NOE** version of the experiment (**NOESY**), Δ must be of the same order of magnitude as the spin–lattice relaxation time T_1. For the chemical-exchange version of the experiment (**EXSY**), Δ must be on the order of the reciprocal of the rate constant. Hence, **NOESY** experiments require T_1 measurements at best or guesses of their values at worst, and **EXSY** experiments require at least a guess of the exchange rate.

For the latter, one usually limits the experiment to a qualitative demonstration that an exchange process is occurring, rather than to a measurement of its rate. Determining the rate by **EXSY** is very difficult and requires numerous time-consuming 2D experiments with variable evolution times.

The ability to demonstrate the existence of through-space dipolar H—H interaction is of considerable practical importance for *conformationally rigid molecules,* as it gives information about stereochemistry. Since the dipole–dipole interaction falls off as the inverse sixth power of the internuclear separation between the two interacting nuclei, one observes **NOESY** cross peaks only for nuclei that are proximate in space (typically less than 3.5 Å apart). **NOESY** spectra are presented and interpreted in a fashion analogous to that of **COSY** spectra. Figure 10.19 shows a **NOESY** spectrum for β-ionone. Correlations may be seen between the 1, 1' CH$_3$ resonance and the H$_8$ and H$_7$ resonances, between the 5 CH$_3$ resonance and the H$_4$, H$_8$, and H$_7$ resonances, and between the 10 CH$_3$ and the H$_8$ and H$_7$ resonances. The reason the olefinic protons H$_7$ and H$_8$ show an NOE with all methyl groups is that the side chain may freely rotate around the C$_6$—C$_7$ bond. It is important in **NOESY** spectroscopy to distinguish cross peaks due to scalar coupling from cross peaks due to dipolar interactions. Consequently, **NOESY** spectra should always be compared with **COSY** spectra of the same sample.

▶ **FIGURE 10.19**
360-MHz ^{1}H **NOESY** spectrum of β-ionone. High-resolution ^{1}H NMR spectra obtained in another experiment are plotted above and on the left-hand side.*

*Adapted from reference 10.

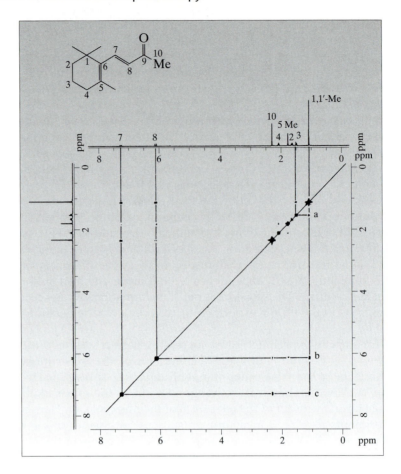

10.7   The Two-Dimensional **INADEQUATE** Experiment

The **INADEQUATE** *(incredible natural abundance double quantum transfer)* experiment provides information on carbon–carbon connectivities by way of $^1J_{CC}$. The general pulse sequence is $90^\circ_{X'}-\tau-180^\circ_{Y'}-\tau-90^\circ_{X'}-t_1-90^\circ_{X'}\,\phi'-\text{FID}(t_2)$. The first part of the 2D pulse sequence, up to and including the second $90^\circ_{X'}$ pulse, serves to establish double quantum coherence of two adjacent ^{13}C spins. During time t_1, this double quantum coherence evolves and is then converted by the 90°_ϕ pulse into single quantum transitions, which can be observed and which correspond to ^{13}C satellites in a normal ^{13}C{^{1}H} experiment. If the value of τ is correctly chosen, the Fourier transformation of the FID with respect to t_2 gives only the satellite spectrum with suppression of the central resonances. Thus, for each pair of directly adjacent ^{13}C nuclei A and X, one obtains two doublets. Mathematical analysis shows that the satellite signals, as functions of t_1, are modulated by the sum of the resonance frequencies of the two coupled C nuclei $(\nu_A + \nu_X)$, which we cannot show diagrammatically.

The experiment consists of n measurements with different t_1 values, with the increment between successive measurements being a few microseconds. The second Fourier transformation of all the F_2 spectra with respect to t_1 gives a two-dimensional spectrum in which the F_1-coordinates of the signals correspond to the double quantum frequencies $(\nu_A + \nu_X)$ and the F_2-coordinates are the frequencies of ν_A and ν_X. However, owing to the nature of the experiment, the frequencies along the F_1-axis do not give $(\nu_A + \nu_X)$ directly, but instead give a frequency $\nu_{\text{measured}} = \nu_A + \nu_X - 2\nu_1$, where ν_1 is the frequency of the pulsed r.f. source. The ν_{measured} values are in a frequency range extending up to a few kilohertz. Since these double quantum frequencies are of no interest to us, we will not consider them further. The important point is that at *one* particular F_1 value we find the satellite spectra of both of the coupled ^{13}C nuclei, so that both doublets appear at the same F_1 value. Usually, the actual value of the coupling constant J_{CC} is not of interest, but rather, the carbon–carbon connectivity is desired. Consider the **INADEQUATE** spectrum shown in Figure 10.20 of menthol. The spectrum at the top is a high-resolution $^{13}C\{^1H\}$ NMR spectrum obtained in a separate experiment. Each carbon nucleus gives a doublet consisting of two satellite signals separated by $^1J_{CC}$. The main signals would be at the center of these doublets.

We begin our analysis at the top and look for horizontal correlation of doublets, as indicated by the dashed lines at constant values of F_1. Thus, we see from the correlated doublets at 23 and 35 ppm that C_5 and C_6 are

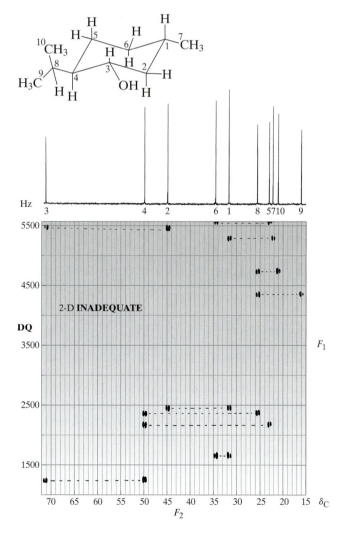

◀ FIGURE 10.20
75-MHz ^{13}C **INADEQUATE** spectrum of menthol with ^1H broadband decoupling. Plotted on top is a high-resolution $^{13}C\{^1H\}$ NMR spectrum obtained in a separate experiment. Coupling pairs of ^{13}C nuclei are connected by dashed lines.

connected. Likewise, we find connections between C_2 and C_3, between C_7 and C_1, between C_{10} and C_8, between C_9 and C_8, between C_1 and C_2, between C_8 and C_4, and between C_4 and C_3. We might also note, for example, that if we start on the upper left and draw a vertical line downward from the C_3 resonance, we find two correlations (i.e., C_3 is connected to two other carbons, C_2 and C_4). Similarly, C_4 is connected to three other carbons, C_8, C_5, and C_3. Other correlations are found in an analogous manner.

Note that the **INADEQUATE** experiment is very powerful, but it is also very time consuming. At present, it is feasible only for very soluble samples such that very high concentrations may be obtained (about 5 to 20 molar), and even then the experiment may take days to perform. It also is not useful for very small molecules, which relax too slowly.

10.8   **HMQC** and **HMBC** Spectra

It is possible to observe heteronuclear correlation between two spin-coupled nuclei by directly detecting either of the two nuclei and indirectly detecting the other. The classical $^1H/^{13}C$ **HETCOR** experiment (Section 10.5) involves direct detection of the less sensitive ^{13}C nucleus and indirect detection of the more sensitive ^1H nucleus. A considerable enhancement of sensitivity may be obtained by directly detecting the ^1H nucleus and indirectly detecting the ^{13}C nucleus. Sensitivity may be a further enhanced by employing an *inverse probe* (Section 2.4) in the experiment. Two such experiments are discussed herein.

A. ^{1}H-Detected Heteronuclear Multiple-Quantum Coherence (**HMQC**) Spectra

The **HMQC** experiment[*] provides a highly sensitive method for determining one-bond proton–carbon shift correlations. The pulse sequence for this experiment is shown in Figure 10.21.

In this pulse sequence, the bilinear rotation-decoupling (**BIRD**) pulse cluster[†] is followed by a delay τ whose magnitude is adjusted so that the inverted magnetizations of protons not bound to the ^{13}C nuclei pass through zero amplitude (as they change from the original to the final positive ampli-

▶ FIGURE 10.21
The pulse sequence of the **HMQC** experiment.

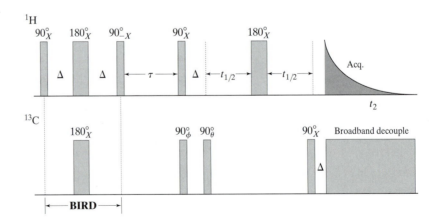

[*] Müller, L., *J. Am. Chem. Soc.*, **1979**, *101*, 4481. Eich, G., Bodenhausen, G., Ernst, R. R., *J. Am. Chem. Soc.* **1982**, *104*, 3731.
[†] Garbow, J. R., Weitkamp, D. P., Pines, A., *Chem. Phys. Lett.* **1982**, *93*, 504.

tude) when the first 90_X° pulse is applied. The relaxation time T_1 of protons directly attached to a ^{13}C nucleus is short because of the presence of adjacent ^{13}C nuclei. Thus, such nuclei relax efficiently, and the time between successive scans can be kept quite brief (about $1.3T_1$ of the fastest-relaxing protons).

The **HMQC** experiment is about 16 or 100 times more sensitive for ^{13}C and ^{15}N detections, respectively, than the conventional **HETCOR** experiment. The sensitivity advantage in the **HMQC** experiment (as well as in the **HMBC** experiment described next) is due to the fact that it relies on the equilibrium magnetizations derived from protons. Since this magnetization is proportional to the larger population difference in proton energy levels (rather than the smaller population difference in ^{13}C energy levels), a stronger NMR signal is obtained. Moreover, since the strength of the NMR signal increases with the frequency of observation, a larger signal will be obtained at the higher proton observation frequencies than at the lower ^{13}C observation frequencies. Consequently, an **HMQC** spectrum can be obtained in the same total time, with an equivalent or better signal-to-noise ratio on a much more dilute sample by a factor of about 200 than in an equivalent **HETCOR** experiment. An **HMQC** spectrum and a **HETCOR** spectrum of the same compound (but not the same concentrations) are compared in Figure 10.22. Usually, the

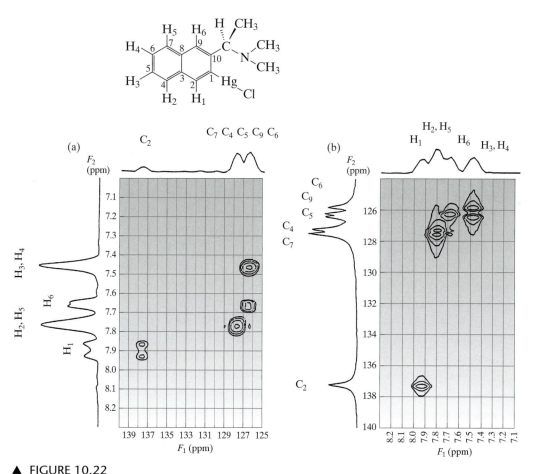

▲ FIGURE 10.22
Expansions of (a) **HMQC** and (b) **HETCOR** spectra of the compound shown obtained at 500-MHz and 125-MHz observation frequencies, respectively.*

*Gül, N., Nelson, J. H., *J. Molec. Struct.* **1999**, *475*, 121.

enhanced signal-to-noise ratio of the **HMQC** experiment makes it preferable to the more time-consuming **HETCOR** experiment, but Figure 10.22 shows an example in which this is not the case. The compound in this example is relatively easily synthesized, and it is quite soluble, so that a concentrated solution is readily obtained. Since in this case there is more dispersion in the proton dimension than in the carbon dimension, the chemical-shift correlations are much more easily made if the better resolved, directly detected F_2 dimension is ^{13}C (**HETCOR**) than if it is ^1H (**HMQC**), despite the greater sensitivity of the latter experiment. This issue will be considered further in Section 10.9.

B.   Heteronuclear Multiple-Bond Connectivity (**HMBC**) Spectra

In contrast to the **HMQC** experiment, which provides connectivity information about directly bonded ^1H—^{13}C interactions (i.e., one-bond coupling), the **HMBC**[*] experiment provides information about long-range ^1H—^{13}C coupling interactions, which are about an order of magnitude smaller. The **HMBC** experiment is also an inverse experiment and is best done with an inverse probe. The pulse sequence used is shown in Figure 10.23. The first 90° ^{13}C pulse serves to suppress one-bond $^1J_{CH}$ correlations, so that cross peaks due to direct connectivities are minimized, allowing the long-range ^1H—^{13}C connectivities $^2J_{CH}$ and $^3J_{CH}$ to be recorded. The second 90° ^{13}C pulse creates zero and double quantum coherences, which are interchanged by the 180° ^1H pulse. After the last 90° ^{13}C pulse, the ^1H signals resulting from ^1H—^{13}C multiple quantum coherence are modulated by ^{13}C chemical shifts and homonuclear proton couplings.

The timing of the **HMQC** and **HMBC** experiments is related to the magnitude of J_{CH}. In the **HMQC** experiment, $\Delta = 1/(2J)$, and since the average value of $^1J_{CH}$ is about 140 Hz, Δ is usually on the order of 3.6 msec, which is a relatively short delay time. In the **HMBC** experiment, Δ is also equal to $1/(2J)$ and Δ_2 is about 60 msec, but now the success of the experiment is strongly dependent upon the guessed magnitude of J_{CH}. If $J_{CH} = 20$ Hz, then $1/(2J) = 25$ ms, but if $J_{CH} = 10$ Hz, then $1/(2J) = 50$ ms, and if $J_{CH} = 5$ Hz, then $1/(2J) = 100$ ms. In practice, signals essentially vanish at long delay times, and a lower limit of $J_{CH} = 10$ Hz seems to apply to this experiment. The

▶ FIGURE 10.23
The pulse sequence of the **HMBC** experiment.

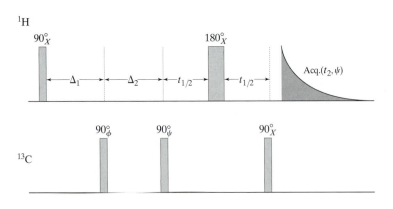

[*]Bax, A., David, D. G., Sarker, S. K., *J. Magn. Reson.* **1985**, 63, 230.

HMQC experiment is more forgiving and easier to use than the **HMBC** experiment. To succeed with the latter, it is often necessary to run several experiments with different values of J_{CH}, or better (if possible), to obtain the values of $^2J_{CH}$ and $^3J_{CH}$ from a proton-coupled ^{13}C spectrum.

10.9   Data Acquisition and Processing Parameters in 2D NMR Spectroscopy

A number of parameters need to be considered in acquiring and processing 2D NMR spectra, including (a) the pulse sequence to be used, which depends upon the experiment to be conducted, (b) the pulse lengths and the delays in the pulse sequence, (c) the spectral widths SW_1 and SW_2 that correspond to the F_1 and F_2 domains, (d) the number of data points or time increments that define t_1 and t_2, (e) the number of transients for each value of t_1, (f) the relaxation delay between each set of pulses that allows an equilibrium state to be reached, (g) the number of preparatory dummy transients (DS) or steady-state pulses per FID required for the establishment of the steady state for each FID, and (h) whether or not to spin the sample and control the temperature. In general, for high-resolution 2D NMR spectroscopy, it is desirable to control the temperature, and more often than not, the sample is not spun.

A.   Pulse Sequence

The choice of the pulse sequence is of fundamental importance. It is prudent to consider in advance exactly what information is required and to choose the right experiment to provide it. For example, if it is necessary to determine all the coupling pathways in a molecule, then a **COSY**-45 is the better experiment than a set of homonuclear $^1H/^1H$ decoupling experiments. However, if it is necessary to determine the values of individual coupling constants, then the latter set of experiments is preferable. Many 2D pulse sequences have been written, but only some are of general utility. Only proven techniques should be routinely chosen to solve structural problems. The software provided with most spectrometers contains pulse sequences for the most widely used experiments such that the user normally does not need to compose them.

B.   Pulse Lengths and Delays

Accurate calibration of pulse lengths for each nucleus and each probe is essential for the success of most 2D NMR experiments. Inaccurate pulse lengths may significantly reduce the sensitivity of the probe and may lead to the appearance of artifacts. In some experiments—particularly inverse ones—inaccurate pulse lengths may lead to failure of the experiment altogether.

C.   Spectral Widths

The spectral width SW_2 relates to the frequency domain. As the evolution period t_1 varies, the intensity and phase of the signals varies as a function of the behavior of the nuclear spins during the evolution and mixing periods. During the detection period, t_2, FIDs are accumulated, each at a discrete value of t_1, and Fourier transformed with respect to t_2, producing a set of spectra in the frequency domain F_2. Arranging these spectra in rows that correspond to rows of the data (matrix) and Fourier transforming at different points across columns of this data matrix creates the second frequency domain F_2. With quadrature detection, the F_2 domain covers the frequency range $SW_2/2$.

Bruker instruments use quadrature detection, with channels A and B being sampled alternately, while in Varian instruments simultaneous sampling occurs, with points A and B taken together as pairs. In both cases, the dwell time is given by

$$\mathrm{DW_2} = \frac{1}{\mathrm{SW_2}} \quad \text{and} \quad t_2 = t_2^0 + n(\mathrm{DW_2})$$

where t_2^0 is the initial value of t_2 and N_2 is the total number of A and B data points.

In quadrature detection, the transmitter offset frequency is positioned at the center of the F_2 domain (i.e., at $F_2 = 0$; in single-channel detection it is positioned at the left edge). Frequencies to the left (or downfield) of the transmitter offset frequency are positive, while those to the right (or upfield) are negative.

The spectral width $\mathrm{SW_1}$ associated with the F_2 frequency domain is $F_1 = \mathrm{SW_1}$. The time increment for the t_1 domain, which is the effective dwell time $\mathrm{DW_1}$ for this period, is $\mathrm{DW_1} = (\mathrm{SW_1})/2$. The time increments during t_1 are kept equal. In successive FIDs, the time t_1 is incremented systematically: $t_1^0 = t_1 + n(\mathrm{DW_1})$, where $n = 0, 1, 2 \ldots N_1 - 1$, N_1 is the number of FIDs, t_1^0 is the value of t_1 at the beginning of the experiment, and $\mathrm{DW_1}$ is the time increment for the t_2 domain.

In homonuclear-shift-correlated experiments, such as **COSY**, the F_1 domain corresponds to the observed nucleus. In heteronuclear-shift-correlated experiments, such as **HETCOR** or **HMQC**, F_1 relates to the indirectly detected or decoupled nucleus. It is, therefore, necessary to set the spectral width $\mathrm{SW_1}$ after considering the 1D spectrum of the nucleus corresponding to the F_1 domain. In 2D J-resolved spectra, the value of $\mathrm{SW_1}$ depends upon the magnitudes of the coupling constants and the type of experiment, **HOM2DJ** or **HET2DJ**. In both of these cases, the width of the largest multiplet, in hertz, determines $\mathrm{SW_1}$, which in turn is related to the homonuclear or heteronuclear coupling constants, respectively. In both of these experiments, the transmitter offset frequency is kept at the center of the F_1 domain (i.e., $F_1 = 0$).

D. Number of Data Points in t_1 and t_2

The effective resolution is determined by the number of data points in each domain, which in turn determines the length of t_1 and t_2. Thus, although digital resolution can be improved by zero-filling,* the basic resolution, which determines the separation of close-lying multiplets and line widths of individual signals, will not be altered by zero-filling.

The effective resolution R for N time-domain data points is given by the reciprocal of the acquisition time AQT: $R = 1/AQT = 2SW/N$. Since, by the Nyquist theorem, there must be at least two data points to define each signal, the digital resolution DR (in hertz per point or the spacing between adjoining data points) must be $\mathrm{DR} = R/\tau = \mathrm{SW_1}/N$. The real part of the spectrum is therefore obtained by the transformation of $2N$ data points, of which N are time-domain points and N are zeros. This level of zero-filling allows the resolution contained in the time domain to be recorded. Additional zero-filling

---

*Bartholdi, E., Ernst, R. R., *J. Magn. Reson.* **1973**, *11*, 9.

will lead only to a smoother appearing spectrum but not to any genuine improvement in resolution.

The number of data points chosen in the F_1 and F_2 domains is also influenced by the available storage in the computer and the time required for data acquisition, transformation, and other instrument operations. In order to avoid wasting valuable instrument time, one should choose the minimum resolution that would yield the desired information. Thus, if peaks are separated by at least 1 Hz, then the desired digital resolution should be $R/2 = 0.5$ Hz, to allow for separation in the F_2 domain. Resolution considerations in the F_2 and F_1 domains are often different. Since the acquisition time is generally small compared to the total pulse-sequence length, it is preferable to choose N_2 (the number of data points in F_2) on the basis of the final digital resolution required *without zero-filling*. The sensitivity and resolution in the F_2 domain will be improved as long as t_2 is equal to or less than the effective relaxation time T_2.

In the t_1 domain, only the number N of data points (number of increments), and not the time involved in the pulse sequence, determines the resolution. Thus, it is advisable to acquire only half the theoretical number of FIDs and to obtain the required digital resolution by zero-filling. Accordingly, the resolution in the F_1 domain is $R_1 = 2SW_1/N_1$ while the resolution in the F_2 domain is $R_2 = 1/AQT = 2SW_2/N_2$. Normally, the resolution in the F_2 domain is much greater than that in the F_1 domain.

E.   Number of Transients

The minimum number of transients required depends upon the phase cycling in the pulse sequence and is usually 4, 8, 16, or 32. The signal-to-noise ratio in the 2D plot is a function of the number of transients,[*] provided that the duty cycle in the pulse sequence is not significantly greater than the relaxation time T_2 or T_2^*. Normally, only the minimal number of transients required to obtain a reasonable signal-to-noise ratio should be used. It is important to recall that doubling the number of transients doubles the overall experiment time, but yields only a factor of 1.4 in increased signal-to-noise ratio.

F.   Data Processing

A completed 2D NMR experiment contains a set of N_1 FIDs composed of N_2 quadrature data points, with $N_2/2$ points from channel A and $N_2/2$ points from channel B, acquired with either alternate (Bruker) or simultaneous (Varian) sampling. The nature of the data processing is critical for a successful outcome. Data processing usually involves the following sequence of steps: (a) dc (direct current) correction (performed automatically by the instrument software), (b) apodization (applying a weighting function) to the t_2 time-domain data, (c) F_2 Fourier transformation and phase correction, (d) apodization of the t_1-domain data and phase correction (unless the spectrum is a magnitude or a power-mode spectrum, in which case no phase correction is made), (e) complex Fourier transformation in F_1, and (f) coaddition of real and imaginary data (if phase-sensitive representation is required), to give a magnitude (M) or a power-mode (P) spectrum. Then, tilting, symmetrization, and calculation of projections or traces may be performed on the resulting spectrum.

---

[*] Levitt, M. H., Bodenhausen, G., Ernst, R. R., *J. Magn. Reson.* **1984,** *58,* 462.

▶ FIGURE 10.24

Fourier transformation across t_2 and t_1 gives signals in v_2 and v_1, both of which have absorption and dispersion components corresponding to real and imaginary parts. This leads to four different combinations of real and imaginary components and the four different line shapes shown below as contour plots, with solid contours (positive) being above the plane of the paper and dashed contours (negative) being below the plane of the paper.*

* Adapted from Reference 6.

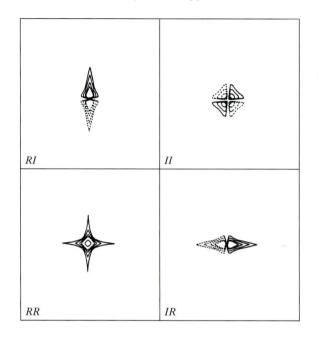

The first set of Fourier transformations in experiments such as **COSY** produces signals in the F_2 dimension. These signals have real (R) and imaginary (I) components. The second set of Fourier transformations across t_1 gives signals in the F_1 domain, which also have real (R) and imaginary (I) components, leading to four different quadrants arising from the four possible combinations of real and imaginary components RR, RI, IR, and II (Figure 10.24). The pure absorption-mode signals correspond to the real-real (RR) quadrant.

There is some contribution of the receiver dc to the signal that needs to be removed. In 1D experiments, the FIDs decay substantially, and the last portion of the FID gives a reasonably good estimate of the dc level. The computer computes the average dc level automatically and subtracts it from each of the two quadrature data sets. This is the baseline correction discussed in Chapter 2. In 2D NMR experiments, since the number of data points is small and the acquisition time is short, the FID does not decay to zero and dc correction is more difficult to do accurately. Phase-cycling procedures are employed in the pulse sequence to remove dc offsets. Since the data points in the F_1 transform arise from frequency-domain spectra, no dc correction is normally required. (There should be no changing dc components at $F_1 = 0$.)

The same type of apodization that is used in 1D NMR spectroscopy is also used in 2D NMR spectroscopy, but with some differences because the line shapes are different. The free-induction decay may be considered to be a complex function of frequency, having real and imaginary coefficients. There are two coefficients in each point of the frequency spectrum, and these real and imaginary coefficients may be described as the cosine and sine terms, respectively. Recall from Eulers formula that $e^{i\theta} = \cos\theta + i\sin\theta$. Each set of coefficients occupies half the memory in the time-domain spectrum. After double Fourier transformation and phasing, the resulting signals may be displayed in the absorption mode. Such *pure 2D absorption peaks* have cross sections parallel to the F_1 and F_2 axes, which are pure 1D Lorentzian lines (Figures 10.25a and b). However, they lack cylindrical or elliptical symmetry, display protruding ridges running down the lines parallel to F_1 and F_2, and have star-shaped cross sections (Figure 10.25c).

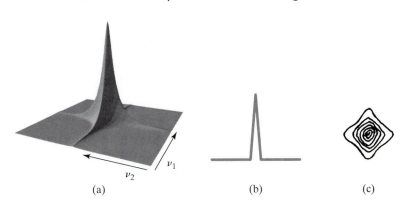

◀ FIGURE 10.25
Different views of pure 2D absorption-peak shapes. (a) a 3D view; (b) a vertical cross section parallel to ν_1 or ν_2; (c) a contour plot.*

*Adapted from Reference 2.

This distortion can be suppressed by Lorentz–Gaussian apodization. The *pure negative 2D dispersion peak shapes* are shown in Figure 10.26. The peaks have positive and negative lines, with vanishing signal contributions as they pass through the baseline. They are broadened at the base, decay slowly, and lead to a poor signal-to-noise ratio. Contour plots of such peaks give the appearance of a Maltese cross. It is desirable to remove their dispersive components in order to improve the signal-to-noise ratio. The *phase-twisted peak shapes* (or mixed absorption–dispersion peak shape) is shown in Figure 10.27. Such peaks arise by the overlapping of the absorptive and dispersive contributions in the peak. The center of the peak contains mainly the absorptive component, while in the direction moving away from the center toward the edge, there is an increasing dispersive component. Mixed phases in peaks reduce the signal-to-noise ratio and can give rise to complicated interference effects when two such peaks lie close together. Overlap between positive regions of two

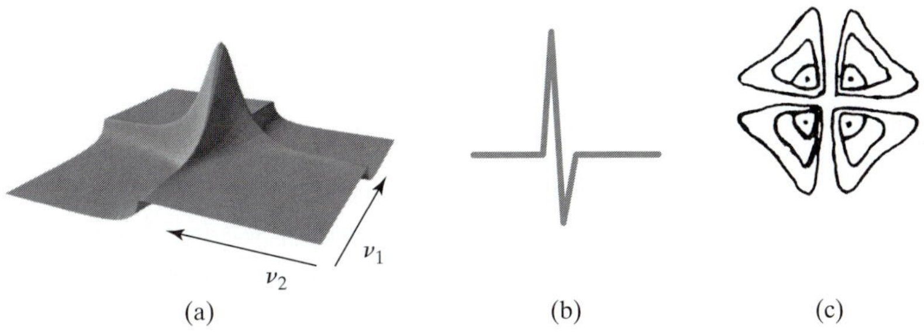

◀ FIGURE 10.26
Different views of pure negative 2D dispersion-peak shapes. (a) a 3D view; (b) a vertical cross section parallel to ν_1 or ν_2; (c) a contour plot.*

*Adapted from Reference 2.

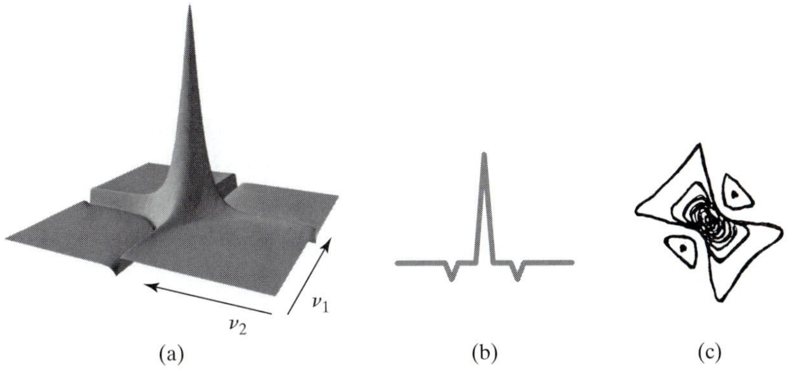

◀ FIGURE 10.27
Different views of an absorption–dispersion peak. (a) a 3D view; (b) a vertical cross section; (c) a contour plot.*

*Adapted from Reference 2.

different peaks can mutually reinforce the lines (constructive interference), while overlap between positive and negative lobes can mutually cancel the signals in the region of overlap (destructive interference).

In part because of the variety of shapes encountered in a peak, 2D spectra are often recorded in the absolute-value mode. The absolute value A is the square root of the sum of the squares of the real (R) and imaginary (I) coefficients:

$$A = \sqrt{R^2 + I^2}.$$

The absolute-value-mode presentation eliminates negative contributions, but it produces broad dispersive components that lead to a broadening of the base of the peak and to tailing effects. The majority of 2D experiments produce phase-twisted peaks that cannot be satisfactorily phased. An absolute-value display is then most convenient, but it requires the application of strong weighting functions. This increases the sensitivity at the expense of resolution. The nature of the weighting function depends upon the line shape, and for absolute-value-mode spectra *sine-bell, sine-bell squared, phase-shifted sine-bell, phase-shifted sine-bell squared*, and *Lorentz–Gauss functions* are all used. Fortunately for the user, most newer software suggests the best type of apodization function to be used for each experiment, with only minor adjustments required.

It is often desirable to tilt or symmetrize the spectrum before it is plotted. For *J*-resolved spectra, a 45° tilt produces orthogonal *J* and *S* axes.* Symmetrization procedures can also be employed for cosmetic improvement of the spectrum when the peaks are symmetrically arranged, such as in *J*-resolved **COSY** and **NOESY** spectra. Symmetrization can be carried out by *triangular multiplication*,[†] in which a pair (a, b) of symmetry-related peaks is replaced by their geometric mean $\sqrt{ab}$. A simpler symmetrization procedure is to replace each pair of symmetry-related peaks by the smaller of the two values a and b.

Additional Reading

1. Software to simulate **HOM2DJ**, **HET2DJ**, and **HETCOR** spectra has been written by Harold T. Bell and is available from him at the Department of Chemistry, Virginia Polytechnic Institute and State University, Blacksburg, Virginia 24061.

2. ATTA-UR-RAHMAN, and CHOUDHARY, M. I., *Solving Problems with NMR Spectroscopy*, Academic, San Diego, 1996.

3. BAX, A., *Two-Dimensional Nuclear Magnetic Resonance in Liquids*, D. Reidel, Dordrecht the Netherlands, 1982.

4. BRAUN, S., KALINOWSKI, H. O., and BERGER, S., *100 and More Basic NMR Experiments*, VCH, New York, 1996. BRAUN, S., KALINOWSKI, H. O., and BERGER, S., *150 and More Basic NMR Experiments*, VCH, New York, 1999.

5. CROASMUN, W. R. and CARLSON, R. M. K., *Two-Dimensional NMR Spectroscopy. Applications for Chemists and Biochemists*, VCH, New York, 2d ed., 1994.

6. DEROME, A. E., *Modern NMR Techniques for Chemistry Research*, Pergamon, New York, 1987.

7. GÜNTHER, H., *NMR Spectroscopy*, Wiley, New York, 2d ed., 1995.

8. KESSLER, H., GEHRKE, M., and GRIESENGER, C., *Two-Dimensional NRM Spectroscopy: Background and Overview of the Experiments*, Angew. Chem. Int. Ed. Engl., 1988, 27, 490.

9. MARTIN, G. E. and ZEKTZER, A. S., *Two-Dimensional NMR Methods for Establishing Molecular Connectivity*, VCH, New York, 1988.

10. NAKANISHI, K., *One-Dimensional and Two-Dimensional NMR Spectra by Modern Pulse Techniques*, University Science Books, Mill Valley, CA, 1990.

11. SANDERS, J. K. M., and HUNTER, B. K., *Modern NMR-Spectroscopy. A Guide for Chemists*, Oxford, New York, N.Y., 2d ed., 1993.

12. SCHRAML, J., and BELLAMA, J. M., *Two-Dimensional NMR Spectroscopy*, Wiley, New York, 1988.

---

*Baumann, R., Kumar, A., Ernst, R. R., Wüthrich, K., *J. Magn. Reson.* **1981**, *44*, 76.

[†]Baumann, R., Wider, G., Ernst, R. R., Wüthrich, K., *J. Magn. Reson.* **1981**, *44*, 402.

CHAPTER 11

Dynamic NMR Spectroscopy

11.1   Introduction

Dynamic NMR spectroscopy (**DNMR**) deals with the effects, in a broad sense, of chemical-exchange processes on NMR spectra and also with the information about changes in the nuclear environment that can be ascertained from the observation of variable-temperature NMR spectra. These changes in environment are due to exchange between sites with different chemical shifts and/or different coupling constants. One of the more important kinds of information obtained in this way concerns the rate constants for the processes studied. The NMR time scale is such that first-order or pseudo-first-order rate constants in the range from $10^5\,\text{s}^{-1}$ to $10^{-1}\,\text{s}^{-1}$ with associated activation energies in the range from 20 to 100 kJmol^{-1} can be measured. Besides this very large time span, the **DNMR** technique has several advantages compared with other kinetic methods. For example, the NMR spectrum gives direct information about the specific parts of the molecule that are affected by the exchange, something that is not normally achieved by other spectroscopic methods. Furthermore, **DNMR** studies are performed on systems in thermodynamic equilibrium, and even degenerate systems, in which the exchange leads to molecules that are indistinguishable from the original ones, can be investigated by this method.

We may ask the general question, What effect will rate processes have on an NMR spectrum? Consider the following hypothetical "compounds" I and II:

If there is no rotation about the $X\text{–}Y$ bond, then the ^{1}H NMR spectra of these two molecules, while both showing AB or AX patterns, will exhibit different coupling constants ($^3J_{\text{HH}}$ will not be the same for the two rotamers) and different chemical shifts (δ_A and δ_B or δ_A and δ_X will not be the same for the two rotamers). On the other hand, if there is free rotation about the $X\text{–}Y$ bond, then compounds I and II cannot exist as two separate entities, as they would be one and the same substance. The ^{1}H NMR spectrum would still be that of an AB or AX spin system, but the chemical shifts of the two protons (H$_1$ and H$_2$) and the coupling constant ($^3J_{\text{HH}}$) would now be averages of the values for the "pure" compounds I and II.

The third possible situation that must be considered is one in which rotation about the $X\text{–}Y$ bond is rapid enough to give a single AB or AX pattern, but the average contributions of the two rotamers I and II are not equal. When this occurs, the chemical shifts and the coupling constant are the result of population-weighted averages that depend on P_I and P_{II}, the fractional populations of species I and II, respectively, as shown in the following equations:

$$J(\text{average}) = J(\text{rotamer I}) \times P_I + J(\text{rotamer II}) \times P_{II}; \qquad \textbf{(11.1)}$$

$$v(\text{average}) = v(\text{rotamer I}) \times P_I + v(\text{rotamer II}) \times P_{II};$$

$$P_I + P_{II} = 1.$$

Thus, in principle at least, it is possible to differentiate between the situation of restricted rotation and the other two situations.

11.2   Determination of Exchange Rates

The simplest possible system to consider is a two-site exchange of a nucleus between sites **A** and **B** under conditions of no spin–spin coupling to other nuclei in either site. For such a system, the residence times in the two sites are inversely related to the first-order rate constants k_f and k_r for the site exchange. Designating the fractional populations of the **A** and **B** sites as P_A and P_B, we find that

$$\frac{\tau_A}{\tau_B} = \frac{P_A}{P_B} = \frac{k_r}{k_f},$$ (11.2)

from which it follows that

$$P_A = \frac{\tau_A}{(\tau_A + \tau_B)} \quad \text{and} \quad P_B = \frac{\tau_B}{(\tau_A + \tau_B)}.$$ (11.3)

The signal line shapes can be described rather simply in four special cases for two-site exchange processes.

A.   Slow Exchange

When exchange is very slow, two separate resonances will be observed. The line widths $\Delta\nu_A$ and $\Delta\nu_B$ are related to the spin–spin relaxation times T_2 and the lifetimes by

$$\pi\Delta\nu_A = \frac{1}{T_{2A}} + \frac{1}{\tau_A}$$ (11.4)

and

$$\pi\Delta\nu_B = \frac{1}{T_{2B}} + \frac{1}{\tau_B}.$$ (11.5)

Exchange thus results in excess broadening, and since $k_a = 1/\tau_a$, it follows that the excess broadening over the natural line width can be used to calculate the exchange rate constant from

$$\pi\Delta\nu_A(\text{excess}) = k_A \quad \text{and} \quad \pi\Delta\nu_B(\text{excess}) = k_B.$$ (11.6)

In Equation 11.6, $\Delta\nu_{A/B}(\text{excess}) = \Delta\nu_{A/B}(\text{experimental}) - \Delta\nu_{A/B}(\text{natural})$. Hence, $\Delta\nu_{A/B}(\text{natural})$ must be either measured or assumed to be negligibly small. Usually, one measures the line width of a nonexchanging internal standard or the line width at the stopped-exchange limit for this purpose. Equation 11.6 applies only when the two signals are not so greatly broadened as to begin to overlap. If the two signals are unequal in intensity, Equation 11.2 suggests that the less intense signal will be the more broadened. Then the following equation obtains:

$$\Delta\nu(\text{excess}) = \frac{4\pi P_A P_B(\delta\nu)^2}{k_A + k_B}.$$ (11.7)

The larger the chemical-shift separation, the greater the rate that can be measured with reasonable accuracy. When the two resonances are well separated, τ_A and τ_B can be obtained without any knowledge of $(\nu_A - \nu_B)$. In fact, if an observed resonance is exchange broadened, the rate of exchange with respect to the observed site can be obtained without any knowledge of the number or nature of the other sites.

For equally populated two-site exchange systems, the lifetime or residence time in either site, τ_A or τ_B, is related to the exchange rate constant, $k_{(exchange)}$, by the equation

$$\frac{2}{\tau_A} = \frac{2}{\tau_B} = 2k_f = 2k_r, \tag{11.8}$$

where $k_{(exchange)}$ is the rate constant in both directions. Since the rates are the same in both directions, the pseudo-first-order or "one way" rate constant k_f or k_r is related to τ by

$$k_f = k_r = \frac{1}{2\tau}. \tag{11.9}$$

In the slow exchange regime, the chemical-shift difference $v_A - v_B$ decreases with increasing temperature. The temperature dependence of the chemical-shift difference is related to the lifetime at any given temperature by

$$\frac{(v_A - v_B) \text{ at temperature of interest}}{(v_A - v_B) \text{ at stopped exchange limit}} = \left[1 - \frac{1}{2\pi^2\tau^2(v_A - v_B)^2}\right]^{\frac{1}{2}}. \tag{11.10}$$

Equation 11.10 is not applicable to situations where τ is small enough to give only a single averaged resonance or where the line widths are not small in comparison with the chemical-shift difference. When it is applicable, the equation may be used together with Equation 11.9 for an equally populated two-site exchange to determine the exchange-rate constant at various temperatures.

B. Fast Exchange

Fast exchange is defined here as a rate such that τ_A and τ_B are small compared to $(v_A - v_B)^{-1}$. When the exchange rate is sufficiently large that only a single resonance is observed, but it is slow enough to still contribute to the exchange-averaged line width, then, for an equally populated two-site exchange, the following equation holds:

$$k = \frac{\pi(v_A - v_B)^2}{2\Delta v_{1/2}}. \tag{11.11}$$

11.3　Coalescence Temperatures and Rate Constants at Coalescence

A classical example of hindered rotation is that observed for N,N-dimethylformamide.* The amide group is planar, and the partial double-bond character of the C—N bond causes exchange of the methyl groups to be slower than might be expected. The rotation is given as follows:

---

*Stewart, W. E., Siddall, T. H., *Chem. Rev.,* **1970**, *70*, 517.

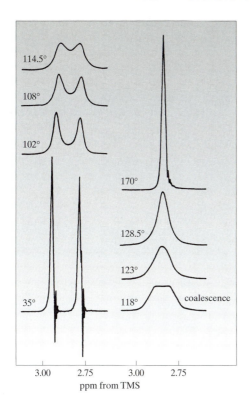

◀ FIGURE 11.1
60-MHz ^{1}H NMR spectrum of neat
N,N-dimethylformamide as a
function of temperature (°C).

The temperature dependence of the 60-MHz ^{1}H NMR spectrum of neat N,N-dimethylformamide is shown in Figure 11.1. At 35°C, two equally intense narrow methyl resonances are observed: δ = 2.79 (CH$_3$ *cis* to CO) and δ = 2.94 (CH$_3$ *trans* to CO). As the temperature is raised, these two sharp resonances begin to broaden, and at 118°C they coalesce into one broad resonance. At still higher temperatures, the single averaged resonance at $\delta_{(avg)} = (\delta\,CH_{3(a)} + \delta\,CH_{3(b)})/2$ progressively narrows as the temperature increases.

Similar behavior is observed in the variable-temperature 60-MHz ^{1}H {^{2}H} NMR spectra of d$_{11}$-cyclohexane in CS$_2$ solution, as illustrated in Figure 11.2. In this case, the averaging process is due to ring inversion, which exchanges equatorial and axial sites for the lone proton.

In both of these examples, the spectra change from the stopped-exchange limit at low temperature progressively through the slow- and intermediate-exchange regimes to coalescence and, finally, fast exchange (Figure 11.2). The two resonances broaden as their chemical-shift difference decreases, until, at coalescence, only one broad resonance is observed. At temperatures higher than the coalescence temperature, this single resonance continues to narrow throughout the fast-exchange regime.

At coalescence in such equally populated two-site exchange systems, the equation

$$k_f = k_r = \pi\frac{(\nu_A - \nu_B)}{\sqrt{2}} = 2.22(\nu_A - \nu_B) \qquad \textbf{(11.12)}$$

▶ FIGURE 11.2
60-MHz ^{1}H{^{2}H} NMR spectrum of cyclohexane-d$_{11}$ 15% vol./vol. in CS$_2$ as a function of temperature (°C).*

*Bovey, F. A., Hood, F. P., Anderson, E. W., Kornegay, R. L., *Proc. Chem. Soc.*, **1964**, *146*; *J. Chem. Phys.*, **1964**, *41*, 2041.

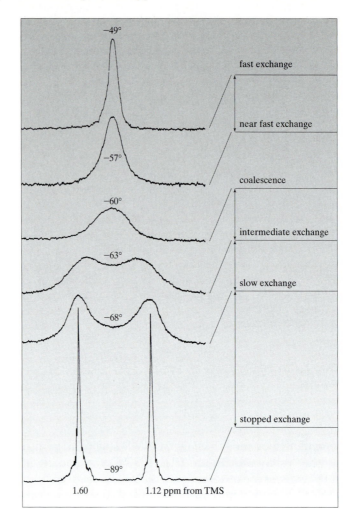

applies, where $\nu_A - \nu_B$ is the chemical-shift difference in Hz at the stopped exchange limit.

SAMPLE PROBLEM 11.1

Calculate the exchange rates at coalescence for (a) the hindered rotation of N,N-dimethylformamide and (b) the chair-to-chair inversion of d$_{11}$-cyclohexane.

Solution

(a) N,N-dimethylformamide

$$k_f = k_r = 2.22(2.94 - 2.79 \text{ ppm})\left(\frac{60 \text{ Hz}}{\text{ppm}}\right) = 19.98 \text{ s}^{-1} = 20 \text{ s}^{-1}$$

(b) d$_{11}$-cyclohexane

$$k_f = k_r = 2.22(1.60 - 1.12 \text{ ppm})\left(\frac{60 \text{ Hz}}{\text{ppm}}\right) = 63.94 \text{ s}^{-1} = 64 \text{ s}^{-1}$$

For an exchange process involving two coupled nuclei in an *AB* spin system, the following equation applies:

$$k_c = 2.22\sqrt{(\nu_A - \nu_B)^2 + 6J_{AB}{}^2}. \tag{11.13}$$

SAMPLE PROBLEM 11.2

From the 60-MHz ^{1}H NMR spectra of 1,2-diphenyldiazetidinone shown at (A) 35°C in acetone-d_6 and (B) −55° in CDCl$_3$ (where J_{AB} = 14 Hz), calculate k_c, the exchange rate at coalescence (−1°C).*

Solution

$$k_c = 2.22\sqrt{(\nu_A - \nu_B)^2 + 6J_{AB}{}^2}$$

$$= 2.22\sqrt{\left[(5.29 - 4.60\,\text{ppm})\left(\frac{60\,\text{Hz}}{\text{ppm}}\right)\right]^2 + 6(14\,\text{Hz})^2}$$

$$= 2.22\sqrt{1713.96 + 1176}\,\text{Hz}$$

$$= 119.34\,\text{s}^{-1} = 119\,\text{s}^{-1}$$

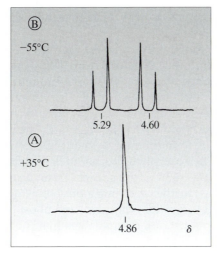

As we see from equations 11.10 to 11.12, k_c is a function of $\nu_A - \nu_B$, which in turn is a function of the field strength of the spectrometer. Consequently,

---

*Adapted from *Basic One- and Two-Dimensional NMR Spectroscopy*, Friebolin, H., VCH, Weinheim, **1993**.

since $\nu_A - \nu_B$ increases with increasing field strength, the coalescence temperature for any exchange process will increase with increasing field strength. Empirically, it is found that doubling the field strength increases the coalescence temperature by about 10°C.

11.4   Complete Line-Shape Analysis

Given values for the chemical shifts and coupling constants in an NMR spin system, NMR spectra may be calculated completely with the aid of a variety of computer programs that have been written for spin-system analysis. Most of these programs are based upon LAOCOON-III[*] and more recent versions of it and are usually available in the software provided by NMR instrument manufacturers. These simulations provide the complete spectral line shapes, including the number of resonance lines, together with their line widths and intensities. When dynamic processes are involved, the line shapes of the resonances also depend on the exchange rate k or the lifetime τ of a nucleus in a particular site. In simple cases, as we have seen, k and τ may be calculated rather easily by hand. In more complicated cases, such as when there is coupling among exchanging nuclei or when multisite exchange is occurring, complete line-shape analysis is required to determine τ and k.

In order to do this, it is first necessary to obtain a number of spectra at different temperatures, ideally encompassing a temperature variation that extends from the stopped-exchange limit up through the fast-exchange regime. Then, the spectra are completely analyzed at the stopped-exchange limit to obtain chemical shifts and coupling constants. Finally, the spectra are calculated at various exchange rates, using a computer program such as DNMR5,[†] and the calculated spectra are either visually matched or iteratively fit by computer to the experimental spectra.

An example of a study of this type for the stereochemical permutation of the trigonal bipyramidal species $HRh[P(OC_2H_5)_3]_4$, as investigated by 90-MHz ^1H NMR spectroscopy, is shown in Figure 11.3. The calculated line shapes were obtained assuming a first-order low-temperature limit spectrum corresponding to C_{3v} symmetry with $|^1J_{HRh}| = 9$ Hz, $^2J_{HPax} = \pm 152$ Hz, and $^2J_{HPeq} = \pm 5$ Hz. At $-46°C$, the spectrum consists of a binomial quintet due to four equal $^2J_{HP}$ couplings, further split into doublets due to $^1J_{RhH}$. The four phosphorus nuclei are thus isochronous on the NMR time scale. At $-134°C$, the spectrum can be crudely described as a doublet due to $^2J_{PaxH}$, further split into a doublet of quartets due to $^1H_{RhH}$ and $^2J_{PeqH}$. Correlation of proton, rhodium, and phosphorus spins throughout the entire temperature range establishes that the exchange process is intramolecular.

[*]Diehl, P., Kellerhals, H., Lustig, E., *Computer Assistance in the Analysis of High-Resolution NMR Spectra,* In *NMR: Basic Priniciples and Progress*, Vol. 6, Diehl, P., Fluck, E., Kosfeld, R., eds. Springer-Verlag, Berlin, **1972**.

[†]Binsch, G., *Band-Shape Analysis in Dynamic Nuclear Magnetic Resonance Spectroscopy,* Jackman, L. M., Cotton, F. A., eds. Academic Press, New York, 1975. Stephenson, D. S., Binsch, G., *J. Magn. Reson*, **1978**, *30*, 145. Ibid. **1978**, *32*, 145.

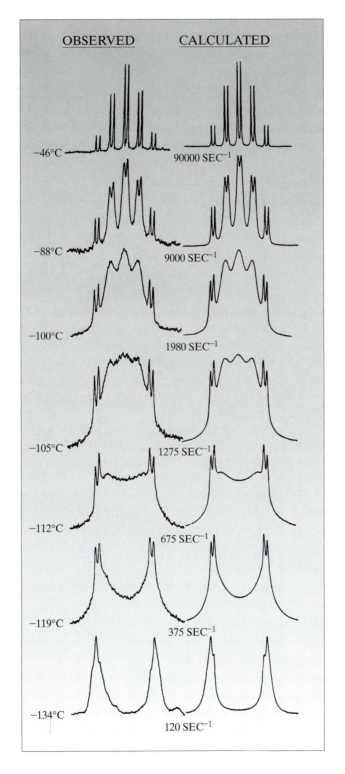

◀ FIGURE 11.3
Observed and calculated 90-MHz
^{1}H NMR spectra in the hydride
region of HRh[P(OC$_2$H$_5$)$_3$]$_4$ in
CHClF$_2$ as a function of
temperature (°C).*

*Meakin, P., Muetterties,
E. L., Jesson, J. P., *J. Am. Chem. Soc.*,
1972, *94*, 5271.

11.5 Activation Parameters

It is possible to determine the Arrhenius activation energy E_A for dynamic processes from a plot of $\ln k$ vs. $1/T$ with a slope of $-E_A/R$, as given by the equation

$$k = k_0\, e^{-E_A/RT}, \quad \text{or} \quad \ln k = \ln k_0 - \frac{E_A}{RT}, \tag{11.14}$$

where T is the temperature in K, k_0 and k are the rate constants at the experimental and infinitely large temperatures, respectively, and R is the universal gas constant (1.978 cal K^{-1} mol^{-1} or 8.3144 J K^{-1} mol^{-1}). It is also possible to determine the activation free energy from the Eyring equation,

$$k = \kappa \frac{k_B T}{h} e^{-\Delta G^\ddagger/RT}, \tag{11.15}$$

which becomes

$$\Delta G^\ddagger = 4.58 T_C \left[10.32 + \log\left(\frac{T_C}{k_C}\right) \right] \text{cal mol}^{-1} \tag{11.16}$$

$$= 19.14 T_C \left[10.32 + \log\left(\frac{T_C}{k_C}\right) \right] \text{J mol}^{-1}$$

after insertion of the values for the Boltzmann constant ($k_B = 3.2995 \times 10^{-24}$ cal K^{-1} or 1.3805×10^{-23} J mol^{-1}), the transmission coefficient (κ, which is usually assumed to be unity), and the Planck constant ($h = 1.5836 \times 10^{-34}$ cal s or 6.6256×10^{-34} Js). Note that the activation energy at coalescence and the coalescence temperature are denoted by $\Delta G_C^\ddagger$ and T_C, respectively. It is now necessary to know only the coalescence rate and the coalescence temperature in order to calculate $\Delta G_C^\ddagger$.

The coalescence temperature for equally populated two-site exchange, as a function of $\Delta G^\ddagger$, for three different ^{1}H observation frequencies is given approximately by Figure 11.4. From this figure, we can see that the practical

▶ FIGURE 11.4
Empirical correlation of coalescence temperature for equally populated two-site exchange (°C) with activation energy $\Delta G^\ddagger$ (kcal/mol) as a function of ^{1}H observation frequency.

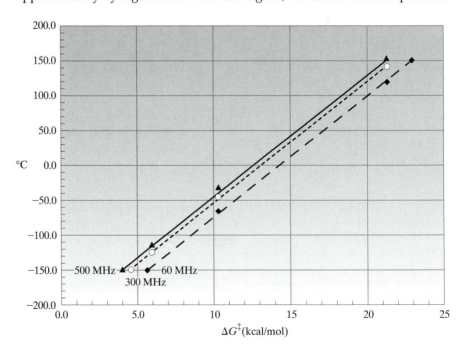

lower limits for the ^{1}H NMR determination of activation free energies go from about 4 kcal/mol at 500 MHz to about 6 kcal at 60 MHz, because about $-150°C$ is usually the lowest obtainable temperature in an NMR spectrometer.

SAMPLE PROBLEM 11.3

Calculate $\Delta G_C^{\ddagger}$ for (a) the hindered rotation of N,N-dimethylformamide and (b) the chair-to-chair inversion of d_{11}-cyclohexane.

Solution

(a) N,N-dimethylformamide:

In Sample Problem 11.1, we found that $k_C = 20\ s^{-1}$ at $T_C = 393\ K$. Thus,

$$\Delta G_C^{\ddagger} = (4.58)(393)(10.32 + \log 393/20) = 20{,}903\ cal\ mol^{-1}$$

$$= 20.9\ kcal\ mol^{-1}$$

$$(87.5\ kJ\ mol^{-1}).$$

(b) d_{11}-cyclohexane:

In Sample Problem 11.1, we found that $k_C = 64\ s^{-1}$ at $T_C = 206\ K$. Thus,

$$\Delta G_C^{\ddagger} = (4.58)(206)(10.32 + \log 206/64) = 10{,}216\ cal\ mol^{-1}$$

$$= 10.2\ kcal\ mol^{-1}$$

$$(42.7\ kJ\ mol^{-1}).$$

The free energy of activation, $\Delta G^{\ddagger}$, is related to the enthalpy of activation, $\Delta H^{\ddagger}$, and the entropy of activation, $\Delta S^{\ddagger}$, by

$$\Delta G^{\ddagger} = \Delta H^{\ddagger} - T\Delta S^{\ddagger}. \tag{11.17}$$

Combining equations 11.15 and 11.17 gives

$$k = \left(\frac{k_B T}{h}\right) e\left(\frac{\Delta S^{\ddagger}}{R}\right) e\left(-\frac{\Delta H^{\ddagger}}{RT}\right) \tag{11.18}$$

$$-R \ln\left(\frac{Kh}{k_B T}\right) = -\Delta S^{\ddagger} + \frac{\Delta H^{\ddagger}}{T},$$

from which it should be clear that a plot of $-R\ln(Kh/k_B T)$ vs. $1/T$ has a slope of $\Delta H^{\ddagger}$ and an intercept of $-\Delta S^{\ddagger}$. After substituting the appropriate values of h and k_B into Equation 11.18, we have

$$\ln\left(\frac{k}{T}\right) = 23.76 - \left(\frac{\Delta H^{\ddagger}}{RT}\right) + \left(\frac{\Delta S^{\ddagger}}{R}\right). \tag{11.19}$$

Thus, a plot of $\ln k/T$ vs. $1/T$ has a slope of $-\Delta H^{\ddagger}/R$ and an intercept of $(23.76 + \Delta S^{\ddagger}/R)$. Unlike the data obtained directly through the use of

Equations 11.14, 11.16, and 11.20, the activation parameters calculated from Equations 11.18 or 11.19 are independent of temperature and should be used to compare the energy profiles of analogous reactions. For unimolecular reactions, the relationship between $\Delta H^{\ddagger}$ and E_A is given by

$$\Delta H^{\ddagger} = E_A - RT. \tag{11.20}$$

The main sources of error in the variable-temperature NMR determination of activation parameters stem from errors in measuring the temperatures T and the coalescence temperatures T_C. Measurements of either are usually no more accurate than about ± 2 K. This leads to uncertainties about the following magnitudes: $\Delta G^{\ddagger}(\pm 0.2\,\text{kcal mol}^{-1})$, $\Delta H^{\ddagger}$ or $E_A(\pm 1\,\text{kcal mol}^{-1})$, and $\Delta S^{\ddagger}(\pm 2$ to $5\,\text{cal mol}^{-1}\,\text{K}^{-1})$.

11.6   Equilibrium Thermodynamics

If two species that are in equilibrium have different thermodynamic stabilities, then variable-temperature NMR spectra for such species, undergoing sufficiently slow exchange that separate resonances for both species may be observed, can be used to determine equilibrium constants K_{eq} for these equilibria. Since the equilibrium constants should be temperature dependent, their measurement (from ratios of integrals in the spectra) as a function of temperature provides a means of measuring equilibrium thermodynamics from the following equations:

$$\Delta G^{\circ} = -RT \ln K_{eq}; \tag{11.21}$$

$$\Delta G^{\circ} = \Delta H^{\circ} - T \Delta S^{\circ}; \tag{11.22}$$

$$\ln K_{eq} = -\frac{\Delta H^{\circ}}{RT} + \frac{\Delta S^{\circ}}{R}. \tag{11.23}$$

Hence, in a plot of $\ln K_{eq}$ vs. $1/T$, the slope is $-\Delta H^{\circ}/R$ and the intercept is $-\Delta S^{\circ}/R$. Analogous equilibrium and activation thermodynamic quantities have similar magnitudes of uncertainties.

11.7   Examples of Dynamic Processes Studied by NMR Spectroscopy

A.   Rotation about C–C Single Bonds

i.   When Both Carbon Atoms Have sp^3 Hybridization

Rotation about the C—C bond in substituted ethanes has been studied extensively. For ethanes with small substituents, the rotation is so rapid that only averaged spectra can be obtained over the entire accessible temperature range. For halogenated and t-butyl substituted ethanes, $\Delta G^{\ddagger}$ in the range 5 to 15 kcal mol^{-1} is observed, with the actual values depending upon the number and nature of the other substituents. We will consider the nature of the spectra that should be observed for two types of substituted ethanes, both of which have three "low-energy" rotamers. Suppose that there is no coupling between the substituents *A–D* and the two hydrogens. When rotation is slow, the NMR spectrum will show three separate *AB* or *AX* patterns:

Case I

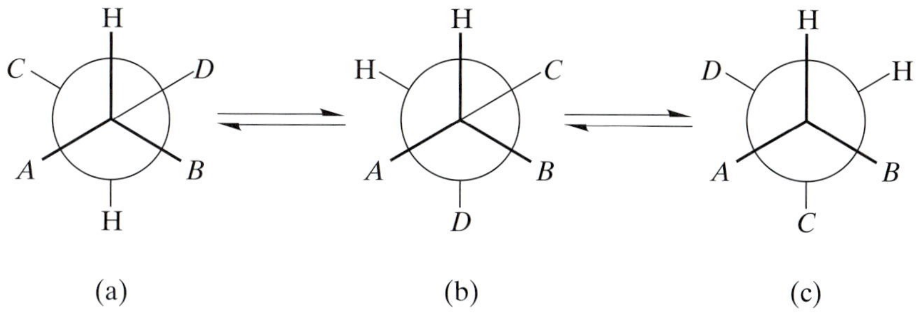

(a)   (b)   (c)

If, however, there is rapid rotation, then only a single *AB* or *AX* pattern will result, independently of whether the three rotamers (*a*–*c*) contribute equally or not. The values of the chemical shifts and the H—H coupling constant will, however, depend upon the relative populations under conditions of fast exchange.

Case II

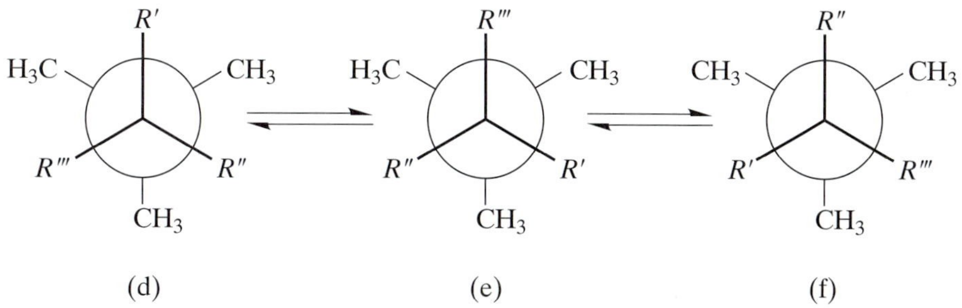

(d)   (e)   (f)

Assume here that no coupling is present. When rotation is slow, the spectrum will show three separate CH_3 resonances, and when rotation is rapid, only one CH_3 resonance will be observed, independently of whether the three rotamers (d)–(f) contribute equally or not. The chemical shift will, however, be a function of the rotamer population under conditions of fast exchange.

ii.   When One Carbon Atom Is sp^2 Hybridized and One Carbon Atom Is sp^3 Hybridized

For a monosubstituted benzene, such as toluene, ethylbenzene, or isopropylbenzene, rotation about the C—C bond is so rapid that it cannot be frozen out, even at temperatures as low as $-150°C$. However, for sterically encumbered systems, such as 1-isopropyl-2-methylnaphthalene, one finds that $T_C = 228$ K and $\Delta G^{\ddagger} = 12.7 \pm 0.2$ kcal mol^{-1} at 60 MHz.

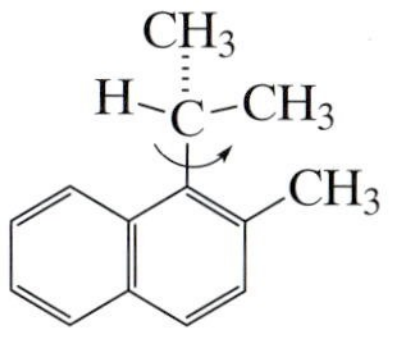

SAMPLE PROBLEM 11.4

Calculate the chemical-shift difference at 60 MHz and 500 MHz for the two diastereotopic methyl resonances of 1-isopropyl-2-methylnaphthalene at the stopped-exchange limit of $T_C = 228$ K and $\Delta G_C^{\ddagger} = 12.4$ kcal mol^{-1},* observed at 60 MHz.

Solution

From Equation 11.16, we have

$$\Delta G_C^{\ddagger} = 4.58\, T_C\left(10.32 + \log\frac{T_C}{k_C}\right) \quad \text{or} \quad \frac{(12400)}{(4.58)(228)} - 10.32 = \log\frac{T_C}{k_C},$$

whence

$$\log\frac{T_C}{k_C} = 1.55 \quad \text{or} \quad k_C = \frac{228}{15/86} = 14.38 \text{ s}^{-1}.$$

Then, from Equation 11.12,

$$k_C = 2.22(\nu_A - \nu_B)$$

or

$$(\nu_A - \nu_B) = \frac{14.38 \text{ s}^{-1}}{2.22} = 6.5 \text{ Hz} = 0.11 \text{ ppm at 60 MHz.}$$

At 500 MHz, $(\nu_A - \nu_B) = (500/60)(6.5 \text{ Hz}) = 54.2$ Hz.

iii. When Both Carbon Atoms Are sp^2 Hybridized

It has long been known that 2,2′-disubstituted biphenyls (shown below) exhibit hindered C—C rotation; hence, atropisomerism and $\Delta G^{\ddagger}$ for such molecules is found in the 14- to 25-kcal mol^{-1} range.[†]

B. Rotation about a Partial Double Bond

Hindered rotations of this type have been observed for amides, thioamides, amidines, enamine derivatives, and aminoboranes. We discuss each in turn.

Amides. The highest activation energy for this class of compounds is that found for N,N-dimethylformamide ($\Delta G^{\ddagger} = 20.9$ kcal mol^{-1}, $T_C = 393$ K). Replacing the formyl hydrogen with any substituent decreases the activation energy (e.g., for $R = C_6H_5$, $\Delta G_{298}^{\ddagger} = 15.0$ kcal mol^{-1}, and for $R = t - C_4H_9$, $\Delta G^{\ddagger} = 12.2$ kcal mol^{-1}). Replacing the N-methyl groups with other substituents

---

*Mannschreck, A., Ernst, L., *Chem. Ber.* **1971**, *104*, 228.
[†]M. Oki, ed., *Applications of Dynamic NMR Spectroscopy to Organic Chemistry*, VCH Publishers, Deerfield Beach, FL, **1985**.

causes only small changes in the activation energy (e.g., for $(C_2H_5)_2NCHO$, $\Delta G^{\ddagger} = 20.4$ kcal mol^{-1}, and for $(CH_3)_3$ CNHCHO, $\Delta G^{\ddagger}_{298} = 19.2$ kcal mol^{-1}).

Thioamides. Thioamides have rotational barriers that are about 2 to 4 kcal mol^{-1} higher than for the equivalent amide. This has been rationalized on the basis of a weaker C—S π bond than the C—O π bond, stabilizing the dipolar canonical form:

Amidines. The barriers to rotation about the C—NR$_2$ bond in amidines are smaller (12 to 14 kcal mol^{-1}) than those found in amides (12 to 21 kcal mol^{-1}), probably due to a reduced contribution of the dipolar canonical form:

Enamines. For normal enamines, the rotational barrier about the C—N bond is too small to be measured by NMR spectroscopy. However, for α, β-unsaturated systems, such as

C—N bond rotation is hindered, and $\Delta G^{\ddagger} \cong 13$ kcal mol^{-1}. For these systems, there is also hindered rotation about the partial C—C double bond, with $\Delta G^{\ddagger} \cong 17$ kcal mol^{-1}.* The substituents R and R' have a large effect on these barriers.[†]

Aminoboranes. Aminoboranes exhibit considerable hinderance to rotation about the B—N bond, due to the large stabilization of the dipolar canonical form. Typical activation energies are 15 to 23 kcal mol^{-1}, similar to those found for amides:[‡]

---

*Dabroski, J., Kozerski, L., *J. Chem. Soc.*, (B) **1971**, 345.

[†] See Jackman, L. M., in *Dynamic Nuclear Magnetic* Resonance Spectroscopy, Jackman, L. M., Cotton F. A., eds., Academic, New York, **1975**, pp. 218–233.

[‡] Friebolin, H., Rensech, R., Wendel, H., *Org. Magn. Reson.*, **1976**, *8*, 287.

C.   Inversions at Stereogenic Centers

While inversion of a chiral carbon atom is so slow that resolution of enantiomers is commonplace, inversion of nitrogen, phosphorus, and sulfur stereocenters are much faster:

Nitrogen. Inversion of amines, in general, is too rapid to be studied by NMR spectroscopy, except in special circumstances. For example, for dibenzylmethyl amine, $k = 2 \times 10^5 \, s^{-1}$ at 25°C, and typical chiral amines cannot be resolved. However, the inversion barrier can be increased by certain substituents. For example, for the amine

(in *n*-hexane), the temperature dependence of the *AB* methylene proton resonances gave a coalescence temperature T_C of 257 K and $\Delta G^{\ddagger}_{257} = 12.7 \, \text{kcal mol}^{-1}$.*

For three-, four-, and six-membered heterocycles containing nitrogen, the inversion rate is slower. We saw previously (Sample Problem 11.2) that for 1,2-diphenyldiazetidinone, $T_C = 272$ K and $\Delta G^{\ddagger} = 13.3 \, \text{kcal mol}^{-1}$. There are two nitrogen atoms in this compound, and N_1 is assumed to be the more slowly inverting nitrogen. For N-chloroaziridine,

coalescence could not be reached before decomposition occurred, and $T_C > 180$°C.

Phosphorus. Phosphorus inversion in phosphines has a high energy barrier typically on the order of 30 kcal mol^{-1}, which is much larger than that of amines. Consequently, chiral phosphines may be resolved. Phosphorus heterocycles, such as phospholes, have lower inversion barriers (15 to 24 kcal mol^{-1}), due to stabilization of the planar transition state. An example is given in Figure 11.5. Three different temperature-dependent resonances are shown in the figure. The methoxy resonance exhibits coalescence of two singlets into one singlet. The pair of doublets for the methine resonance coalesces into a single doublet, and the pair of doublets of doublets for the ring methyl resonance coalesces into a single doublet of doublets. The coalescence temperatures are different for the three *reporter groups*, because the chemical-shift differences of these groups for the two diastereomers in the slow exchange limit are different. Nonetheless,

---

*Griffith, D. L.; Roberts, J. D., *J. Am. Chem. Soc.* **1965**, *87*, 4089.

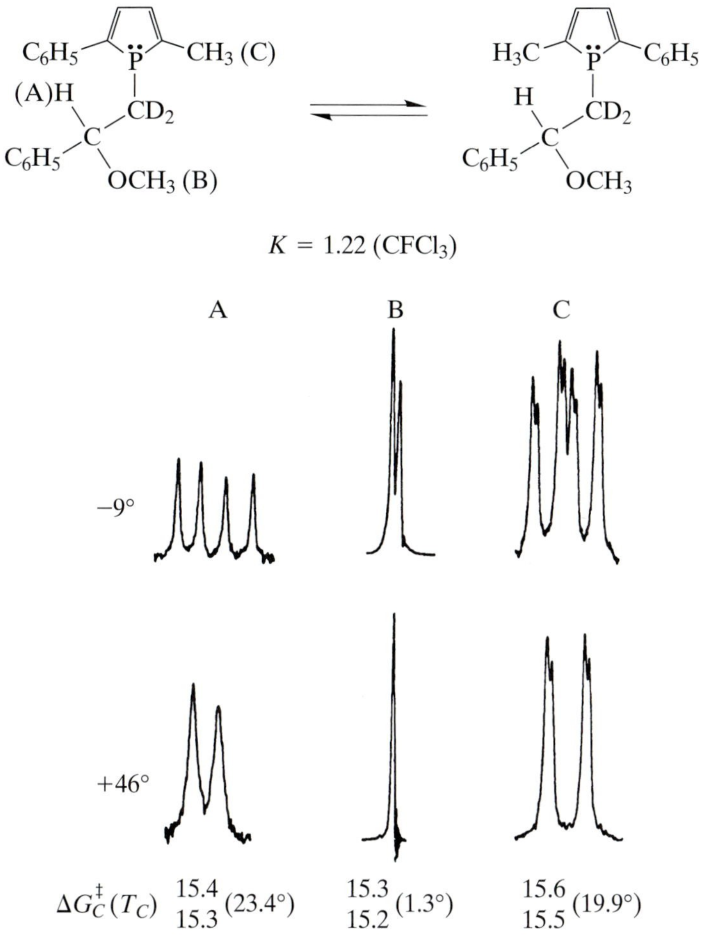

$$K = 1.22 \ (CFCl_3)$$

◄ FIGURE 11.5

Expansions of the 60-MHz ^{1}H spectra of the phosphole molecule shown in CFCl$_3$. At $-9°$C, exchange is slow, and at 46°C, exchange is rapid. Listed below the resonances of each of the reporter groups are the activation energies and the coalescence temperatures.

Egan, W., Tang, R., Zou, G., Mislow, K., *J. Am. Chem. Soc.,* **1971**, *93*, 6205.

the activation energies, as determined from the three separate coalescence temperatures, are all within the experimental uncertainty (± 0.2 kcal mol^{-1}), as they must be if the data analysis is correct.

Sulfur. For chiral sulfoxides and sulfonium ions, the inversion barriers (15 to 43 kcal mol^{-1}) are usually sufficiently large to allow for their resolution. In contrast, for unsymmetrically substituted dialkyl-, diaryl-, or alkyl-aryl-sulfides coordinated to a transition metal, the activation energy is so reduced that inversion is often rapid at ambient temperatures and below. For example, for *cis*-[(C$_2$H$_5$)$_2$S]$_2$PtCl$_2$, $T_C = 333$ K and $\Delta G^{\ddagger} = 16.8$ kcal mol^{-1}, and for *trans*-[(C$_2$H$_5$)$_2$S]$_2$PtCl$_2$, $T_C = 274$ K and $\Delta G^{\ddagger} = 13.9$ kcal mol^{-1}.*

D. Ring Inversion

The ring inversions of cyclic molecules were among the earliest dynamic processes to be studied by **DNMR**. We have already considered the ring inversion of cyclohexane-d$_{11}$. Four-membered carbocyclic rings undergo a butterfly inversion, cyclopentanes undergo pseudorotation, and both

---

*Orgell, K. G., *Coord. Chem. Rev.* **1989**, *96*, 1.

▶ FIGURE 11.6
25-MHz $^{13}C\{^1H\}$ NMR spectra of
cis-decalin in toluene-d_8 at
(a) −80°C and (b) +50°C.*

*Mann, P. E., *J. Magn. Reson.*,
1976, *21*, 18.

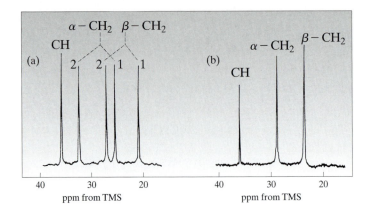

processes are too rapid to measure by NMR spectroscopy. Inversion of rings larger than cyclohexane is generally faster than that of cyclohexane. Inversion of cycloheptane is too fast to measure, but for cyclooctane-d_{14}, $\Delta G^{\ddagger} = 8.1$ kcal mol^{-1}, and for cyclononane, cyclodecane, and cyclododecane, $\Delta G^{\ddagger} \cong 6$ to 7 kcal mol^{-1}. The inclusion of double bonds, carbonyl groups, or bulky substituents reduces the rate of inversion. The inversion of cycloheptatriene (shown below) can be measured by observing the partial collapse of the methylene multiplet; $\Delta G^{\ddagger} = 6.1$ kcal mol^{-1}. The ring inversion of *cis*-decalin (shown below) cannot be studied by ^{1}H NMR spectroscopy, because the spectrum shows very little temperature dependence. However, the $^{13}C\{^1H\}$ NMR spectrum (Figure 11.6) is quite

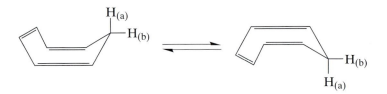

temperature dependent. At temperatures above 40°C, the cyclohexane rings in *cis*-decalin invert so rapidly that only two CH$_2$ resonances are observed. At −80°C, ring inversion is stopped, and now four CH$_2$ resonances are observed. The bridgehead carbons are chemical-shift equivalent for all

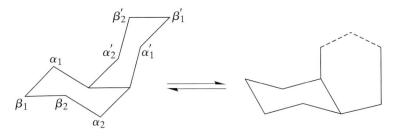

conformations, so their resonance is independent of temperature. Analysis of the rate constants as a function of temperature gave $\Delta G^{\ddagger}_{298} = 12.6$ kcal mol^{-1}, $\Delta H^{\ddagger} = 13.6$ kcal mol^{-1}, and $\Delta S^{\ddagger} = 3.5$ eu.

E. Valence Tautomerism

The valence tautomerism (bond-shift isomerism) of bullvalene has been studied by **DNMR**. Bullvalene has the structure of a trivinylcyclopropane in which there are two kinds each of aliphatic and olefinic protons. The ^{1}H NMR spectrum of bullvalene at 120°C consists of a single sharp resonance at δ 4.5 ppm. On cooling the sample to room temperature the line width of this resonance increases to several hundred Hz. Upon further cooling, the broad resonance splits, and at −59°C there are two multiplets with relative intensities of 2 to 3 at δ ≅ 2 and δ ≅ 5.6 ppm, respectively.* This temperature-dependent behavior of the ^{1}H NMR spectrum is explained on the basis of the following Cope rearrangement, which occurs so rapidly at elevated temperatures that all the protons "experience" an averaged environment and the spectrum consists of a single resonance:

At low temperatures, the Cope rearrangement is slow, and separate resonances are observed for both types of proton. This valence tautomerism can also be investigated by ^{13}C{^{1}H} NMR spectroscopy. At 141°C in the 25-MHz ^{13}C{^{1}H} NMR spectrum, a single resonance is observed at δ 86.4, whereas at −60°C, there are four resonances: two for the olefinic carbons at δ = 128.3 and δ = 128.5, one for the cyclopropane carbons at δ = 21.0, and one for C_4 at δ = 31.0 ppm. In the coalescence region (about 40 to 50°C), $\Delta\nu_{1/2} \cong 4000$ Hz!†

F. Keto-Enol Tautomerism

β-Diketones undergo keto-enol tautomerism as illustrated in the following figure:

keto   enol

This equilibrium is easily studied by ^{1}H NMR spectroscopy, as is exemplified by the ^{1}H NMR spectrum of 2,4-pentanedione (Figure 11.7). Since the interconversion between the tautomers is generally slow on the ^{1}H NMR time scale, separate resonances are observed for the keto and enol forms. The olefinic resonance of the enol form (δ = 5.5) and that of the CH_2 resonance

---

* See Schröder, G., Oth, J. F. M., Merenyi, R., *Angew. Chem. Int. Ed. Engl.*, **1965**, *4*, 752.
† Oth, J. F. M., Müllen, K., Gilies, J-M., Schröder, G., *Helv. Chim. Acta.*, **1974**, *57*, 1415.

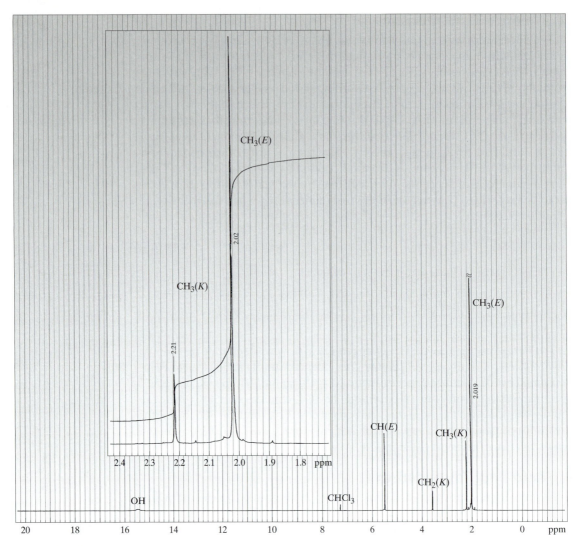

▲ FIGURE 11.7
500-MHz ^{1}H NMR spectrum of 2,4-pentanedione in CDCl$_3$ at about 25°C. The inset shows an expansion of the methyl region. (E = enol, K = keto form.)

in the keto form ($\delta = 3.5$) are readily assigned. Comparing their integrated intensities shows that the equilibrium mixture contains more molecules in the enol (80%) than in the keto (20%) form. From this, assignment of the two methyl resonances follows directly. The resonance of the strongly acidic hydrogen-bonded OH proton is found far downfield ($\delta = 15.5$). Increasing the temperature or adding a base causes the rate of the interconversion of the tautomers to increase, and only a time-averaged spectrum is then observed. For ethyl acetoacetate, in contrast to 2,4-pentanedione, the keto form predominates in the equilibrium mixture (approximately 90% keto to 10% enol). This is generally true for β-ketoesters, compared to β-diketones, where the keto form predominates in the former and the enol form predominates in the latter.

G.   Intermolecular Proton Exchange

The kinetics of the proton exchange of acids and bases have been widely studied by ^{1}H NMR spectroscopy. Under normal conditions, protons directly bonded to oxygen or nitrogen atoms undergo rapid exchange. Proton exchange may be studied by ^{1}H NMR spectroscopy in three ways:

a. By adding D_2O or other reagents containing labile deuterons, the exchange with substrate protons may be measured analytically by observing the disappearance of the substrate proton resonances. This method does not exploit the rate-dependent characteristics inherent in NMR spectroscopy, but is useful for relatively slow exchange processes (k_c about $10^{-2}\,s^{-1}$).

b. By observing the broadening and possible coalescence of substrate and solvent peaks, which requires little further elaboration. (See Section 11.2.)

c. By observing the collapse of spin multiplets. Consider, for example, the ^{1}H NMR spectra of ethanol (Figure 11.8). At room temperature, the lifetime τ of the OH proton is so long that coupling is observed between the OH and CH_2 protons. The OH resonance appears as a triplet. But the resonances are broad, showing the onset of an exchange process. A separate resonance is observed for the residual trace of H_2O in the sample.

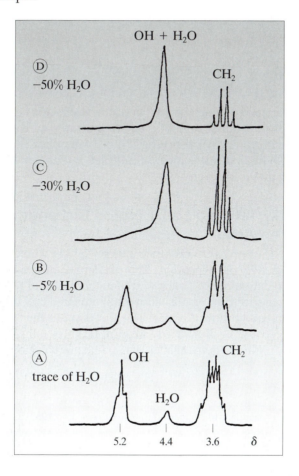

◀ FIGURE 11.8
60-MHz ^{1}H NMR spectra of ethanol (A) and three different ethanol–H_2O mixtures (B–D) at 25°C. Only the OH and CH_2 resonances are shown. The CH_3 triplet at $\delta = 1.2$ remains unchanged.

Adding H_2O increases the rate of exchange, as can be clearly seen in spectra B to D. For a water concentration of 50%, the protons on the alcohol and water oxygen atoms exchange so rapidly that one sees only an averaged resonance at $\delta \cong 4.6$ (Figure 11.8), and the coupling to the CH_2 protons is lost. Raising the temperature increases the exchange rate and has a similar effect on the resonances.

Acidifying the sample increases the exchange rate by many orders of magnitude by virtue of the following equilibrium:

$$CH_3CH_2OH + H_3O^+ \rightleftharpoons CH_3CH_2OH_2^+ + H_2O.$$

For this reason, the spectra of carboxylic acids and phenols in aqueous solution always contain only one average resonance for all the exchangeable protons. (See Section H.)

The proton exchange process in the dibenzylmethylammonium ion illustrated in the following equilibrium, has been used to measure the nitrogen inversion rate.[*] In a concentrated HCl solution at 25°C, coupling is observed between the NH and CH_2 protons (H_a and H_b), showing that exchange of the NH proton is slow under these conditions.

Upon reducing the HCl concentration, this coupling disappears. The NH protons must now be exchanging rapidly. The exchange process may be monitored by observing the resonance for the CH_2 protons. At slow exchange rates it is an *AB* multiplet, but at fast exchange rates a single resonance is observed.

H.   Intermolecular Proton Exchange in Carboxylic Acids

The chemical shifts of various nuclei in carboxylic or amino acids often are strongly pH dependent. This pH dependence of the chemical shift can be used to determine pK_a values. Under conditions of fast proton exchange, an average spectrum results from the protonated and deprotonated species that are in equilibrium. An acid–base titration can then be performed by NMR spectoscopy,[†] as illustrated in Figure 11.9.

For the general acid HA in aqueous solution, the equilibrium reaction

$$HA + H_2O \rightleftharpoons H_3O^+ + A^-$$

---

[*] Saunders, M., Yamada, F., *J. Amer. Chem. Soc.*, **1963**, *85*, 1982.

[†] Popov, A. I., in *Modern NMR Techniques and Their Application in Chemistry*, Popov, A. I. and Hallenga, K., eds., Marcel Dekker, New York, **1991**.

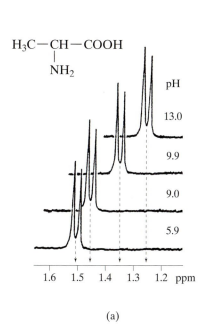

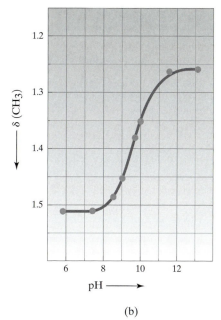

◀ FIGURE 11.9
pH dependence of the ^{1}H CH$_3$
resonance of alanine in aqueous
solution. (a) CH$_3$ doublet
resonance at selected pH values.*
(b) CH$_3$ chemical shift at various
pH values.

* Adapted from Reference 1.

(a)   (b)

obtains. Assuming that A^- does not contain exchangeable protons and that exchange is fast (as is usually the case for these equilibria), the observed ^{1}H chemical shift is given by

$$\delta_{obs} = \chi_{H_3O^+}\delta_{H_3O^+} + \chi_{HA}\delta_{HA} + \chi_{H_2O}\delta_{H_2O}. \qquad \textbf{(11.24)}$$

Note that the χ's do not denote the mole fractions of the respective species, but rather represent *the mole fractions of exchangeable protons that they contain.* Calculations can be considerably simplified if the water signal is taken as the internal reference. In this case $\delta H_2O = 0$, and Equation 11.24 simplifies to

$$\delta_{obs} = \chi_{H_3O^+}\delta_{H_3O^+} + \chi_{HX}\delta_{HX}. \qquad \textbf{(11.25)}$$

For a monoprotic acid,

$$K_a = \frac{[HA]}{[H_3O^+][A^-]}, \text{ or}$$

$$pK_a = pH - \log\frac{[HA]}{[A^-]}. \qquad \textbf{(11.26)}$$

When the acid is half neutralized, $[HA] = [A^-]$ and $pK_a = pH$. Hence, the inflection point in an NMR titration curve represents this halfway point, and the pK_a can be determined from it. For the example shown in Figure 11.9, pK_a is about 9.5, or $K_a \cong 3 \times 10^{-10}$.

For alanine, $pK_a = 9.87$, as measured by other techniques. Alanine is in fact a complicated example, as more than two forms of it are present in solution. The pH dependence of the chemical shifts of other reporter nuclei, such as ^{13}C and ^{19}F, can be similarly used to determine pK_a values for carboxylic acids, as illustrated in Sample Problem 11.5.

SAMPLE PROBLEM 11.5

Determine the pK_a value for fluoroacetic acid (FCH_2CO_2H) in aqueous solution from the pH dependence of its ^{19}F chemical shift shown in the following graph:

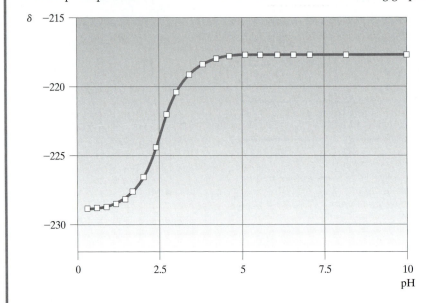

Solution

The inflection point in the curve comes at pH $= 2.50 = pK_a$.

Additional Reading

1. GÜNTHER, H., *NMR Spectroscopy: An Introduction*, Wiley, New York, 1980, 2d ed., 1995.

2. JACKMAN, L. M., and COTTON, F. A., eds., *Dynamic Nuclear Magnetic Resonance Spectroscopy*, Academic Press, New York, 1975.

3. EMSLEY, J. W., FEENEY, J., and SUTCLIFFE, L. H., *High Resolution Nuclear Magnetic Resonance Spectroscopy*, Pergamon, Oxford, Vols. I & II, 1965.

4. MACOMBER, R. S., *A Complete Introduction to Modern NMR Spectroscopy*, Wiley, New York, 1998.

5. POPLE, J. A., SCHNEIDER, W. G., and BERNSTEIN, H. J., *High Resolution Nuclear Magnetic Resonance*, McGraw-Hill, New York, 1959.

6. SANDSTROM, J., *Dynamic NMR Spectroscopy*, Academic, New York, 1982.

7. For a discussion of activation energies for unequally populated two-site exchange see: SHANAN-ATIDI, H., BAR-ELI, K. H., *J. Phys. Chem.* 1970, 74, 961.

CHAPTER 12

Lanthanide-Shift Reagents

12.1   Introduction

In Chapter 7, it was noted that the presence of paramagnetic impurities in a sample shortens nuclear relaxation times through dipolar relaxation leading to line broadening. In Chapter 8, it was noted that it is sometimes desirable (e.g., for quantitative ^{13}C NMR measurements), to intentionally add a paramagnetic reagent such as chromium acetylacetonate [$Cr(AcAc)_3$], in order to suppress the nuclear Overhauser effect. This reagent is sometimes called a *shiftless relaxation agent*, because, when it is present in moderate concentrations, it has very little effect on chemical shifts. In contrast, many paramagnetic reagents may behave as Lewis acids and coordinate with molecules in solution which contain donor atoms such as O and N or sometimes S and P that may function as Lewis bases. In these situations, the paramagnetic reagent produces large changes in chemical shifts of the nuclei in the coordinated molecule, due to the magnetic moment of the unpaired electron(s).[*] Such shifts often simplify spectra and make spectral analysis easier.

Paramagnetic nickel and cobalt complexes were among the first shift reagents to be used, but they produce severe line broadening. In 1969, Hinckley[†] discovered that the *tris* β-diketonate chelates of lanthanides, Ln (β-diketonate)$_3$—particularly those of europium, praseodymium, and ytterbium, gave rise to considerable shifts without significant line broadening. Typical β-diketones that have been used in lanthanide-shift reagents (**LSR**s) are listed in Table 12.1.

These reagents are soluble in CCl_4, $CDCl_3$, and other polar solvents and thus can be added directly to the solution being observed by NMR spectroscopy; the resulting chemical-shift changes are known as lanthanide-induced shifts (**LIS**s).

Table 12.1   Typical β-diketones that Have Been Used in Lanthanide-Shift Reagents; Ln (β-diketonate)$_3$[*]

			$RCOCH_2COR''$
R	*R'*	Abbreviation	Name
t-butyl	*t*-butyl	DPMH(thdH)	Dipivalomethane or 2,2,6,6-tetramethyl-3,5-heptanedione
t-butyl	C_3F_7	FODH	6,6,7,7,8,8,8-heptafluoro-2,2-dimethyl-3.5-octanedione
t-butyl	CF_3	PTAH	1-pivaloyl-3,3,3-trifluoroacetone or 2,2-dimethyl-6,6,6-trifluoro-3,5-hexanedione
C_3F_5	C_2F_5	FHDH	1,1,1,2,2,6,6,7,7,7-decafluoro-3,5-heptanedione
	t-butyl	PCAMH	3-pivaloyl-d-camphor
	C_3F_7	FBCAMH	3-heptafluorobutanoyl-d-camphor
	CF_3	TFCH	3-trifluoroacetyl-d-camphor

[*]*Nuclear Magnetic Resonance Shift Reagents*, Sievers, R. E., ed., Academic, New York, N.Y., **1973**.

---

[*]LaMar, G. N., Horrocks, W., Holm, R. H., *NMR of Paramagnetic Molecules: Principles and Applications*, Academic, New York, N.Y., **1973**.
[†]Hinckley, C. C., *J. Amer. Chem. Soc.* **1969**, *91*, 5160.

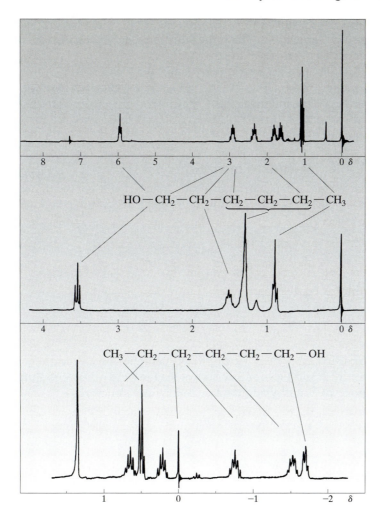

220-MHz ^{1}H NMR spectra of 1-hexanol (25 μL, 0.2×10^{-3} M) in 0.5 mL of CDCl$_3$ (middle) and after the addition of 14 mg (1.3×10^{-5} M) of Eu(FOD)$_3$ (top) and of 30 mg (2.9×10^{-5} M) of Pr(FOD)$_3$ (bottom). The OH resonance is not shown.

The europium and ytterbium complexes give rise mainly to downfield shifts, while prasedymium reagents give rise to upfield shifts. Figure 12.1 shows, as an example, the 220-MHz ^{1}H NMR spectrum of 1-hexanol, together with the spectra obtained after the addition of Eu(FOD)$_3$ and Pr(FOD)$_3$, respectively. Even at 220 MHz, the normal spectrum resolves only the α- and β-methylene and methyl resonances. After the addition of the shift reagents, the multiplets from all of the CH$_2$ groups become clearly resolved, giving rise to essentially first-order spectra. Note the presence of the *tert*-butyl proton resonances of the shift reagents that occur at about δ = 0.42 for Eu(FOD)$_3$ and δ = 1.40 ppm for Pr(FOD)$_3$.

12.2   Theory of Shift Reagent–Substrate Interaction

The coordinately unsaturated shift reagents complex reversibly to the substrate via the reaction

$$n[\text{substrate}] + [\text{shift reagent}] \underset{}{\overset{K_{eq}}{\rightleftharpoons}} [\text{adduct}]. \qquad \textbf{(12.1)}$$

Since the equilibrium is rapid on the NMR time scale, time-averaged spectra are observed in which the substrate chemical shifts are the population-weighted averages of the free substrate and the lanthanide-adduct chemical shifts. As increased amounts of shift reagent are added, the equilibrium of Equation 12.1 is shifted to the right, and for small amounts of added reagent, the substrate chemical shifts are directly proportional to the molar ratios of substrate to shift reagent. Two cases will be considered for the equilibrium reaction (Equation 12.1).

A. 1:1 Adduct Formation ($n = 1$)

Let the initial concentration of substrate be S, the concentration of added shift reagent be E, and the concentration of adduct formed be A. It is then possible to write

$$K = \frac{A}{[S - A][E - A]} \tag{12.2}$$

which is readily rearranged to

$$E = A + \frac{A}{K[S - A]} \tag{12.3}$$

Figure 12.2a shows the predicted behavior of 1:1 adducts obeying Equation 12.2, and Figure 12.2b shows the Eu(DPM)$_3$-induced shifts of the methyl group of 2-methyladamantan-2-ol, whose structure is

in CCl$_4$ solution.

The agreement between Figures 12.2a and b is very good apart from the initial "lag," which is clearest in the most dilute case and is believed to be due to residual water in the solvent competing with the alcohol for coordination to Eu(DPM)$_3$.

B. 2:1 Adduct Formation ($n = 2$)

In this case, the operative equation is

$$E = A + \frac{A}{K[S - 2A]^2} \tag{12.4}$$

Plots of Equation 12.4 are similar to Figure 12.2, except that the scale of the horizontal axis is expanded by a factor of 2 (i.e., maximum curvature occurs at a mole ratio $[E]/[S] = 0.5$) and the curvature is more marked than for the 1:1 adduct case.

For both the 1:1 and 2:1 adduct cases, one expects to see a limiting value in which the substrate shift is independent of the amount of lanthanide-shift reagent, and this can be observed for large shift reagent concentrations. It is conventional to define ΔM values as the (usually extrapolated) shifts obtained for the 1:1 mole ratio of substrate and shift

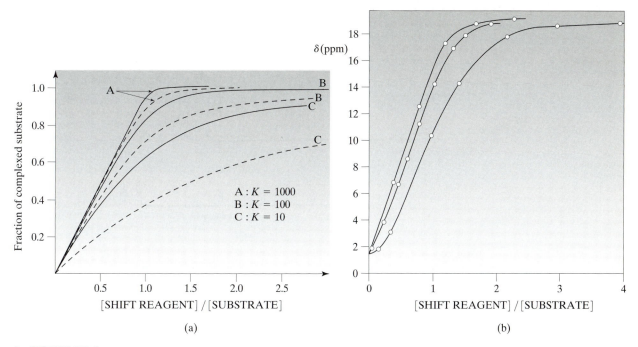

▲ FIGURE 12.2

(a) Behavior* of substrates forming 1 : 1 adducts, as predicted by Equation 12.3. The fraction of substrate complexed is $[A]/[S]$, and the molar ratio is $[E]/[S]$. Solid lines represent 0.5 M substrate, broken lines 0.1 M substrate. (b) Shifts induced in the methyl resonance of 2-methyladamantan-2-ol on addition of Eu(DPM)$_3$. Concentrations of 2-methyladamantan-2-ol are 0.024, 0.061, and 0.084 M in CCl$_4$ solution. The more concentrated solutions give the larger shifts.

*Sanders, J. K. M., Hanson, S. W., Williams, D. H., *J. Am. Chem. Soc.*, **1972**, *94*, 5325.

reagent. For the equilibrium of Equation 12.1 and a large equilibrium constant, the following equation results:

$$\Delta M = \delta(A) - \delta(S). \tag{12.5}$$

SAMPLE PROBLEM 12.1

Calculate ΔM for the methyl resonance of 2-methyladamantan-2-ol from the data presented in Figure 12.2b.

Solution

Extrapolation back from high values of the ratio $[E]/[S]$ gives $\delta(A) = 19$ ppm and $\delta(S) = 1.5$ ppm, whence $\Delta M = 19 - 1.5 = 17.5$ ppm.

Figure 12.3 shows, as another example, the 100-MHz ^{1}H NMR spectrum of *cis*-4-*tert*-butylcyclohexanol in CDCl$_3$ solution for various ratios of $[E]/[S]$. From this figure, we see that all the resonances are shifted increasingly to low field, with an increase in the $[E]/[S]$ ratio. It is often useful to plot, in a single diagram, the chemical shifts for all the resonances against the concentration

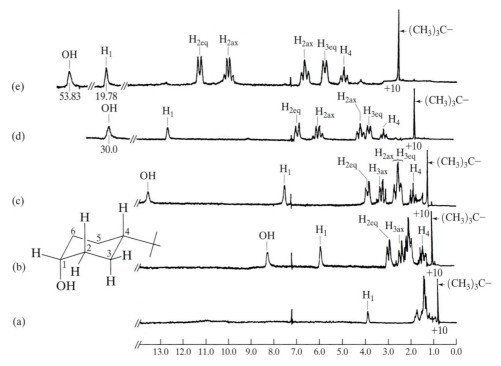

▲ **FIGURE 12.3**

100-MHz ^{1}H NMR spectra* of *cis*-4-*tert*-butylcyclohexanol (20 mg, 1.28 × 10^{-4} mol) in CDCl$_3$ (0.4 mL) containing various amounts of Eu(DPM)$_3$: (a) 0.0 mg; (b) 10.3 mg; (c) 16.0 mg; (d) 31.1 mg; (e) 60.2 mg.

*Demarco, P. V., Elzey, T. K., Lewis, R. B., Wenkert, E., *J. Am. Chem. Soc.*, **1970**, *92*, 5734.

(or amount) of the added shift reagent, as in Figure 12.4. Since the shifts are proportional to the amount of LSR added, the experimental points should lie on straight lines.

From such a plot, it is possible to determine whether any of the resonances cross over as the LSR concentration is increased.

Also, by extrapolating back to zero LSR concentration, one can obtain the chemical shifts and assignments if these were not already apparent from the spectrum in the absence of the LSR.

The ability of many classes of compound to coordinate with lanthanide-shift reagents gives the latter widespread application in NMR spectroscopy. The strength of the interaction, and therefore ΔM, depends on the ability of the substrate to act as a Lewis base and the lanthanide complex to act as a Lewis acid. Typical ΔEu values, [Eu(DPM)]$_3$ for methylene protons attached to various donor groups, are listed in Table 12.2. The data in the table illustrate the relationship between the magnitude of ΔEu and the donor ability of the functional group; ΔEu increases with increasing donor ability. Approximate

Table 12.2   LIS ΔEu, [Eu(DPM)$_3$] of R CH$_2X$ Protons

X	NH$_2$	OH	COR	CHO	OCH$_2$R	CO$_2$Me	CN	NO$_2$
ΔEu(ppm)	35	25	15	11	10	7	5	<1

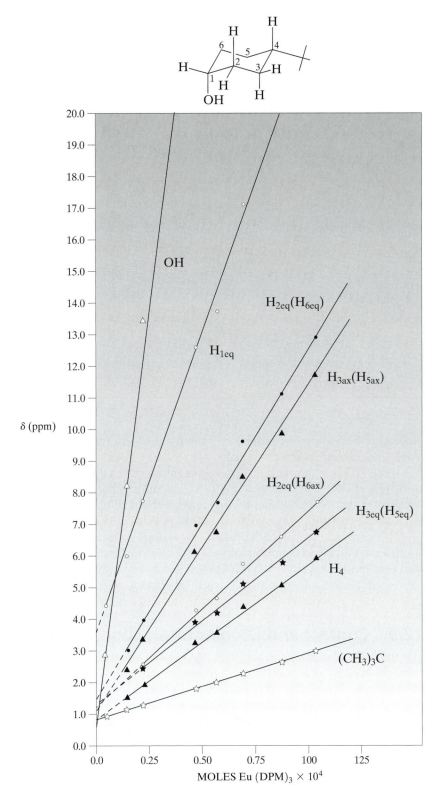

Variations in the chemical shifts for the different protons of *cis*-4-*tert*-butylcyclohexanol (1.28×10^{-4} mol) in 0.4 mL of $CDCl_3$ with increasing concentration of $Eu(DPM)_3$. Straight lines shown are derived by least squares.

Table 12.3 Some Selected ΔEu, [Eu(DPM)$_3$] Values (ppm)

Compound	Nucleus	1	2	3	4	5	6
$\overset{1}{H}O\overset{1}{C}H_2\overset{2}{C}H_2\overset{3}{C}H_2\overset{4}{C}H_3$	^{1}H 90.6(OH)	24.5	13.9	9.7	4.6		
$H_2N\overset{1}{C}H_2\overset{2}{C}H_2\overset{3}{C}H_2\overset{4}{C}H_3$	^{1}H	32.5	20.2	11.8	6.1		
	^{13}C	97.0	−18.6	10.6	9.3		
HN(CH₂—CH₂)(CH₂—CH₂)	^{1}H	37.7	16.6				
	^{13}C	143	−8.2				
4-tert-butylcyclohexanone	^{1}H eq		13.2	3.5			
	^{1}H ax		10.3	5.1	4.3	—	1.3
	^{13}C	32.2	8.6	8.1	4.8	2.7	1.7
2-adamantanol	^{1}H		17.2	5.2	4.2(eq) 6.0(ex)		
	^{13}C	53.7	22.6	9.6	8.4		
pyridine	^{1}H	31.0	10.7	9.7			
	^{13}C	90.0	−0.9	30.2			

binding affinities (donor abilities) are primary amine > hydroxyl > ketone > aldehyde > ether > ester > nitrile.

Table 12.3 lists additional ΔEu values for some selected compounds for both ^{1}H and ^{13}C nuclei. Because the **LIS**s are comparable for the two nuclei, but the ^{13}C chemical shift dispersion range is about 20 times that of the proton, the effect of shift reagents on ^{13}C NMR spectra is less dramatic than for proton spectra. Indeed, accidental chemical-shift equivalence in ^{13}C NMR spectra is relatively uncommon for medium-sized molecules. Thus, the use of shift reagents to simplify ^{13}C NMR spectra is not very common. However, the combined use of shift reagents, ^{13}C and ^{1}H spectra, and ^{13}C/^{1}H **HETCOR** spectroscopy is very powerful for simplifying and assigning both spectra.

12.3   Contact and Dipolar Interactions

The unpaired electron(s) on the shift reagent can affect the substrate chemical shifts in either of two distinct processes: *contact* and *dipolar (pseudocontact)* interactions. The contact term arises from the delocalization of the shift reagents' unpaired spin(s) onto the substrate atoms. The magnitude of this shift is, therefore, directly proportional to the value of the unpaired spin density at the observed nucleus and will necessarily be different for different nuclei. Thus, ^{13}C contact shifts will usually be much larger than ^{1}H contact shifts, as there is always a much greater spin density on the carbons than on corresponding protons. Although this electron delocalization can take place through single or double bonds, in saturated systems only nuclei one or two bonds

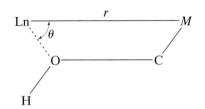

◄ FIGURE 12.5
Geometric parameters for dipolar
shifts in a lanthanide–alcohol
complex.

removed from the shift reagent are usually affected. In contrast, in conjugated
systems, extensive delocalization can occur, giving more widespread contact
shifts.

The dipolar term, which is often called the *pseudocontact term*, results simply from the direct magnetic field of the unpaired electrons on the shift reagent
at the substrate nuclei. This shift is given by the basic dipole equation

$$\Delta M = \frac{D}{r^3}(3\cos^2\theta - 1)\text{ ppm,}\qquad\qquad \textbf{(12.6)}$$

where D is the magnetic moment of the lanthanide, r is the distance of the nucleus from the lanthanide, and θ is the angle between the magnetic axis of the
shift reagent and r, as illustrated in Figure 12.5.

This simple form of the basic dipole equation, in which the shift reagent
is assumed to have effective axial symmetry along the Ln—O bond, gives a
quantitative picture of dipolar shifts. The magnetic field produced at the observed nucleus M by the lanthanide magnetic moment is, of course, independent of the nature of M. Thus, the same shift would be obtained when M
is either ^{1}H or ^{13}C. Obviously, in a real molecule, the carbons and hydrogens
of the substrate are not in the same place, but Equation 12.5 may be used to
calculate the dipolar shifts of all the nuclei in the substrate molecule.

In general, the dipolar term dominates, and as a consequence, the ΔM
values reflect the geometrical predictions. Particularly for acyclic compounds
(*cf.* Table 12.3), in which conformational averaging removes the angle-
dependent term, the shifts decrease with increasing distance from the donor
atom in rough agreement with the r^{-3} dependence. Also, if one excludes the
α- and β-carbons, the carbon and proton shifts are in reasonable agreement,
as predicted.

For cyclic compounds, the angular dependence is clearly seen, and it is
possible to observe negative ΔEu values (i.e., upfield shifts) and correspondingly positive ΔPr values (i.e., downfield shifts) for those nuclei in the molecule for which $3\cos^2\theta - 1$ is less than zero.

12.4   Determination of Molecular Geometry

An important utility of lanthanide-shift reagents lies not only in their ability to
simplify otherwise unresolvable spectra, but also in the possibility of taking advantage of the geometrically dependent term of Equation 12.6 to obtain the
geometry of the substrate molecule. However, for this property to be of use, it
is necessary to show that the actual shifts of molecules with known geometries are dipolar in origin and not contact shifts. The problem of separating the
two contributions is that sufficient experimental data are required to determine the unknown molecular parameters of the complex *before* the dipolar contribution may be calculated. These are, for the alcohol complex considered, the

▶ FIGURE 12.6

Calculated (in parentheses) and observed* ΔM values for ^{13}C and ^1H shifts of isoborneol in $CDCl_3$. Upper numbers for $Pr(FOD)_3$, lower numbers for $Eu(FOD)_3$.

*Hawkes, G. E., Leibfritz, D., Roberts, D. W., Roberts, J. D., *J. Am. Chem. Soc.*, **1973**, *95*, 1659.

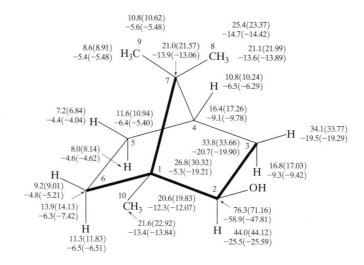

O—Ln bond length, the C—O—Ln bond angle, the C—C—O—Ln dihedral angle, and the effective magnetic moment D of the complex. Thus, four different ΔM values are required merely to define the unknowns in Equation 12.6, even when there is no contact contribution at all!

However daunting this problem may be, some systematic studies along these lines have been made, and two examples will be described. Figure 12.6 shows the observed ΔM values for all the ^1H and ^{13}C nuclei of isoborneol for both $Eu(FOD)_3$ and $Pr(FOD)_3$, together with the calculated dipolar shifts for a given geometry. (For $Eu(FOD)_3$, O—Eu = 2.51 Å, angle C—O—Eu = 129.5°, and angle C_3—C_2—O—Eu = 86.0°.) The praseodymium geometry is similar, except that the O—Pr distance is 2.70 Å.

As there are nine carbon and 11 proton values, the geometry is considerably overdetermined. There is essentially complete agreement between the calculated and observed shifts, except for carbons C_1 and C_2. In contrast, all the proton shifts give good agreement. This suggests that only C_1 and C_2 have contact contributions to their ΔM values.

In aromatic compounds more widespread contact shifts are to be expected, and thus the separation of the dipolar and contact contributions is more difficult. One method of resolving this problem is to study a given substrate with a number of different shift reagents. Although the contact and dipolar terms will vary for each shift reagent, the geometrical relationship of the substrate nuclei, and therefore their relative dipolar shifts, will be unchanged in the different cases, and this provides a method by which the contact and dipolar terms may be separated. Such an approach has been applied to quinoline, with structure

and the contact and dipolar contributions to the ΔM values for the carbon and hydrogen atoms in the molecule are given in Table 12.4. The dipolar shifts are consistent with an Eu—N distance of 3.87 Å and an Eu—N—C_2 angle of 117.3°.

Table 12.4   Contact and Dipolar Contributions to the Eu Shifts for Quinoline → Eu(DPM)$_3$

		2	3	4	5	6	7	8	8a	4a
Position										
^{13}C										
	contact shift	39.8	−17.6	1.4	−4.0	−1.7	4.4	15.9	0.8	−23.5
	dipolar shift	44.7	21.6	17.4	11.9	9.1	11.0	30.6	43.4	21.4
^{1}H										
	contact shift	−13.3	−6.4	−3.8	−3.0	0.0	1.6	−8.3		
	dipolar shift	39.6	13.6	11.0	8.3	5.3	3.9	34.8		

Here, in contrast to the isoborneol case, the contact contributions are appreciable for almost every atom in the molecule. Indeed, in some cases the two components are almost equal in magnitude, but opposite in sign, leading to a very small final ΔEu value. For example, C$_3$ has a ΔEu value of 4.0 ppm, whereas the two individual contributions are +21.6 and −17.6 ppm. Note that the protons exhibit considerable contact contributions. Thus, the interpretation of these data on the basis of dipolar shifts alone could lead to erroneous geometrical conclusions.

Because multiple equilibria are involved, the interpretation of lanthanide-induced shifts (**LIS**s) for di- and polyfunctional molecules is complex, and the results should be interpreted with great caution.

12.5   Chiral Lanthanide-Shift Reagents

Enantiomers are indistinguishable by NMR spectroscopy. Consequently, it is not possible by conventional NMR spectroscopy alone to determine the optical purity of a substance. However, if diasteromers are formed by reaction with an enantiomerically pure substance, then the diastereomers give rise to different NMR spectra. It is not necessary that strong bonds be formed in these diastereomers, but only that a complex be formed in solution, even if it exchanges rapidly on the NMR time scale. Thus, chiral lanthanide-shift reagents are of considerable utility in the determination of optical purity.

Assume that a substrate S is present as a racemic mixture of $S(+)$ and $S(-)$. Then, adding a pure enantiomer of a lanthanide-shift reagent, say, $E(+)$, will result in the formation of two diastereomeric adducts (A):

$$S(+) + E(+) \rightleftharpoons A(+, +); \qquad \textbf{(12.7)}$$

$$S(-) + E(+) \rightleftharpoons A(-, +).$$

Figure 12.7 gives an example. Here, the racemic substrate is 1-phenylethylamine, and the enantionmerically pure lanthanide-shift reagent is Eu(TFC)$_3$. In the absence of the chiral-shift reagent, only single CH and CH$_3$ resonances are present in the ^{1}H NMR spectrum of the racemate. Upon adding the pure D enantiomer of Eu(TFC)$_3$, both these resonances are split into two separate resonances. For the racemic mixture, the two CH quartets and the two overlapping CH$_3$ doublets each have equal integrated intensities, as they should

▶ FIGURE 12.7
250-MHz ^{1}H NMR spectra in CDCl$_3$ of (A) racemic 1-phenylethylamine, (B) racemic 1-phenylethylamine in the presence of D-Eu(TFC)$_3$, and (C) 4:1::L:D-1-phenylethylamine in the presence of D-Eu(TFC)$_3$.*

*Adapted from *Basic One- and Two-Dimensional NMR Spectroscopy*, Friebolin, H., VCH, Weinheim, **1993**.

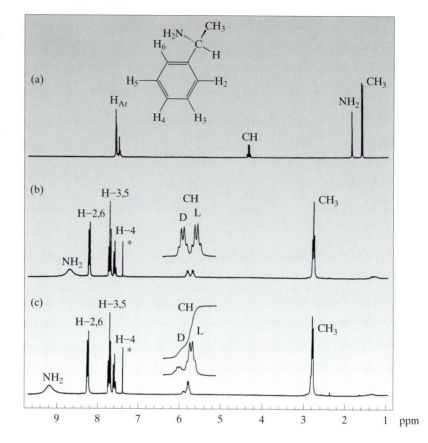

for a 1:1 mixture of diastereomers that forms from the racemic substrate. The optical purity of a partially resolved mixture can be assessed from the integrals, as illustrated in Figure 12.7c. It is not yet possible to predict which of the two possible diastereomers will exhibit the larger **LIS**. Hence, it is not yet possible to determine absolute configurations in this way.

The determination of the optical purity of a wide variety of substances by ^{1}H, ^{13}C, and ^{31}P NMR spectra of diastereomers has been discussed in several reviews.*

12.6 Practical Considerations

Because the phenomenon of lanthanide-induced shifts results from a fast-exchange equilibrium, the line widths of the resonances are given by

$$\Delta \nu \approx \frac{\pi (\Delta \delta)^2}{2k},$$

(12.8)

where $\Delta\delta$ is the chemical-shift difference (in Hz) for the nuclide in question and k is the rate of exchange. Equation (12.8) is called the *fast-exchange*

Asymmetric Synthesis, Morrison, J. D., ed., Academic Press, San Diego, CA, Vol. 1, **1983**.
Parker, D., *Chem. Rev.*, **1991**, *91*, 1441.
Offerman, W., Mannschreck, A., *Org. Magn. Reson.*, **1984**, *22*, 355.
Nelson, J. H., *Coord. Chem. Rev.*, **1995**, *139*, 245.

approximation. Since chemical-shift anisochrony ($\Delta\delta$) is directly proportional to the field strength $\mathbf{H_0}$, the resonance line width increases with the square of the field strength. This problem is especially severe when $\Delta\delta$ is large, as is often the case for lanthanide-induced shifts. Since broad resonances are generally undesirable, there is an advantage to using low-field spectrometers in studying lanthanide-shift reagents. Because such spectrometers (≤ 100 MHz) are becoming increasingly rare, it is useful to note that there is also a temperature dependence hidden in Equation 12.8. Since k is related to the activation energy for the chemical exchange process, the following proportionality obtains:

$$\Delta v \; \alpha \mathbf{H_0^2} \exp\left(\frac{\Delta G^{\ddagger}}{\mathrm{RT}}\right). \qquad \textbf{(12.9)}$$

Equation 12.9 shows that the line width is proportional not only to the square of the field strength, but also to the temperature. Although increasing the field strength of the spectrometer broadens the line width of the resonance, raising the temperature tends to counteract this effect. In practice, it is wise to conduct a LIS analysis on the lowest field spectrometer available; if line broadening is a problem, warming the sample may help to alleviate the problem.

The interaction of a ligand with a lanthanide complex may result in a change in chemical shift, $\Delta\delta$, for some of the nuclides of the ligand, especially those that are in close proximity to the coordinating atom. If the shift reagent and the ligand are chiral, there may be different lanthanide-induced shifts for corresponding nuclei in the two enantiomers of the ligand (nuclides that are enantiotopic by external comparison). This induced anisochrony (chemical-shift difference) is $\Delta\Delta\delta$. Since $\Delta\Delta\delta$ is a function of concentration, temperature, and ligand, a comparison of "resolving power" among different reagents is difficult. The three ligands found in the most common commercially available chiral-shift reagents are TFC, FBCAM (HFC), and dicampholylmethane (DCM). The TFE and FBCAM complexes are sold as europium, ytterbium, or praseodymium complexes, while only Eu(DCM)$_3$ is commercially available. For 1-phenylethanol and 1-phenylethylamine, Eu(DCM)$_3$ was found to give rise to the largest $\Delta\Delta\delta$'s, while there was little difference between Eu(TFC)$_3$ and Eu(FBCAM)$_3$. In choosing a shift reagent, it should be recalled that europium and ytterbium induce downfield shifts, while praseodymium induces upfield shifts. Additionally, the three metals may also cause line broadening to differing extents. For 1-phenylethanol and 1-phenylethylamine, Pr(FBCAM)$_3$ induces larger shifts than Eu(FBCAM)$_3$ and does so at a lower concentration. Still, the consensus appears to be that no single lanthanide-shift reagent is superior for all possible ligands.

Fraser* recommends the following experimental protocol:

1. Try as many as four chiral-shift reagents, the approximate order of capacity being Eu(DCM)$_3$ > Pr(FBCAM)$_3$ $\approx$ Yb(FBCM)$_3$ > Eu(FBCM)$_3$.

---

*Fraser, R. R. in *Asymmetric Synthesis*; Morrison, J. P., ed., Academic: Orlando, **1983**; Vol. 1, pp. 173–196.

2. Try changing the temperature. A lower temperature can have a substantial influence on lanthanide-induced shifts, while a higher temperature may sharpen resonances.

3. If you are still unsuccessful, try derivatizing the ligand to form a stronger, harder Lewis base.

Before conducting a lanthanide-shift reagent study, the experimentalist should consult Sullivan's review* for detailed experimental guidelines. Briefly, the guidelines suggest drying the substrate, the solvent, and the shift reagent (by sublimation if they are prepared fresh, or over P_2O_5 *in vacuo* if they are purchased); keeping the substrate concentration low (~0.1–0.25 M); adding the shift reagent (either as a solid or as a concentrated solution) in small increments; filtering the solution after each addition (the molar ratio needed for a good induced shift is rarely greater than 1:1, and too much lanthanide can broaden lines and even cause the induced shifts to decrease); reshimming the spectrometer after the shift reagent is added, in order to compensate for the presence of the paramagnetic ions; and checking for paramagnetic precipitates after the sample has been spinning for several minutes. In addition, recall that the method is usually more effective at low fields.

---

*Sullivan, G. R. in *Topics in Stereochemistry*, Eliel, E. L., Allinger, N. L., eds.; Wiley-Interscience: New York, **1978**; Vol. 10, pp 287–329.

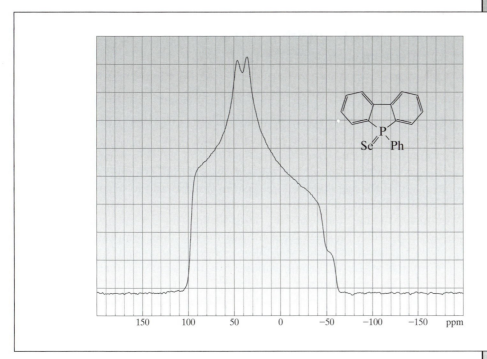

CHAPTER 13

^{31}P CP NMR spectrum of 1-phenyldibenzophosphole selenide.

NMR of Solids

13.1   Introduction

T he NMR spectra of solids are generally characterized by broad, feature-less resonances that are nearly devoid of structural information. The main cause of line broadening is *"dipolar broadening."* In recent years, significant advances have been made that allow solid-state spectra to be obtained with a resolution which approaches that obtained for solutions. This chapter will focus on the methods employed to obtain high-resolution solid-state spectra.

13.2   Proton Dipolar Broadening

A.   General Considerations

Consider a single CH spin system in a solid organic compound. The local magnetic field at the C—13 nucleus is given by

$$\mathbf{B}_{\text{local}} = \pm h\gamma_{\text{H}}(3\cos^2\Theta_{\text{CH}} - 1)/4\pi r_{\text{CH}}^3. \tag{13.1}$$

In this equation, γ_{H} is the proton gyromagnetic ratio, r_{CH} is the carbon–proton internuclear separation (approximately 0.95 Å for a directly bonded proton), and Θ_{CH} is the angle of the CH vector with respect to $\mathbf{B}_0$ (Figure 13.1). The $\pm$ sign results from the fact that the local field may either add to or subtract from the laboratory field $\mathbf{B}_0$, depending upon whether the proton dipole is aligned with or opposed to the direction of $\mathbf{B}_0$, states of affairs that are almost equally proba-ble. If r_{CH} and Θ_{CH} were fixed throughout the bulk sample, as for isolated $^{13}\text{C}-^1\text{H}$ pairs in a single crystal, this interaction would result in a splitting of the ^{13}C resonance into two equal components with a separation that would depend upon Θ. The effect can be large. If the $^{13}\text{C}-^1\text{H}$ bonded pairs make an angle $\Theta = 0°$, Equation 13.1 predicts that the splitting will be on the order of 40 kHz.

Most organic solids are amorphous or microcrystalline, such that in a bulk sample there is a random orientation resulting in a summation over many values of Θ_{CH} and r_{CH}. As a consequence, the proton dipolar broadening in a typical solid sample will be many kilohertz. Rapid reorientation of the $^{13}\text{C}-^1\text{H}$ internuclear vectors, especially if they are isotropic or nearly so, will result in a narrowing of the broadened lines if the rate of reorientation ex-ceeds the line width in Hz. If the reorienting $^{13}\text{C}-^1\text{H}$ vectors sample all Θ_{CH} angles at a rate that is faster than the reciprocal of the dipolar coupling (i.e., $\gg 5 \times 10^4\,\text{s}^{-1}$), dipolar broadening is reduced to a small value, as in liquids. Such rapid tumbling is not possible for solids, but since the $(3\cos^2\Theta_{\text{CH}} - 1)$ term in Equation 13.1 is zero when $\Theta = 54.7°$ (the so-called magic angle), it was long ago recognized that spinning solid samples at this angle with re-spect to $\mathbf{B}_0$ would be effective in reducing the line width.

It is not mechanically feasible to design rotors that can spin faster than about $10^4\,\text{Hz}$, well short of the 40 or 50 kHz required. At higher speeds, all suitable ma-terials disintegrate under the extreme gravitational forces generated by such rapid spinning. However, while magic-angle spinning (**MAS**) is not useful for removing proton dipolar broadening, it is very effective for collapsing the less-er broadening arising from chemical-shift anisotropy. (See Section 13.3.)

B.   High-Power Proton Decoupling

A more practical way of removing the proton dipolar broadening is to em-ploy a high-power proton-decoupling field (i.e., dipolar decoupling). This is analogous to heteronuclear broadband proton decoupling in a standard $^{13}\text{C}\{^1\text{H}\}$ experiment, except that, because the dipolar coupling has a much greater magnitude than scalar coupling, much greater power is required for

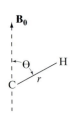

▲ FIGURE 13.1

Illustration of the variables in Equation 13.1. The C—H vector of length *r* makes an angle Θ with respect to the external magnetic-field direction $\mathbf{B}_0$.

dipolar decoupling. Rather than the approximately 1-gauss decoupling field used for scalar decoupling in solutions, a field of about 10 gauss is necessary for dipolar decoupling. Because of this high power, the decoupler is turned on for only a short period (10^{-3} s) called the *contact time*.

13.3   Chemical-Shift Anisotropy

In isotropic mobile liquids, because of rapid molecular tumbling, only the symmetry properties of individual nuclei within molecules are of importance in determining chemical shifts. All nuclei of the same element that are related by a symmetry operation have the same chemical shift. The chemical shift of a nucleus is related to the electron density at the observed nucleus by the relationship

$$\mathbf{B}_{local} = \mathbf{B}_0(1 - \sigma). \tag{13.2}$$

The screening constant is a directionally dependent second-rank tensor* given by

$$\sigma = \begin{bmatrix} \sigma_{11} & \sigma_{12} & \sigma_{13} \\ \sigma_{21} & \sigma_{22} & \sigma_{23} \\ \sigma_{31} & \sigma_{32} & \sigma_{33} \end{bmatrix} [\lambda_{ij}^2]. \tag{13.3}$$

By appropriate choice of coordinate system, Equation 13.3 may be diagonalized to

$$\sigma = \sigma_{11}\lambda_{11}^2 + \sigma_{22}\lambda_{22}^2 + \sigma_{33}\lambda_{33}^2, \tag{13.4}$$

where λ_{ii} are the direction cosines of the principal axes of the three principal screening constants (σ_{ii}) with respect to the external magnetic field (Figure 13.2).

Figure 13.3 shows a graphical representation of the experimental ^{13}C chemical-shift tensor in calcium formate. Note that, unlike the situation in Figure 13.3, in this case none of the principal components of the tensor lie along bond axes. (Some authors use δ_{ii} instead of σ_{ii} to describe the tensor components.) The use of δ_{ii} are to be preferred as they are the measurable quantities. In fact, shielding and chemical shift cannot be used interchangeably. Since the chemical shift δ is defined as

$$\delta \equiv \frac{(\nu - \nu_{ref})}{\nu_{ref}} \quad \text{then,} \quad \delta \equiv \frac{(\nu - \nu_{ref})}{\nu_{ref}} = \frac{(\sigma_{ref} - \sigma)}{(1 - \sigma_{ref})} \simeq (\sigma_{ref} - \sigma).$$

If the molecule is oriented so that the C—X bond is along the $\mathbf{B}_0$ direction, the observed chemical shift will correspond to the screening constant σ_{33}; similar statements apply to σ_{11} and σ_{22}. For any arbitrary orientation, the chemical shift

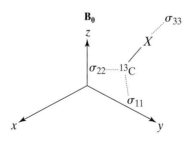

◀ FIGURE 13.2
The principal values σ_{11}, σ_{22}, and σ_{33} of the carbon-13 screening tensor, represented along three mutually perpendicular axes in a laboratory frame for a molecule containing a ^{13}C—X bond. One of these axes, corresponding to σ_{33}, is shown as coinciding with the C—X bond, although it is generally not necessary that any of the axes coincide with chemical bonds.

---

*A tensor is a three-dimensional vector.

▶ FIGURE 13.3
Graphical representation of the ^{13}C chemical-shift tensor* in calcium formate; views from above (left) and within the plane (right) of the formate ion.

*Ackerman, J. L., Tegenfeldt, J., Waugh, J. S., *J. Amer. Chem. Soc.,* **1974**, *96*, 6843.

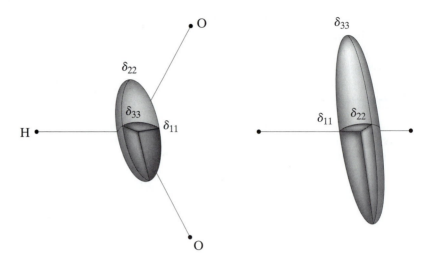

is prescribed by Equation 13.4. In solution, the screening constant is given by the isotropic average of that equation. Since this is $\frac{1}{3}$ for each λ_{ii}^2, we have

$$\sigma_{iso} = \frac{1}{3}(\sigma_{11} + \sigma_{22} + \sigma_{33}) \quad \text{and similarly} \quad \delta_{iso} = \frac{1}{3}(\delta_{11} + \delta_{22} + \delta_{33}). \quad \textbf{(13.5)}$$

The quantity in parenthesis is the *trace* of the tensor (Equation 13.3)—that is, the sum of its diagonal elements.

Most solids are polycrystalline, so that, in a bulk sample, the microcrystals and their constituents are oriented randomly with respect to $\mathbf{B_0}$. Under these circumstances, the screening constant—or chemical shift—takes on a continuum of values, forming the line shapes shown in Figure 13.4. The lighter shapes represent the theoretical line shapes, and the bolder curves represent the experimental line shapes in which a Lorentzian broadening function has been added. Figure 13.4a shows the situation for a site with cubic symmetry, where $\delta_{11} = \delta_{22} = \delta_{33}$. Figure 13.4b shows the situation for axial symmetry, where $\delta_{11} = \delta_{22} \neq \delta_{33}$, and Figure 13.3c shows the situation for a site with lower symmetry.

If $\delta_{22} = \delta_{33}$, as may happen, then Figure 13.4b will be reversed from left to right. Note that δ_{33} is customarily taken as corresponding to the largest shielding

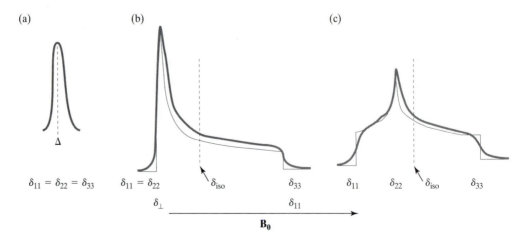

▲ FIGURE 13.4
Schematic powder line shapes for the shielding tensor with (a) cubic symmetry, (b) axial symmetry, and (c) lower symmetry. The dashed line represents $\delta_{iso} = \frac{1}{3}(\delta_{11} + \delta_{22} + \delta_{33})$.

value (the most upfield, or smallest, chemical shift), but the observation of such a pattern does not actually tell us the orientation of the principal axes of the tensor with respect to the molecular framework. In order to determine this, it is necessary to study the solid-state spectra of oriented single crystals whose structure is known. The principal values δ_{11}, δ_{22}, and δ_{33} are usually expressed in ppm and can be measured directly from the singularities or discontinuities in the spectrum, as indicated in Figure 13.4, and Sample Problem 13.1.

From the shielding tensor components, one normally defines the shielding anisotropy and the shielding asymmetry factor

$$\Delta\delta = \delta_{33} - \frac{1}{2}(\delta_{11} + \delta_{22})$$

$$\Delta\delta = \left(\frac{3}{2}\right)(\delta_{11} - \delta_{iso}) \quad \text{or} \quad \Delta\delta = \left(\frac{3}{2}\right)(\delta_{33} - \delta_{iso})$$

(13.6)

and

$$\eta = \frac{(\delta_{22} - \delta_{11})}{(\delta_{33} - \delta_{iso})} \quad \text{or} \quad \eta = \frac{(\delta_{22} - \delta_{33})}{(\delta_{11} - \delta_{iso})}$$

(13.7)

respectively. The choice depends upon which component (δ_{11} or δ_{33}) is further from δ_{iso}.

SAMPLE PROBLEM 13.1

Analyze the powder pattern shown to obtain δ_{11}, δ_{22}, δ_{33}, δ_{iso}, $\Delta\delta$, and η.

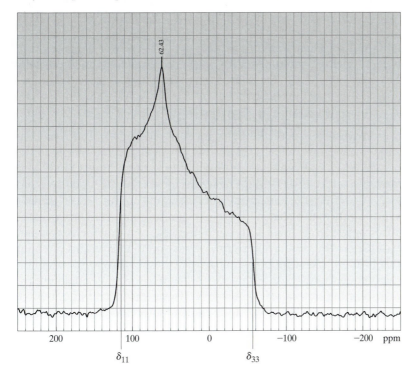

Solution

As indicated, $\delta_{11} = 115$ ppm, $\delta_{22} = 62.4$ ppm, and $\delta_{33} = -57$ ppm, from which it follows that $\delta_{iso} = \frac{1}{3}[115 + 62.4 - 57] = 40.1$ ppm. Also, $\Delta\delta = -57 - \frac{1}{2}(115 + 62.4) = -145.7$ ppm, and $\eta = (62.4 - 115)/(-57 - 40.1) = -52.6/-97.1 = 0.54$ ppm.

Table 13.1   Typical ^{13}C Chemical-Shift Tensor Values for Various Carbon Functionalities

Group	δ_{11}	δ_{22}	δ_{33}	δ_{iso}	$\Delta\delta$	η
$-CH_3$	38	29	3	23.3	-30.5	0.44
$-CH_2-$	54	41	16	37	-31.5	0.62
$-\overset{\vert}{\underset{\vert}{CH}}$	40	32	28	33.3	-8.0	1.51
$-CH_2OH$	80	70	25	58.3	-50	0.30
$-\overset{\vert}{CHOH}$	90	83	42	71.7	-44.5	0.24
$-OCH_3$	83	71	12	55.3	-65.0	0.28
$\diagdown C=C \diagup$	223	127	34	128	-141	1.02
$-C(O)H$	276	229	90	198.3	-162.5	0.43
$RC(O)R$	279	245	85	203	-177	0.29
$-C(O)OH$	247	171	106	174.7	-103	1.11
$-C(O)O^-$	239	188	105	177.3	-108.5	0.71
$-C(O)NH_2$	243	179	94	172	-117	0.82
$-C\equiv C-$	151	151	-79	74.3	-230	0
$-C\equiv N$	227	208	-89	115.3	-306.5	0.09

Table 13.1 contains some typical chemical-shift tensor data for a number of different types of commonly occurring carbon functionalities, presented in order to give the reader some idea of the magnitudes of these parameters.

13.4   Magic-Angles Spinning

The carbon anisotropies (Table 13.1) are of much interest and repay careful experimental and theoretical study. However, in all but the simplest of molecules, there will be several overlapping patterns producing a broad and un-interpretable spectrum. Under these circumstances, it usually becomes necessary to sacrifice the anisotropy information in order to retain the isotropic chemical shifts with a reasonable degree of resolution. The carbon-13 shielding anisotropies $\Delta\sigma$ in Table 13.1—or better, $|\delta_{33} - \delta_{11}|$, which expresses the width of the powder pattern in ppm, correspond to line widths of about 800 to 5000 Hz at an observation frequency of 100 MHz (proportionately higher at higher frequencies). These frequencies are within the range attainable by mechanical spinning with air-driven rotors. Under such rotation, the direction cosines in Equation 13.4 become time dependent in the rotor period. Taking the time average under rapid rotation, we find that Equation 13.4 becomes

$$\sigma = \frac{1}{2}\sin^2\beta(\sigma_{11} + \sigma_{22} + \sigma_{33}) + \frac{1}{2}(3\cos^2\beta - 1) \qquad \textbf{(13.8)}$$

$$\times \text{ (function of direction cosines)}.$$

The angle β is now the angle between the sample rotation axis and the magnetic field direction $\mathbf{B_0}$. When β is the magic angle, $54.7°$, $\sin^2\beta = \frac{2}{3}$ and

$\cos^2 \beta = \frac{1}{3}$, so that equation 13.8 reduces to the isotropic value $\sigma_{iso} = \frac{1}{3}(\sigma_{11} + \sigma_{22} + \sigma_{33})$. Thus, under magic-angle rotation, the chemical-shift pattern collapses to the isotropic average.

13.5   Cross Polarization

A single 90° observation pulse applied at the resonant frequency produces a free-induction decay signal. This situation was described previously for liquid samples, and the same holds true for solid samples as well. In solids, however, molecular motion is limited, leading to very inefficient spin–lattice relaxation and long T_1 values. Cross polarization (CP) takes advantage of the fact that proton spin diffusion generally causes all of the protons in a solid to have the same T_1 value and the fact that the proton T_1 (abundant spin) is usually short compared to the carbon T_1 (or that of any other dilute spin). Cross polarization works by effectively forcing an overlap of the proton and carbon (or other dilute spin) energies in the rotating frame. The means of doing this was demonstrated in 1962 by Hartmann and Hahn.* Energy transfer between nuclei with widely differing Larmor frequencies can be made to occur when

$$\gamma_X \mathbf{B}_{1X} = \gamma_{1H} \mathbf{B}_{1H}, \tag{13.9}$$

where γ_X and $\mathbf{B}_{1X}$ are, respectively, the gyromagnetic ratio and applied field strength of the dilute spin (such as ^{13}C), and γ_{1H}, and $\mathbf{B}_{1H}$ are those of the proton. Equation 13.9 expresses the *Hartmann–Hahn condition*. Since γ_{1H} is four times γ_C, the Hartmann–Hahn match occurs when the strength of the applied carbon field ($\mathbf{B}_{1C}$) is four times the strength of the applied proton field ($\mathbf{B}_{1H}$). When the proton and carbon rotating-frame energy levels match, polarization is transferred from the abundant protons to the dilute ^{13}C nuclei. Because polarization is being transferred from the protons to the carbons (or other dilute spins), it is the shorter proton T_1 that dictates the repetition rate for data acquisition in a Fourier-transform experiment.

Figures 13.5 and 13.6 show, respectively, the pulse sequence for cross polarization and the spin-vector diagrams for the ^1H and ^{13}C rotating frames. First, a 90°_X pulse is applied to the protons, tipping the proton spins onto the y'-axis. The phase of the proton $\mathbf{B}_1$ field is then shifted by 90°. In this way, the protons are *spin locked* along the y'-axis, and for the duration

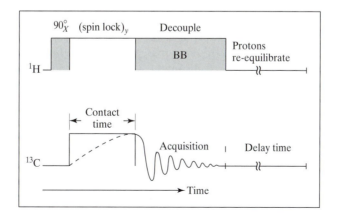

◀ FIGURE 13.5
The ^1H ^{13}C cross-polarization pulse sequence for solid-state observation.

---

*Hartmann, S. R., Hahn, E. L., *Phys. Rev.*, **1962**, *128*, 2042.

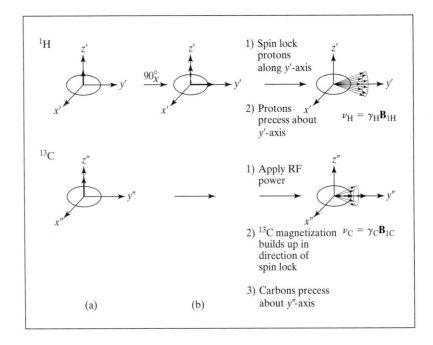

of the spin-locking pulse they are forced to precess about the y'-axis of their rotating frame with a frequency $\nu_H = \gamma_{1H}\mathbf{B}_{1H}$. Meanwhile, the carbons are put into contact with the protons.

Proton-carbon contact is accomplished by turning the carbon field $\mathbf{B}_{1C}$ on during the spin-lock time, causing the carbon magnetization to build up in the direction of the spin-lock field. The carbon magnetization now precesses about its y'-axis with a frequency $\nu_C = \gamma_C\mathbf{B}_{1C}$. The proton magnetization is now precessing about the $\mathbf{B}_{1H}$ field with a frequency $\nu_H = \gamma_{1H}\mathbf{B}_{1H}$, and since the Hartmann–Hahn match has been established, $\nu_C = \nu_H$. Therefore, the z components of both the proton and carbon magnetizations have the same time dependence. Because of this common time dependence, mutual spin flip-flops can occur between protons and carbons. The process can be visualized as a flow of polarization, or *coolth*, from the abundant proton spins to the rare carbon (or other dilute) spins. It can be shown that under these circumstances there is an enhancement of the rare-spin signal intensity by as much as the ratio of the gyromagnetic ratios of the abundant and rare spins. In the ^{1}H^{13}C cross-polarization case, this factor is $\gamma_H/\gamma_C = 4$, which is greater than the maximum Overhauser enhancement. This is in addition to the great time advantage of not having to deal with long ^{13}C spin–lattice relaxation times in spectrum accumulation.

To achieve minimum line width in magic-angle spectra, it is necessary that the spinning axis be set to within 0.5% of the magic angle in order to reduce the anisotropy pattern to within 1% of its static value. At high magnetic-field strengths, the anisotropy pattern may be as much as 10 kHz wide. Since a spinning rate of 10 kHz is difficult to achieve, one must accept the presence of spinning sidebands located on each side of the isotropic chemical-shift position by a distance equal to the spinning frequency (Figure 13.7). These spinning sidebands are not necessarily a nuisance and may in fact be "used" at relatively low spinning rates to trace out complex anisotropy patterns that

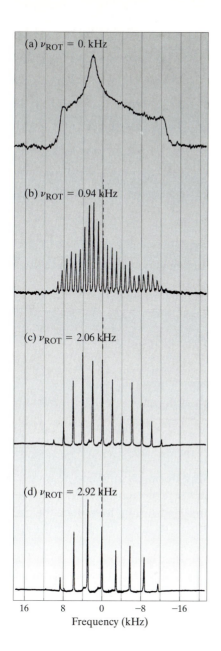

◄ FIGURE 13.7
CP/MAS $^{31}P\{^1H\}$ NMR spectra* of barium diethyl phosphate as a function of spinning frequency (ν_{ROT}).
*Herzfield, J., Berger, A. E., *J. Chem. Phys.*, **1980**, *73*, 6021.

are difficult to disentangle in the static spectra. Note that while the positions of the spinning sidebands are dependent on the spinning rate, the position of the isotropic chemical shift (indicated by dashed lines in the figure) is not. This fact is often used to identify the isotropic chemical shift.

13.6   Recapitulation

The total cross-polarization magic-angle-spinning (**CP/MAS**) experiment involves the use of the following techniques:

1. **Cross-polarization techniques,** to increase the signal-to-noise ratio and bypass the long T_1 relaxation process of the dilute nuclei being observed. The technique has no effect on the line width.

▶ FIGURE 13.8
60.75-MHz CP ^{31}P NMR spectra*
of [(PhDBP)CuCl]$_4$ as a function of
observation conditions. (a) Static
(no **MAS**), ^{1}H-decoupled; note that
the static ^{1}H-coupled spectrum is a
featureless, almost flat line;
(b) **MAS**, ^{1}H-coupled, (c) **MAS**,
^{1}H-decoupled.

*Attar, S., Bowmaker, G. A., Alcock,
N. W., Frye, J. S., Bearden, W. H.,
Nelson, J. H., *Inorg. Chem.*, **1991**,
30, 4743.

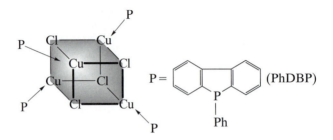

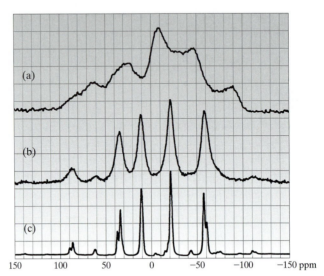

2. **High-power proton decoupling** during acquisition, to remove the dipolar interactions involving the protons.

3. **Continuous magic-angle sample spinning**, to remove the chemical-shift anisotropy.

The effects of these techniques are illustrated in Figure 13.8. The spectra in this figure are characterized by a single phosphorus chemical shift that is both scalar and dipolar coupled to ^{63}Cu and ^{65}Cu, both of which have $I = \frac{3}{2}$, giving rise to two overlapping asymmetric 1:1:1:1 four-line multiplets.

13.7 Crystallographic Equivalence, Chemical-Shift Equivalence, and Magnetic Equivalence in CP/MAS NMR Spectra

Two nuclei of the same element are crystallographically equivalent in a crystalline solid if they are related by a symmetry operation of the space group. As a result, crystallographically equivalent nuclei have identical isotropic chemical shifts. In addition, the principal components of their respective chemical-shift tensors are of identical magnitude. However, the orientations of their chemical-shift tensors are not generally coincident. In order to be magnetically equivalent, the nuclei must be crystallographically equivalent, *and the orientations of the principal axes of both tensors must coincide.* Generally, this requirement is fulfilled only if the two nuclei are related in the solid state by a center of inversion or a translation.

An example* is provided by the **CP/MAS** ^{31}P NMR spectra of $Na_4P_2O_7 \cdot 10H_2O$, where the two ^{31}P nuclei in the pyrophosphate anion are crystallographically equivalent, but magnetically distinct. The line shapes are dependent on the spinning rate. The single resonance observed at either slow or rapid spinning rates splits into two resonances at intermediate spinning rates. Another interesting solid-state effect occurs when the two crystallographically equivalent, but magnetically different, spins of a spin pair are also J coupled. Under such circumstances, the homonuclear J interaction can be reintroduced in the **CP/MAS** NMR spectra, even though the two nuclei have identical isotropic chemical shifts. That is, for a pair of crystallographically equivalent spins, up to four lines may be observed in **CP/MAS** NMR spectra. Moreover, the four-line spectra of such spin systems may be unusual compared to well-known AB spectra observed in solution. In particular, the J-recoupled ^{31}P **CP/MAS** NMR spectra of some bis-phosphine metal complexes exhibit spectra in which $^2J(^{31}P, ^{31}P)$ is given by the separations between alternate lines in the four-line multiplet, as opposed to the outer splittings observed in solution NMR spectra.†

As an example of the spinning-rate dependence, the ^{31}P **CP/MAS** NMR spectra of $[Ir(PPh_2Me)_2(COD)]PF_6$ are shown for two field strengths in Figure 13.9.

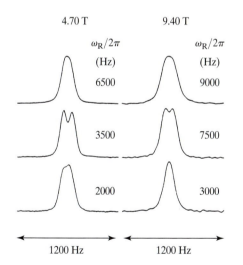

4.70 T   9.40 T

$\omega_R/2\pi$ $\omega_R/2\pi$

(Hz)   (Hz)

6500   9000

3500   7500

2000   3000

1200 Hz   1200 Hz

◄ FIGURE 13.9

Expansions of the isotropic region of the ^{31}P **CP/MAS** NMR spectra‡ of [Ir(PPh$_2$Me)$_2$(cyclooctadiene)]PF$_6$ as a function of spinning rate ($\omega_R/2\pi$) at two field strengths (4.70 and 9.40 T).

‡Wu, G., Wasylishen, R. E., *Inorg. Chem.*, **1994**, *33*, 2774.

---

*Hayashi, S., Hayamizu, K. *Chem. Phys. Lett.*, **1989**, *161*, 158;
Kabo, A., McDowell, C. A. *J. Chem. Phys.*, **1990**, *92*, 7156.
†Wu, G., Eichele, K., Wasylishen, R. E., in *Phosphorus-31 NMR Spectral Properties in Compound Characterization and Structural Analysis*, Quin, L. D., Verkade, J. G., eds., VCH, New York, **1994**, pp. 441–450.

13.8   Molecular Motion in the Solid State

The occurrence of molecular motion in the solid state will give a partial averaging of the chemical-shift anisotropy pattern. For simple, well-defined motions in molecules for which the orientation of the principal chemical-shift tensor elements is known or may be inferred, the change in the anisotropy pattern caused by the motion may be used to deduce the nature of the motional process. For example, the cross-polarization ^{13}C NMR anisotropy pattern of the ring carbons of polycrystalline hexamethylbenzene at $-186°$C is anisotropic. At ambient temperature, an axially symmetric pattern is observed. In the change, δ_{11} and δ_{22} average to give $\delta_\perp$, but δ_{33} is unchanged and is called $\delta\|$. From the known orientation of the principal shielding elements derived from single-crystal studies, it was concluded that the motion involves in-plane rotation of the aromatic ring.

More detailed information allowing the deduction of the mechanism of the reorientation process in favorable cases may be obtained from a consideration of the detailed line-shape changes occurring during the motional averaging. An example is provided by the variable-temperature ^{31}P powder spectra of white phosphorus (P$_4$) (Figure 13.10). Thus, Spiess et al. were able to deduce that reorientation of white phosphorus must be described by

▶ FIGURE 13.10
(a) Calculated and
(b) experimental powder line shapes* for a random-jumping tetrahedron with jump rates $k = \tau^{-1}$. The ^{31}P spectra were obtained at 92 MHz on the β-phase of white phosphorus at the indicated temperatures. The values of the jump rates for the experimental spectra were obtained from T_1 data.

*Spiess, H. W., *Chem. Phys.*, **1974**, *6*, 217; Spiess, H. W., Groseau, R., Haeberlen, U., *Chem. Phys.*, **1974**, *6*, 226.

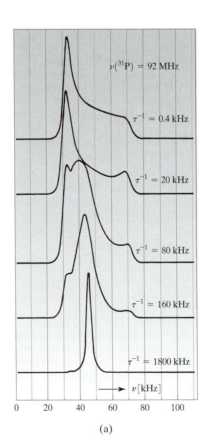

(a)

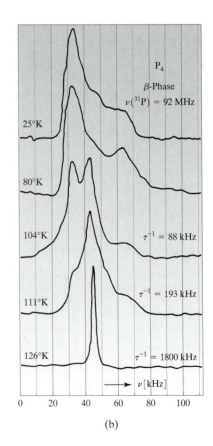

(b)

a "solid" model in which the molecule jumps between fixed positions. The spectrum at low temperatures is axially symmetric, as expected, but at high temperatures it is a single (averaged) isotropic line due to the motion.

The agreement between the observed and calculated line shapes during the transition is strong support for the proposed mechanism, although isotropic (free) rotation would give the same average spectrum.

Additional Reading

1. ABRAHAM, A., *The Principles of Nuclear Magnetism*, Oxford University Press, London, 1970.

2. AXELSON, D. E., *Solid State Nuclear Magnetic Resonance of Fossil Fuels: An Experimental Approach*, Multiscience Publications, Toronto, Canada, 1985.

3. CHANDRAKUMAR, N., and SUBRAMANIAN, S., *Modern Techniques in High-Resolution FT-NMR*, Springer-Verlag, New York, 1987, Chapter 6.

4. DUNCAN, T. M., *A Compilation of Chemical Shift Anisotropies*, Farragut Press, Chicago, 1986.

5. ERNST, R. R., BODENHAUSEN, G., and WOKAUN, A., *Principles of Nuclear Magnetic Resonance in One and Two Dimensions*, Oxford University Press, New York, 1987.

6. FUKUSHIMA, E., and ROEDER, S. B., *Experimental Pulse NMR: A Nuts and Bolts Approach*, Addison-Wesley, Reading, PA, 1981.

7. FYFE, C. A., *Solid State NMR for Chemists*, C.F.C. Press, Guelph, Ontario, Canada, 1983.

8. GERSTEIN, B. C., *Anal. Chem.*, **1983**, *55*, 781A, 899A.

9. GERSTEIN, B. C., and DYBROSKI, C. R., *Transient Techniques in NMR of Solids, an Introduction to Theory and Practice*, Academic, New York, 1985.

10. GRIFFIN, R. G., *Anal. Chem.*, **1977**, *49*, 951A.

11. HAEBERLEN, U., *High Resolution NMR in Solids, Advances in Magnetic Resonance, Supplement 1*, Academic, New York, 1976.

12. HARRIS, R. K., *Nuclear Magnetic Resonance Spectroscopy*, Pitman, London, 1983, Chapter 6.

13. HARRIS, R. K., and PACKER, K. J., *Eur. Spectrosc. News*, **1978**, *21*, 37.

14. LYERLA, J. R., YANNONI, C. S., Fyfe, C. A., *Acc. Chem. Res.*, **1982**, *15*, 208.

15. MEHRING, M., *High Resolution NMR in Solids*, 2d ed., Springer-Verlag, Berlin, **1983**.

16. YANNONI, C. S., *Acc. Chem. Res.*, 1982, *15*, 201.

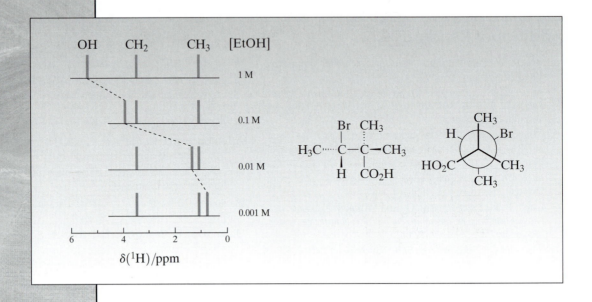

Problems

This chapter is a collection of problems that were written to illustrate the principles discussed in the text. They appear in a sequence that corresponds roughly with the sequence of topics presented. Where data for the problems were taken from the literature, a citation is given to the original literature. An associated instructor's guide contains detailed answers to these problems.

1. How many orientations with respect to an external magnetic field, and consequently, how many different energy states, are possible for a nucleus that possesses $I = \frac{9}{2}$?

2. How many ways can one assign n identical nuclei $+\frac{1}{2}(\alpha)$ spin?

3. The formula for the relative receptivities of two nuclei is

$$R_x^y = \frac{a_x \gamma_x^3 I_x(I_x + 1)}{a_y \gamma_y^3 I_y(I_y + 1)},$$

   where a_x and a_y are the natural abundances of nuclei X and Y and the other symbols have their usual meaning. Calculate the receptivities of ^{31}P, ^{119}Sn, ^{183}W, and ^{195}Pt relative to ^1H and ^{13}C. Compare your results with the appropriate D_H and D_C values given in Table 1.1. $R_{195Pt}^{13C} = 19.5$ means that a platinum signal should be 19.5 times more intense than a ^{13}C signal for the same molarity. What time ratios would be necessary in order to achieve the same signal-to-noise ratios in ^{31}P, ^{119}Sn, ^{183}W, and ^{195}Pt spectra compared with ^1H spectra of substances containing the same number of both nuclei, assuming that duty cycles for observation of all these nuclei are the same length?

4. Calculate the resonance frequency for a ^{31}P nucleus in a field of 7.02 tesla.

5. What are the fundamental differences between the two ways in which continuous-wave NMR spectra may be obtained?

6. What are the ordinate and abscissa on an NMR spectrum? Explain.

7. (a) At 25°C, what fraction of ^1H nuclei in a 2.35-tesla field are in the upper and lower spin states? (b) Under the same conditions, what fraction of ^{13}C nuclei are in the upper and lower spin states?

8. What is the *Larmor frequency*? Why is it important to an NMR experiment?

9. Is it generally more desirable to obtain a spectrum of a concentrated sample on a low-field spectrometer or a dilute sample on a high-field spectrometer? Are there situations where one or both would be desirable?

10. Describe how the magnitude of $\mathbf{B_0}$ affects the chemical-shift separation of two resonances. Does it also affect the magnitude of coupling constants?

11. What properties should an *ideal* NMR probe possess?

12. Deuterium (^2H) is the most common lock signal, but ^1H, ^7Li, and ^{19}F have also been employed for this purpose. Is it possible to use a nucleus as lock signal while observing a spectrum of the same nuclei?

13. Based upon the right-hand thumb rule, and assuming that only an equilibrium magnetization directed upon the z'-axis exists, draw the orientations of the magnetization vectors after the application of

(a) a 90°_X pulse, (b) a 180°_X pulse, (c) a 90°_{-Y} pulse, and (d) a 90°_Y pulse followed by a 90°_X pulse.

14. In a decaying signal (FID), why does the amplitude decay asymmetrically toward zero while the precessional frequency remains constant?

15. Poor digital resolution will result in the loss of some of the fine structure and truncated line shapes of an NMR signal. To increase the digital resolution, one needs to either maintain the same number of data points, but reduce the spectral width, or, alternatively, to maintain the spectral width, but increase the number of data points. Which is preferable for achieving a better signal-to-noise ratio? Why?

16. If the spectral width is insufficient to cover the entire spectrum, then one or more of the resonances will appear as a *fold-over*, or *aliased*, signal. In general, how could such a signal be identified? Could you use the same criteria if only one signal is present in the spectrum?

17. Define *sensitivity* and *resolution* in NMR spectroscopy.

18. Summarize the common methods for enhancing sensitivity in NMR spectroscopy.

19. A typical narrow-bore NMR tube has an outside diameter of 5.00 mm and an inside diameter of 4.20 mm. It is usually filled to a height of about 2.5 cm. (a) What volume of solution is necessary to fill the tube to this height? (Assume that the bottom of the tube is flat.) (b) What is the wall thickness of the glass tube? (c) If the sample cavity (inside the receiver coils) is 0.050 cm^3, what are the dimensions (radius, height, and volume) of the volume that is actually occupied by the sample in the receiver coils?

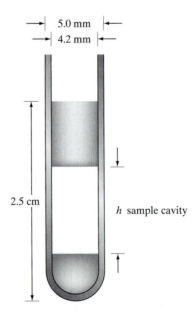

20. A single NMR scan of a certain highly dilute sample exhibits an S/N ratio of 1.9. If each scan requires 5 sec, what is the minimum time required to obtain a spectrum with an S/N ratio of 19?

21. Differentiate between *tuning* and *shimming* an NMR spectrometer. How does probe tuning affect the quality of an NMR spectrum?

22. The proton NMR signal of CH_3NO_2 occurs at 259.8 Hz to high frequency of that of TMS for a spectrometer operating at 60 MHz. Calculate δ for CH_3NO_2 at 60 MHz and the separation, in Hz, of the CH_3NO_2 and TMS resonances at 100, 300, and 500 MHz.

23. What is the difference between the proton resonance frequencies of benzene and acetone (7% in $CDCl_3$) at a constant magnetic field of $B_0 = 2.35$ tesla? What is the difference in the shielding constant between the two molecules? Hence, what is the difference in the magnetic fields actually experienced by the protons of these two molecules?

24. ^1H NMR spectra of toluene at 60 MHz and 80 MHz each show two resonances. At 60 MHz, there are resonances 140 Hz and 430 Hz downfield of TMS. At 80 MHz, these resonances occur at 187 Hz and 573 Hz downfield of TMS. Assign the two resonances. Show that the chemical shifts (δ) are field independent.

25. How does magnetization transfer from ^1H to a spin-coupled ^{13}C nucleus affect the signal intensity of the ^{13}C resonance?

26. Using Shoolery's additivity rules, predict the ^1H chemical shifts of the CH_2 groups in the following compounds:

   (a) CH_2Cl_2 (b) $BrCH_2C(O)Ph$ (c) $PhCH_2C(O)CH_3$

   Similarly, predict the ^1H chemical shifts for the methine and methylene resonances of $HC(OCH_2CH_3)_3$.

27. Using Shoolery's rules, predict the chemical shift positions for those protons attached to carbon atoms in (a) N-methylacetamide ($CH_3C(O)NHCH_3$), (b) 2-butanone ($CH_3C(O)CH_2CH_3$), (c) ethylpropanote ($CH_3CH_2C(O)OCH_2CH_3$), (d) vinylacetate ($H_2C=CHOC(O)CH_3$), and (e) n-propylacetate ($CH_3C(O)OCH_2CH_2CH_3$). Find the spectra of these compounds in the Aldrich library, and compare the experimental results with your predictions.

28. Using Beauchamp and Marquez's modified Shoolery rules, predict the chemical shifts for those protons attached to carbon atoms for the compounds given in Problem 27. Which set of predictions, those using Shoolery's rules or those using Beauchamp and Marquez's modified rules, is closer to the experimental data?

29. Predict the proton chemical shifts of the vinyl hydrogens in (a) styrene, (b) methylvinylketone, (c) vinylbromide, and (d) acrylonitrile. Compare your predictions with experimental values, and cite your source of the experimental values.

30. Predict the proton chemical shifts of the phenyl-group hydrogens in (a) nitrobenzene, (b) toluene, (c) phenol, and (d) analine. Compare your predictions with experimental values, and cite your source of the experimental values.

31. Explain the observations that in dilute CCl_4 solutions the OH protons of phenol and methysalicylate resonate at $\delta = 4.24$ and $\delta = 10.58$ ppm, respectively. The OH resonance position for one of these compounds is quite concentration dependent, whereas, for the other compound, there is little concentration dependence. Why?

32. The ^1H NMR resonances of the following two compounds occur at quite different chemical shifts:

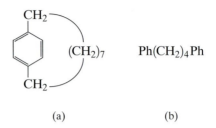

(a)   (b)

Which has the lower set of δ values? Why?

33. If the ^1H NMR spectrum of ketene dimer, $(CH_2=C=O)_2$, shows only two resonances of equal intensity, which of the following isometric structures is correct for this species?

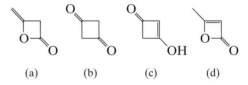

(a)   (b)   (c)   (d)

34. The ^{93}Nb NMR spectrum of $[Et_4N]^+[Nb(PF_3)_6]^-$ consists of a septet of nonadecets. Explain. What does this description of the multiplet imply about the relative magnitudes of $^1J(^{31}P^{93}Nb)$ and $^2J(^{19}F^{93}Nb)$? (See Rehder, D., Bechthold, H-C., Paulsen, K., *J. Magn. Reson.*, **1980**, *40*, 305.)

35. The ^{19}F NMR spectrum of $[Me_4N][BiF_6]$ consists of a 10-line multiplet. Explain. What are the relative intensities of the lines in this multiplet? Is it proper to call this a decet? Why or why not? (See Morgan, K., Sayer, B. G., Schrobilgen, G. J., *J. Magn. Reson.* **1983**, *52*, 139.)

36. Predict the appearance of the ^{19}F NMR spectrum of $[Et_4N]^+[AsCIF_5]^-$. (See Dore, M. F. A., Sanders, J. C. P., Jones, E. L., Parkin, M. J., *J. Chem. Soc. Chem. Commun.* **1984**, 1578.)

37. Explain the ^{119}Sn NMR spectrum of $[SnH_{3-n}D_n]^-$ shown in the following figure (see Wasylishen, R. E., Burford, N., *Can. J. Chem.* **1987**, *65*, 2707):

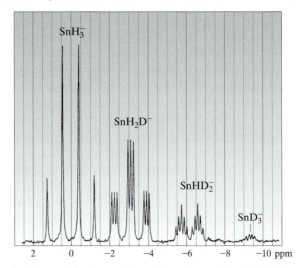

38. The ^{19}F NMR spectra of SF_4 and PF_5 both show single resonances at room temperature, but two resonances at low temperature. Explain.

39. Predict the chemical shifts for the $^{13}C\{^1H\}$ NMR spectra of (a) methylvinylketone ($CH_3C(O)CH=CH_2$), (b) 1-bromobutane-2ol ($BrCH_2C(OH)HCH_2CH_3$), (c) methylcyclohexane ($CH_3C_6H_{11}$), (d) 1-bromo-3-nitrobenzene, and (e) pentane. Find data for these compounds in the literature, and compare the experimental results with your predictions.

40. Predict the ^{15}N chemical shifts of 2-, 3-, and 4-methlpyridine. Find data in the literature, and compare them with your predictions. (Hint: Look for a book by Witanowski and Webb.)

41. Predict the $^{31}P\{^1H\}$ NMR chemical shifts of Me_3P, $Me_3PMo(CO)_5$, and *cis*- and *trans*-$(Me_3P)_2PdCl_2$. Find values in the literature, and compare them with your predictions.

42. Explain in detail how you could use NMR spectroscopy to study equilibria of the type

$$(R_3P)_2PtCl_2 + (R_3P)_2PtBr_2 \rightleftharpoons 2(R_3P)_2PtBrCl.$$

43. Describe how ^1H NMR spectroscopy could be used to distinguish among the following isomeric compounds:

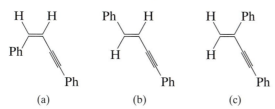

(a)   (b)   (c)

(See: Yi, C. S., Liu, N. *Organometallics* **1996**, *15*, 3968.)

44. For what conformation of the carbene complex shown in the following figure will the CH_2 protons be enantiotopic or diastereotopic?

$(CO)_5W=C$ OCH_2—◯—CH_3 CH_3O

How will the ^1H NMR line shapes of the CH_2 resonance(s) reflect this? (See Amin, S. K., Jayaprakash, K. N., Nandi, M., Sathe, K. M., Sarker, A., *Organometallics* **1996**, *15*, 3528.)

45. Explain the origins of the ^1H chemical-shift differences for the indicated protons in the following pairs of compounds:

(a)   *H*   vs.   *H* H‖

δ 8.65 ppm   δ 10.35 ppm

(b)

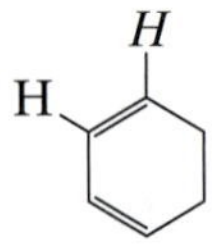

δ 5.9 ppm   vs.   δ 7.16 ppm

(c)

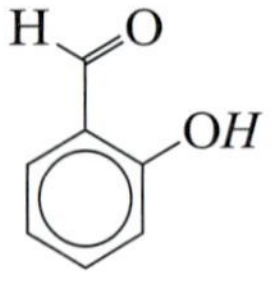

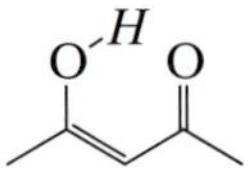

δ 11.3 ppm   vs.   δ 15.2 ppm

46. If the ^{17}O NMR spectrum of $Co_4(CO)_{12}$ shows four equally intense resonances, what is the structure of $Co_4(CO)_{12}$?

47. State whether the indicated (**H** or **H′** pairs) in the following molecules are enantiotopic, diastereotopic, or homotopic and whether they would be isochronous or anisochronous:

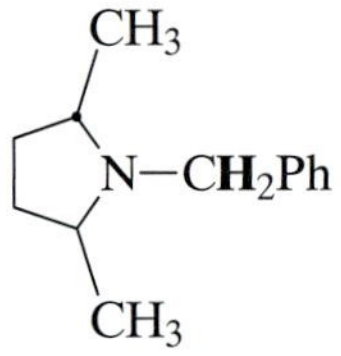

(a)

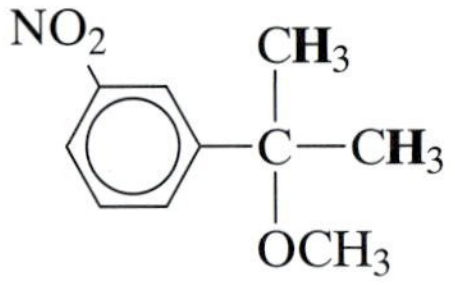

(b)

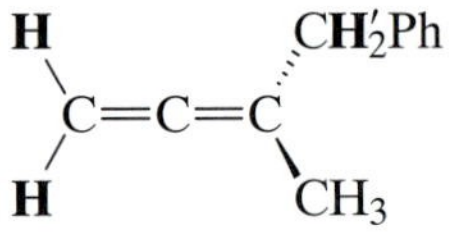

(c)

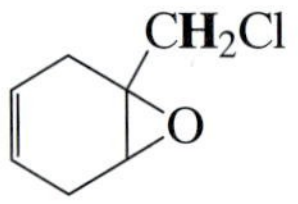

(d)

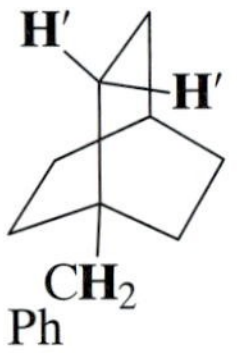

(e)

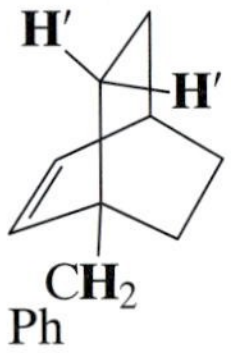

(f)

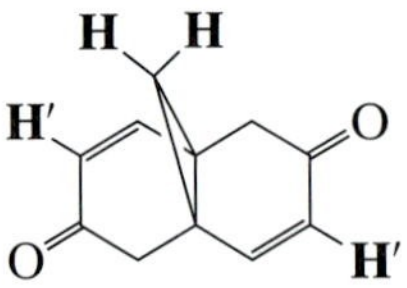

(g)

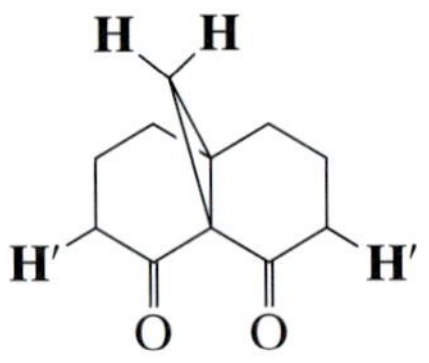

(h)

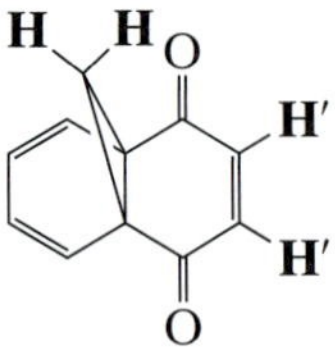

(i)

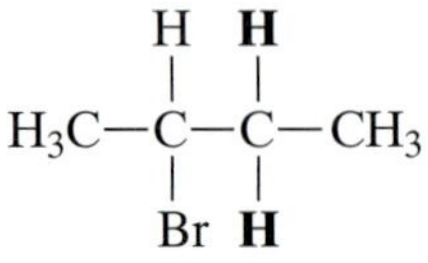

(j)

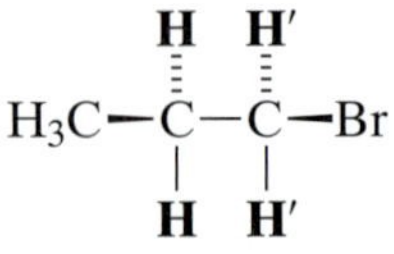

(k)

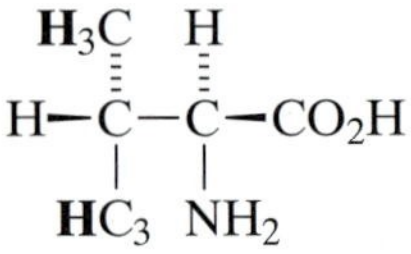

(l)

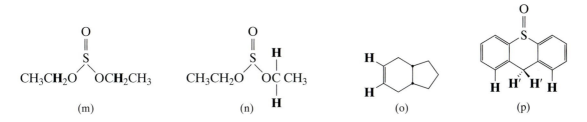

(m)   (n)   (o)   (p)

48. Give the spin-system designations that are appropriate for the following line shapes if all the nuclei have $I = \frac{1}{2}$:

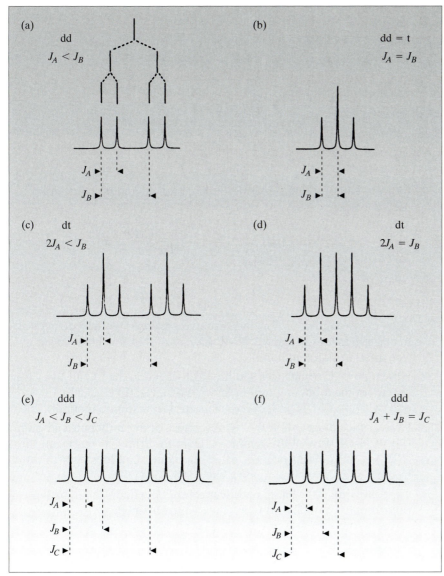

49. Suggest spin-system designations for the protons in the following molecules:

a. p-bromochlorobenzene

b. cyclopropylchloride

c. the chair form of cyclohexane

d. o-dichlorobenzene

e. ethylbromide

f. thiophene

50. Explain the concentration effect on the OH chemical shift of ethanol in CCl_4 solution illustrated in the following figure:

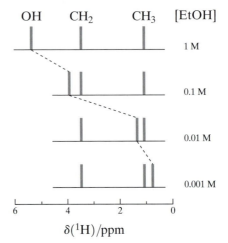

51. Indicate all homotopic, enantiotopic, and diastereotopic nuclei for the molecule:

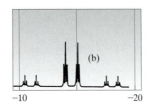

How many distinct proton and carbon resonances would you expect to observe in the ^1H and ^{13}C NMR spectra of this molecule? Assume free rotation about all bonds.

52. How many ^{13}C resonances should be observed in the ^{13}C NMR spectra of the following molecules: (a) H_3CCH_3; (b) H_3CCH_2Cl; (c) $H_3CC(O)CH_2CH_3$; (d) cyclohexanone; (e) cyclopropanone?

53. Draw a stick diagram for the ^{13}C resonance of the carbonate carbon in $(Ph_2PCH_2CH_2PPh_2)Pt(CO_3)$ if $\delta = 166$ ppm, $^2J(PtC) = 66$ Hz, and $^3J(PC) = 4$ Hz. (See Andrews, M. A., Gould, G. L., Klooster, W. T., Koenig, K. S., Voss, E. J., *Inorg. Chem.* **1996**, *35*, 5478.)

54. The following 100-MHz ^1H NMR spectra in the hydride region were unfortunately not labeled when they were obtained:

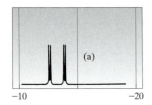

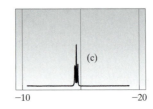

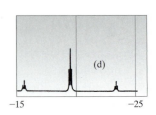

Nonetheless, the spectra are known to correspond to the following complexes: (a) *trans*-PtHBr(PEt$_3$)$_2$, (b) PdH(CN)(Ph$_2$PCH$_2$CH$_2$PPh$_2$), (c) [PtH(PEt$_3$)$_3$]ClO$_4$, and (d) *trans*-PdHCl(PEt$_3$)$_2$. Assign each spectrum to the appropriate complex and give reasons to support your assignment. Estimate the values of ^1J(Pt-H) for the two platinum complexes.

55. Given the following parameters for the ^{1}H NMR spectrum of 2-furaldehyde, draw a stick diagram of the spectrum:

$\delta H_A = 7.81$ ppm $J_{AM} = 1.65$ Hz

$\delta H_M = 6.67$ ppm $J_{AR} = 0.80$ Hz

$\delta H_R = 7.23$ ppm $J_{AX} = 0.81$ Hz

$\delta H_X = 9.66$ ppm $J_{MR} = 3.62$ Hz

$J_{MX} = J_{RX} = 0$

Use a computer program to simulate the 300-MHz spectrum, and compare your simulated spectrum with that given in the Aldrich catalog.

56. The ^{19}F NMR spectrum of ClF$_3$ at low temperature consists of a doublet considerably to high frequency (low field) of a 1:2:1 triplet. The total intensities of the two multiplets are in a 2:1 ratio, and the splittings are all the same. Explain the spectrum and relate it to that which should be observed for this compound if its structure obeys the VSEPR prediction.

57. From the 470.3-MHz ^{19}F NMR spectra of CFCl$_3$ shown in the following figure, calculate the magnitudes (in ppm) of the carbon and chlorine primary isotope effects on the ^{19}F chemical shift in this molecule:

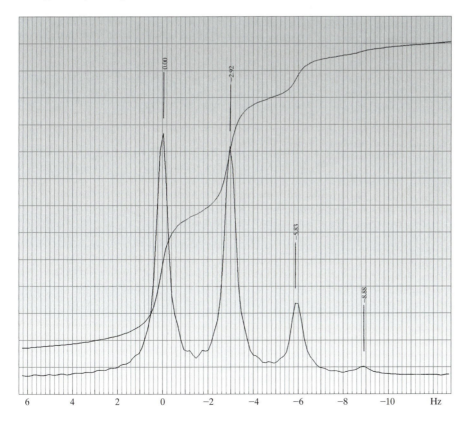

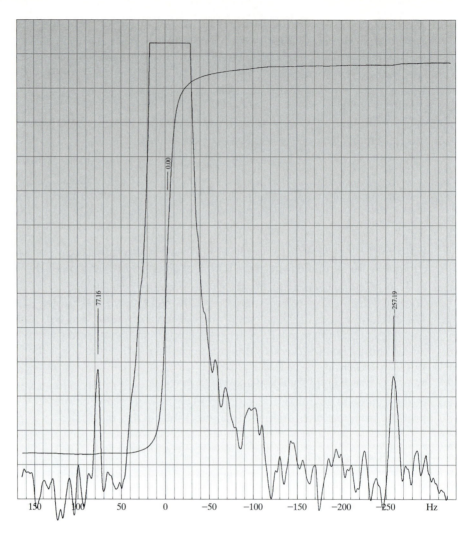

Identify the isotopomers that give rise to each chemical shift. What is the magnitude of $^1J\,(^{19}F^{13}C)$?

58. The following NMR data were reported for the characterization of 3,4-difluoropyrrole (Woller, E. K.; Smirnov, V. V.; Dimagno, S. G., *J. Org. Chem.* **1998**, *63*, 5706): ^1H (500 MHz, CDCl$_3$) δ = 7.23 (bt, 1H, $J^{14}NH$ = 50 Hz), 6.35 (dd, 2H, J_1 = 3.5, J_2 = 1.2 Hz, H$_{2,5}$); $^{13}C\{^1H\}$ (125 MHz, CDCl$_3$) δ = 139.1 (dd, J_1 = 23 Hz, J_2 = 12 Hz), 100.4 (dd, J_1 = 21 Hz, J_2 = 3.5 Hz); ^{19}F (470 MHz, CDCl$_3$) δ = −181.54 (dd, J_1 = 2 Hz, J_2 = 1 Hz). Are these data consistent with the structure of the molecule and the spin systems that would be expected?

59. Give spin-system designations for the aryl protons of the molecules illustrated in the figures that follow. Which of the spectra that are shown are first-order spectra?

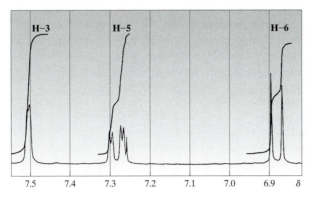

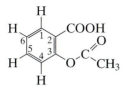

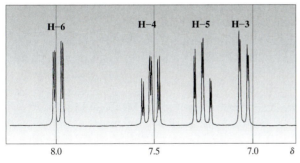

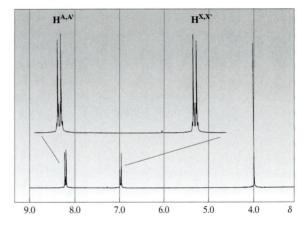

60. For the following AB spectrum, calculate J and the two δ values if the spectrum were obtained at 40 MHz:

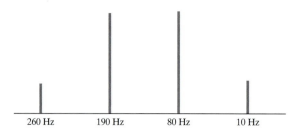

260 Hz   190 Hz   80 Hz   10 Hz

What would the appearance of this spectrum be at 500 MHz? Give the spin-system designation for the 500-MHz spectrum. Simulate the two spectra, using a computer program for spin-system analysis. What are the relative line intensities in your simulations?

61. Given the following information, construct stick diagrams for the ^1H, ^{13}C, ^{14}N, ^{15}N, and ^{129}Xe NMR spectra of $HC\equiv{}^{15}NXeF^+$ and $HC\equiv{}^{14}NXeF^+$:

Chemical Shifts (ppm)	
$\delta\ ^{129}Xe = -1552$	$\delta\ ^{15}N = 234.5$
$\delta\ ^{19}F = -198.7$	$\delta\ ^{13}C = 104.1$
$\delta\ ^{14}N = -235.1$	$\delta\ ^1H = 4.70$

Coupling Constants (Hz)	
$^1J(^{129}Xe^{19}F) = 6161$	$^2J(^{15}N^{19}F) = 23.9$
$^1J(^{129}Xe^{14}N) = 332$	$^2J(^{15}N^1H) = 13.0$
$^1J(^{129}Xe^{15}N) = 471$	$^3J(^{19}F^{13}C) = 18.0$
$^1J(^{14}N^{13}C) = 22$	$^3J(^{129}Xe^1H) = 24.7$
$^1J(^{13}C^1H) = 308$	$^4J(^{19}F^1H) = 2.6$

(See Emara, A. A. A., Schrobilgen, G. J., *Inorg. Chem.* **1992**, *31*, 1323.)

62. Predict the appearances of the ^1H (hydride region only), $^{31}P\{^1H\}$, $^{119}Sn\{^1H\}$, and $^{195}Pt\{^1H\}$ NMR spectra of *trans*- $HPt(Et_3P)_2SnCl_3$. For the ^{31}P, ^{119}Sn, and ^{195}Pt spectra, consider decoupling only the ethyl-group protons. (See Albinati, A., Von Guten, U., Pregosin, P. S., Ruegg, H. J., *J. Organomet. Chem.* **1985**, *295*, 239.)

63. Use a computer program to simulate an AB spectrum at 60 MHz with $J_{AB} = 10$ Hz and $\Delta\nu_{AB} = 120, 60, 20, 10$, and 3 Hz. The first spectrum should resemble an AX spectrum, and the last is what is termed a tightly coupled AB spectrum.

64. The following NMR data have been reported for $Me_3N \rightarrow BF_2Cl$:
$\delta\ ^1H = 2.71$ ppm, $^3J(^1H^{11}B) = 1.8$ Hz, $\delta\ ^{11}B = -4.5$ ppm, $\delta\ ^{19}F = 143.4$ ppm, $^1J(^{11}B^{19}F) = 44.8$ Hz, and $^4J(^1H^{19}F) = 0.7$ Hz. Draw stick diagrams of the ^1H, ^{11}B, and ^{19}F NMR spectra of the species, ignoring contributions from the ^{10}B isotope. Predict the values of

$^3J(^1H^{10}B)$ and $^1J(^{10}B^{19}F)$. (See Benton–Jones, B., Davidson, M. E. A., Hartmann, J. S., Klassen, J. J., Miller, J. M., *J. Chem. Soc. Dalton. Trans.* **1972**, 2603, Miller, J. M.; Jones, T. R. B., *Inorg. Chem.* **1976**, *15*, 284, and Binder, H., Fluck, E., *Z. Anorg. Allg. Chem.* **1971**, *381*, 123.)

65. Interpret the 94.1-MHz ^{19}F and 100-MHz ^1H NMR spectra (shown below) of $F_4P^{15}NH_2$ in $HCCl_2F$ solution. What is the spin system for this molecule? Estimate the coupling constants if $^1J(^{31}P^{19}F_e) = 936$ Hz and $^1J(^{31}P^{19}F_a) = 760$ Hz (see Cowley, A. H., Schweiger, J. R., *J. Am. Chem. Soc.* **1973**, *95*, 4179)

a. upfield half of $^{19}F_a$ resonance
b. upfield half of $^{19}F_e$ resonance
c. 100-MHz ^1H NMR spectrum

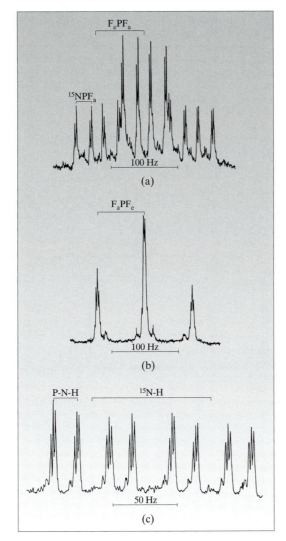

Reproduced with permission from the reference above, copyright 1973 American Chemical Society.

66. Completely interpret the following 300-MHz ^{1}H and 75-MHz ^{13}C NMR spectra of Na$^+$ H$_3$^{13}C^{13}CH$_2$CO$_2$— in D$_2$O solution:

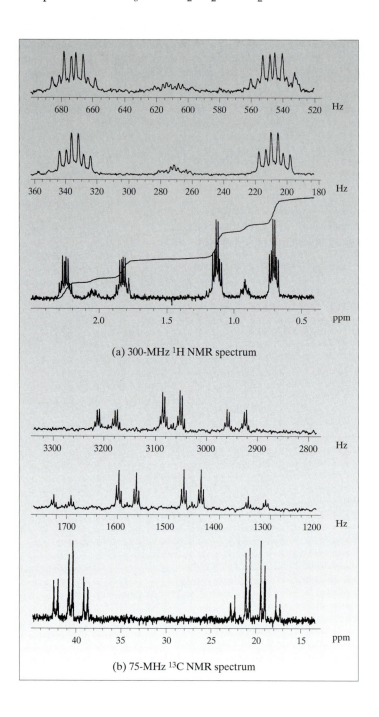

(a) 300-MHz ^{1}H NMR spectrum

(b) 75-MHz ^{13}C NMR spectrum

In your interpretation, calculate the chemical shifts and coupling constants. From the ^{1}H NMR spectrum, determine the ratio of the ^{13}C to ^{12}C species present (i.e., the level of ^{13}C labeling).

67. Complete the following table:

Spin System	Observables	Spectrum Order	Maximum Number of Lines
A_2	1δ	first	1
AB	$1\delta; 1J$	second	4
AM			
AX			
A_3			
A_2B	$2\delta; 2J$	second	9(8)
A_2X			
ABC			
ABX			
AMX			
$AA'X=[A]_2X$			
A_4			
A_3X			
A_2MX			
$ABMX$			
$AA'MX=[A]_2MX$			
A_2X_2			
$AA'XX'=[AX]_2$			
$AA'BB'=[AB]_2$			

68. Predict the ^{13}C chemical shifts for the three isomers of hydroxyacetophenone:

para-hydroxyacetophenone   ortho   meta

Compare your predictions with experimental results, and cite your source of the experimental results.

69. Predict the ^1H and ^{13}C chemical shifts for (a) hexane, (b) 1-hexene, and (c) 1-hexyne. Compare your predictions with experimental results, and cite the source(s) of the experimental results.

70. A neutral compound with the molecular formula $C_{10}H_{12}O_2$ showed ^{13}C chemical shifts of $\delta = 22, 68, 128, 129, 131, 132,$ and 166 ppm. An APT spectrum showed that the signal at $\delta = 22$ arises from a CH_3 group, those at $\delta = 131$ and 166 are due to quarternary carbons, and all other signals are due to CH carbons. Deduce the structure of the compound.

71. The ^{13}C NMR spectrum of an unknown compound ($C_6H_{10}O$) exhibited the following resonances: $\delta = 20(q)$, $(27q)$, $31(q)$, $124(d)$, $154(s)$, and $197(s)$. Deduce the structure of this compound, and support your conclusion by predicting the ^{13}C chemical shifts from additivity relationships.

72. From the following values of $^1J(XF)$, calculate the values of $^1K(XF)$:

	$^{13}CF_4$	$^{15}NF_3$	$^{17}OF_2$
$^1J(XF)$	−257	160	300
	$^{29}SiF_6^{2-}$	$^{31}PF_6^{-}$	$^{33}SF_6$
$^1J(XF)$	110	−706	251.6
	$^{73}GeF_6^{2-}$	$^{75}AsF_6^{-}$	$^{77}SeF_6$
$^1J(XF)$	98	−932	−1421
	$^{119}SnF_6^{2-}$	$^{121}SbF_6^{-}$	$^{125}TeF_6$
$^1J(XF)$	1601	−1938	3717

Is there any discernible trend in the $^1K(XF)$ values?

73. Completely interpret the 499.84-MHz ^1H and 125.70-MHz ^{13}C NMR spectra of disodium malate (Na^+ $^-O_2CCH_2CH(OH)CO_2^-$ Na^+) in D_2O. Calculate all chemical shifts and coupling constants. Simulate the ^1H NMR spectrum at 60-MHz and 500-MHz observation frequencies, and give spin-system designations for both. The spectra are as follows:

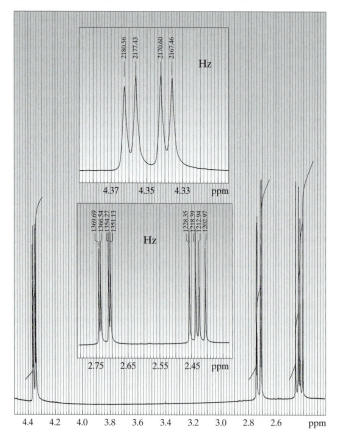

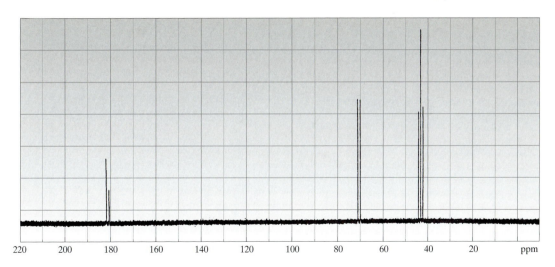

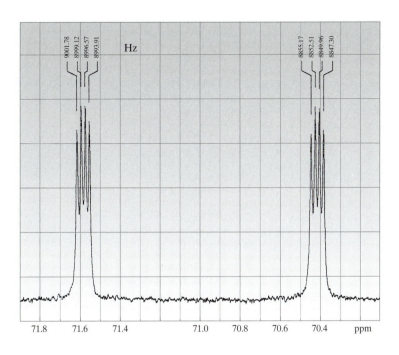

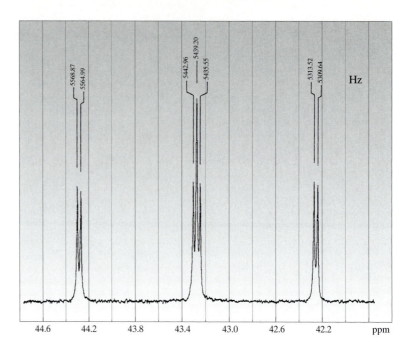

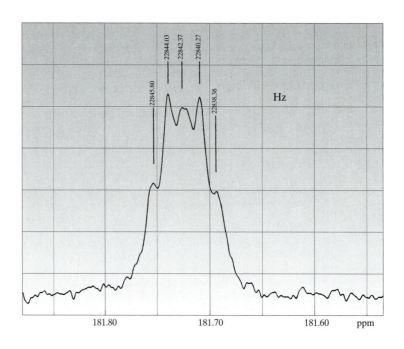

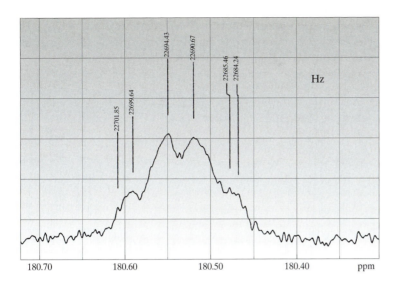

74. There is a linear relationship between $^1J(CH)$ and the percent s character in the carbon–hydrogen bond: $^1J(CH) = 5.7$ (% s) - 14.8 Hz. Use this equation to calculate $^1J(CH)$ for ethane, ethylene, and acetylene, and compare your calculated values with experimental results. Would you expect a similar relation to obtain for $^1J(CH)$ in the series CH_3NH_2, $H_2C{=}NH$, and $HC{\equiv}N$? (Consult Breitmaier and Voelter, page 135, for the data.)

75. Given the information in the following table, which nuclide, ^{115}Sn, ^{117}Sn, or ^{119}Sn, is most favorable for direct observation and why?

Isotope	I	Natural abundance, %	NMR reson freq, MHz (^{1}H = 100 MHz)	γ (10^7 rad s^{-1} T^{-1})
^{115}Sn	1/2	0.35	32.864	−8.8014
^{117}Sn	1/2	7.61	35.626	−9.589
^{119}Sn	1/2	8.58	37.272	−10.0318

76. Predict the ^{1}H and ^{19}F NMR spectra of $CF_3SO_3CH_2CH_2CF_3$. (See Tsushima, T., Kawada, K., Ishara, S., Uchida, O., Higaki, J., Hirata, M., *Tetrahedron* **1988**, *44*, 5375.)

77. ^{15}N labeled ammonium chloride (^{15}NH$_4$Cl) was dissolved in a mixture of H$_2$O and D$_2$O and strongly acidified. Interpret as fully as possible the following 400-MHz ^{1}H NMR spectrum that results:

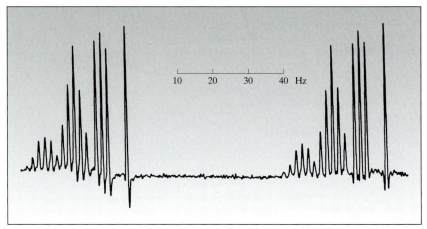

How many ^{15}N resonances should be observed for this mixture? (See Sanders, K. M., Hunter, B. K., Jameson, C. J., *Chem. Phys. Letters* **1988**, *143*, 471.)

78. The 500-MHz ^{1}H NMR spectrum of C$_2$H$_5$OC(O)CH=CH(CF$_3$) in CDCl$_3$ shows the following reasonances: δ = 6.78(*dq*, $J_{(HH)}$ = 15.5 Hz, $J_{(HF)}$ = 6.5 Hz, 1H); 6.49(*dq*, $J_{(HH)}$ = 15.5 Hz, $J_{(HF)}$ = 2.0 Hz, 1H); 4.28(*q*, $J_{(HH)}$ = 7.1 Hz, 2H), and 1.33(*t*, $J_{(HH)}$ = 7.1 Hz, 3H). The ^{13}C{^{1}H} NMR spectrum of the same sample shows the following resonances: δ = 163.86 (*s*, quarternary carbon), 131.21(*q*, *J*(CF) = 35.5 Hz, CH), 128.88(*q*, *J*(CF) = 6.2 Hz, CH), 122.03 (*q*, *J*(CF) = 269.9 Hz, and no attached protons). Completely assign the ^{1}H and ^{13}C{^{1}H} NMR spectra, and deduce whether the molecule is the *cis* or the *trans* isomer.

79. An alkene with the molecular formula C$_3$H$_4$F$_2$ shows a doublet of doublets (*J* = 45 and 10 Hz) and a douplet of quartets (*J* = 45 and 8 Hz) in its ^{19}F NMR spectrum. What is the structure of this compound?

80. From the ^{11}B NMR spectrum of B$_9$H$_{14}^-$ shown in the following figure, deduce the structure of the anion (see also Figure 3.7):

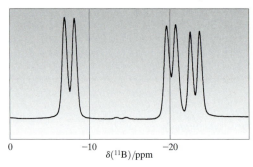

81. The following linear relationship obtains between $^1J_{CC}$ and the product of the carbon percent *s* characters in the C—C bond:

$$^1J_{CC} = \frac{7.3(\%s_x)(\%s_y)}{100} - 17 \text{ Hz}.$$

Use this equation to calculate the $^1J_{CC}$ coupling constants for ethane, ethylene, and acetylene, and compare your calculated values with experimental results.

82. Suggest a structure for a compound $C_5H_{13}N$ whose $^{13}C\{^1H\}$ NMR spectrum exhibits three resonances at δ = 54.4, 30.4, and 10.4 ppm. Assign each signal to a carbon in your structure, and then use additivity relationships to predict the chemical shifts of each carbon.

83. Propose a structure for a compound C_8H_6 whose $^{13}C\{^1H\}$ NMR spectrum exhibits six resonances at δ = 77.4, 83.8, 122.4, 128.3, 128.7, and 132.2 ppm. Assign each signal to a carbon in your structure, and use additivity relationships to predict the chemical shifts of the four most downfield resonances.

84. Suggest structures for three of the isomers of C_7H_8O whose $^{13}C\{^1H\}$ NMR chemical shifts are given in the following table:

A	141.8	129.1	128.1	127.7	65.1
B	159.9	129.5	120.7	114.1	54.8
C	152.6	130.5	130.2	115.3	20.6

Assign each resonance to one or more carbons in your structures, and use additivity relationships to predict the chemical shifts.

85. For each set of condensed spectral data in the following table, suggest a structure that is consistent with the data:

Molecular Formula	Nucleus	δ(ppm)(relative intensity)
a. C_7H_8O	^1H	2.43(1H), 4.58(2H), 7.28(5H)
b. $C_3H_5Cl_3$	^1H	2.20(3H), 4.02(2H)
c. $C_{14}H_{12}O$	^1H	3.88(2H), 7.38(10H)
		4.37(2H), 7.20(10H)
d. $C_{10}H_{14}O$	^{13}C	31.5(86), 33.7(10), 114.9(55),
		125.9(36), 141.6(7), 154.7(7)
e. C_5H_7N	^{13}C	35.6(18), 108.3(88), 121.6(58)
f. $C_7H_{12}O_2$	^{13}C	25.5(60), 25.9(70), 29.0(66),
		43.1(38), 183.0(25)

Correlate observed chemical shifts with predicted values.

86. Draw a stick diagram for the ^1H NMR spectrum of $\begin{array}{c} H_3C \\ \\ H_3C \end{array} \! \! C{-}CH_2OH$ with H. Compare your predictions with data from the literature, and cite your source of the data.

87. (a) In the hydrogen molecule, $^1J_{HH}$ is +280 Hz. Predict the value (sign and magnitude) of $^1J_{HD}$ in HD. (b) The 21.4-MHz $^{13}C\{^1H\}$ NMR spectrum of $CDCl_3$ consists of three resonances, at 75.5, 77.0, and 78.5 ppm. Calculate $^1J_{CD}$ for $CDCl_3$ and predict $^1J_{CH}$ for $CHCl_3$.

88. What is the hybridization of the CH bond in $CHCl_3$ if $^1J_{CH}$ is 209 Hz?

89. The complex ions $M(C{\equiv}C{-}C{\equiv}N)_4{}^{2-}$, where $M{=}Ni$, Pd, Pt, recently were prepared (Zhou, Y., Arif, A. M., Miller, J. S., *Chem. Commun.* **1996**, 1881). From the following data, assign the ^{13}C chemical shifts:

	Ni	Pd	Pt	$^nJ_{PtC}$
	107.0	107.1	115.3	1034.5 Hz
$\delta\ ^{13}C(ppm)$	80.0	76.1	74.5	311.2 Hz
	126.6	122.1	107.8	35.9 Hz

What are the structures of these ions, T_d or D_{4h}?

90. Do the following data correlate with those shown in Figure 6.1? For $^{13}C(CH_3)_4$ $^1J_{CC}$ = 36.9 Hz, $\gamma^{13}C$ = 6.7283, and for $^{15}N(CH_3)_4{}^+$ $^1J(^{15}N^{13}C)$ = 5.8 Hz, $\gamma^{15}N$ = −2.712. Explain why or why not.

91. Cyclohexane and its derivatives usually exist in a chair conformation such as the following:

For a rigid chair conformation, predict the values of $^3J(H_xH_e)$, $^3J(H_xH_a)$, and $^2J(H_aH_e)$ for a monosubstituted cyclohexane.

92. Consider the following molecule:

The 60-MHz ^1H NMR spectrum of this compound shows, in addition to two nine-proton singlets for the two $(H_3C)_3C-$ groups, a set of resonances for H_a, H_b, and H_c as follows:

H	ν from TMS	J(Hz)
H_a	163	ab = −18
H_b	195	ac = 2
H_c	263	bc = 10

a. Why are H_a and H_b anisochronous?

b. Calculate $\Delta\nu/J$ for each coupling interaction.

c. What type of spin system do these nuclei constitute?

d. Draw a stick diagram for a first-order spectrum.

e. What do the magnitudes of $^3J_{ac}$ and $^3J_{bc}$ indicate about the preferred conformation for the molecule? Draw it in Newman projection.

f. Use a computer program to simulate the spectrum at 60, 100, and 300 MHz.

93. Consider the following data given for vinylmercuryhydride, with structure shown (Guillemin, J.-C., Bellec, N., Sze'tsi, S. K., Nyula'szi, L., Veszpre'mi, T., *Inorg. Chem.* **1996**, *35*, 6586):

$$H_c, H_b, C=C, H_d, H_g, H_a$$

5a

^{1}H NMR (CDCl$_3$, = −40°C): δ = 5.47 (ddd, 1H, CH$_2$); ^3J(HH) *trans*) = 20.6 Hz, ^{2}H(HH gem) = ^4J(HH) = 3.6 Hz, ^3J(HgH) = 1.63 Hz), 6.08 (ddd, 1H, CH$_2$), ^3J(HH *cis*) = 13.7 Hz, ^4J(HH) = 9.6 Hz, ^2J(HH gem) = 3.6 Hz, ^3J(HgH) = 302 Hz), 6.84 (ddd, 1H, CH), ^3J(HH *trans*) = 20.6 Hz, ^3J(HH *cis*) = 13.7, ^3J(HH) = 2.7 Hz, ^2J(HgH) = 139 Hz), 14.4 (ddd, 1H, HgH), ^4J(HH) = 9.6 Hz, ^4J(HH) = 3.6 Hz, ^3J(HH) = 2.7 Hz, ^1J(HgH) = 2800 Hz.

a. Why are these resonances designated as ddd when four coupling constants are given?

b. Use the structural drawing to show the assignments of the resonances.

c. Is the preceding interpretation consistent with the following spectra?

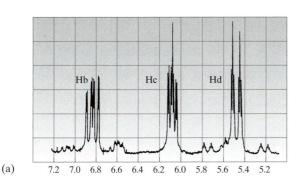

(a)   7.2 7.0 6.8 6.6 6.4 6.2 6.0 5.8 5.6 5.4 5.2

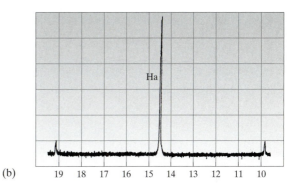

(b)   19 18 17 16 15 14 13 12 11 10

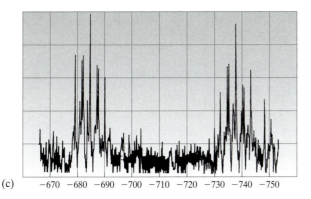

(c)   −670 −680 −690 −700 −710 −720 −730 −740 −750

^{1}H (300-MHz) and ^{199}Hg (123.2-MHz) NMR spectra of 5a. (a) ^{1}H NMR spectra (Hb, Hc, and Hd); (b) ^{1}H NMR spectra (Ha); (c) ^{199}Hg NMR spectrum (gate decoupling). Reproduced with permission from above reference, copyright 1996, American Chemical Society.

d. On the basis of the magnitudes of ^3J(HH) and ^2J(HH), what is mercury's electronegativity?

94. Explain the appearances of the 3-MHz ^{14}N NMR spectra for various ammonium ions in the following figure (Ogg, R. A., Day, J. D., *J. Chem. Phys.* **1957**, *26*, 1340):

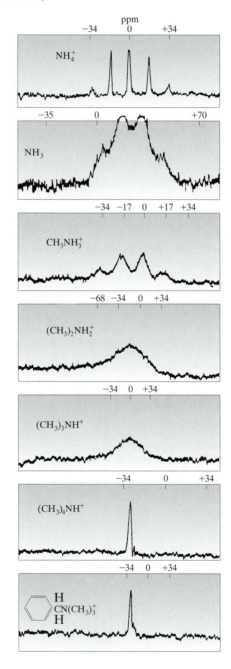

Reproduced with permission from above reference, copyright 1957, American Institute of Physics.

What is the field strength of the magnet, in tesla?

95. The 40-MHz ^{19}F NMR spectra of (a) BF_3/BCl_3, (b) BF_3/Br_3, and (c) $BF_3/BCl_3/BBr_3$ mixtures are as follows (Coyle, T. D., Stone, F. G. A., *J. Chem. Phys.* **1960**, *32*, 1892):

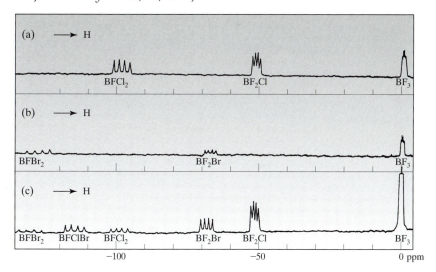

Reproduced with permission from above reference, copyright 1960, American Institute of Physics.

   a. What kind of equilibria do these spectra illustrate?

   b. Convert the chemical shifts given in the following table to $CFCl_3$ as a reference:

Compound	$\delta_{BF_3}^{int}$	$\delta CFCl_3$	$\delta_{CF_2CO_2H}^{ext}$	$^1J^{11}B^{19}F$	$^1J^{10}B^{19}F$
BF_3	0	+127	+48.4	15	
BF_2Cl	−51.5		−3.1	34	
BF_2Br	−68.4		−20	56	
$BFCl_2$	−99.0		−50.6	74	
$BFClBr$	−144.8		−66.4	92	
$BFBr_2$	−130.4		−82.0	108	

   c. Calculate $^1J^{10}B^{19}F$ values for each compound.

   d. Explain the variation in ^1JBF values.

   e. Is there an additivity relationship for $^1J^{11}B^{19}F$?

   f. Is there an additivity relationship for $\delta\,^{19}F$?

96. What should be the appearance of the ^{31}P NMR spectra of (a) PF_5 (room temperature and low temperature), (b) PF_6^-, (c) PF_4O^-, (d) $PF_2O_2^-$, and (e) PO_3^-?

97. The 470.306-MHz ^{19}F NMR spectrum of an old sample of $NaPF_6$ in D_2O is shown in the following figure:

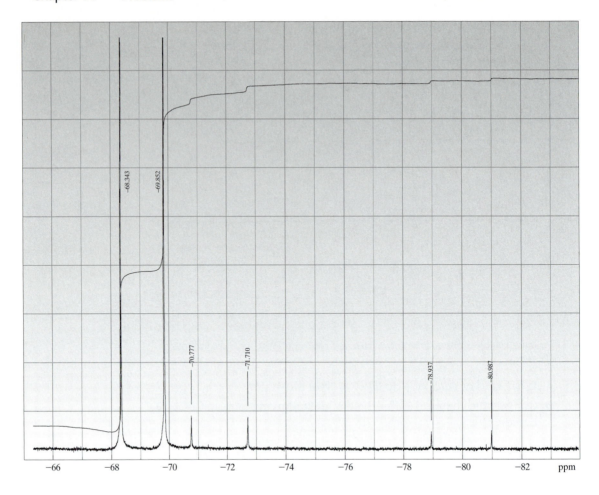

Assign the resonances, calculate the chemical shifts and coupling constants, and write chemical reactions that indicate how the additional solvolysis products are formed. Estimate the relative amounts of the solvolysis products. Predict the appearance of the ^{31}P NMR spectrum of this sample.

98. Give spin-system designations for the three isomers of dichlorobenzene:

How could ^{1}H and ^{13}C{^{1}H} NMR spectroscopy distinguish these three compounds? Find data in the literature to support your arguments, and cite your literature source(s).

99. Draw coupling trees for the ^1H NMR spectrum of $CH_3CH_2SPF_2$:

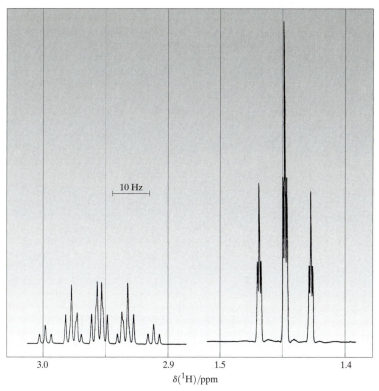

Label the spin system and describe the multiplets.

100. Draw coupling trees for the ^1H and $^1H\{^{15}N\}$ NMR spectra of $PF_2{}^{15}NHSiH_3$:

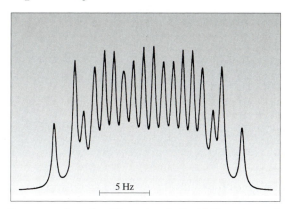

Resonances from the SiH_3 protons in the NMR spectrum of $PF_2{}^{15}NHSiH_3$. The pattern is a doublet of doublets of doublets of triplets, but because of the relationships between coupling constants, many lines overlap: $^3J_{PH} = 8$, $^3J_{HH} = 4$, $^2J_{NH} = 3$ and $^4J_{FH} = 2$ Hz approximately.

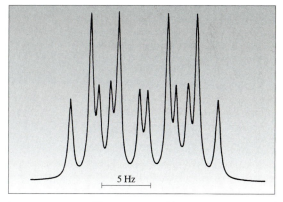

^1H spectrum of $PF_2{}^{15}NHSiH_3$, with ^{15}N noise decoupling.

What should be the relative intensities of the observed lines?

101. The normal 60-MHz ^{1}H NMR spectrum (a) and the ^{1}H{^{14}N} NMR spectrum (b) of formamide are as follows:

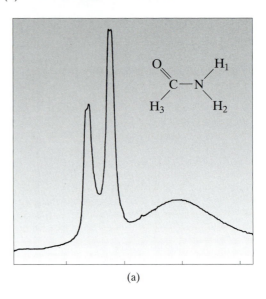

(a)

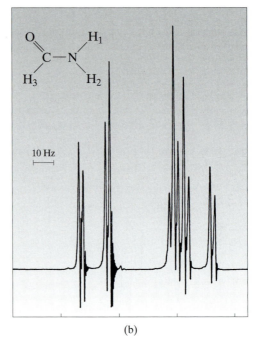

10 Hz

(b)

Analysis of the latter gives J_{12} = 2.8 Hz, J_{13} = 13.1 Hz, and J_{23} = 2.8 Hz. Determine the relative chemical shifts (in Hz) of the three interacting protons, and simulate the ^{1}H{^{14}N} spectrum as an *ABC* spin system, using a computer program for spin-system analysis.

102. Given the following information from Slawin, A. M. Z., Smith, M. B., and Wooling J. D., (*J. Chem. Soc., Dalton Trans.*, **1996**, 3659), predict the appearance of the 36.2-MHz ^{31}P{^{1}H} NMR spectrum of compound 6: $\delta(P_E)$ = 15.0, $\delta(P_O)$ = 26.9, $^2J(P_EP_O)$ = 3.3 Hz, $^2J(PtP_E)$ = 116.5 Hz,

$^2J(PtP_O) = 30.8$ Hz, and $^1J(PSe) = 486$ Hz. Is the following spectrum consistent with your predictions?

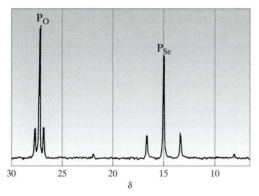

5 E = S
6 E = Se

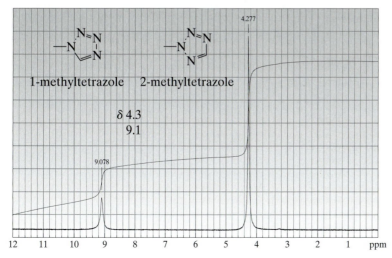

^{31}P–{^{1}H} NMR spectrum (36.2 MHz) of compound 6. Reproduced with permission from above reference, copyright 1996, Royal Society of Chemistry.

103. Alkylation of sodium tetrazolate with CH_3I could produce two products: 1-methyltetrazole and 2-methyltetrazole. From the ^{1}H, ^{13}C{^{1}H}, and ^{15}N{^{1}H} NMR spectra shown in the following figure, deduce what was formed in the reaction and assign all the chemical shifts (see Nelson, J. H., Takach, N. E., Henry, R. A., Moore, D. W., Tolles W. N., Gray, G. A., *Magn. Reson. Chem.*, **1986**, *25*, 984; Naumenko, V. N., Koren, A. O., Gaponik, P., *Magn. Reson. Chem.*, **1992**, *30*, 558; and Sveshnikov, N. N., Nelson, J. H., *Magn. Reson. Chem.* **1997**, *35*, 209):

499.875-MHz ^{1}H NMR spectrum in CDCl$_3$.

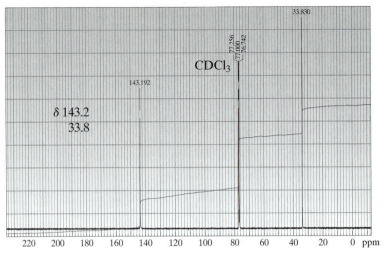

125.707-MHz $^{13}C\{^1H\}$ NMR spectrum in $CDCl_3$.

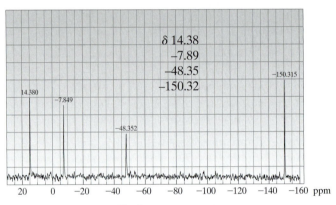

30.46-MHz $^{15}N\{^1H\}$ NMR spectrum in $CDCl_3$.

104. The reaction of sodium 5-benzyltetrazolate with CH_3I produces a mixture of 1-methyl-5-benzyltetrazole and 2-methyl-5-benzyltetrazole. Assign the ^1H, $^{13}C\{^1H\}$, and $^{15}N\{^1H\}$ NMR spectra shown in the following figures, and determine the ratio of the isomers (see Nelson, J. H., Takach, N. E., Henry, R. A., Moore, D. W., Tolles, W. M., Gray, G. A., *Magn. Reson. Chem.*, **1986**, *24*, 984; Naumenko, V. N., Koren, A. O., Gaponik, P., *Magn. Reson. Chem.*, **1992**, *30*, 558; Sveshnikov, N. N., Nelson, J. H., *Magn. Reson. Chem.*, **1997**, *35*, 209):

1-methyl-5-benzyltetrazole   2-methyl-5-benzyltetrazole

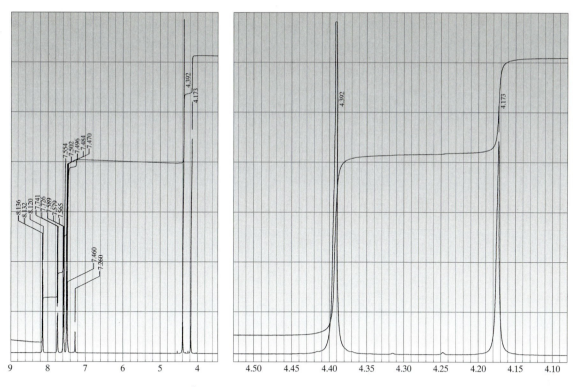

499.869-MHz ^{1}H NMR spectrum in CDCl$_3$.

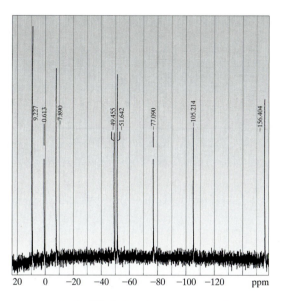

50.667-MHz ^{15}N{^{1}H} NMR spectrum in CDCl$_3$.

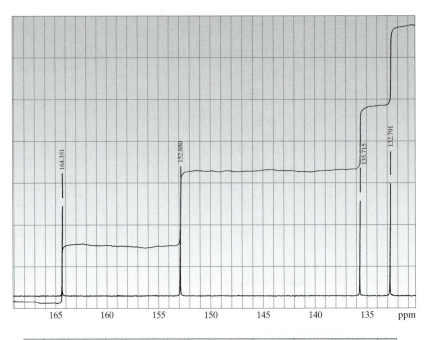

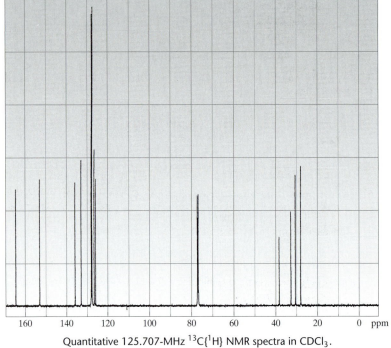

Quantitative 125.707-MHz ^{13}C{^{1}H} NMR spectra in CDCl$_3$.

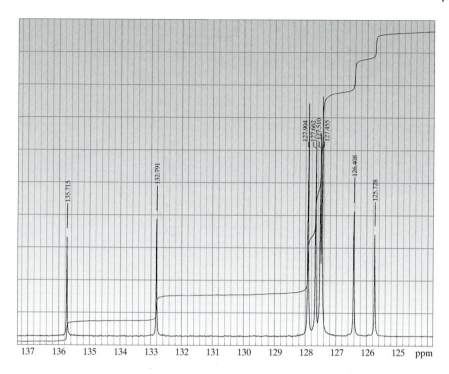

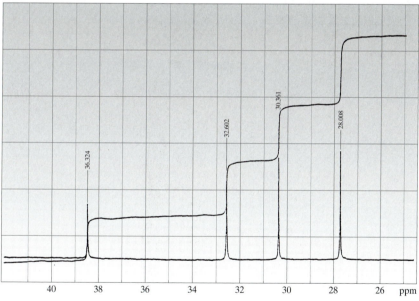

105. Explain the following ^1H, ^{13}C, and $^{13}C\{^1H\}$ NMR spectra of

acetaldoxime,

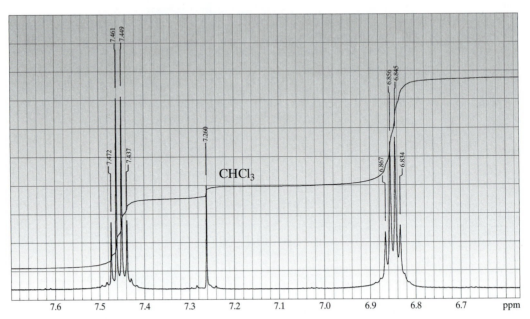

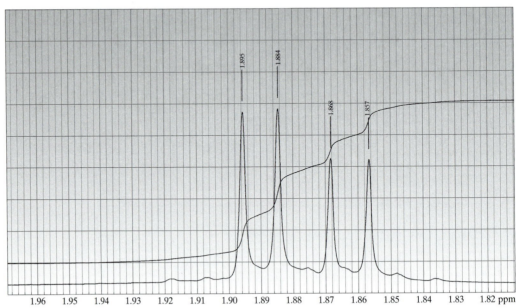

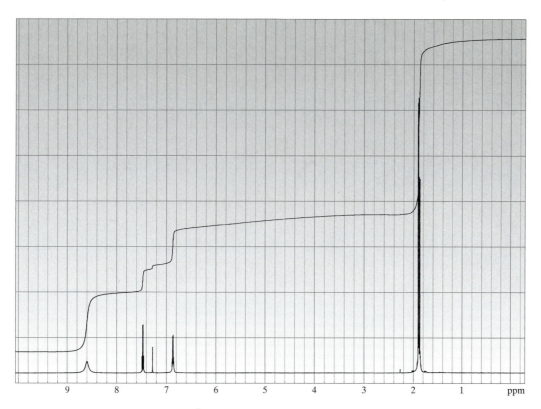

499.868-MHz ^{1}H NMR spectrum in CDCl$_3$.

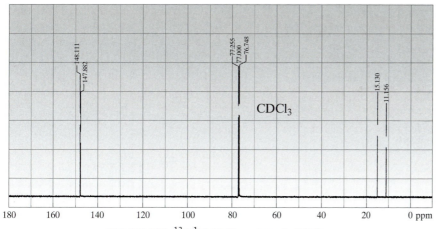

125.702-MHz ^{13}C{^{1}H} NMR spectrum in CDCl$_3$.

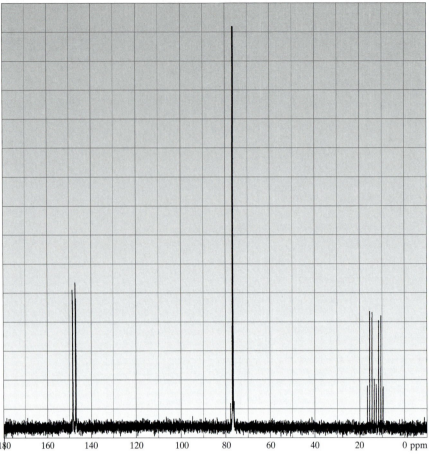

125.702-MHz ^{13}C NMR spectrum in CDCl$_3$.

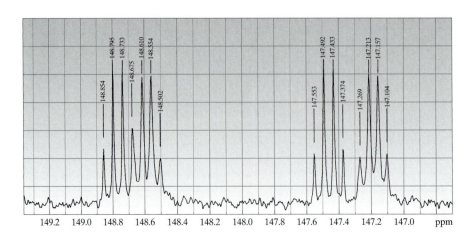

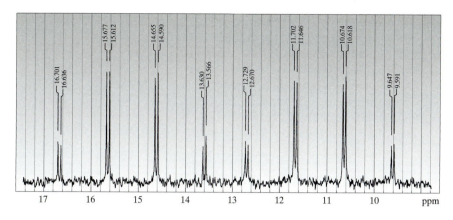

Would you expect these spectra to be dependent on temperature?
Why or why not?

106. Assign the $^{31}P\{^1H\}$ NMR resonances shown in the following figure:

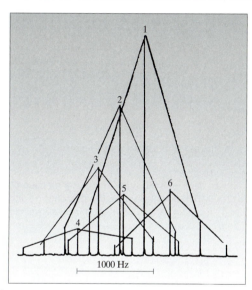

81.0-NHz $^{31}P\{^1H\}$ NMR spectrum of the products formed in the reaction of trans-
[$PtCl_4(PEt_3)_2$] with trans- [$PtBr_4(PEt_3)_2$]. The six complexes formed have from four
chlorines and no bromines to four bromines and no chlorines. There are two isomers
of the complex, with two atoms of each halide. Platinum satellites are indicated by
the lines connecting them to the central resonances.

Draw structures that correspond to the resonances labeled 1–6.

107. Why does the hydride resonance of $T\overset{*}{p}$ $RuH(H_2)_2$, where $T\overset{*}{p}$ = hydridotris (3,5-dimethylpyrazolyl) borate, dissolved in C_6D_6 change with time as shown in the following figures? (See Moreno, B., Sabo–Etienne, S., Chaudret, B., Rodriguez, A., Jalon, F., Trofimenko, S., *J. Am. Chem. Soc.*, **1995**, *117*, 7441.)

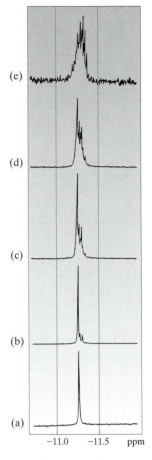

Modification of the high field ^{1}H NMR signal of 5a as a function of time in C_6D_6 at room temperature (250 MHz): (a) initial signal; (b) after 30 s; (c) after 5 min; (d) after 10 min; (e) after 24 h. Reproduced with permission from above reference, copyright 1995, American Chemical Society.

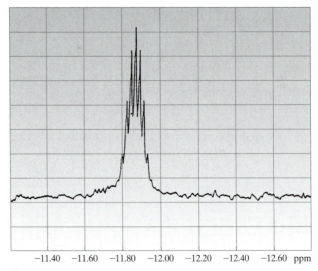

High-field ^{1}H NMR signal of Tp' $RuHD_4$ (5b-d_4) showing the expected nonet pattern (250 MHz, C_6D_6).

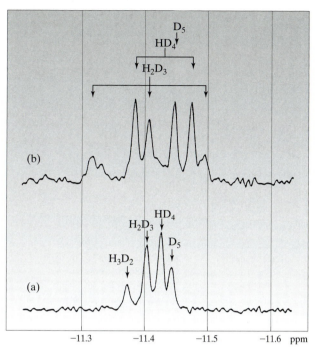

^{2}D NMR spectra (61.42 MHz, C_6H_6) of a mixture of 5a-d_5, 5a-d_4, 5a-d_3, and 5a-d_2: (a) {^{1}H}^{2}D NMR spectrum; (b) ^{2}D NMR spectrum.

Do the ^{2}H NMR spectra support your explanation?

108. Explain the satellite intensities in the ^{119}Sn and ^{195}Pt NMR spectra shown in the following figures (see Nelson, J. H., Wilson, W. L., Cary, L. W., Alcock, N. W., Clase, H. J., Jas, G. S.; Ramsey–Tassin, L., Kenney, J. W. III, *Inorg. Chem.*, **1996**, *35*, 883. Holt, M. S., Wilson, W. L., Nelson, J. H., *Chem. Rev.*, **1989**, *89*, 11):

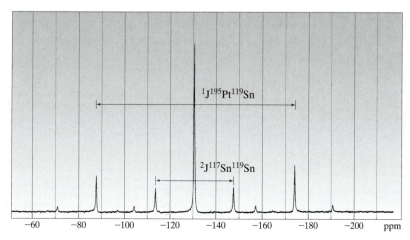

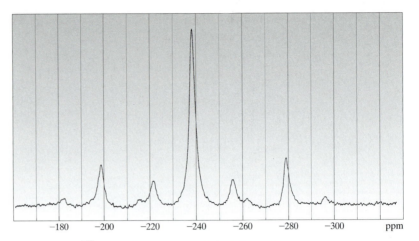

186.36-MHz ^{119}Sn NMR spectra of $[PhCH_2PPh_3]_3[Pt(SnCl_3)_5]$ (upper) and $[PhCH_2PPh_3]_3[Pt(SnBr_3)_5]$ (lower) in acetone-d_6 at 298 K. For these two spectra, the line widths at half-height ($\Delta\nu_{1/2}$) are 76 and 486 Hz, respectively.

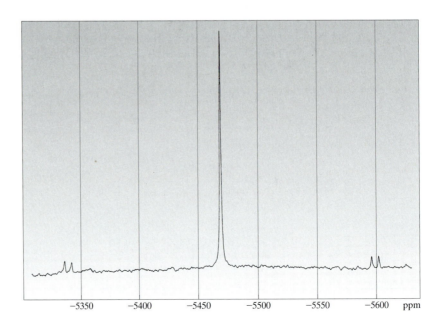

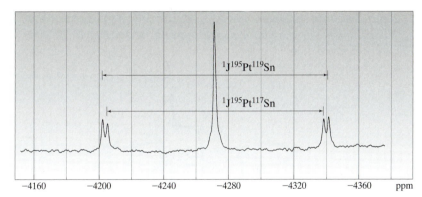

107.43-MHz ^{195}Pt NMR spectra of $[PhCH_2PPh_3]_2–[PtBr_3(SnBr_3)]$ (upper) and $[PhCH_2PPh_3]_3[Pt(SnBr_3)_5]$ (lower) in acetone-d_6 at 298 K. Reproduced with permission from above reference, copyright 1996, American Chemical Society.

109. Predict the values of $^2J_{HH}$ and $^3J_{HH}$ in vinylchloride, methylvinylketone, and vinyltrimethysilane.

110. The 24.29-MHz ^{31}P NMR spectrum of the diphosphite ion is shown in the following figure:

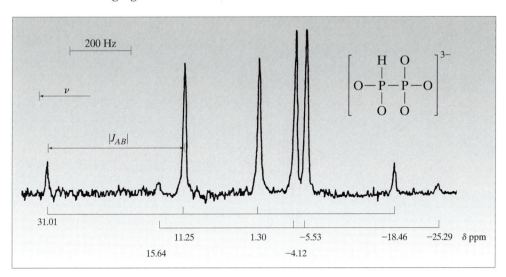

Analyze this spectrum, which is the *AB* part of an *ABX* (*A, B* = ^{31}P, *X* = ^{1}H) spin system, as completely as possible, and predict the appearance of the ^{1}H NMR spectrum of this species if δ ^{1}H is about 5 ppm and ^1J(PH) = 444 Hz.

111. a. State what you would expect for the relative magnitudes of the ^{35}Cl^{17}O quadrupole coupling constants in the following species, and explain your reasoning: (a) ClO^-, (b) ClO_2^-, (c) ClO_3^-, and (d) ClO_4^-.

b. Which species would exhibit the broader ^{35}Cl NMR resonance in a homogeneous solution, CCl_4 or NaCl? Why?

112. Explain the line shapes in the $^{31}P\{^1H\}$ and $^{51}V\{^1H\}$ NMR spectra of $[V(NPPh_3)_4]Cl$ (see Aistars, A., Doedens, R. J., Doherty, N. M., *Inorg. Chem.*, **1994**, *33*, 4360):

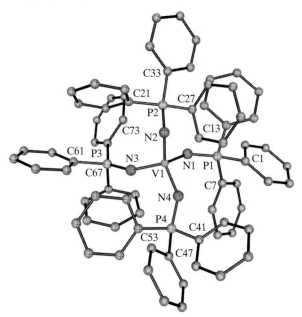

Drawing of the cation $[V(NPPh_3)_4]^+$ in the structure of $[V(NPPh_3)_4]Cl \cdot 4MeCN$ with 50% thermal ellipsoids for nonhydrogen atoms.

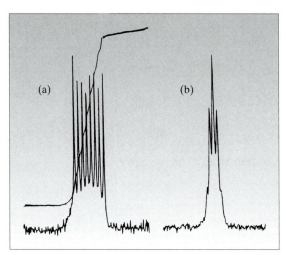

(a) ^{31}P NMR and (b) ^{51}V NMR spectra of $[V(NPPh_3)_4]$ Cl in CD_2Cl_2. Reproduced with permission from above reference, copyright 1994, American Chemical Society.

113. Given the following information, construct the stick diagram for the ^1H NMR spectrum of styrene:

$^3H_{(HH')} = 10.6$ Hz, $^3H_{(HH'')} = 17.4$ Hz

$^2J_{(H'H'')} = -1.4$ Hz

$\delta\, H' = 5.32$ ppm, $\delta\, H'' = 5.74$ ppm, $\delta\, H = 6.71$ ppm

$\delta\, C_6H_5 = 7.3$ ppm

Label the spin system for the interacting nuclei. Simulate the spectrum at 60 MHz with $^2J(H'H'')$ both positive and negative, and compare your results with the experimental spectrum that can be found in the Aldrich collection of spectra.

114. The ^1H and ^2H NMR spectra of n-butyl acetate, $CH_3C(O)OCH_2CH_2CH_2CH_3$, are as follows:

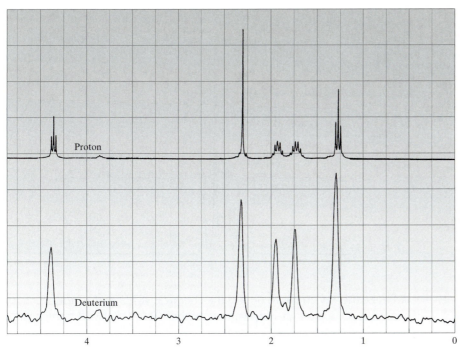

Would you expect the ^1H and ^2H chemical shifts to be the same? Explain. While ^1H—^1H coupling is manifested in the ^1H spectrum, ^2H—^2H coupling is not observed in the ^2H spectrum. The ^2H line widths are larger than the ^1H line widths. Explain these observations.

115. The 108.7-MHz ^{31}P NMR spectrum of $(CH_3O)_3P$ is as follows:

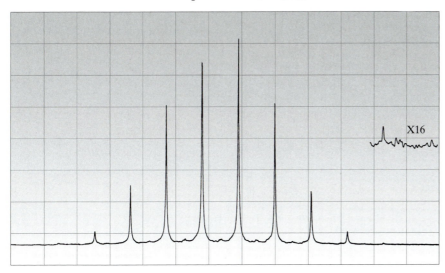

Is the spectrum first order? What is the spin system? What should be the relative intensities of the lines in this spectrum? What should be the appearance of the $^{31}P\{^1H\}$ and ^1H NMR spectra of this compound?

116. The 60.74-MHz CP/MAS $^{31}P\{^1H\}$ NMR spectrum of $(n\text{-}Bu_3P)_2HgCl_2$ exhibits a central *AB* quartet flanked by ^{199}Hg satellites that make up an *ABX* spin system:

Chemical Shifts (ppm)	
76.51	31.12
73.70	28.33
69.29	−2.73
66.42	−5.51
36.87	−7.07
34.08	−9.83

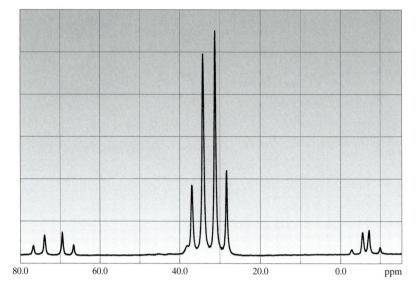

Analyze this spectrum to obtain the chemical shifts and coupling constants. Calculate the appearance of the $^{199}Hg\{^1H\}$ NMR spectrum for this compound. (See Bowmaker, G. A., Clase, H. J., Alcock, N. W., Kessler, J. M., Nelson, J. H., Frye, J. S., *Inorg. Chim. Acta*, **1993**, *210*, 107.)

117. The following is the $^{13}C\{^1H\}$ NMR spectrum of a commercial sample of $D_3{}^{13}CCO_2{}^-\ Na^+$ in D_2O:

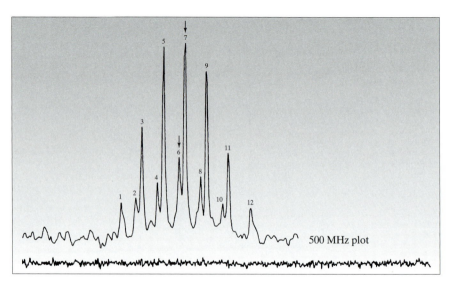

No	Freq (Hz)	PPM	Int%
1	689.69	27.536	922
2	676.26	27.000	1031
3	670.16	26.756	2764
4	656.73	26.228	1414
5	650.44	25.976	4786
6	637.28	28.448	2025
7	631.10	25.196	4774
8	617.67	24.668	1555
9	612.79	24.465	4188
10	598.14	23.880	892
11	593.26	23.685	2133
12	573.73	22.986	785

Explain the line shapes of the ^{13}C resonances. Calculate the relative amounts of $D_3^{13}CCO_2^-$, $HD_2^{13}CCO_2^-$, and $H_2D^{13}C^{13}CO_2^-$ present in the sample? What is the ^{13}C observation frequency? Do these spectra illustrate a primary isotope effect on the ^{13}C chemical shift?

118. Consider the compounds in the following figure. Why is the magnitude of $^3J_{(PP)}$ for **A** (26.91 Hz) greater than that for **B** (15.30 Hz), which is in turn greater than that for **C** (7.50 Hz) (see Maitra, K., Catalano, V. J., Nelson, J. H., *J. Organomet. Chem.*, **1997**, *529*, 409 and Crumbliss, A. L.; Topping, R. J. in Verkade, J. G., Quin, L. D. (eds.) *Phosphorus-31 NMR Spectroscopy in Stereochemical Analysis*, VCH, Deerfield Beach, Florida, **1987**, pp 531–557)?

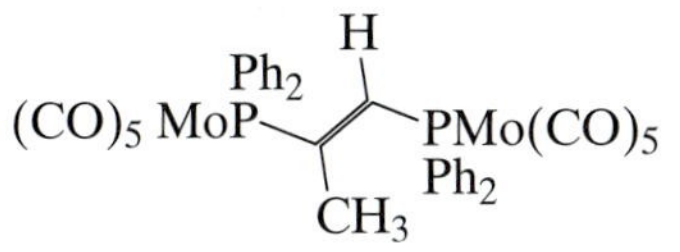

(a)

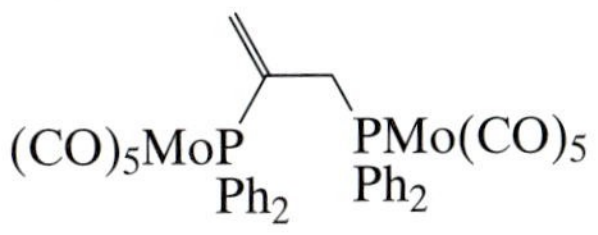

(b)

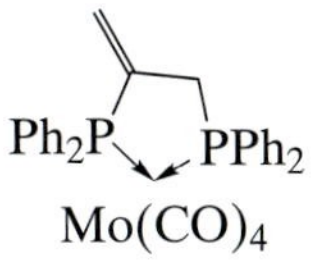

(c)

119. The ^1H and ^{19}F NMR spectra of $C_2HF_3Br_2$ are shown in the following figure:

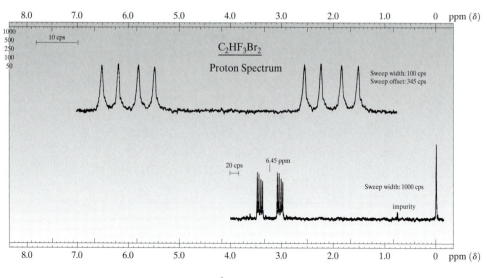

^1H NMR Spectrum.

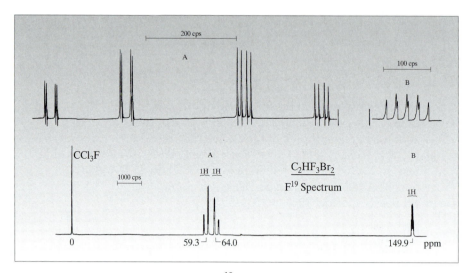

^{19}F NMR Spectrum.

What is the spin system for this molecule? Are the ^1H or ^{19}F spectra first order? Are two of the flourines homotopic, enantiotopic, or diastereotopic? Can the chemical shifts and coupling constants be obtained without the aid of a computer simulation? Describe how you would proceed to analyze these spectra. What is the ^1H observation frequency?

120. Explain the 25.2-MHz $^{13}C\{^1H\}$, $^{13}C\{^{19}F\}$, and $^{13}C\{^{19}F, {}^1H\}$ NMR spectra of 2,2,3,3-tetrafluoropropanol shown in the following figures:

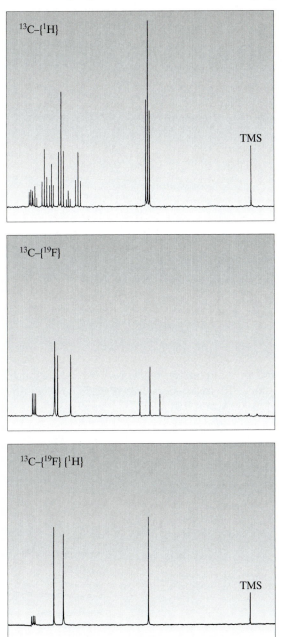

Assign the chemical shifts, including that due to the solvent C_6D_6, and identify the coupling constants.

121. In a Fourier-transform experiment, the signal observed following the pulse was found to be a maximum for $t = 50$ μs. Over what range of NMR transition frequencies would this pulse be effective? What would be the observed effect of a pulse of 100 μs duration?

122. The results of an experiment to determine a 90° observation pulse width are as follows:

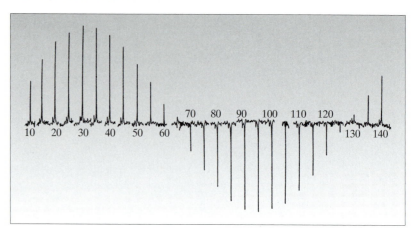

From these data, what are the 45°, 90°, 180°, and 360° pulse widths? The numbers on the figure are in μs.

123. From the following inversion-recovery ^{103}Rh NMR spectra, estimate the value of T_1 for ^{103}Rh in $(\eta^5-C_5H_5)Rh(\eta^4-C_6H_8)$ (delay times are 60, 30, 16, 8, 4, 2, 1, 0.5, 0.1, and 0.01 sec):

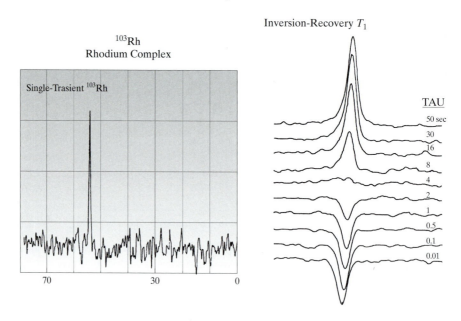

Would you expect the ^{103}Rh T_1 to be field dependent? The data were obtained at 12.6 MHz. What is the field strength of the spectrometer, in tesla?

124. From the following inversion-recovery ^{95}Mo NMR spectra of Na$_2$MoO$_4$, calculate the value of T_1 for ^{95}Mo in the compound:

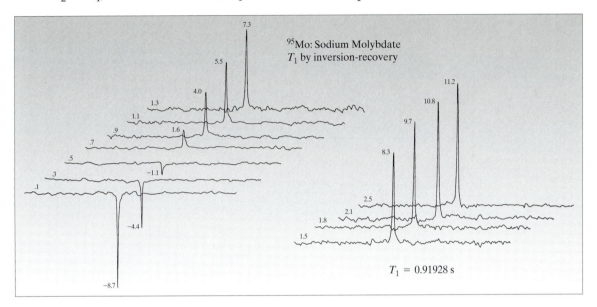

^{95}Mo: Sodium Molybdate
T_1 by inversion-recovery

$T_1 = 0.91928$ s

#	Time (s)	Intensity	#	Time (s)	Intensity
1	0.10	−8.70	7	1.30	+7.30
2	0.30	−4.40	8	1.50	+8.30
3	0.50	−1.10	9	1.80	+9.70
4	0.70	+1.60	10	2.10	+10.80
5	0.90	+4.00	11	2.50	+11.20
6	1.10	+5.50			

Is the answer given in the data provided on the spectrum? Do you agree with this answer?

125. An inversion-recovery T_1 (180-τ-90) ^{13}C{^{1}H} experiment performed on ethylbenzene with τ values of 120.0, 1.0, 2.5, 5.0, 10.0, 12.5, and 15.0 seconds gave the following results:

τ(sec)	120	1	2.5	5.0	7.5	10.0	12.5	15.0
				Intensities %				
6.55	4045	−2939	−1835	−486	939	1805	2202	2406
19.98	4574	−2432	−1961	−1172	581	1399	1760	1978
δ 116.24	5071	−3133	−2418	−1016	788	1557	2034	1981
118.28	7769	−4417	−4168	−1966	787	2666	3672	3680
118.92	9730	−5450	−4339	−2828	798	2692	3429	3795
134.52	198	−462	−438	−288	180	311	500	852

Calculate the T_1 values. Assign the chemical shifts. Based upon the carbon type and the dominant mechanism(s) for ^{13}C relaxation, do the T_1 values meet with your expectations?

126. The following is a 125.698-MHz $^{13}C\{^1H\}$ DEPT spectrum of menthol:

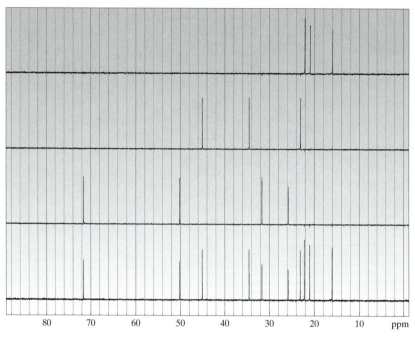

Label the spectrum, and show how it is consistent with the structure of menthol, shown.

127. Completely assign the following 500-MHz ^1H **HOMO 2DJ** NMR spectrum in the vinyl region of $[(\eta^6-C_6H_6)Ru(Ph_2PCH=CH_2)(CH_3CN)Cl]PF_6$:

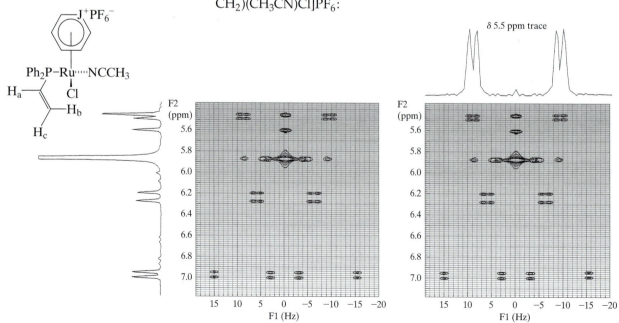

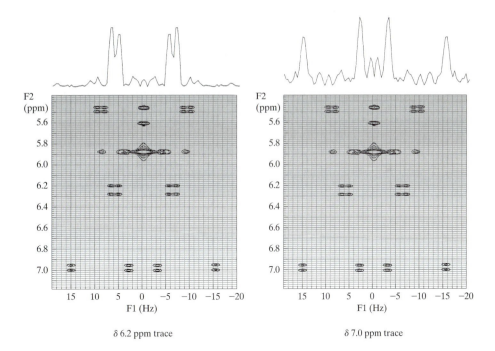

δ 6.2 ppm trace   δ 7.0 ppm trace

Identify the resonance at δ = 5.9 ppm.

128. 300 MHz **¹H HOMO 2DJ** (in the aliphatic region only). ¹H, **COSY**, and 75 MHz ¹³C/¹H **HETCOR** (in the aliphatic region only) spectra of the molecule $C_{22}H_{24}P_2S_2$ shown on the right, together with a coupling tree for the **HOMO 2DJ** spectrum and chemical shift assignments for all spectra. Show that these assignments are consistent with the structure.

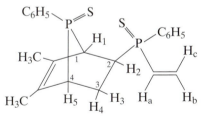

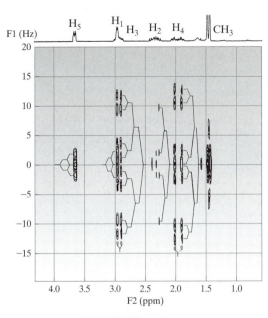

HOMO 2DJ spectrum.

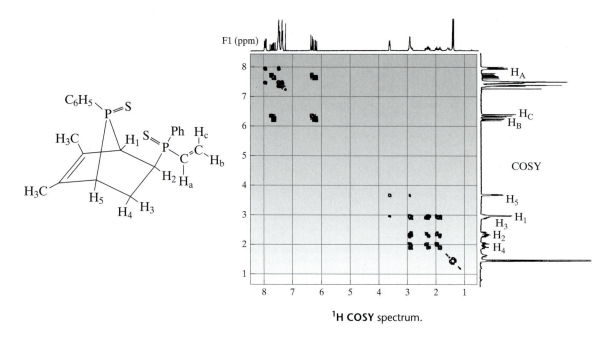

¹H **COSY** spectrum.

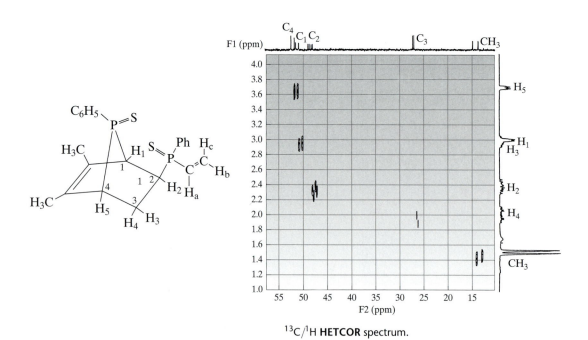

¹³C/¹H **HETCOR** spectrum.

129. The 400-MHz ^{1}H, ^{1}H(**COSY**) and 100-MHz ^{13}C/^{1}H **HETCOR** spectra of 2-*exo*-bromonorbornane are as follows (Abraham, R. J., Fisher, J., *Mag. Reson. Chem.* **1985**, 23, 862):

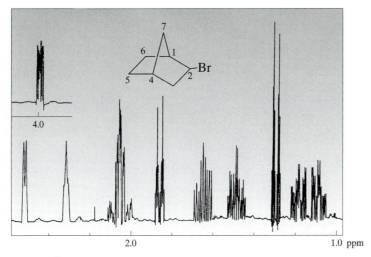

The 400-MHz ^{1}H NMR spectrum of 2-*exo*-bromonorbornane (0.13M in CDCl$_3$ solution).

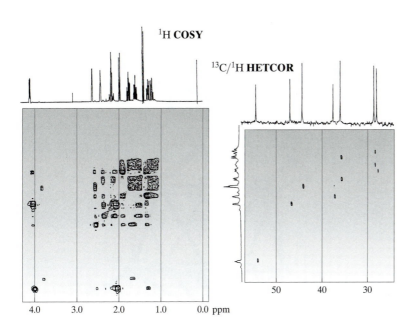

From these figures, assign both the ^{1}H and ^{13}C spectra.

130. Fully interpret the $^{11}B\{^1H\}–^{11}B\{^1H\}$ **COSY** of $[SMe_3][B_{12}H_{10}(SMe_2)]$ and $1,7(SMe_2)_2B_{12}H_{10}$ shown in the following figures (Hamilton, E. J. M., Jordan, G. T.IV, Meyers, E. A., Shore, S. G., *Inorg. Chem.*, **1996**, *35*, 5335):

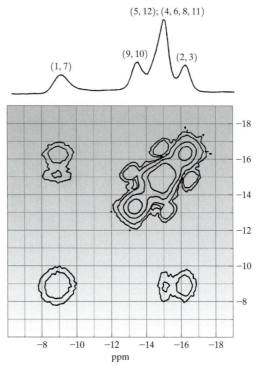

$^{11}B\{^1H\}–^{11}B\{^1H\}$ (COSY) spectrum of $1,7\text{-}(SMe_2)_2B_{12}H_{10}$ showing all peak assignments.

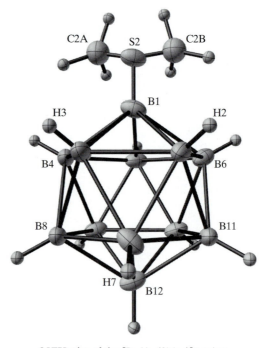

ORTEP plot of the $[B_{12}H_{11}(SMe_2)]^-$ anion.

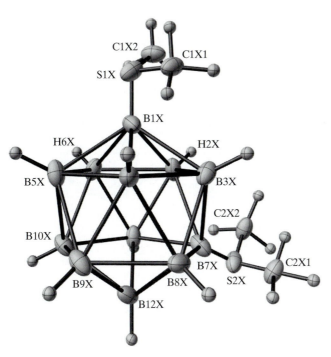

ORTEP plot of molecule X of 1,7-$(SMe_2)_2B_{12}H_{10}$, showing eclipsed conformation of the two SMe_2 units.

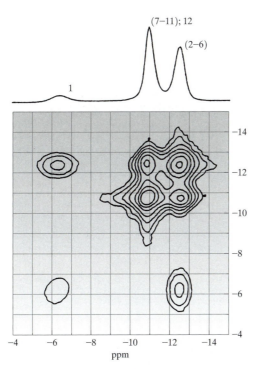

$^{11}B\{^1H\}$–$^{11}B\{^1H\}$ (COSY) spectrum of $[SMe_3][B_{12}H_{11}$–$(SMe_2)]$ showing all peak assignments. Reproduced with permission from above reference, copyright 1996, American Chemical Society.

131. Interpret the following ^{11}B, $^{11}B\{^1H\}$, and $\{^{11}B-^{11}B\}$ **COSY** spectra for $C_4B_{18}H_{22}$ (Janousek, A., Stibr, B., Fontine, X. L. R., Kennedy, J. D., *J. Chem. Soc., Dalton Trans.*, **1996**, 3813):

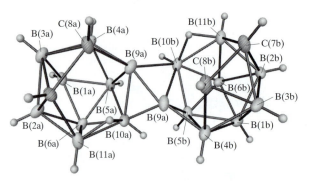

An ORTEP drawing of the crystallographically determined molecular structure of molecule 1 of $C_4B_{18}H_{22}$. Two crystallographically independent near-identical molecules occur in the unit cell. Boron and carbon atoms are represented as ellipsoids at the 50% probability level, and hydrogen atoms as circles, each with an arbitrary small radius.

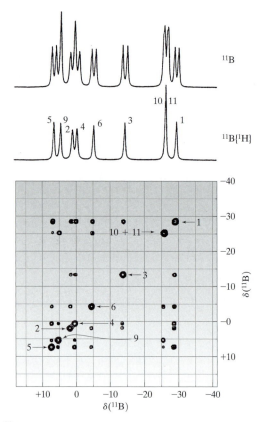

The 128-MHz ^{11}B NMR spectra of $C_4B_{18}H_{22}$ in CD_2Cl_2 solution at 297 K. The top trace is the straightforward ^{11}B spectrum, and under it is an equivalent spectrum recorded with complete $\{^1H\}$ decoupling. The bottom diagram is a "two-dimensional" contour plot arising from a $[^{11}B-^{11}B]$ correlation-spectroscopy (COSY) experiment, also conducted with complete $\{^1H\}$ decoupling. Reproduced with permission from above reference, copyright 1996, Royal Society of Chemistry.

132. The $^{13}C\{^1H\}$ NMR chemical shifts of the carbonyl carbon for $PhCH_2{}^{13}CO_2H$ (1.5×10^{-2} M in H_2O) as a function of pH are listed in the following table (Holmes, D. L., Lightner, D. A., *Tetrahedron*, **1995**, *51*, 1607):

pH	δ ^{13}C	pH	δ ^{13}C	pH	δ ^{13}C
2.26	177.100	4.00	178.350	6.55	181.065
2.49	177.100	4.26	178.956	7.04	181.083
2.75	177.153	4.61	179.755	7.59	181.086
3.00	177.220	4.91	180.291	8.04	181.090
3.26	177.359	5.38	180.787	8.84	181.090
3.50	177.565	5.76	180.960	9.54	181.093
3.75	177.903	6.08	181.037	10.03	181.093

From these data, calculate pk_a for phenylacetic acid, and compare your calculated value with the value obtained from the literature.

133. For the compound

calculate pk_a from the following ^{19}F chemical shifts as a function of pH (200 mM in DMSO-d_6):

pH	δ ^{19}F	pH	δ ^{19}F	pH	δ ^{19}F
0.30	−188.56	3.35	−182.42	6.05	−180.91
1.70	−188.86	3.80	−181.56	6.54	−180.91
2.01	−187.08	4.21	−181.21	7.04	−180.92
2.40	−185.86	4.62	−181.05	8.12	−180.92
2.71	−184.69	5.08	−180.96	10.03	−180.92
3.01	−183.53	5.54	−180.94		

134. From the following ^{19}F chemical shifts of *o*-, *m*-, and *p*- fluoropbenzoic acids (2.00 mM in 5% DMSO/H$_2$O) calculate the pK_a values for the three acids:

Ortho	pH	δ ^{13}C	pH	δ ^{13}C	pH	δ ^{13}C
	0.30	−112.77	3.35	−114.89	6.05	−116.80
	1.70	−112.86	3.80	−115.75	6.54	−116.80
	2.01	−113.00	4.21	−116.27	7.04	−116.80
	2.40	−113.28	4.62	−116.55	8.12	−116.81
	2.71	−113.68	5.08	−116.70	10.03	−116.82
	3.01	−114.19	5.54	−116.77		

Meta	pH	δ ^{13}C	pH	δ ^{13}C	pH	δ ^{13}C
	0.30	−113.92	3.35	−114.20	6.05	−114.74
	1.70	−113.93	3.80	−114.39	6.54	−114.75
	2.01	−113.94	4.21	−114.55	7.04	−114.75
	2.40	−113.96	4.62	−114.64	8.12	−114.76
	2.71	−114.01	5.08	−114.71	10.03	−114.76
	3.01	−114.08	5.54	−114.75		

Para	pH	δ ^{13}C	pH	δ ^{13}C	pH	δ ^{13}C
	0.30	−106.60	3.35	−107.55	6.05	−110.97
	1.70	−106.62	3.80	−108.42	6.54	−110.99
	2.01	−106.66	4.21	−109.28	7.04	−111.00
	2.40	−106.75	4.62	−110.06	8.12	−111.03
	2.71	−106.90	5.08	−110.57	10.03	−110.99
	3.01	−107.11	5.54	−110.82		

135. The data in question 134 show that, at low pH, the ^{19}F chemical shifts for the three isomers of fluorobenzoic acid fall in the order δ ^{19}F *meta* > δ ^{19}F *ortho* > δ ^{19}F *para*, while at high pH, the order δ ^{19}F *ortho* > δ ^{19}F *meta* > δ ^{19}F *para* is observed. Explain these results.

136. The ^{1}H and ^{13}C{^{1}H} NMR spectra of *meso*-tetraphenylporphyrin (TPP) are temperature dependent, as illustrated in the following figures:

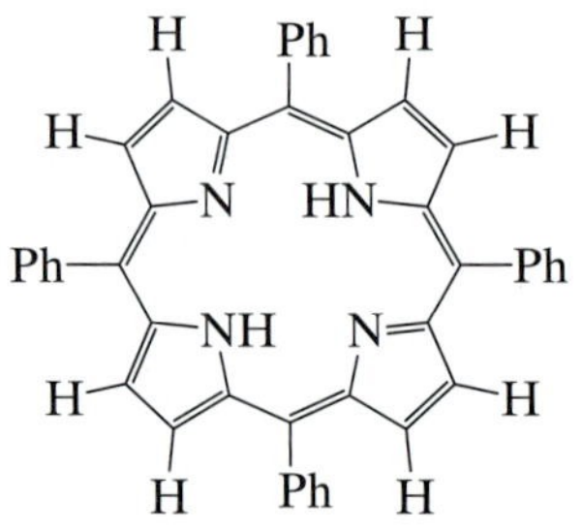

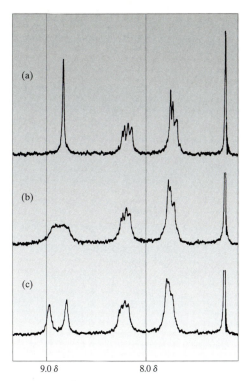

The 100-MHz ^{1}H spectrum of *meso*-tetraphenylpor-phyrin (TPP) 0.03M in CDCl$_3$ at (a) −12°C, (b) −46°C, and (c) −60°C. (From Abraham, Hawkes, and Smith. *Tetrahedron Letters*, 1974, 71–74, Pergamon Press, Oxford.)

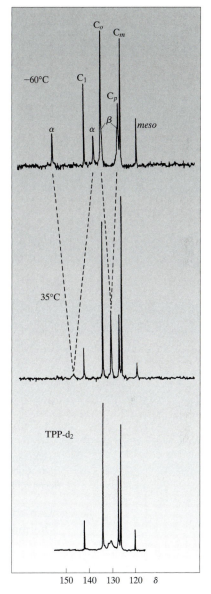

The 25.2-MHz ^{13}C–{^{1}H} spectrum of TPP in CDCl$_3$ at 35°C (middle). −60°C (upper), and of the dideuteriat-ed species (lower). (From Abraham, Hawkes, and Smith. *J. Chem. Soc., Perkin Trans. 2*. 1974, 627–634.)

Explain why the spectra are temperature dependent.

137. Why are the ^{1}H NMR spectra of compound (1a) temperature dependent (see Brunner H., Oeschey, R., Nuber, B. O., *Organometallics*, **1996**, *15*, 3616)?

^{1}H NMR spectra between 2.00 and 1.55 ppm (methyl doublet region) of (1a) in methanol-d_4 at various temperatures. The horizontal offset of the spectra is 0.015 ppm. Reproduced with permission from above reference, copyright 1996, American Chemical Society.

Can one obtain either or both equilibrium and activation thermodynamic parameters from data of this type?

138. Explain the following ^{13}C{^{1}H} and VT ^{19}F NMR spectra for [Tl(tpp)(O$_2$CCF$_3$)] (see Chou, L.-F., Chen, J.-H., *J. Chem. Soc., Dalton Trans.*, **1996**, 3787):

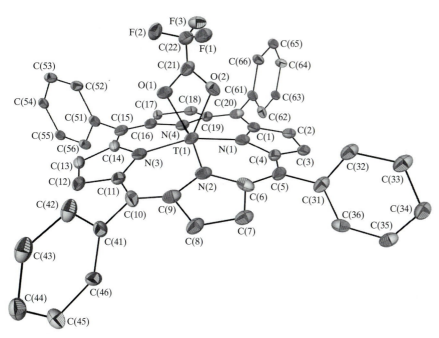

Molecular configuration and atom-labelling scheme for [Tl(tpp)(O$_2$CCF$_3$)] · CH$_2$Cl$_2$. Hydrogen atoms and solvent C(71)H(71A)H(71B)—Cl(1)Cl(2) are omitted for clarity.

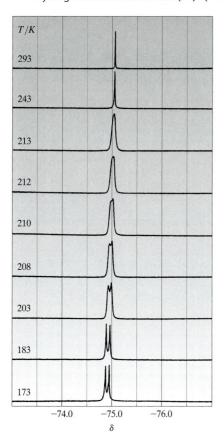

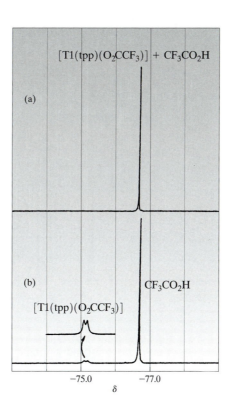

The 376.50-MHz ^{19}F NMR spectra for [Tl(tpp)(O$_2$CCF$_3$)] in [^{2}H$_8$]thf at various temperatures.

The 282.40-MHz ^{19}F NMR spectra for [Tl(tpp)(O$_2$CCF$_3$)] ≈ 0.05 cm^3 CF$_3$CO$_2$H in [^{2}H$_8$]thf at (a) 20 and (b) −100°C.

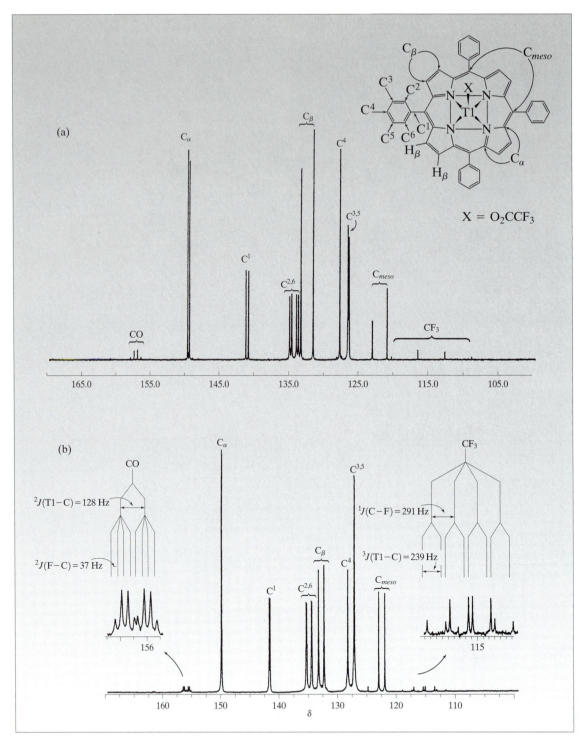

The ^{13}C broadband NMR spectra for [Tl(tpp)(O_2CCF_3)]: (a) in $CDCl_3$ at 20°C (75.47 MHz), (b) in [$^2H_8$]thf at −100°C (150.92 MHz). Reproduced with permission from above reference, copyright 1996, Royal Society of Chemistry.

139. The structure of compound 1 is as follows:

$P_o = OP(OPh)_2$

Why are the following $^{31}P\{^1H\}$ NMR spectra of compound 1 temperature dependent?

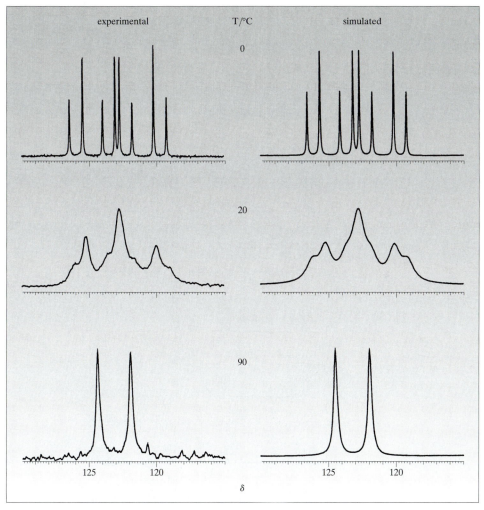

Variable-temperature 121.6-MHz $^{31}P\{^1H\}$ NMR spectra. Reproduced with permission from reference below, copyright 1996, Royal Society of Chemistry.

What are the spin systems under slow and rapid exchange conditions?
(See Van De Beuken, E. K., de Lange, W. G. J., van Leeuwen, P. W. N. M., Veldman, N.; Spek, A. L., Feringa, B. L., *J. Chem. Soc., Dalton Trans.*, **1996**, 3561.)

140. Show how the following VT $^{31}P\{^1H\}$ NMR spectra are consistent with a temperature-dependent equilibrium involving three rhodium diphosphine complexes (see Kadyrov, R., Freier, T., Heller, D., Michalik, M., Selke, R., *J. Chem. Soc., Chem. Commum.*, **1995**, 1745):

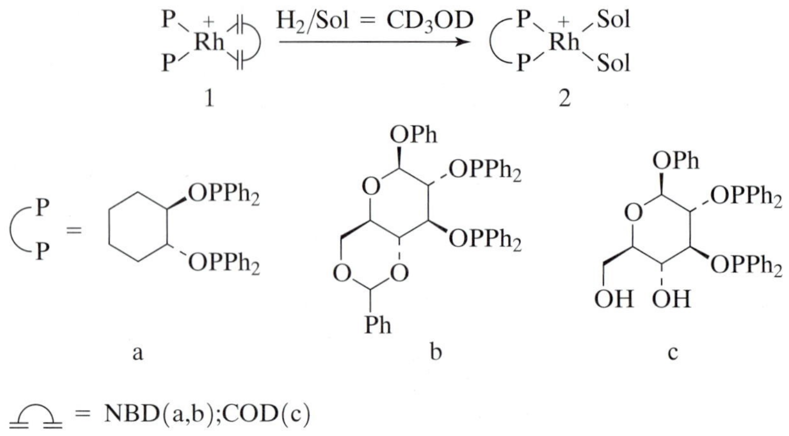

Scheme 1

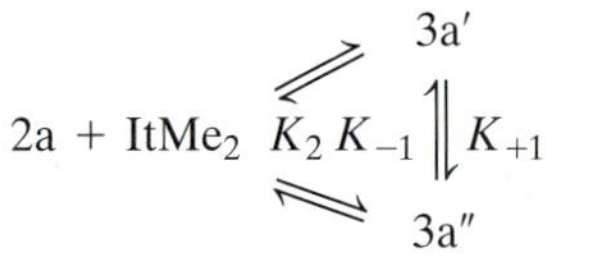

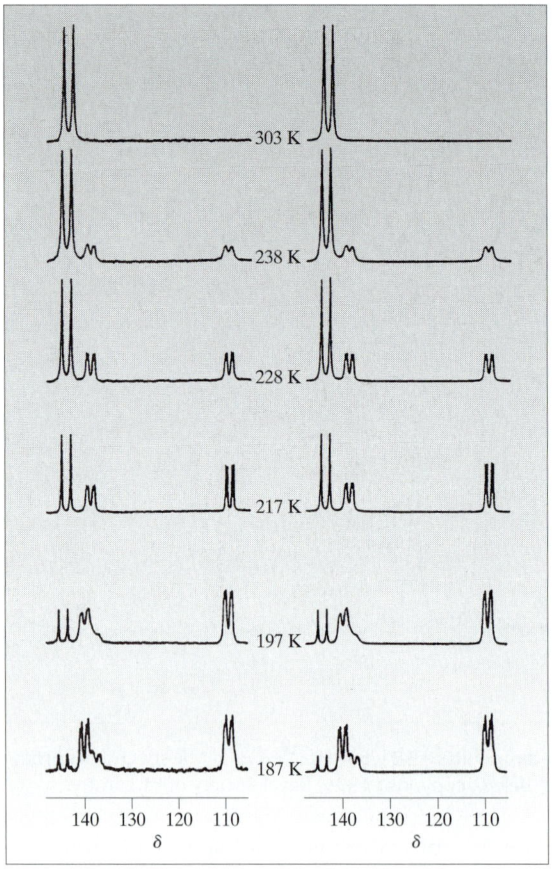

Experimental and calculated $^{31}P\{^1H\}$ NMR spectra of the complexes 2a, 3a', and 3a" in CD$_3$OD. Reproduced with permission from above reference, copyright 1995, Royal Society of Chemistry.

141. The N—H resonance of 1,2,4-triazole appears as a singlet at ambient temperature both in the pure liquid and in dry dimethylsulfoxide–d_6. While this resonance remains a singlet even at low temperatures, the singlet C—H proton resonance present at ambient temperature becomes "two" resonances at temperatures below 0°C. Explain. Predict the temperature dependence of the $^{13}C\{^1H, {}^{14}N\}$ NMR spectrum. (See Creagh, L. T., Truitt, P., *J. Org. Chem.*, **1968**, *33*, 2956.) The structure of 1,2,4-triazole is as follows:

142. Under slow exchange conditions, the following equation is valid:

$$\frac{J \text{ at small } \tau}{J \text{ at large } \tau} = \left[1 - \frac{1}{2\pi^2\tau^2(J \text{ at large } \tau)^2} \right]^{\frac{1}{2}}.$$

At 46.8° $^3J(H, CH_3)$ in the N,N-dimethylanilinium ion is 5.11 Hz, while at 26.8° and below, it is 5.26 Hz. The structure of the ion is

Calculate the N—H exchange rate at 46.8°C. Given further information that the coupling constant is 5.02 Hz and 4.67 Hz at 50.6°C and 55.5°C, respectively, estimate the activation energy for the exchange reaction. (See Reynolds, W. F., Schaefer, T., *Can. J. Chem.*, **1963**, *41*, 540.)

143. In the exchange process $A \underset{k_b}{\overset{k_a}{\rightleftharpoons}} B$, $\delta_A = 4.1$ ppm, $\delta_B = 3.7$ ppm, $T_{2A}^{-1} = 3 \text{ s}^{-1}$, and $T_{2B}^{-1} = 2 \text{ s}^{-1}$, with a 200-MHz spectrometer.

   a. What would be the observed shift when $k_a = 10^4 \text{ s}^{-1}$ and $k_b = 5 \times 10^3 \text{ s}^{-1}$?

   b. What would be the observed line width when $k_a = 3 \text{ s}^{-1}$ and $k_b = 1.5 \text{ s}^{-1}$?

144. Suppose a tetra-substituted ethane molecule were limited to just two conformations, A (where $^3J_{HH} = 13$ Hz) and B (where $^3J_{HH} = 4$ Hz). If there is rapid interconversion of A and B and the observed $^3J_{HH} = 7$ Hz, calculate the average fraction of the molecular population in each conformation.

145. Suppose you are investigating the dynamic behavior of the exchange process of dimethylformamide. At the stopped exchange limit, the 60-MHz ^1H NMR spectrum shows singlets for the methyl hydrogens at $\delta = 2.98$ and 2.81 ppm, while the 20-MHz $^{13}C\{^1H\}$ NMR spectrum shows resonances for the methyl carbons at 36.2 and 31.1 ppm. (a) What is the value of $\Delta\nu_0$ for each spectrum? (b) Which spectrum will exhibit coalescence at the higher temperature?

146. Given the following ranges for T_2 for the nuclei listed, estimate the natural line widths for their resonances:

^1H: 0.1 to 10 sec; ^{13}C: 2 to 100 sec;

^{15}N: 0.5 to 170 sec; ^{11}B: 0.01 to 1 sec;

^{19}F: 1 to 25 sec; ^{195}Pt: 0.03 to 1.3 sec.

147. By consulting Green, M. L. H., Wang, L.-L., Selta, A., *Organometallics*, **1992**, *11*, 2660, derive equations to express the rate constants that one could obtain by a variable-temperature NMR study of an AB_2 spin system undergoing chemical exchange.

148. Explain the reason(s) for the differences in the 60-MHz ^1H NMR specta of methylamine hydrochloride in D_2O and in D_2O containing excess HCl shown in the following figures (see Glaros, G., Cromwell, N. H., *J. Chem. Educ.*, **1971**, *48*, 202):

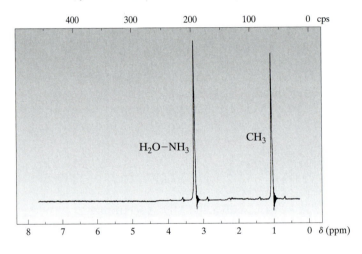

NMR spectrum of a solution of 200 mg methylamine hydrochloride in 0.3 ml water.

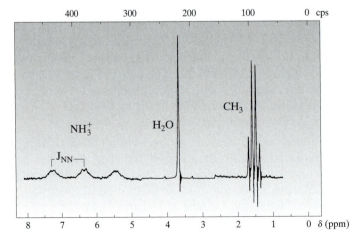

NMR spectrum of a solution of 200 mg methylamine hydrochloride in 0.3 ml water with one drop of dilute hydrochloric acid added. Reproduced with permission from above reference, copyright 1971, Division of Chemical Education, Inc.

149. 2,4-Pentanediol has been separated into two diastereomers (Pritchard, J. G., Vollmer, R. L., *J. Org. Chem.*, **1963**, *28*, 1545). One diastereomer has a low melting point (50–52°C), and the other forms an oil. The methylene resonance of one diastereomer is a triplet, while the methylene resonance of the other diastereomer is a triplet of doublets. Which of these spectra belongs to the *dl*- and which to the *meso*-diastereomer?

(*meso*)   (*dl*)

150. Predict the ^{1}H and ^{1}H **HOMO 2DJ** spectra of both diastereomers of 2,4-pentanediol, and compare your predictions with the ^{1}H spectra found in the Aldrich catalog.

151. The 25.14-MHz ^{13}C{^{1}H} NMR spectrum of *cis*- decalin is as follows (Dalling, D. K., Grant, D. M., Johnson, L. F., *J. Am. Chem. Soc.*, **1971**, *93*, 3678):

Temp. °C	Carbons	$\Delta\nu_{1/2}$	ref $\Delta\nu_{1/2}$	k, sec^{-1}
45.5	1,5,4,8	7.5	2.8	10,000
45.5	2,6,3,7	7.0	2.8	8,990
27.1	1,5,4,8	26.8	2.5	1,930
27.1	2,6,3,7	24.6	2.5	1,710
15.6	1,5,4,8	56.2	3.0	884
15.6	2,6,3,7	51.5	3.0	778
8 ± 5	1,5,4,8		coalescence	384
5 ± 5	2,6,3,7		coalescence	344
0.8	1,5	80	3.3	241
0.8	2,6	73.5	3.3	221
−12.2	1,5,4,8	22.6	4.0	58.4
−12.2	2,6,3,7	27.2	4.0	72.9
−29.0	1,5,4,8	6.8	3.9	9.1
−29.0	2,6,3,7	6.8	3.9	9.1

From the data given in the above table showing the temperature dependence of the spectrum, calculate the activation thermodynamic parameters $\Delta G^\ddagger$, $\Delta H^\ddagger$, and $\Delta S^\ddagger$ for the exchange process.

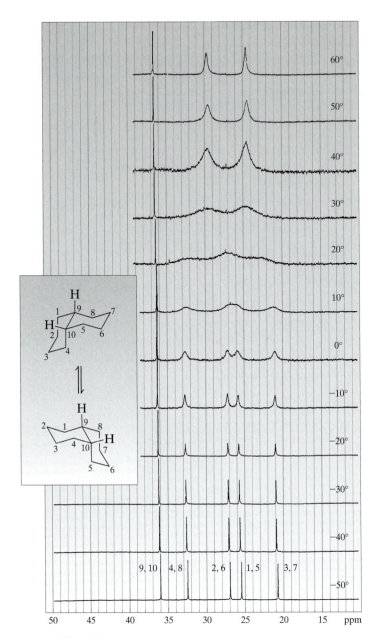

The 125-MHz $^{13}C-\{^{1}H\}$ spectrum of *cis*-decalin at the temperatures indicated.

Describe the nature of the exchange process.

152. The Cope rearrangements of homotropilidene and semibulvalene may be studied by variable-temperature $^{13}C\{^{1}H\}$ NMR spectroscopy.

Suggest what you would expect to observe in the spectra as a function of temperature.

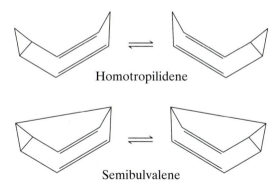

Homotropilidene

Semibulvalene

(See Bicker, R., Keisler, H., Steigel, A., Strohrer, W. D., *Chem. Ber.*, **1975**, *108*, 2708; Cheng, A. K., Anet, F. A. L., Mioduski, J., Meinwald, J., *J. Am. Chem. Soc.*, **1974**, *96*, 2887.)

153. The *cis-trans* isomerization of $[(4-Cl-C_6H_4)P(CH_3)_2]_2PdCl_2$ in nitrobenzene, as a function of temperature, has been studied by NMR spectroscopy (Verstuyft, A. W., Cary, L. W., Nelson, J. H., *Inorg-Chem.*, **1975**, *14*, 1495; Verstuyft, A. W., Nelson, J. H., *Inorg-Chem.*, **1975**, *14*, 1501). From the following data, calculate ΔG_{205}, $\Delta H°$, and $\Delta S°$ for this isomerization process:

$T(°C)$	$K_{eq} \left(\dfrac{trans}{cis} \right)$
−15	0.0647
−03	0.1042
09	0.1967
24	0.2856
39	0.4543
57	0.8680

154. The *cis-trans* isomerization of dibromobis (1-n-butyl-3,4-dimethylphosphole) palladium(II) in $CDCl_3$ has been studied by NMR spectroscopy (MacDougall, J. J., Mathey, F., Nelson, J. H., *Inorg. Chem.* **1980**, *19*, 1400). From the following data, calculate ΔG_{305}, $\Delta H°$, and $\Delta S°$ for this isomerization process:

$T(°C)$	$K_{eq} \left(\dfrac{trans}{cis} \right)$
29.5	0.304
34	0.359
38	0.381
45	0.487
51	0.523
59	0.681
62	0.736

155. The temperature dependence of the phosphine dissociation equilibrium of $(n\text{-Bu}_3P)_2CdBr_2$ has been studied by 121.66-MHz $^{31}P\{^1H\}$ NMR spectroscopy (Kessler, J. M., Reeder, J. M., Vac, R., Yeung, C., Nelson, J. H., Frye, J. S., *Mag. Reson. Chem.*, **1991**, *29*, S94). Consider the data given in the following table:

Temp, K	J_{obs}/J_{stop}	$\Delta\nu_{1/2}(observed)(Hz)$
193	0.995	4.649
203	0.988	7.311
213	0.983	11.281
223	0.977	15.915
233	0.968	22.354
243	NA	30.676
253	NA	47.222
263	NA	62.924
273	NA	85.749
283	NA	120.803

From these data, calculate the equilibrium thermodynamic parameters ΔG°_{298}, ΔH°, and ΔS° and the activation thermodynamic parameters $\Delta G^{\ddagger}_{298}$, $\Delta H^{\ddagger}$, and $\Delta S^{\ddagger}$ for the dissociation equilibrium reaction

$$(n\text{-Bu}_3P)_2CdBr_2 \rightleftharpoons (n\text{-Bu}_3P)CdBr_2 + n\text{-Bu}_3P.$$

The data were collected in the slow-exchange regime, so that $k_{ex} = \pi\Delta\nu_{cor}$ and $\Delta\nu_{cor} = \Delta\nu_{obs}$ − the natural linewidth, which in this case is 4.347 Hz. Since the observed $^1J(^{113}Cd^{31}P)$ is a population-weighted average, one can calculate the population of $(n\text{-Bu}_3P)_2CdBr_2$ from the ratio J_{obs}/J_{stop}, where J_{stop} is the coupling constant in the absence of exchange. Then $K_{eq} = \dfrac{(1 - J_{obs}/J_{stop})^2}{(J_{obs}/J_{stop})}$. Which should have the larger value, $^1J(^{111}Cd^{31}P)$ or $^1J(^{113}Cd^{31}P)$? Why?

156. H. Rüegger and P. S. Pregosin (*Inorg. Chem.*, **1987**, *26*, 2912) have studied the reaction

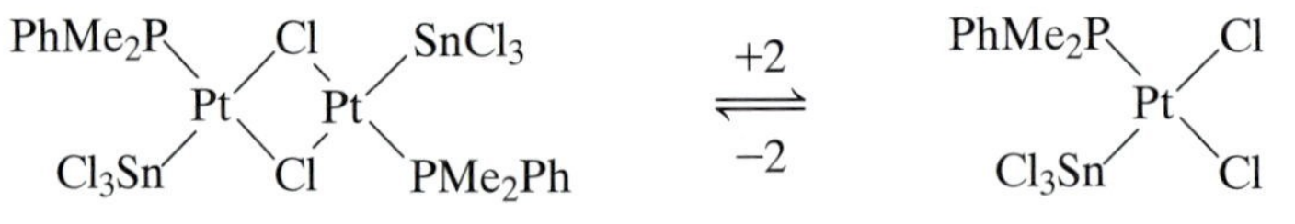

where (**2**) is F—⟨⟩—CN by ^{31}P and ^{19}F exchange spectroscopy, as illustrated in the following figure:

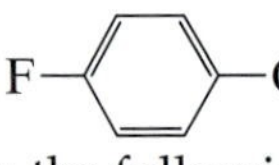

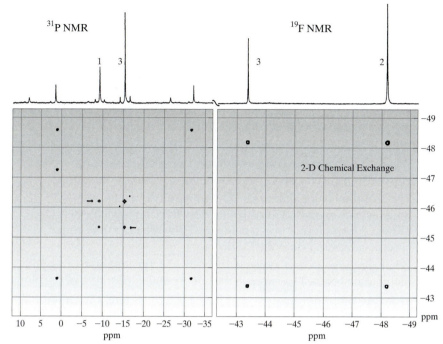

Two-dimensional exchange spectra for the reaction of 1 with 2: (left) ^{31}P; (right) [19]F. The arrows (left) indicate the cross peaks. The high-field ^{19}F signal arises from uncoordinated nitrile. The chemical-shift scale in the figure is defined relative to trifluorotoluene. Reproduced with permission from above reference, copyright 1987, American Chemical Society.

Explain how these spectra indicate that (**1**) and (**3**) are undergoing an exchange process involving gain and loss of (**2**). What is the usual acronym for this type of two-dimensional spectroscopy?

157. Consult Benn, R., Cibura, K., Hoffmann, P., Jonas, K., and Rufinska, A., (*Organometallics*, **1985**, *4*, 2214), and explain how the [13]C **EXSY** spectrum gave information about the chemical exchange shown in the following figure:

158. Describe what two-dimensional NMR spectroscopy technique could be used to
  a. determine what protons are spin coupled to each other,
  b. separate homonuclear (J_{HH}) from heteronuclear (J_{XH}) spin–spin coupling,
  c. determine what protons are proximate in space,
  d. separate carbon chemical shifts from carbon–hydrogen coupling constants,
  e. correlate proton and carbon chemical shifts,
  f. determine carbon–carbon connectivities, and
  g. establish what nuclei are undergoing chemical exchange.

159. When there is only one time variable in a 2D NMR experiment (i.e., t), why is it necessary to process the data with *two* Fourier transformation operations?

160. Why is accurate calibration of pulse widths and delays essential for the success of a 2D NMR experiment?

161. What factors dictate the choice of the number of data points in both dimensions in 2D NMR spectroscopy?

162. What factors govern the minimum number of transients in a 2D NMR spectrum?

163. What are *dummy* or, as they are sometimes called, *steady-state* scans, and why are a number of them often acquired before actual data acquisition?

164. What is *DC* or *baseline* correction?

165. Why is it not necessary to perform DC correction before the second Fourier transformation when processing a 2D NMR data set?

166. The two-dimensional data set $S(t_1, t_2)$ requires two Fourier transformation operations. Explain why the time variable t_2 is almost always Fourier transformed before t_1. How are magnitude-mode spectra computed from the FID?

167. What are the sources of noise in 2D NMR spectra, and how can symmetrization be used to improve the quality and appearance of the plot?

168. What are projection spectra, and how are they different from normal 1D NMR spectra?

169. How many types of plots are generally used in 2D NMR spectroscopy? What are they called?

170. Explain the 10.14-MHz $^{15}N\{^{1}H\}$ and ^{15}N NMR spectra of formamide shown in the following figure:

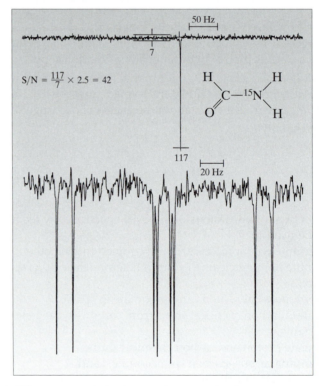

Decoupled ^{15}N spectrum of 90% formamide. Accumulated 10 times (10 × 100 sec.). Nondecoupled ^{15}N spectrum of 90% formamide, with NOE. Accumulated 400 times (27 min.).

Why is the ^{15}N resonance upside down? Explain the pulse sequences that were used to obtain these spectra. Is the ^{15}N spectrum first order? Describe the multiplet in the ^{15}N spectrum and identify the spin system. What are the approximate magnitudes of the N—H coupling constants?

171. In assigning the chemical shifts and coupling constants for the $^{13}C\{^1H\}$ NMR spectrum of $(\eta^6-C_6H_6)Ru(Ph_2PCH{=}CH_2)Cl_2$, the resonances for the *ipso* phenyl carbons and the α- and β-vinyl carbons were difficult to assign. State what NMR experiment(s) would aid in the assignment of these resonances and how you would interpret the data.

172. Explain the following **2D EXSY** ^1H NMR spectra of $[Ru(H)_2(CO)(PPh_3)_3]$ (also shown in the figure) in the hydride region (Ball, G. E., Mann, B. E., *J. Chem. Soc. Chem. Commun.*, **1992**, 561):

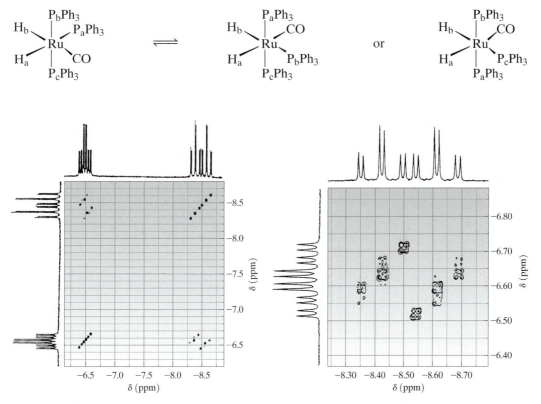

(a) A **2D EXSY** spectrum of the hydride resonances of $[Ru(H)_2(CO)(PPh_3)_3]$ at 49.1°C in $[^2H_8]$ toluene. A mixing delay of 0.2 s was used. (b) An expansion of the signal from (a) over the cross peaks centered at $\delta = -8.44$ (x-axis) and $\delta = -6.54$ (y-axis). Reproduced with permission from above reference, copyright 1992, Royal Society of Chemistry.

173. Explain the following 300-MHz ^{1}H NOE difference spectra:

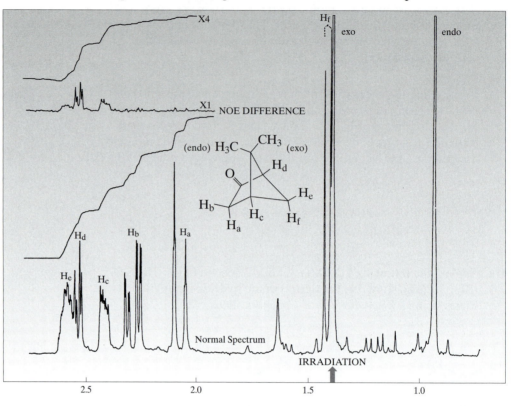

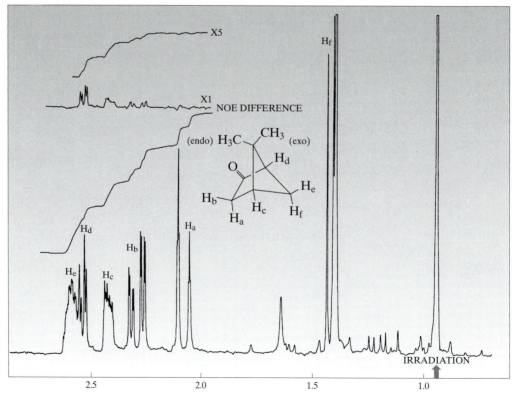

174. Explain the main use of ^{1}H NOE difference spectroscopy. Why does it involve a mathematical subtraction of the normal ^{1}H NMR spectrum from the NOE-enhanced ^{1}H NMR spectrum?

175. Explain the following 300-MHz ^{1}H **COSY** spectrum:

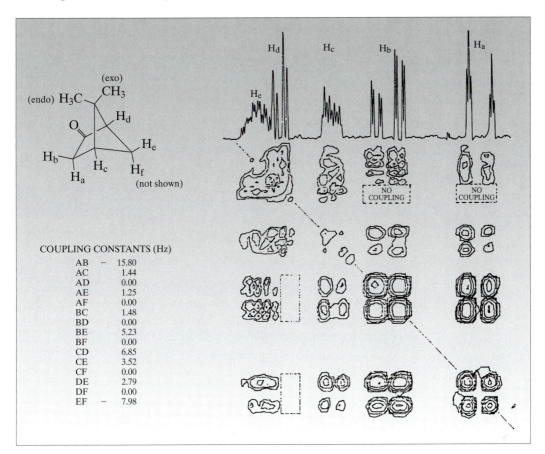

Note that the methyl resonances are not shown in this expansion. From the spectra shown in Problem 173, can you conclude whether or not there is any coupling to the methyl protons?

176. What is the difference between homo- and heteronuclear **2DJ-RESOLVED** spectroscopy?

177. Is it possible to obtain proton–proton coupling information from a 1D ^{1}H NMR experiment?

178. What are the advantages of a **COSY-45** over a **COSY-90** experiment?

179. Interpret the following 75-MHz ^{13}C 2D **INADEQUATE** spectrum of cedrol:

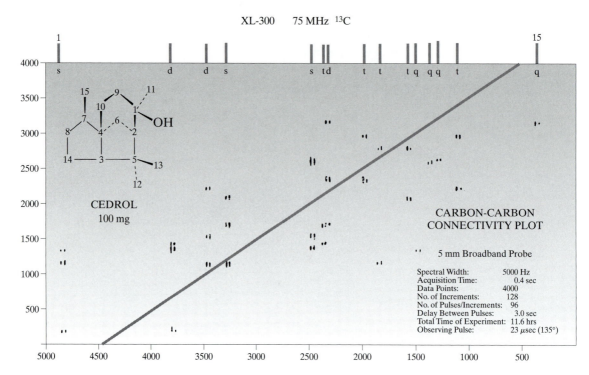

XL-300   75 MHz ^{13}C

CEDROL
100 mg

CARBON-CARBON
CONNECTIVITY PLOT

5 mm Broadband Probe

Spectral Width: 5000 Hz
Acquisition Time: 0.4 sec
Data Points: 4000
No. of Increments: 128
No. of Pulses/Increments: 96
Delay Between Pulses: 3.0 sec
Total Time of Experiment: 11.6 hrs
Observing Pulse: 23 μsec (135°)

The letters on the stick-diagram ^{13}C spectrum on top of the figure indicate the line shapes of the proton-coupled ^{13}C spectrum. Give the chemical formula for the molecule. What other NMR spectra would you obtain in order to completely confirm the structure of cedrol?

180. Explain the following ^{13}C 2D **INADEQUATE** spectrum of α-picoline:

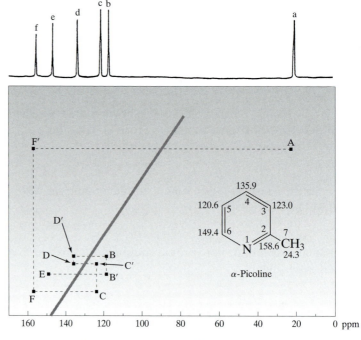

Two-dimensional **INADEQUATE** spectrum of α-picoline.

181. Completely assign the following ^{13}C 2D **INADEQUATE** spectrum of 3-methyltetrahydrofuran:

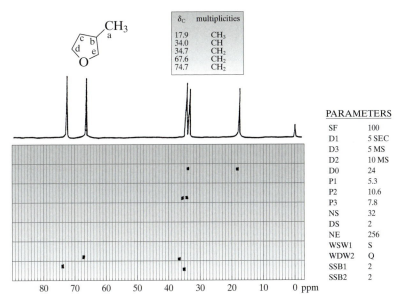

δ_C	multiplicities
17.9	CH_3
34.0	CH
34.7	CH_2
67.6	CH_2
74.7	CH_2

PARAMETERS

SF	100
D1	5 SEC
D3	5 MS
D2	10 MS
D0	24
P1	5.3
P2	10.6
P3	7.8
NS	32
DS	2
NE	256
WSW1	S
WDW2	Q
SSB1	2
SSB2	2

2D **INADEQUATE** spectrum of methyl tetrahydrofuran.

182. Completely interpret the following 75.57-MHz ^{13}C **INADEQUATE** spectrum of menthol (see text for structure and carbon numbering scheme):

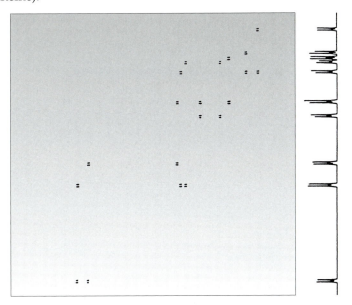

183. In addition to a singlet at 16.50 ppm, the ^{1}H NMR spectrum of 2,3-dibromopropionic acid shows three resonances: $\delta_A = 4.49$, $\delta_M = 3.92$, and $\delta_X = 3.68$ ppm, with $J(AM) = 11.00$ Hz, $J(AX) = 4.50$ Hz, and $J(MX) = 10.00$ Hz. Draw a stick diagram of the spectrum, and predict the appearance of the **COSY** and **2D HOJ** spectra for the compound.

184. Explain completely the following 125.705-MHz ^{13}C and ^{13}C\{^{1}H\} NMR spectra of 2,4-pentanedione:

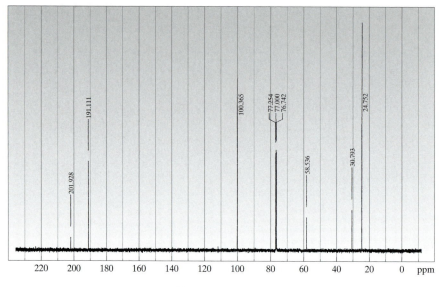

125.705-MHz ^{13}C\{^{1}H\} NMR spectrum in CDCl$_3$.

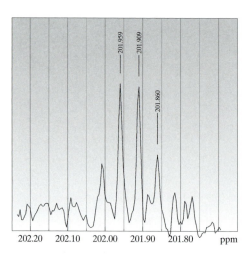

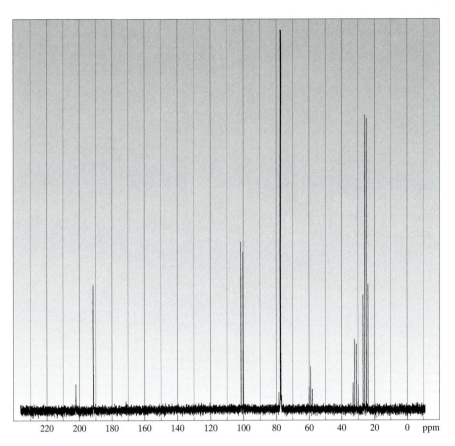

125.705-MHz ^{13}C NMR spectrum in CDCl$_3$.

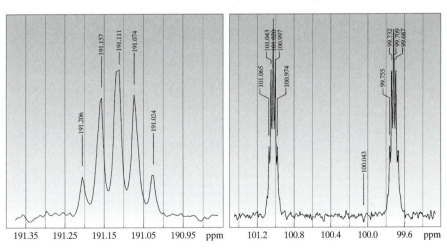

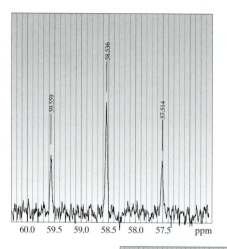

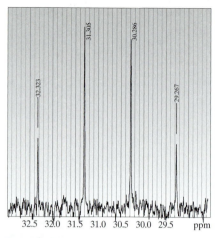

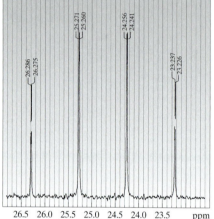

In your explanation, assign all chemical shifts and calculate the CH coupling constants.

185. The 99.54-MHz ^{1}H NMR spectra of $[(PhCH_2)_3P]_2Pd(CNS)_2$ in the absence (bottom) and presence (top) of the lanthanide-shift reagent Eu(fod)$_3$ were used to assign the CH$_2$ resonances of the three different isomers that are present in a CDCl$_3$ solution at ambient temperature (MacDougall, J. J., Verstuyft, A. W., Cary, L. W., Nelson, J. H., *Inorg. Chem.*, **1980**, *19*, 1036):

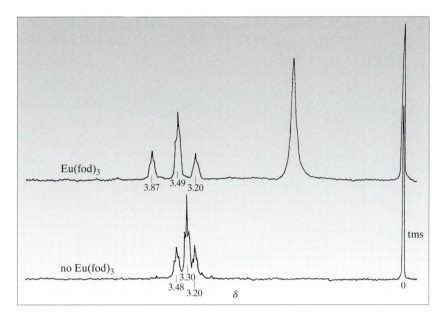

Eu(fod)₃

3.87 3.49 3.20

no Eu(fod)₃

3.30
3.48 3.20

δ

tms

0

Reproduced with permission from above reference, copyright 1980, American Chemical Society.

Explain how this was done.

186. Draw the structure of Eu(AcAc)₃. How many isomers are possible for Eu[PhC(O)CHC(O)CH₃]₃? Would one have to use an isomerically pure compound of the latter type as a shift reagent? Why or why not?

187. *M*(tfac)₃ complexes, where tfac is trifluoroacetylacetonate, have been studied by ^{1}H and ^{19}F NMR spectroscopy (Fay, R. C., Piper, T. S., *J. Am. Chem. Soc.*, **1963**, *85*, 500). Describe what one expects to observe in the ^{1}H and ^{19}F NMR spectra of the *cis* and *trans* isomers of these compounds:

cis   *trans*

188. Two dimensional ^{31}P NMR spectroscopy has been used to obtain both phosphorus–phosphorus and phosphorus–platinum coupling constants for complexes such as Pt₂(dppm)₂Cl₂. Consult Krevor, Simonis, Karson, Castro, and Aliakbar (*Inorg. Chem.*, **1992**, *31*, 312), and explain how this was accomplished.

189. Explain completely the following 500-MHz ^1H NMR and 125.7-MHz $^{13}C\{^1H\}$ NMR spectra for $PhC(O)CH_2C(O)CH_3$ dissolved in $CDCl_3$ at ambient temperature.

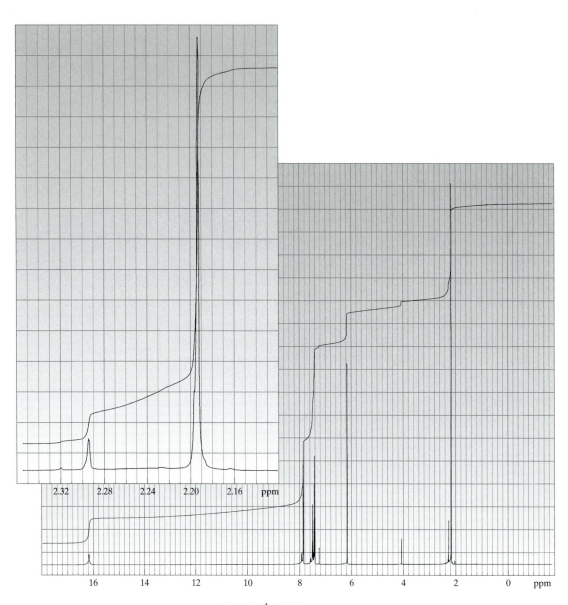

500-MHz ^1H NMR spectrum.

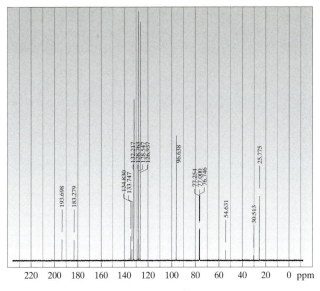

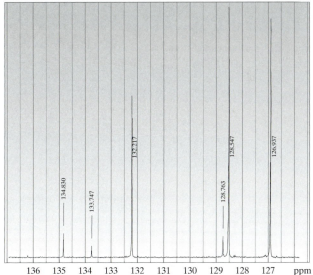

125.7-MHz ^{13}C{^{1}H} NMR spectrum.

Would you expect these spectra to be temperature or solvent dependent? Why or why not?

190. The selective-excitation (sometimes called DANTE) 22.5-MHz ^{13}C NMR spectra for nicotine shown in the following figures can be used to completely assign the ^{13}C NMR spectrum of the substance in CDCl$_3$ solution:

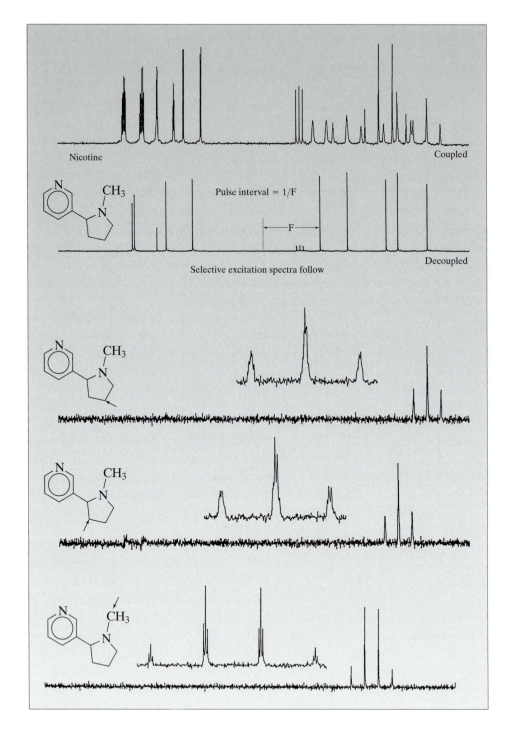

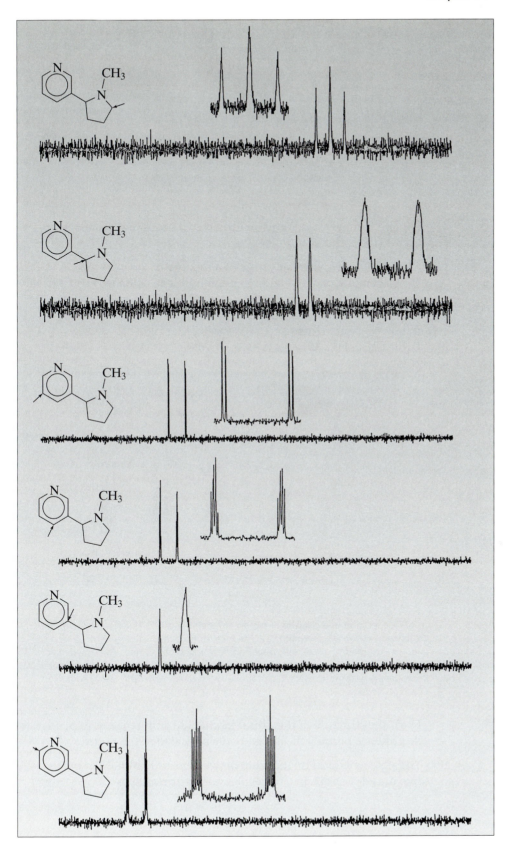

Selective-excitation spectra follow.

Explain the line shapes and identify the CH coupling constants in each of the carbon multiplets.

191. The 60-MHz ^{1}H NMR spectrum of a mixture of the *meso-* and *dl-* diastereomers of *cis-* and *trans-* 2,3-butylene oxide (4a) and (4b) in the presence of the chiral-shift reagent tris-[3-trifluoromethylhydroxy-methylene)-d-camphorato]europium (III), (2a), is shown in the following figure (Kainosho, M.; Ajisaka, K.; Pirkle, W. H.; Beare, S. D., *J. Am. Chem. Soc.*, **1972**, *94*, 5924).

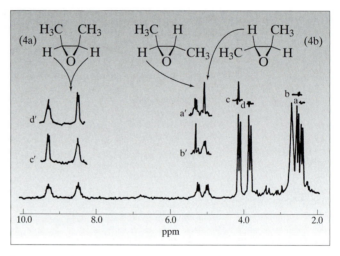

^{1}H NMR spectrum of a mixture of *meso-* (4a) and *dl* (4b) 2,3-butylene oxide in the presence of europium reagent (2a). The inset double-resonance spectra were obtained by irradiating at (a) and observing decoupling at (a'), etc. The drawings of the *trans* isomers (4b) are not meant to depict absolute configurational assignments. Reproduced with permission from above reference, copyright 1972, American Chemical Society.

Draw the structure of this chiral-lanthanide-shift reagent. Explain how its presence permits the assignments indicated in the figure.

192. The stereochemical relationship between a lanthanide-shift probe and two nuclei, **1** and **2**, in a molecule can be expressed as

$$\frac{\delta_1}{\delta_2} = \frac{r_2^3 \cos^2 \theta_1 - 1}{r_1^3 3 \cos \theta_2 - 1},$$

where r_1 and r_2 are the distances between the lanthanide and the nucleus, and θ_1 and θ_2 are the angles to the principal axis of the lanthanide. The geometry of the probe and the nuclei is as follows:

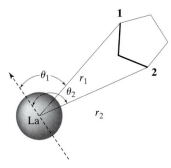

What shift ratio do you expect when $r_1 = r_2$, $\theta_1 = 30°$, and $\theta_2 = 60°$? What happens when $\theta_2 = 54.73°$?

193. Interpret the 121-MHz CP/MAS ^{31}P{^{1}H} NMR spectra of *cis*-(Ph$_3$P)$_2$PtCl$_2$, and *trans*-(*n*-Bu$_3$P)$_2$PtI$_2$ shown in the following figures (see Rahn, J. A., Baltusis, L., Nelson, J. H., *Inorg. Chem.*, **1990**, *29*, 750):

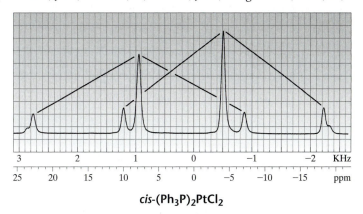

cis-(Ph$_3$P)$_2$PtCl$_2$

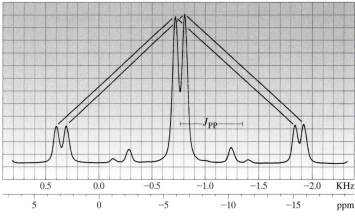

trans-(*n*-Bu$_3$P)$_2$PtI$_2$

Calculate the chemical shifts and coupling constants.

194. Interpret the following 81.02-MHz, $^{31}P\{^1H\}$ CP/MAS NMR spectra of PhDBP = Se:

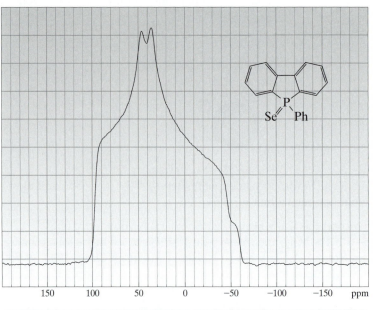

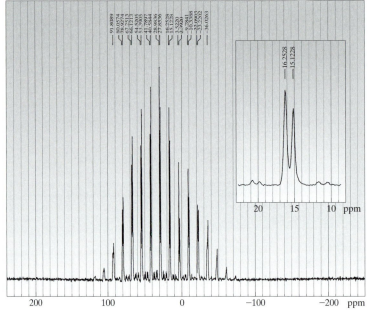

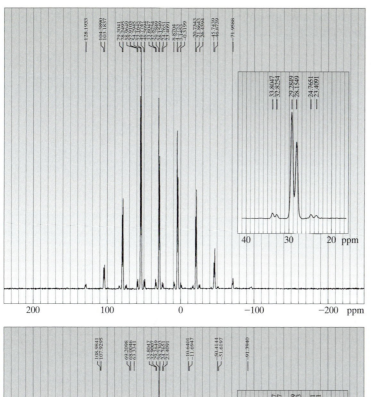

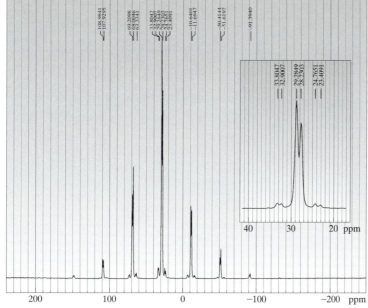

Note that for this compound there are two inequivalent molecules in the unit cell. (See Eichele, K., Wasylishen, R. E., Kessler, J. M., Solujic', L. J., Nelson, J. H., *Inorg. Chem.*, **1996**, *35*, 3904; Alyea, E. C., Ferguson, G., Malito, J., Ruhl, B. L., *Acta Crystallogr.*, **1986**, *C42*, 882.) In your interpretation, calculate the chemical shifts, the coupling constants, and the spinning rates for each spectrum. Calculate the chemical-shift tensor quantities δ_{11}, δ_{22}, δ_{33}, δ_{iso}, $\Delta\delta$, and η.

195. Interpret the following 81.02-MHz $^{31}P\{^1H\}$ CP/MAS NMR spectra of *trans*- $(PhDBP)_2PdCl_2$:

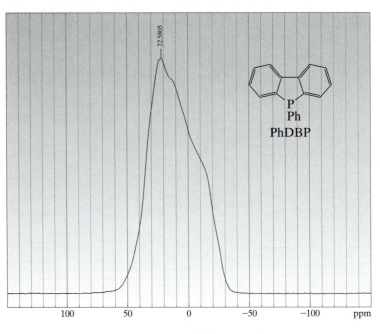

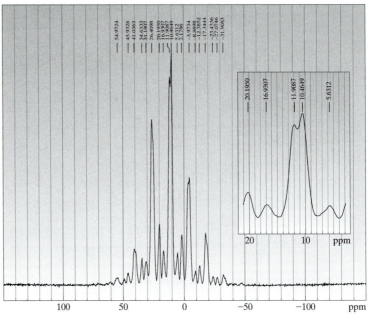

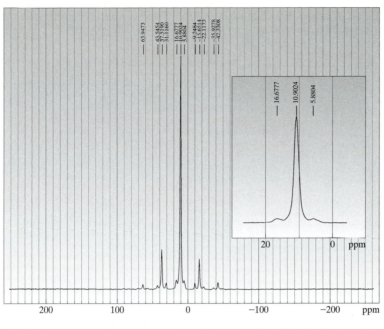

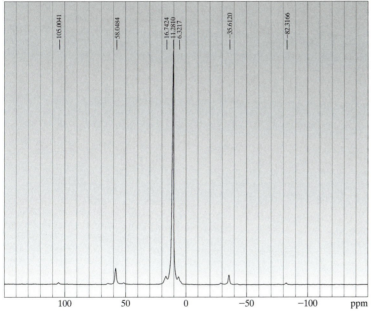

Note that this is a tightly coupled *AB* spin system, as there is no inversion center in the crystal structure that relates the two phosphorus nuclei in the molecule. Is the value of ^2J(PP) consistent with the *trans* geometry? Estimate δ_{11}, δ_{22}, and δ_{33} for the two molecules. Calculate the spinning rates for each of the spectra. (See Eichele, K., Wasylishen, R. E., Kessler, J. M., Solujic', Lj., Nelson, J. H., *Inorg. Chem.*, **1996**, *35*, 3904.)

196. Interpret the following 81.02-MHz $^{31}P\{^1H\}$ CP/MAS NMP spectra of (PhDBP)Cr(CO)$_5$:

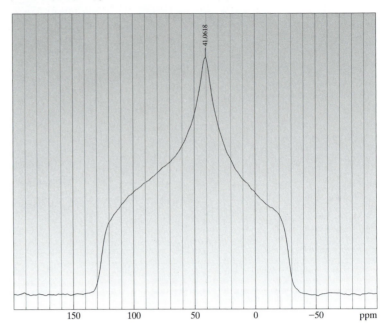

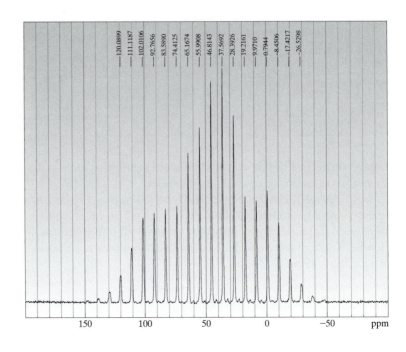

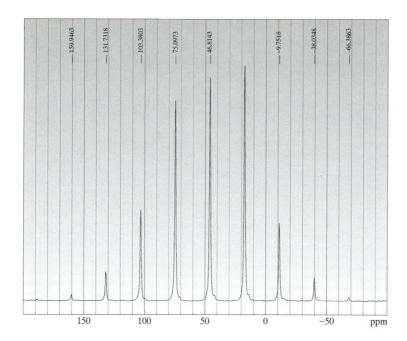

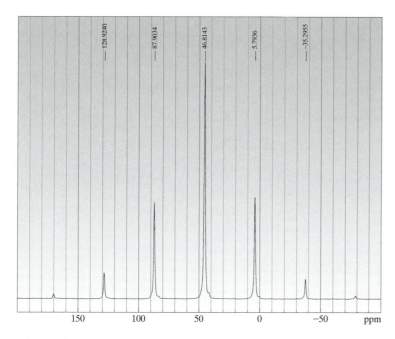

What is the symmetry of the chemical-shift tensor? Calculate δ_{11}, δ_{22}, δ_{33}, δ_{iso}, $\Delta\delta$, and η. Calculate the spinning rates for each of the spectra.

197. There is more than one way that chemical-shift anisotroscopy $\Delta\delta$ and asymmetry η have been defined in the literature:

$$\Delta\delta = \delta_{33} - \frac{1}{2}(\delta_{11} + \delta_{22}) \quad \text{and} \quad \Delta\delta = (\delta_{33} - \delta_{11});$$

$$\eta = \frac{\delta_{22} - \delta_{11}}{\delta_{33} - \delta_{iso}} \quad \text{and} \quad \eta = \delta_{22} - \delta_{11}.$$

    a. When will $\Delta\delta$ be equal under the two definitions?

    b. What is the range of η?

    c. What are the values of η for spherically and axially symmetric cases?

198. Interpret the following 81.02-MHz $^{31}P\{^1H\}$ CP/MAS NMR spectrum of *cis*-$(PhDBP)_2Mo(CO)_4$:

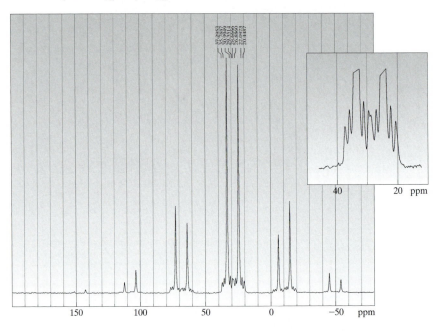

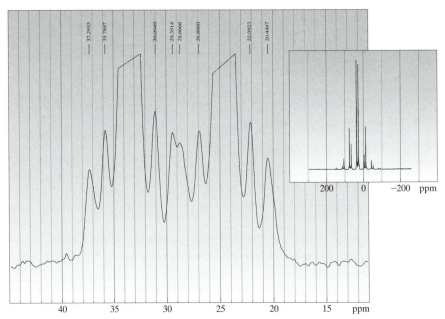

Is there any symmetry element that relates the two phosphorus nuclei in this molecule? Is $^2J(PP)$ small or large? What would you expect it to be? Calculate the value of $^1J(^{95}Mo^{31}P)$. Calculate the spinning rate.

(See Affandi, S., Nelson, J. H., Alcock, N. W., Howarth, D. W., Alyea, E. C., Sheldrick, G. M., *Organometallics*, **1988**, 7, 1724.)

199. Interpret the following 81.02-MHz $^{31}P\{^{1}H\}$ CP/MAS NMR spectra of *cis*- $(PhDBP)_2W(CO)_4$:

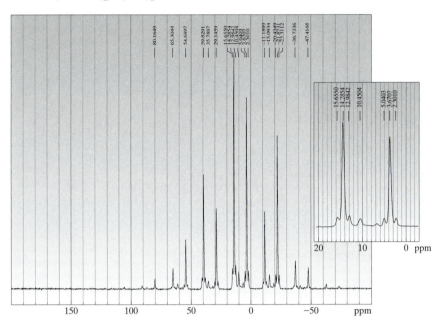

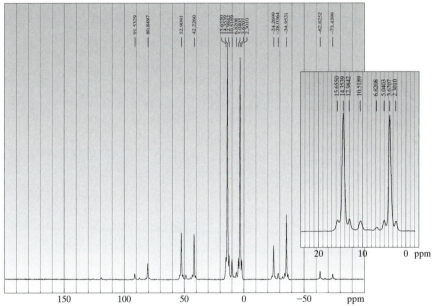

Is there any symmetry element that relates the two phosphorus nuclei in this molecule? In the crystal? Is $^{2}J(PP)$ small or large? What would you expect it to be? Calculate the value of $^{1}J(^{183}W^{31}P)$. Calculate the spinning rates for the two spectra. (See Affandi, S., Nelson, J. H., Alcock, N. W., Howarth, D. W., Alyea, E. C., Sheldrick, G. M., *Organometallics*, **1988**, 7, 1724.)

200. A molecule with the chemical formula C_3H_5Br was dissolved in $CDCl_3$, and the following spectra were obtained on the resulting solution:

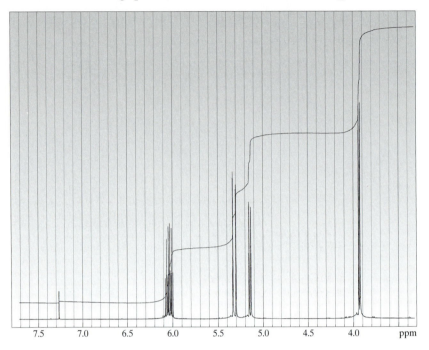

499.869-MHz ^{1}H NMR spectrum.

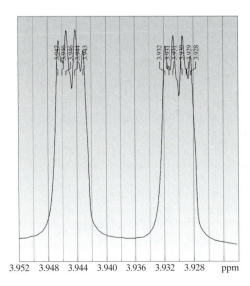

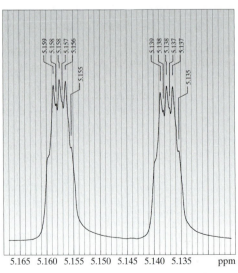

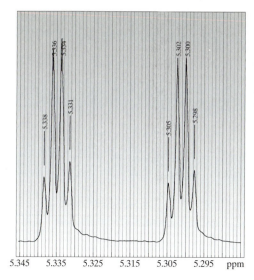

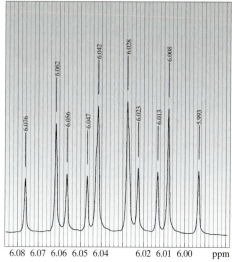

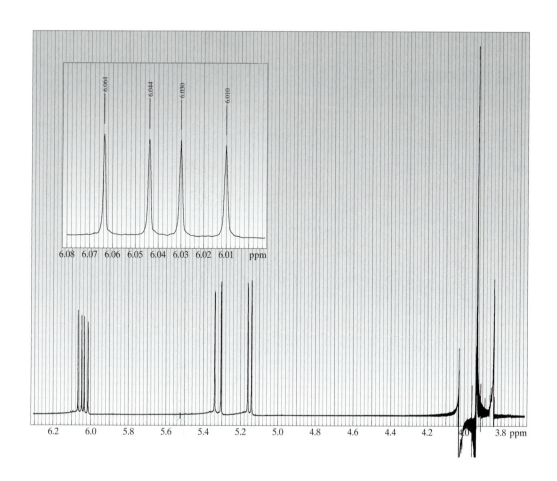

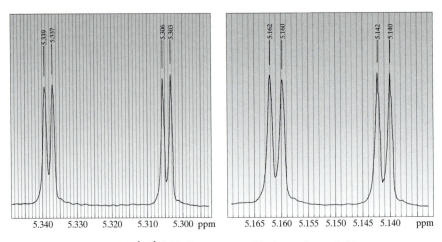

499.469-MHz ^{1}H{^{1}H} NMR spectrum, with decoupling at 3.94 ppm.

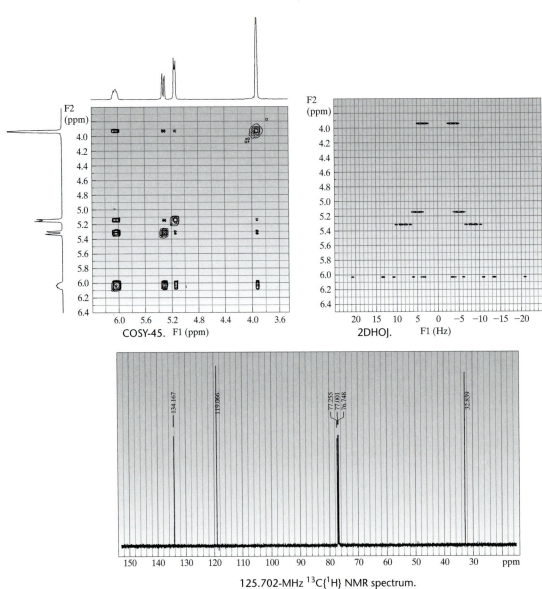

COSY-45. F1 (ppm)

2DHOJ. F1 (Hz)

125.702-MHz ^{13}C{^{1}H} NMR spectrum.

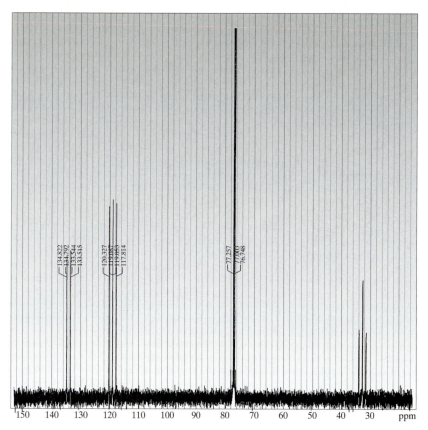

125.702-MHz ^{13}C NMR spectrum.

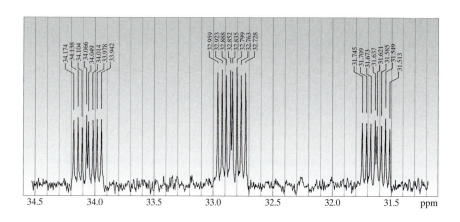

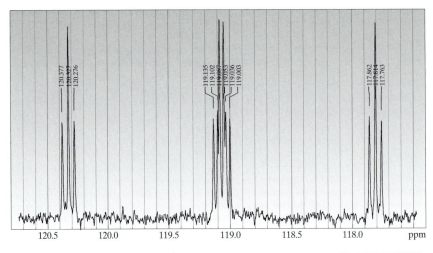

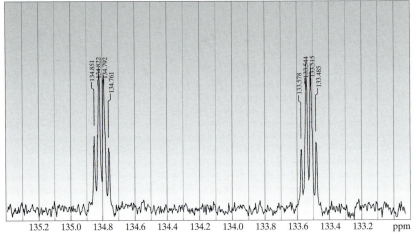

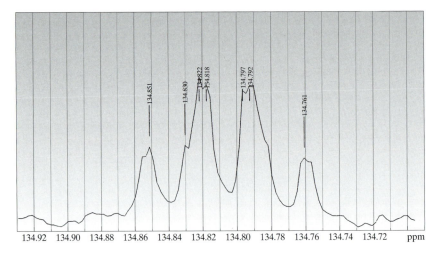

Describe the nature of each spectrum and interpret the spectra. What is the structure of the molecule? Proton spectra were obtained at 499.869 MHz, and carbon spectra were obtained at 125.702 MHz. Calculate all chemical shifts and coupling constants.

201. The 500-MHz ^{1}H NMR spectrum of di-2-pyrrolylketone in CDCl$_3$ at 25°C shows an apparent *dt* (δ = 6.35 ppm, *J* = 4.0, 2.5, 2.5 Hz), an apparent *td* (δ = 7.08 ppm, *J* = 2.5, 2.5, 1.5 Hz), a *ddd* (δ = 7.15 ppm, *J* = 4.0, 2.5, 1.5 Hz), and a broad singlet at 9.5 ppm ($\Delta\nu_{1/2}$ = 35 Hz). Explain these data and suggest an experiment that would confirm your explanation. The structure of di-2-pyrrolylketone is as follows:

di-2-pyrrolylketone

202. Interpret the following 499.867-MHz proton, 125.702-MHz carbon, and various two-dimensional spectra for a CDCl$_3$ solution of camphor (see inset):

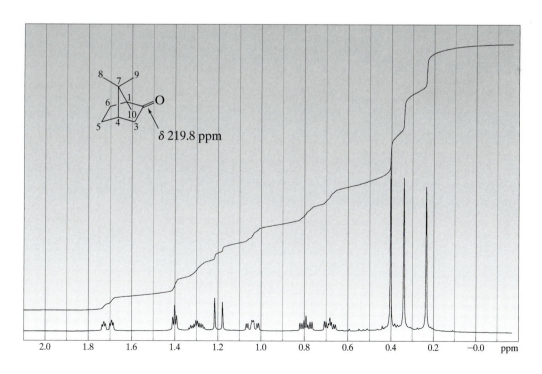

δ 219.8 ppm

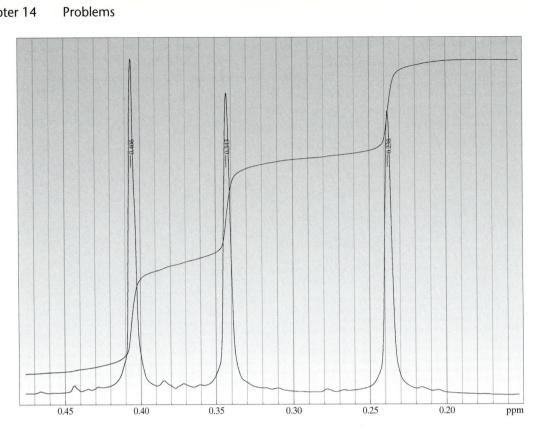

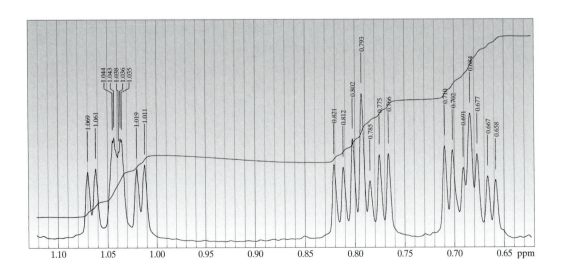

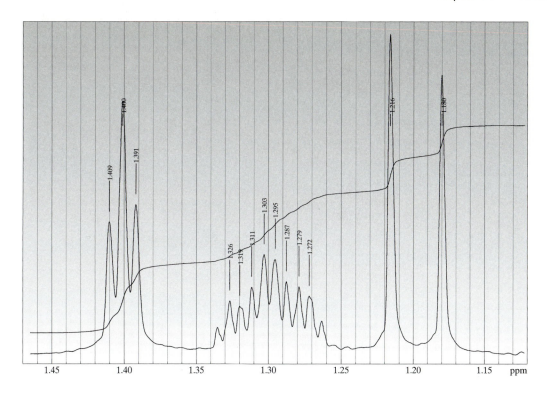

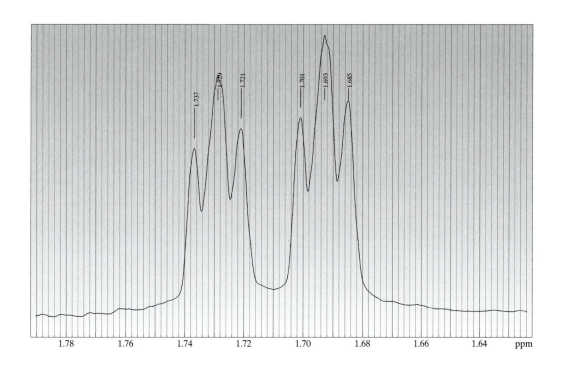

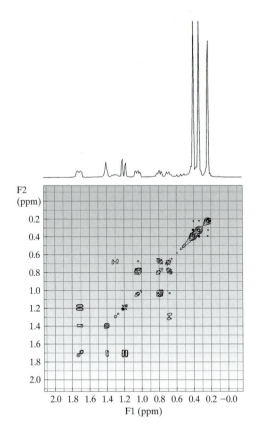

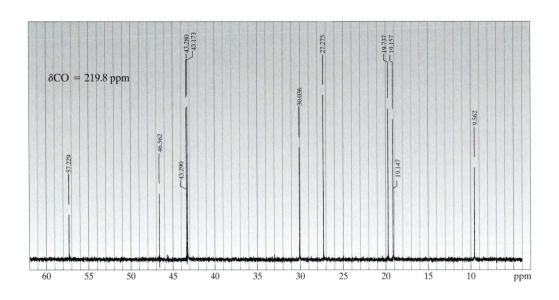

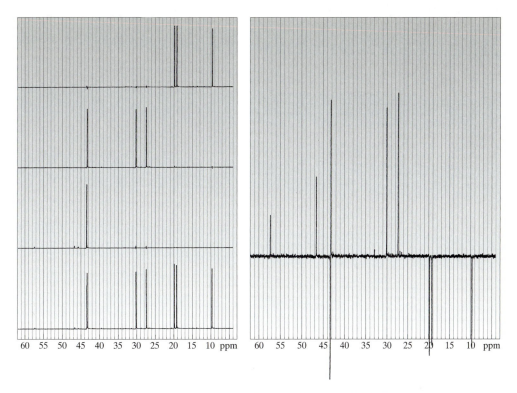

Which of the foregoing spectra are **APT** and which are **DEPT** spectra?

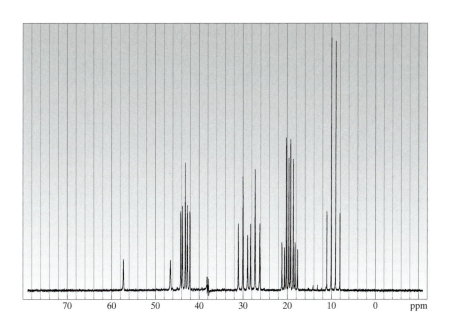

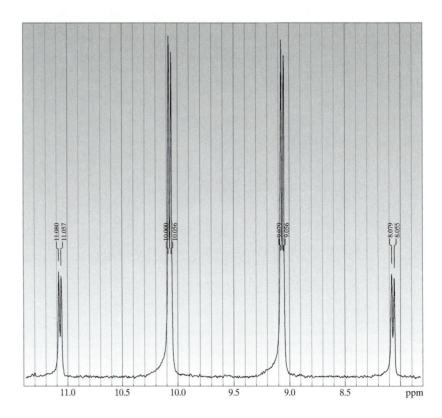

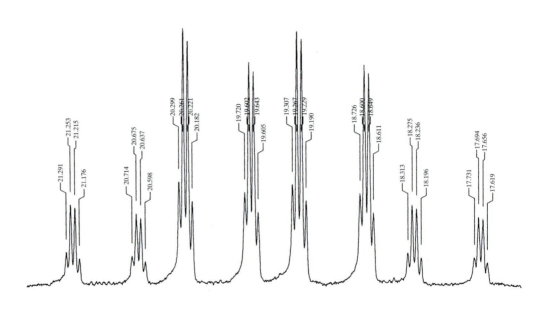

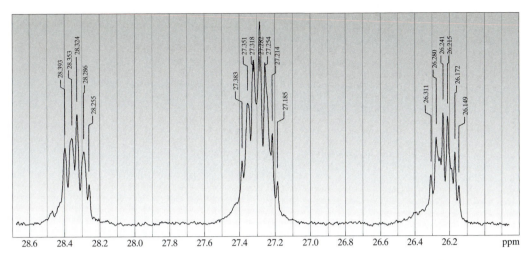

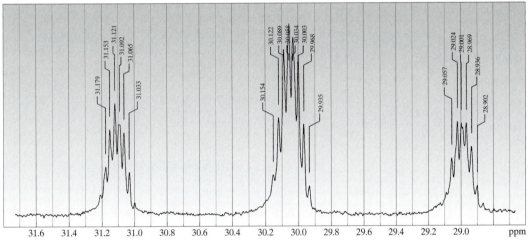

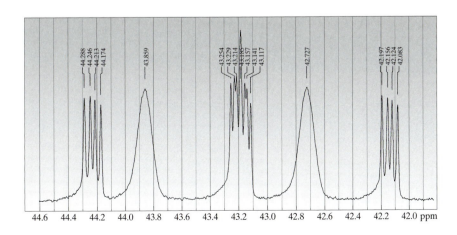

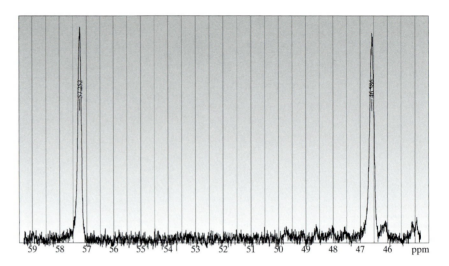

Calculate all the CH coupling constants from the preceding expansions of the ^{13}C NMR spectra. Are any of them unusually large or small?

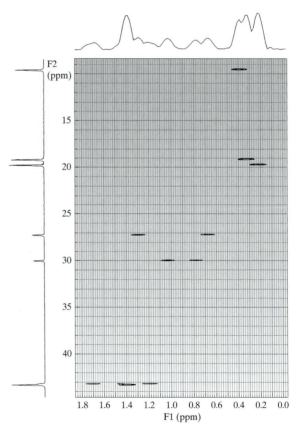

Explain the appearance of the preceding **HETCOR** spectrum in the 43 ppm region (lower left).

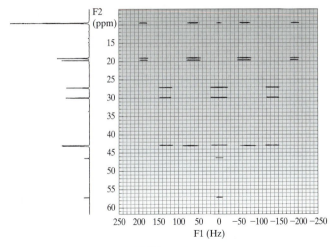

HET2DJ spectrum.

203. The 499.867-MHz proton and 125.695-MHz carbon spectra that follow were obtained on a CDCl$_3$ solution of a chiral compound with the formula C$_4$H$_9$Br. Identify the compound, assign the resonances, and determine all the chemical shifts and coupling constants.

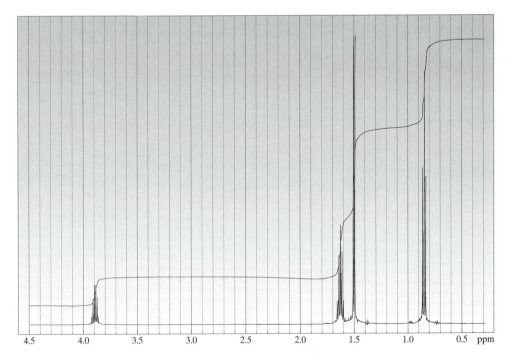

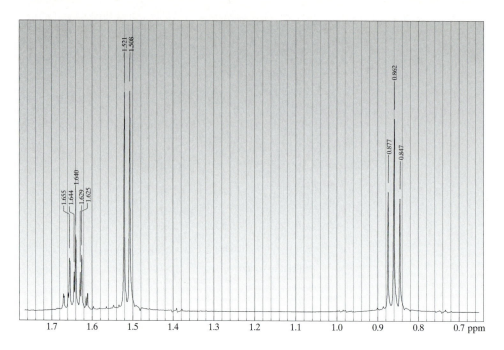

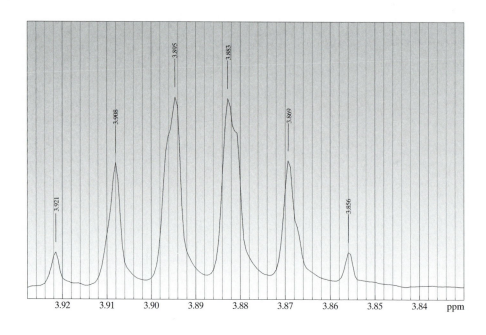

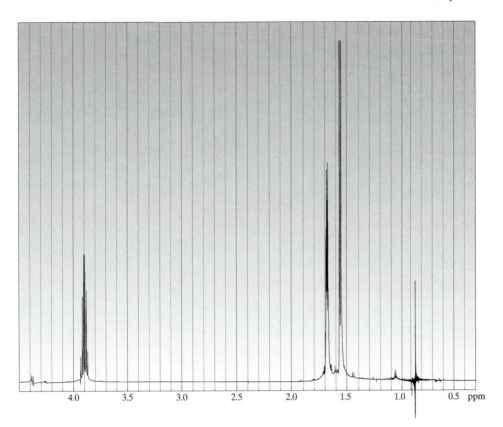

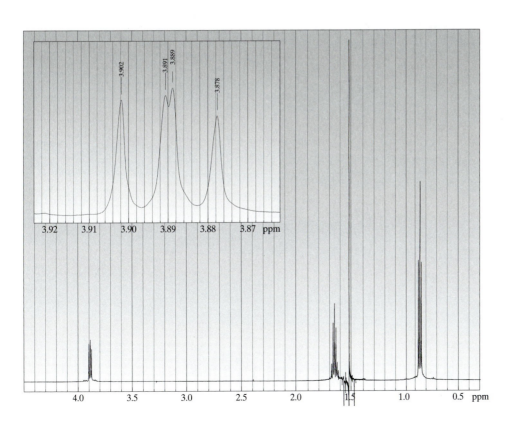

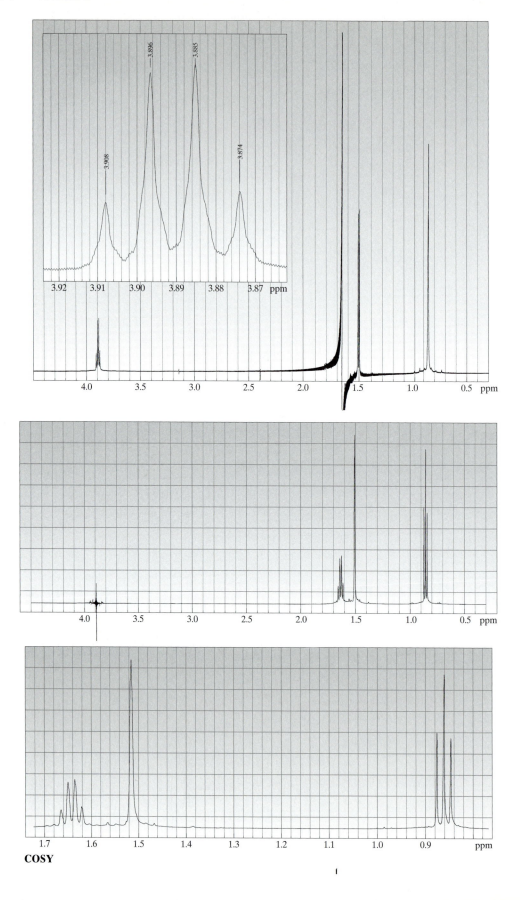

COSY

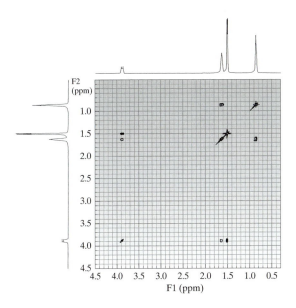

Is the **COSY** spectrum consistent with the results of the preceding homonuclear decoupling experiments?

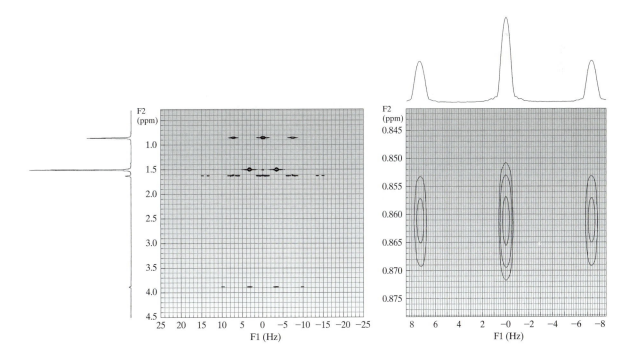

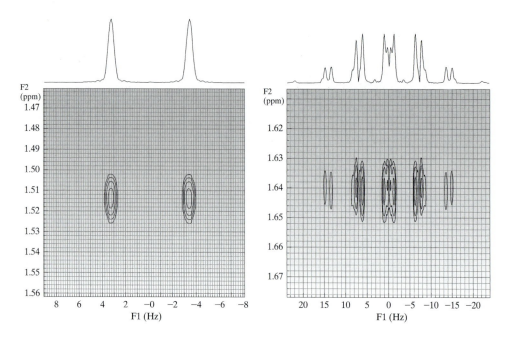

Why is the 1.64-ppm multiple *t* so complex in the **HOMO2DJ** experiment?

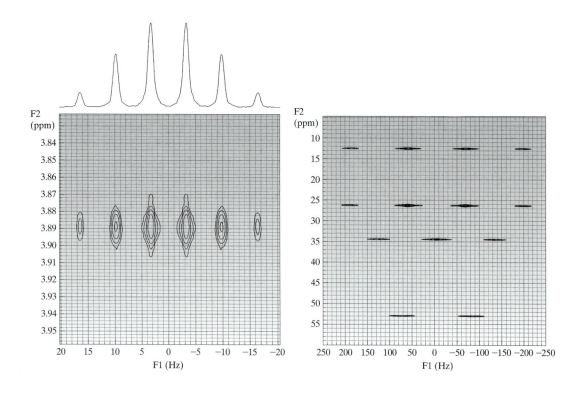

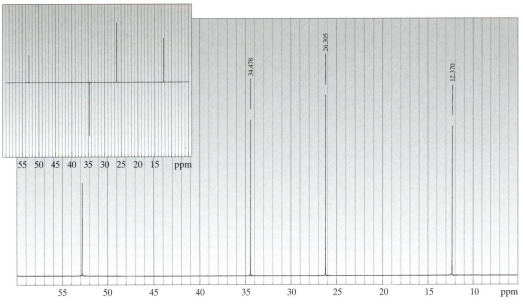

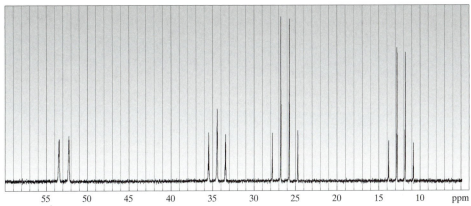

What was done to obtain the preceding spectrum?

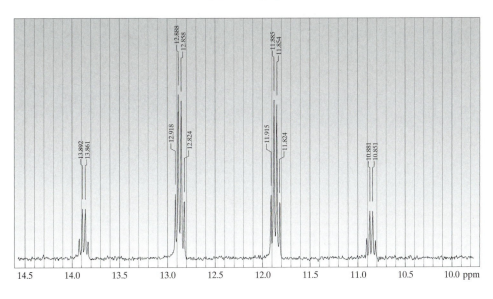

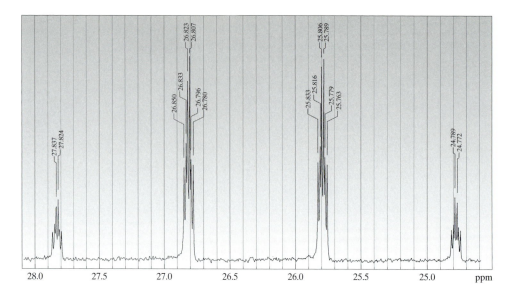

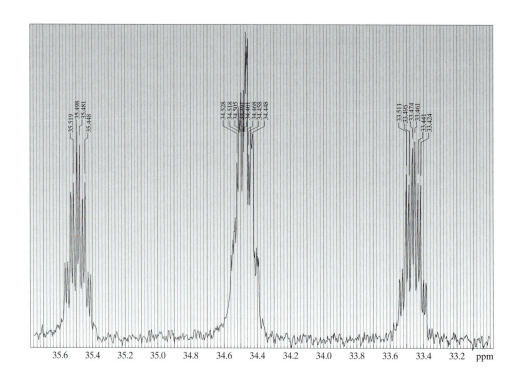

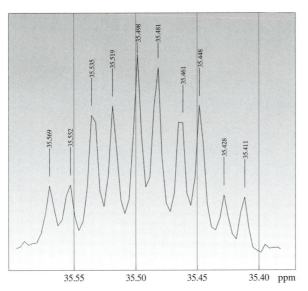

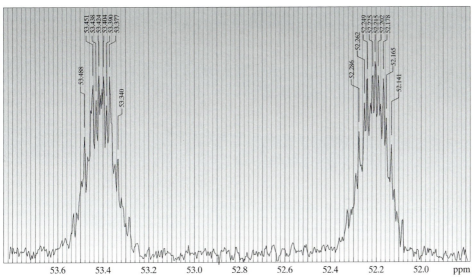

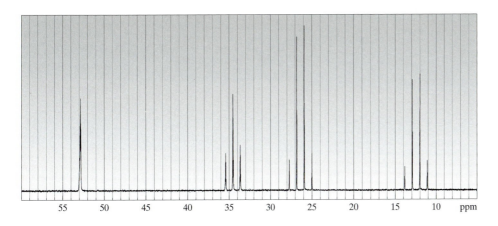

What was done to obtain the foregoing spectrum?

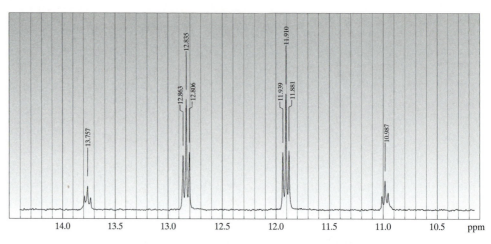

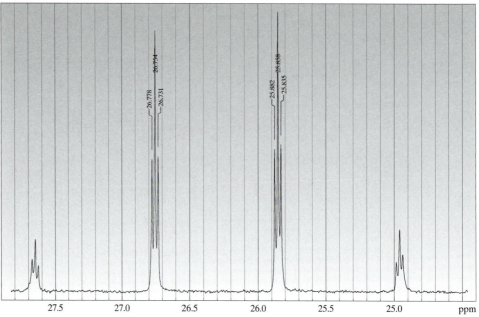

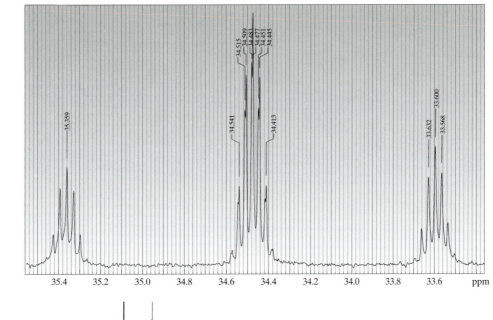

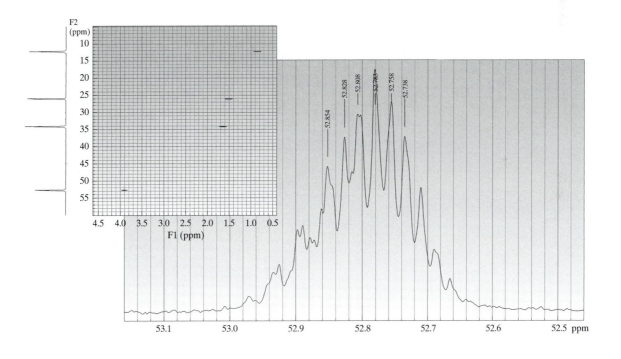

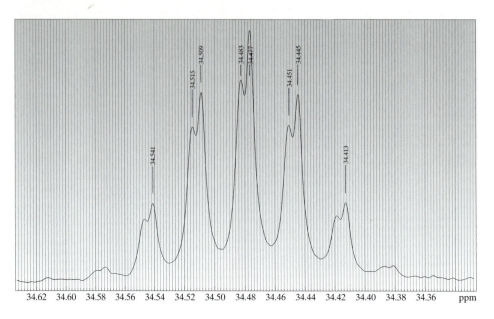

204. Explain, as completely as possible, the **DEPT**, **COSY**, and **HETCOR** spectra of acrylic acid *n*-butyl ester that follow. Identify which spectrum is which.

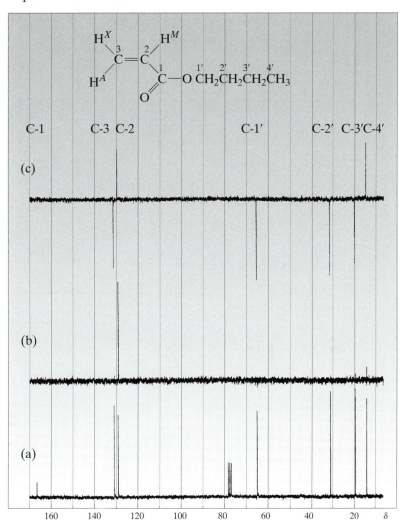

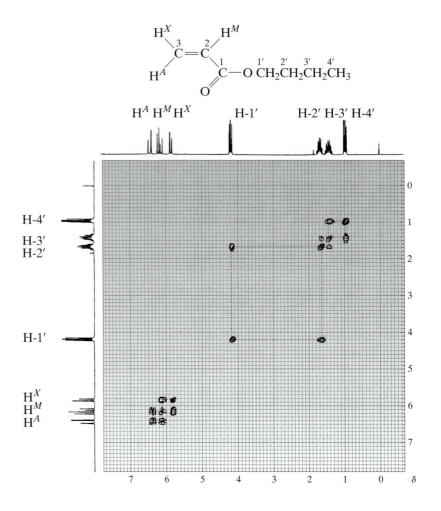

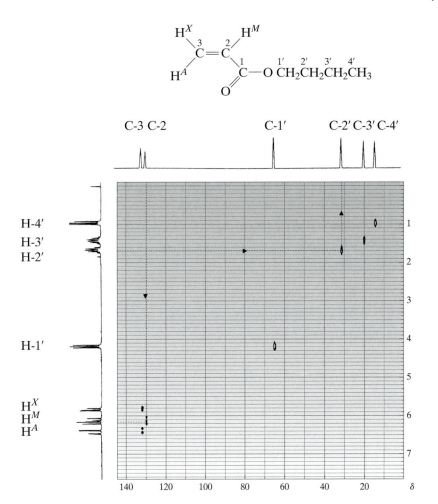

Can you justify the indicated chemical-shift assignments? Are they unambiguous?

205. Using the spectra that follow, completely assign the 499.868-MHz proton and 125.705-MHz carbon NMR spectra of (n-BuDBP)Mo(CO)$_5$ in CDCl$_3$ solution. (See Nelson, J. H., Affandi, S. A., Gray, G. A., Alyea, E. C., *Magn, Reson. Chem.*, **1987**, *25*, 774.)

$$CH_2(\alpha)$$
$$CH_2(\beta)$$
$$CH_2(\gamma)$$
$$CH_3(\delta)$$

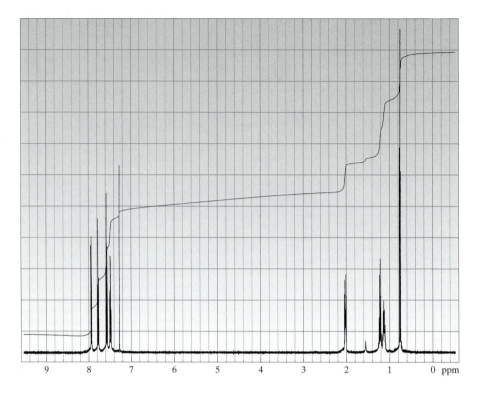

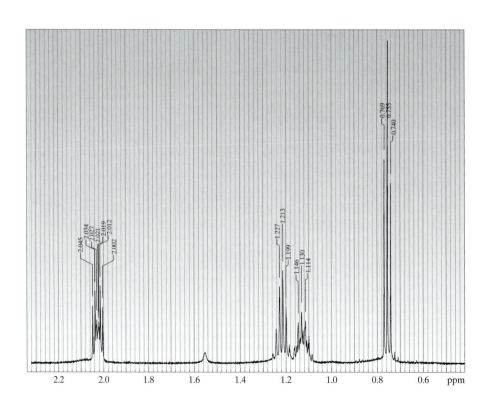

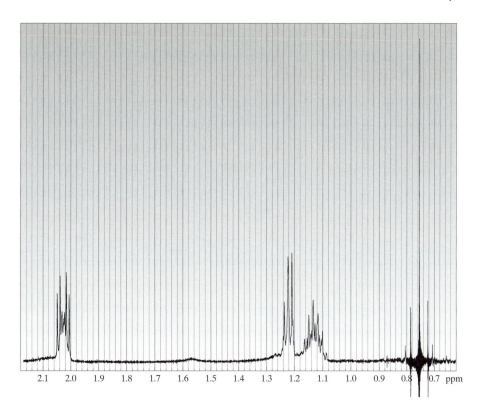

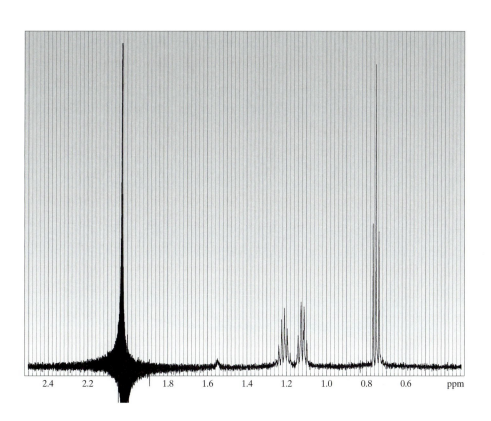

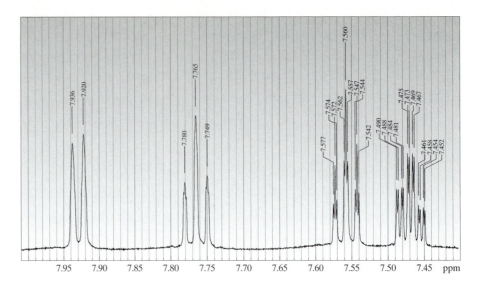

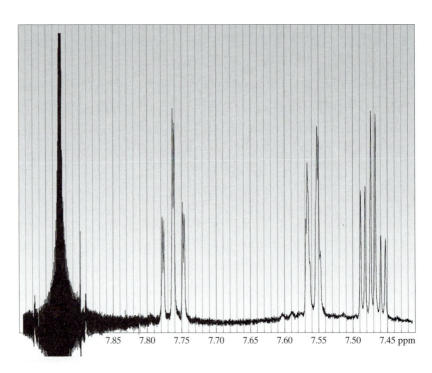

Is the **COSY** spectrum consistent with the results of the homonuclear decoupling experiments?

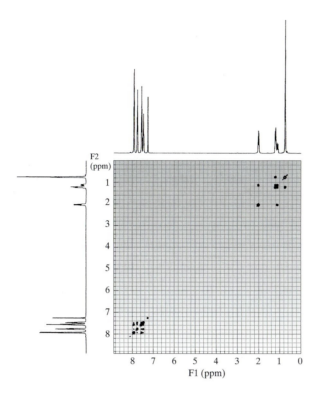

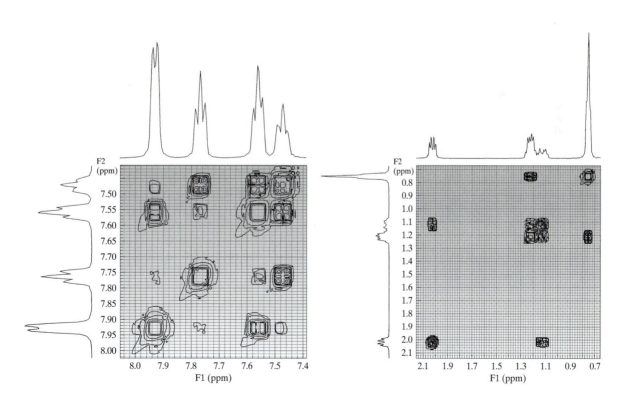

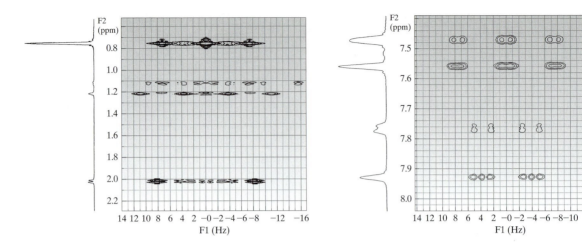

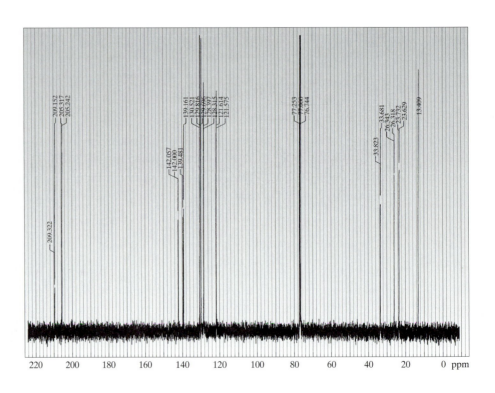

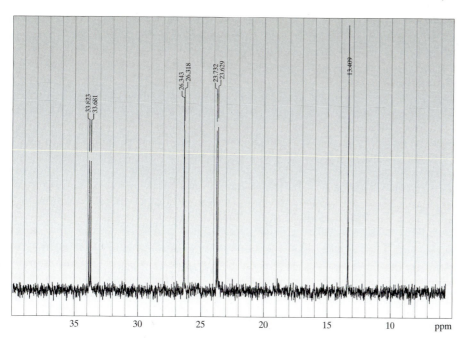

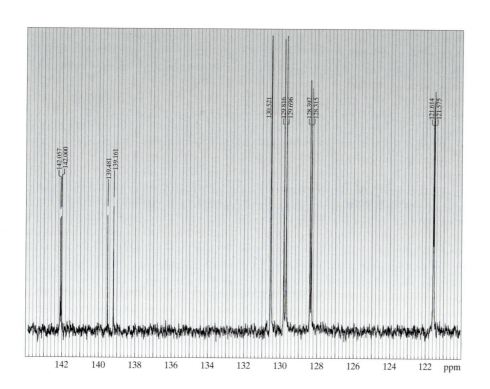

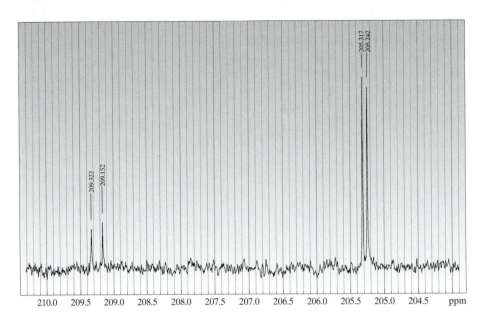

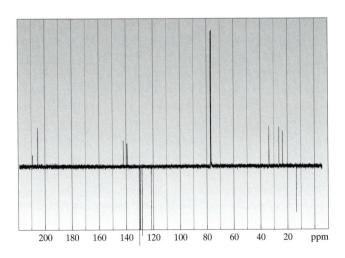

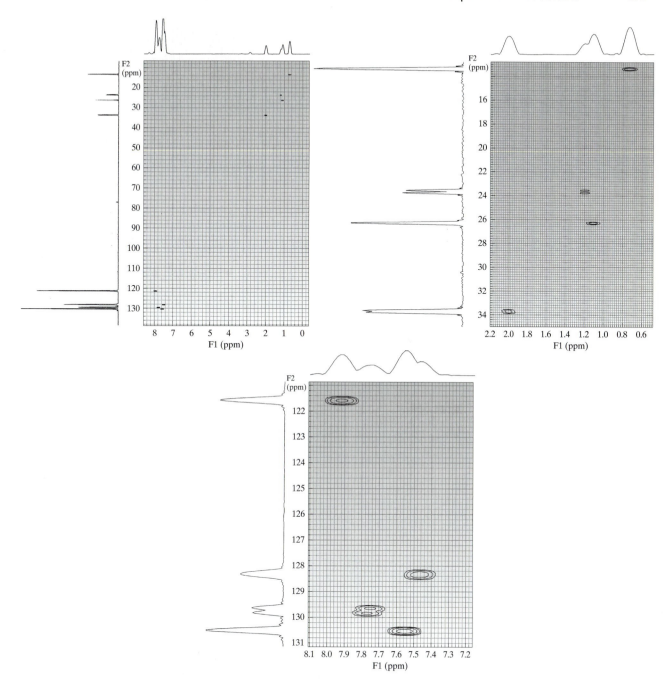

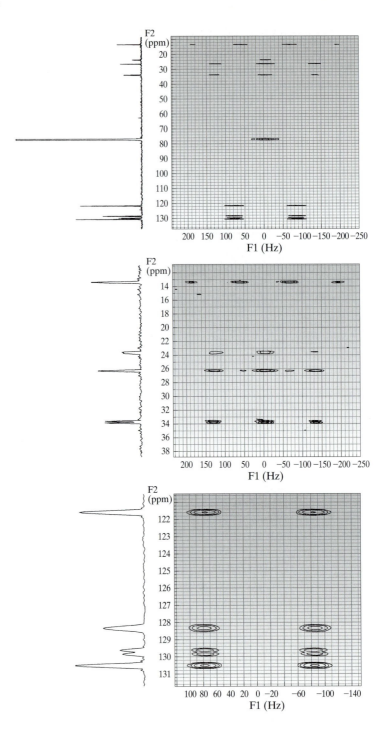

206. Assign the spectra that follow for a mixture of 1-methyl (minor) and 2-methyl (major) naphthalene, and determine the ratio of the two isomers. (See Breitmaier, E., Voelter, W., *Carbon-13 NMR Spectroscopy*, VCH, New York, 1989, for assistance.) Calculate all chemical shifts and coupling constants.

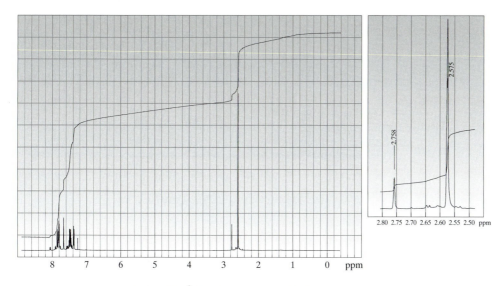

499.868-MHz ^{1}H NMR spectrum in CDCl$_3$.

1-methylnaphthalene   2-methylnaphthalene

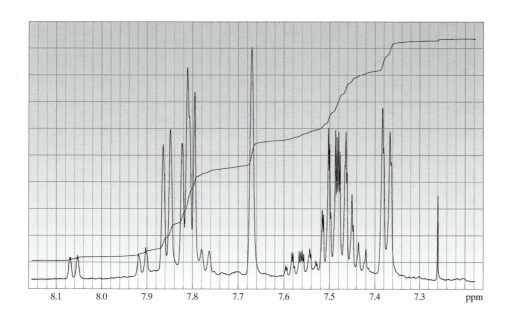

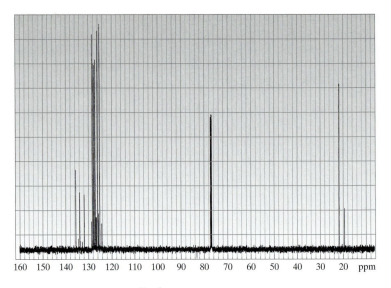

125.702-MHz ^{13}C{^{1}H} NMR spectrum in CDCl$_3$.

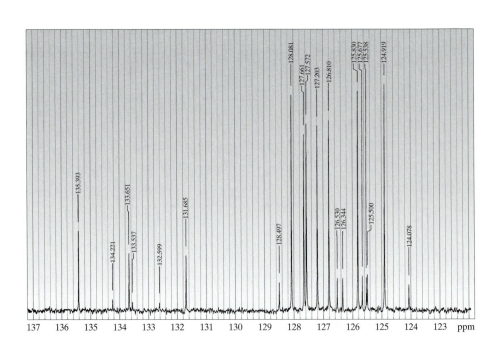

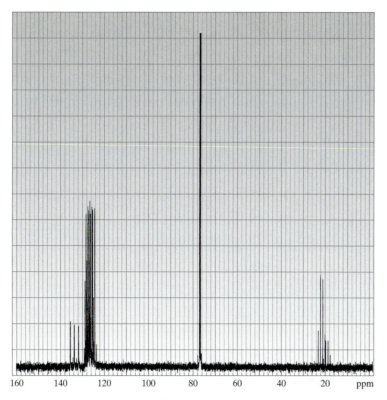

125.701-MHz ^{13}C NMR spectrum in CDCl$_3$.

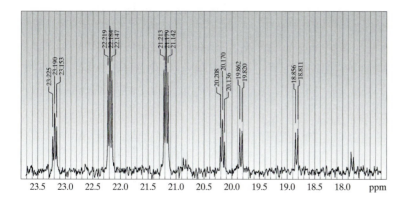

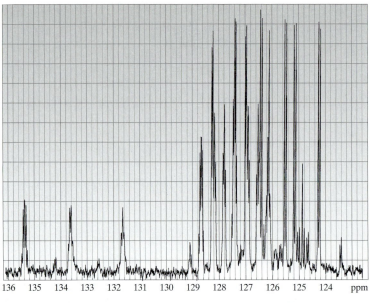

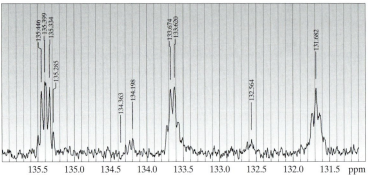

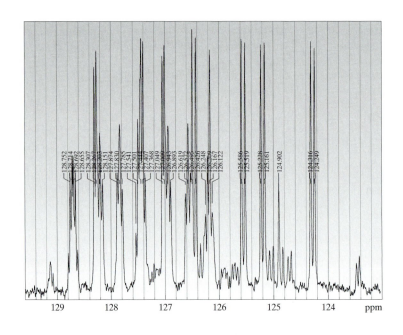

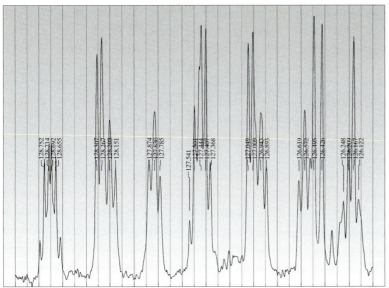

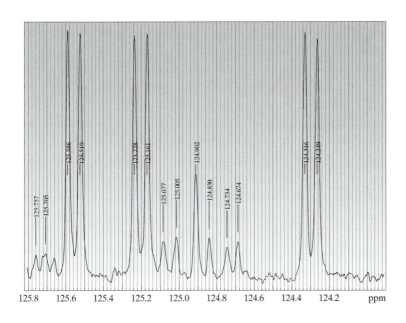

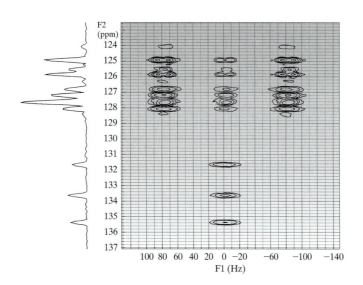

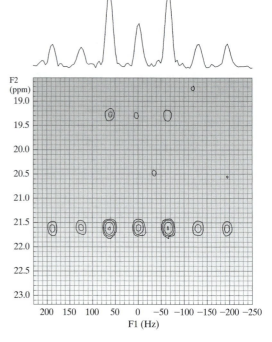

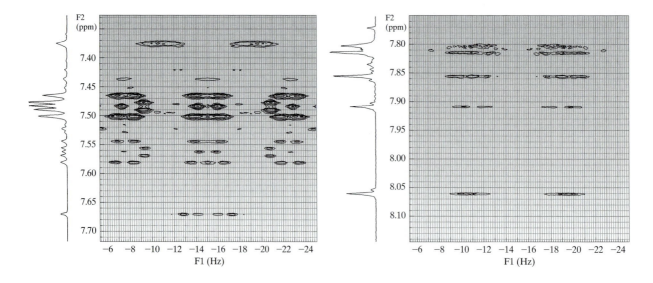

What is wrong with the preceding two spectra?

207. From the following spectra, deduce whether $(DMPP)_2PdCl_2$ (see inset) exists as the *cis* isomer, the *trans* isomer, or both in $CDCl_3$ solution:

P
Ph
DMPP

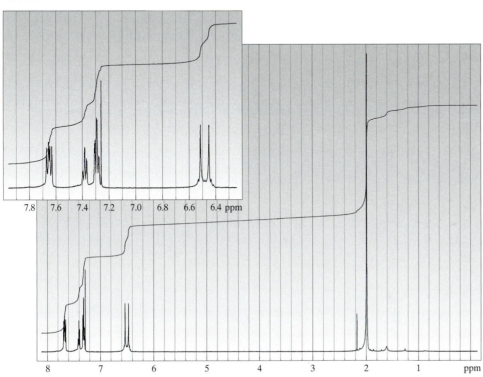

499.868-MHz ^{1}H NMR spectrum in $CDCl_3$.

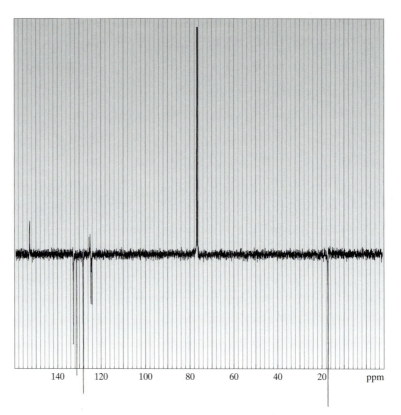

125.701-MHz **APT** spectrum in CDCl$_3$.

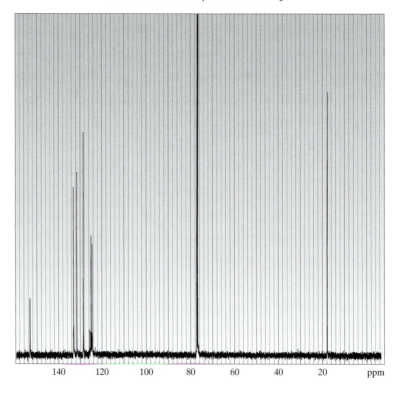

125.701-MHz ^{13}C{^{1}H} NMR spectrum in CDCl$_3$.

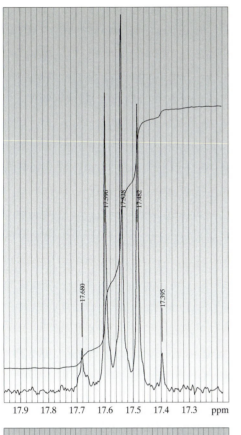

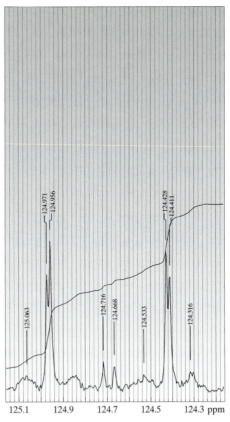

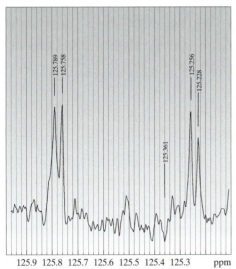

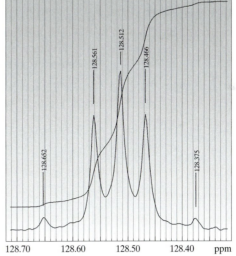

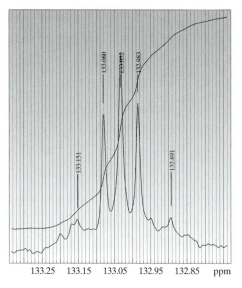
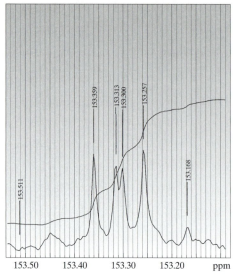

Calculate $^2J(PP)$ from an analysis of the $AA'X$ multiplets in the $^{13}C\{^1H\}$ NMR spectrum. Is the magnitude of $^2J(PP)$ consistent with your conclusion? (See Nelson, J. H. and coworkers, *Inorg. Chem.*, **1980**, *19*, 709, 1400.)

208. Describe how variable-temperature $^{13}C\{^1H\}$ CP/MAS NMR spectroscopy was used to study the motion of adamantane in the solid state. (See Resing, H. A., *J. Chem. Phys.*, **1965**, *43*, 1828 and *Mol. Cryst. Liq. Cryst.*, **1969**, *9*, 101.)

209. Consult Rothwell and Waugh (*J. Chem. Phys.*, **1981**, *74*, 2721), and describe how activation energies for molecular motion in the solid state may be determined from $^{13}C\{^1H\}$ CP/MAS NMR line widths.

210. Consult Wu and Wasylishen (*Inorg. Chem.*, **1992**, *31*, 145), and describe the utility of **2DHOJ** CP/MAS ^{31}P NMR spectroscopy.

211. CP/MAS $^{13}C\{^1H\}$ NMR spectra of naphthazarin B at 25° and −160° are shown in the following figure:

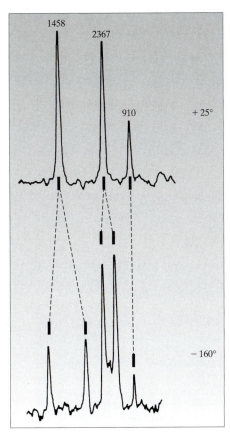

Naphthazarin B

Reproduced with permission from reference below, copyright 1980, American Chemical Society.

Peak assignments, which correspond to numbered carbon atoms in the structure, are as follows: carbons 1,4,5, and 8: δ = 173; carbons 2,3,6, and 7: δ = 134; carbons 9 and 10: δ = 112 (+25°C). Account for the dramatic change that occurs in the spectrum when the temperature is lowered to −160°C. (See Shiau, W.-I., Duesler, E. N., Paul, I. C., Curtin, D. Y., Blann, W. G., Fyfe, C. A., *J. Am. Chem. Soc.*, **1980**, *102*, 4546.)

212. Variable-temperature ^{31}P and ^{27}Al NMR spectra have been reported for the complex $[Al(H_3ppma)_2](NO_3)_3 \cdot 2H_2O$ (Lowe, M. P., Rettic, S. J., Orvig, C., *J. Am. Chem. Soc.*, **1996**, *118*, 10446), whose structure is as follows:

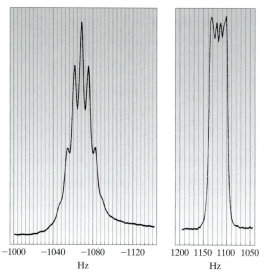

Explain the appearance of the following ^{27}Al and $^{31}P\{^1H\}$ NMR spectra:

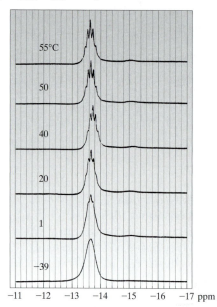

^{27}Al (78.2 MHz, 50°C)(left) and ^{31}P (81.0 MHz, room temperature)(right) NMR spectra of $[Al(H_3ppma)_2](NO_3)_3$ in CD_3OD.

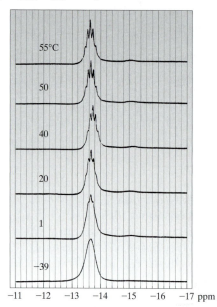

^{27}Al NMR spectra (78.2 MHz) in the variable temperature study of $[Al(H_3ppma)_2](NO_3)_3$ in CD_3OD. Reproduced with permission from above reference, copyright 1996, American Chemical Society.

Why does the ^{27}Al line width increase with decreasing temperature?

213. Consult Shurko, R. W., Wasylishen, R. E., and Nelson, J. H., (*J. Phys. Chem.*, **1996**, *100*, 8057), and describe the origin of the temperature dependence of the ^{31}P CP/MAS spectrum of $PBu_3Co(DH)_2Cl$:

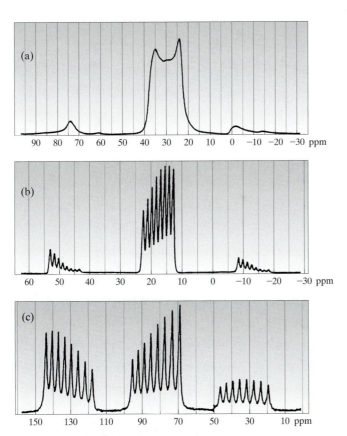

	PR_3	X
I	PBu_3	Cl
II	PPh_3	CH_3
III	$P(OEt)_3$	N_3

Schematic illustration of a phosphorus-substituted cobaloxime. The axial ligands are PR_3, where R = *n*–Bu, Ph, OEt, and X = anionic species (e.g., Cl^-, CH_3^-, N_3^-).

^{31}P CP/MAS NMR spectra of (a) $PBu_3Co(DH)_2Cl$ (**I**), ν_{rot} = 5001 Hz; (b) $PPh_3Co(DH)_2CH_3$ (**II**), ν_{rot} = 5004 Hz; and (c) $P(OEt)_3Co(DH)_2N_3$ (**III**), ν_{rot} = 7973 Hz. Outer peaks are spinning sidebands. All spectra were acquired at 9.4 T at room temperature.

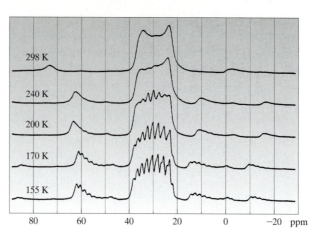

Variable-temperature ^{31}P CP/MAS NMR spectra of PBu$_3$Co(DH)$_2$Cl (**I**) obtained at 9.4 T, ν_{rot} = 3927 Hz. Reproduced with permission from above reference, copyright 1996, American Chemical Society.

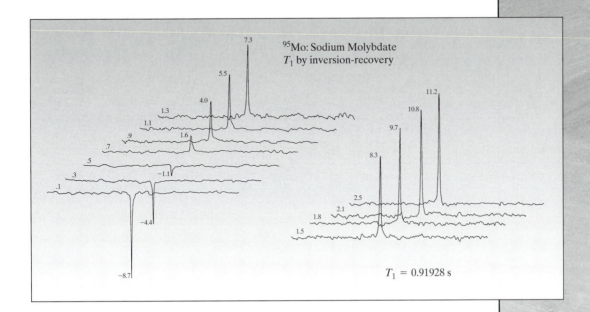

^{95}Mo: Sodium Molybdate
T_1 by inversion-recovery

$T_1 = 0.91928$ s

Selected Values of Longitudinal Relaxation Times for Some Nuclei

A. ^{1}H Relaxation Times (T_1) of Selected Aliphatic and Aromatic Compounds[*]

Aliphatic Compounds

Compound	T_1(s)	Compound	T_1(s)	Compound	T_1(s)
CH_3OH	5.4–9.0	CH_3Br	22	CH_3I	12.4
CH_3CN	16.1	CH_3COCH_3	15.6	CH_3SOCH_3	2.9
CH_3CO_2H	4.3	CH_2Cl_2	30	CH_2Br_2	14.2–16.3
CH_2I_2	5.9–6.5	$CHCl_3$	78–92	$CHBr_3$	22
C_6H_{12}	7.1	cyclohexanone	8.5	$C_6H_{11}C_6H_5$	1.5
$C_6H_{11}(CH_2)_nC_6H_{11}$	0.15–0.8	H_2O	3.5–3.9		

Aromatic Compounds

Compound	T_1(s)	Compound		T_1(s)	Compound		T_1(s)
C_6H_6	18.8–23.6	$C_6H_5CH_3$	CH	16–22	p-$C_6H_4(CH_3)_2$	CH	16.7
			CH_3	9–12		CH_3	7.3
C_6H_5F	28.6	$C_6H_5OCH_3$	CH	16.0	C_5H_5N		17.6
			CH_3	5.8			
C_6H_5Cl	80	C_4H_4S		24.0	C_4H_4O		40.0

B. ^{13}C Relaxation Times (T_1) of Selected Compounds

Aliphatic Compounds

Compound	T_1(s)	Compound	T_1(s)	Compound	T_1(s)
CH_3Br	8.8	CH_3I	14.1	CH_3CCl_3	12.5
$(CH_3)_2CHCl$	21.6	CH_3OH	17.5	$CHCl_3$	32.4
$(CH_3)_2C{=}O$	36.1	$CHBr_3$	1.65	CS_2	36
HCO_2H	10.1	$(CH_2OH)_2$	2.0	C_6H_{12}	17–21

Compound	C_1	C_2	C_3	C_4	C_5
$\overset{1}{C}H_3\overset{2}{C}N$	20.4	46			
$\overset{1}{C}H_3\overset{2}{C}O_2H$	9.8	30			
$\overset{1}{C}H_3\overset{2}{C}O_2\overset{3}{C}H_3$	17.4	29.0	18.3		
$H\overset{2}{C}O_2\overset{1}{C}H_3$	14.5	18.0			

(continues on next page)

---

[*]These data are for neat liquids; the T_1's for dilute solutions may be appreciably longer as they are quite sensitive to viscosity.

Aliphatic Compounds (*cont.*)

Compound	C_1	C_2	C_3	C_4	C_5
$\overset{1}{C}H_2OH\overset{2}{C}HOHCH_2OH$	0.06	0.1			
$\overset{1}{C}H_3\overset{2}{C}H_2\overset{3}{C}H_2\overset{4}{C}H_2OH$	3.0	3.9	3.6	4.2	
$(\overset{1}{C}H_3\overset{2}{C}H_2\overset{3}{C}H_2\overset{4}{C}H_2)_3P$	4.2	3.6	2.9	2.1	
$(CH_3CH_2CH_2CH_2)_3As$	4.8	4.4	3.8	2.9	
$(CH_3CH_2CH_2CH_2)_3Sb$	5.1	4.8	4.1	3.1	
$(CH_3CH_2CH_2CH_2)_3PNi(CO)_3$	3.7	2.6	1.8	1.3	
$CH_3CH_2CH_2CH_2NH_2$	12.1	15.0	13.4	13.4	
$CH_3CH_2CH_2CH_2NH_3{}^+CF_3CO_2{}^-$	1.5	2.3	3.1	3.4	
$H_2NCH_2CO_2{}^-$	2.9	79			
decane	7.1	7.4	6.3	5.4	5.1
1-decanol	0.65	0.77	0.77	0.8(4–6)	
	1.1(7)	1.6(8)	2.2(9)	3.1(10)	

Cycloalkanes	C_1	C_2	C_3	C_4	C_5	C_6
R = H	11.6	21.7	13.7	14.0	12.7	
trans-R = C(CH$_3$)$_3$	3.0	17.2	3.2	3.5	6.2	
	12.2	15.2	13.5			
	30.5	42.8	31.0	22.6		
	17.1	34.2	26.3	27.9		
dioxane	5.8					
	8.0	6.9	7.8	7.9	6.9	6.8
1-methyladamantane		9.6	16.3	8.2	(CH$_3$, 6.0)	
cyclobutane	36	cyclooctane 10.3				

Compounds Containing a C=C or C=N Double Bond

Compound	C_1	C_2	C_3	C_4	C_5
H_3C^1, CH_3 / $C^2=C$ / H, H (cis-2-butene)	23.4	25.1			
H_3C^1, H / $C^2=C$ / H, CH_3 (trans-2-butene)	27.0	14.5			
H_3C, H / $C^2=C^3$ / H_3C^1, H	25.8	40.0	20.6		
H_3C^1, CH_3 / $C^2=C$ / H_3C, CH_3	24.2	79.3			
H_3C^1, O / N–C^3 / H_3C^2, H	18.6	11.1	20.2		
H_3C^1, S / N–C^3 / H_3C^2, H	14.7	9.7			
H_3C^1, OH / $C=N$ / H_3C^2	10.0	4.5			
H_3C^1, OCH_3^4 / $N^{\oplus}=C^3$ / H_3C^2, H, $SO_4^5CH_3^-$	4.9	2.6	2.0	3.0	4.5

Aromatic Compounds	C_1	C_2	C_3	C_4	C_5	C_6
C_6H_6	29.3					
H3C— (1,2,3,4,5)	16.3	58	20	21	15	
H3CCH2— (1,2,3,6,4,5)	9.4	15.8	79.3	13.5	13.5	10.8
H3C— / H3C— (1,2,3,4)	12.4	38	13.2	12.4		
H3C— with CH3, CH3 (1,2,3)	11.9	41	7.8			
HC≡C— (1,2,3,6,4,5)	8.5	53	56	13.2	13.2	9.0
O2N— (1,2,3,4)	56	6.9	6.9	4.8		
HO— (1,2,3,4)	16.7	2.4	2.4	1.4		
H2N— (1,2,3,4)		11.5	11.7	8.8		
biphenyl (1,2,3,4)	3.2	5.9	5.9	61		
biphenyl—CH3 (1,2,3,4,5)	3.1	5.9	5.9	6.0	5.0	
biphenyl—NH2 (1,2,3,4,5)	1.4	2.8				
CH3 naphthalene (1,2)	5.0	5.0				

C. ^{14}N, ^{15}N Relaxation Times (T_1) of Selected Compounds

Compound	^{15}N T_1(s)	^{14}N T_1(ms)
PhCN	Ca420	
PhNO$_2$	450[a]	6.2
ND$_3$	413	25
NH$_3$	186	35
NH$_4$NO$_3$	140	
NH$_4$NO$_3$	37	
CH$_3$CN	90	
C$_5$H$_5$N	85	1.65
trans-PhN = NPh	56	
K$_3$[Co(^{13}CN)$_6$]	35	
NaNO$_2$	27	
HCONHCH$_3$	23	
n-BuONO	24	1.6
KCN	21	
HCONH$_2$	21–25	
H$_3$CCONH$_2$	14	
PhNH$_2$	31.5	
NH$_4$Cl	4	1600
HOCH$_2$CH$_2$CH$_2$NH$_2$	3.6	
HOCH$_2$CH$_2$NH$_2$	4.3	
quinoline	220	

[a] 170s at 7.42 tesla

D. ^{19}F Relaxation Times (T_1) of Selected Compounds

Compound	T_1(s)	Compound	T_1(s)
CIF	4	F$_3$CC≡CCF$_3$	1.8–2.8
FCIO$_3$	0.1–2	C$_6$H$_5$CF$_3$	2.6–3.0
CF$_4$	2–3.6	2,6-C$_5$Cl$_3$F$_2$N	15.7
CFCl$_3$	0.7–8.4		
C$_6$D$_5$F	0.6–24.8		
1,3,5-C$_6$D$_3$F$_3$	3.8–8		
NaPF$_6$	5.9		
Na$_2$PO$_3$F	3.4		
CHFCl$_2$	1.4		

E. ^{31}P Relaxation Times (T_1) of Selected Phosphorus Compounds

Compound	T_1(s)	Compound	T_1(s)
PCl$_3$	3.5	PBr$_3$	6.0
POCl$_3$	9.5	H$_3$PO$_4$	0.14
H$_4$P$_2$O$_7$	7.6	NaPF$_6$	6.6
NaPO$_3$F	5.1	P$_4$O$_6$	15.9
C$_6$H$_5$PH$_2$	1.3	PPh$_3$	30.8
PEt$_3$	10.4	Ppr$_3$	15.4
Ptol$_3$	26.7	P(OMe)$_3$	6.0
P(OEt)$_3$	12.4	OP(Me)$_3$	11.5
OP(OBu)$_3$	9.3	SP(OMe)$_3$	11.4
[PBu$_4$]$^+$Br$^-$	2.3	[PPh$_4$]$^+$Br$^-$	11.8
[PEt$_4$]$^+$I$^-$	9.0	[Ph$_3$PMe]$^+$Br$^-$	6.0
trans-(PEt$_3$)$_2$PdCl$_2$	14.7	trans-(Ppr$_3$)$_2$PdCl$_2$	9.5
trans-(PBu$_3$)$_2$PdCl$_2$	6.0	trans-(Ptol$_3$)$_2$PtCl$_2$	14.0
trans-[P(C$_6$H$_{11}$)$_3$]$_2$PdCl$_2$	4.6	trans-(Me$_2$PPh)$_2$PdCl$_2$	14.6
cis-(Me$_2$PPh)$_2$PdCl$_2$	18.5		

F. ^{195}Pt Relaxation Times (T_1) of Selected Platinum Compounds[a]

Compound	T_1(s)	Compound	T_1(s)
Na$_2$PtCl$_6$	0.3–1.35	H$_2$PtCl$_6$	1.28
H$_2$PtBr$_6$	0.32	H$_2$PtI$_6$	1.11
Na$_2$PtCl$_4$	0.46–0.7	H$_2$PtCl$_4$	0.46
H$_2$PtBr$_4$	0.37	[Pt(en)$_2$]Cl$_2$	0.28
K$_2$[Pt(SCN)$_4$]	0.83	K$_2$[Pt(CN)$_4$]	1.09
(Bu$_4$N)[PtCl$_3$(C$_5$H$_5$N)]	14–35 ms	Pt(1,5-COD)$_2$	0.15
(R$_3$P)$_2$Pt	0.009–0.12	(R$_3$P)$_3$Pt	0.6–4.2
(R$_3$P)$_4$Pt	1.4–8.3		

---

[a] At 2.11 tesla, ^{195}Pt T_1 values are highly field dependent decreasing significantly with increasing field strength.

Index

Isotope	I	Natural abundance (%)	μ/μ_N	$\gamma/10^{7}\ \mathrm{rad\,s^{-1}\,T^{-1}}$	$Q/10^{-28}\ \mathrm{m^2}$	Ξ (MHz)	ℓ	D^{P}	D^{C}
87Rb[k]	3/2	27.85	3.5502	8.7807	0.13	32.823	2.3×10^{-2}	4.92×10^{-2}	2.79×10^{2}
87Sr	9/2	7.02	-1.208	-1.163	0.3	4.349	6.7×10^{-3}	1.91×10^{-4}	1.08
91Zr	5/2	11.23	-1.5415	-2.4959	-0.21[l]	9.3298	1.4×10^{-2}	1.06×10^{-3}	6.04
93Nb	9/2	100	6.818	6.564	-0.22	24.54	3.6×10^{-3}	0.487	2.77×10^{3}
95Mo	5/2	15.72	1.081	1.750	-0.022	6.542	1.5×10^{-4}	5.14×10^{-4}	2.92
(97Mo)	5/2	9.46	-1.104	-1.787	0.255	6.679	2.1×10^{-2}	3.29×10^{-4}	1.87
99Tc[k]	9/2	—	6.281	6.046	0.3	22.60	6.7×10^{-3}	—	—
99Ru	5/2	12.72	-0.7623[m]	-1.234[m]	7.6×10^{-2}	4.614[m]	1.8×10^{-3}	1.46×10^{-4}	0.827
101Ru	5/2	17.07	-0.8544[m]	-1.383[m]	0.44	5.171[m]	6.2×10^{-2}	2.75×10^{-4}	1.56
105Pd	5/2	22.23	-0.760	-1.23	0.8	4.60	0.20	2.52×10^{-4}	1.43
(113In)	9/2	4.28	6.1058	5.8782	0.82	21.973	5.0×10^{-2}	1.50×10^{-2}	85.0
115In[k]	9/2	95.72	6.1190	5.8908	0.83	22.020	5.1×10^{-2}	0.337	1.91×10^{3}
121Sb	5/2	57.25	3.9747	6.4355	-0.28	24.056	2.5×10^{-2}	9.30×10^{-2}	5.27×10^{2}
(123Sb)	7/2	42.75	2.8876	3.4848	-0.36	13.026	1.8×10^{-2}	1.98×10^{-2}	1.13×10^{2}
127I	5/2	100	3.3238	5.3817	-0.79	20.117	0.20	9.20×10^{-2}	5.39×10^{2}
131Xe[i]	3/2	21.18	0.8918	2.206	-0.12	8.245	1.9×10^{-2}	5.94×10^{-4}	3.37
133Cs	7/2	100	2.9231	3.5277	-3×10^{-3}	13.187	1.2×10^{-6}	4.82×10^{-2}	2.73×10^{2}
(135Ba)	3/2	6.59	1.080	2.671	0.18	9.984	4.3×10^{-2}	3.28×10^{-4}	1.86
137Ba	3/2	11.32	1.208	2.988	0.28	11.17	0.10	7.89×10^{-4}	4.47
139La	7/2	99.911	3.150	3.801	0.22	14.210	6.6×10^{-3}	6.05×10^{-2}	3.42×10^{2}
177Hf	7/2	18.50	0.8960	1.081	4.5	4.042	2.8	2.57×10^{-4}	1.46
179Hf	9/2	13.75	-0.705	-0.679	5.1	2.54	1.9	7.42×10^{-5}	0.421
181Ta	7/2	99.988	2.66	3.22	3	12.0	1.2	3.65×10^{-2}	2.07×10^{2}
(185Re)	5/2	37.07	3.753	6.077	2.3	22.72	1.7	5.13×10^{-2}	2.91×10^{2}
187Re[k]	5/2	62.93	3.791	6.138	2.2	22.94	1.5	8.81×10^{-2}	5.00×10^{2}
189Os[i]	3/2	16.1	0.8475	2.096	0.8	7.836	0.85	3.87×10^{-4}	2.20
(191Ir)	3/2	37.3	0.1877	0.4643	1.1	1.735	1.6	9.77×10^{-6}	5.54×10^{-2}
193Ir	3/2	62.7	0.2044	0.5054	1.0	1.889	1.3	2.11×10^{-5}	0.120
197Au	3/2	100	0.18701	0.46254	0.59	1.729	0.46	2.58×10^{-5}	0.147
201Hg[i]	3/2	13.22	-0.71871	-1.7776	0.44	6.645	0.26	1.94×10^{-4}	1.10
209Bi[k]	9/2	100	4.511	4.342	-0.38	16.23	1.1×10^{-2}	0.141	8.01×10^{2}

a Excluding the lanthanides.

b See footnote b to Table 1.1.

c Loc. cit. in footnote d to Table 1.1, except where otherwise stated.

d Loc. cit. in footnote c to Table 1.1.

e It should be noted that reported values of Q may be in error by as much as 20–30%.

f Calculated from the quoted value of γ. (Therefore, diamagnetic corrections are included and the frequency quoted is not with respect to TMS.)

g $\ell = (2I + 3)Q^{2}/[I^{2}(2I - 1)]$.

h See footnote (f) to Table 2.1.

i A useful isotope of $I = \frac{1}{2}$ exists.

j R. Neumann, F. Träger, J. Kowalski, & G. zu Putlitz, Z. Physik A279, 249 (1976).

k Radioactive, with a long half-life. Other radioactive nuclei (e.g., ^{40}K, ^{41}Ca) have been examined by NMR, but are unimportant for chemical studies.

l S. Büttgenbach, R. Dicke, H. Gebauer, R. Kuhnen, & F. Träber, Z. Physik A286, 125 (1978).

m Derived from data in C. Brevard & P. Granger, J. Chem. Phys. 75, 4175 (1981).

n S. Büttgenbach, R. Dicke, H. Gebauer & M. Herschel, Z. Physik A280, 217 (1977).